LOGISTICS OF WARFARE IN THE AGE OF THE CRUSADES

Logistics of Warfare in the Age of the Crusades

Proceedings of a Workshop held at the Centre for Medieval Studies, University of Sydney, 30 September to 4 October 2002

Edited by

JOHN H. PRYOR
University of Sydney, Australia

ASHGATE

Published by
Ashgate Publishing Limited
Wey Court East
Union Road
Farnham
Surrey, GU9 7PT
England

Ashgate Publishing Company
110 Cherry Street
Suite 3-1
Burlington
VT 05401-3818
USA

Ashgate website: http://www.ashgate.com

British Library Cataloguing in Publication Data
Logistics of warfare in the age of the crusades : proceedings of a workshop held at the Centre for Medieval Studies, University of Sydney, 30 September to 4 October 2002
1. Logistics – History – To 1500 – Congresses 2. Military art and science – History – Medieval, 500–1500 – Congresses 3. Crusades – Congresses
I. Pryor, John H., 1947–
355.4'11'0902

Library of Congress Cataloging-in-Publication Data
Logistics of warfare in the Age of the Crusades / edited by John H. Pryor.
p. cm.
Includes bibliographical references and index.
ISBN 0-7546-5197-5 (alk. paper)
1. Crusades. 2. Logistics. 3. Military art and science—History—Medieval, 500–1500. 4. Military history, Medieval. I. Pryor, John H., 1947–

U37.P79 2006
909.07—dc22

2005032504

ISBN-13: 978-0-7546-5197-0
ISBN-10: 0-7546-5197-5

MIX
Paper from responsible sources
FSC® C004959
FSC
www.fsc.org

Printed and bound in Great Britain
by Printondemand-worldwide.com

Contents

Maps, Figures, and Tables ix

Preface xi

Abbreviations xiii

Note on citations of primary sources, on Arabic, and on Geographical names xvii

Maps xix

Chapter 1: Introduction: modelling Bohemond's march to Thessalonikē 1
John H. Pryor
Centre for Medieval Studies, University of Sydney, Australia

Chapter 2: The logistics of the Mongol-Mamlūk war, with special reference to the battle of Wādī 'l-Khaznadār, 1299 C. E. 25
Reuven Amitai
Department of Islamic and Middle Eastern Studies, The Hebrew University of Jerusalem, Israel

Chapter 3: Crusader logistics: from victory at Nicaea to resupply at Dorylaion 43
Bernard S. Bachrach
Department of History, University of Minnesota, USA

Chapter 4: Ship types and fleet composition at Genoa and Venice in the early thirteenth century 63
John E. Dotson
Department of History, University of Southern Illinois at Carbondale, USA

Chapter 5: Logistics and the Second Crusade 77
John France
Department of History, University of Wales at Swansea, UK

Chapter 6: Harbours and facilities along the eastern Mediterranean sea lanes to *Outremer* 95
Ruthy Gertwagen
Oranim Academic College, Israel

Chapter 7: Provisioning Peter the Hermit: from Cologne to Constantinople, 1096 119
Charles R. Glasheen
Department of History, University of Northern Florida, USA

Chapter 8: Roads and communications in the Byzantine Empire: wagons, horses, and supplies 131
John Haldon
Centre for Byzantine, Ottoman, and Modern Greek Studies, University of Birmingham, UK

Chapter 9: Reflections on maps, Crusading, and logistics 159
Benjamin Z. Kedar
The Institute for Advanced Studies, The Hebrew University of Jerusalem, Israel

Chapter 10: Infantry in Muslim armies during the Crusades 185
Yaacov Lev
Department of Middle Eastern Studies, Bar-Ilan University, Israel

Chapter 11: Food and the Fourth Crusade: a new approach to the "Diversion Question" 209
Thomas F. Madden
Department of History, St Louis University, USA

Chapter 12: Money and logistics in the forces of the First Crusade: coinage, bullion, service, and supply, 1096-99 229
Alan V. Murray
Institute for Medieval Studies, University of Leeds, UK

Chapter 13: The Northern Crusaders: the logistics of English and other Northern Crusader fleets 251
Richard W. Unger
Department of History, University of British Columbia, Canada

Chapter 14: Digest 275
John H. Pryor

Glossary of Technical Terminology 293

Selective Gazetteer 297

Consolidated Bibliography 305

Index 343

Maps, Figures, and Tables

Maps (in order of usage)

Map 1	Bohemond's march to Thessalonikē	xix
Map 2	Greater Syria	xx
Map 3	From Constantinople to Dorylaion	xxi
Map 4	The Central and Eastern Mediterranean	xxii
Map 5	The Ionian Sea	xxiii
Map 6	The Southern Aegean and Lycia	xxiv
Map 7	Palestine and Syria	xxv
Map 8	Byzantine Asia Minor	xxvi
Map 9	The Atlantic and Western Mediterranean	xxvii
Map 10	From Clermont to Constantinople	xxviii
Map 11	The Balkans	xxix
Map 12	Egypt and Palestine	xxx
Map 13	The Fourth Crusade	xxxi
Map 14	The Baltic Sea	xxxii

Figures

1.1	Stretch of abandoned road, probably covering the *Via Egnatia* near Struga at the northern end of the Lake of Ohrid	5
9.1	Syria according to al-Iṣṭakhrī (1193 C.E.)	164
9.2	Sketch of al-Iṣṭakhrī's map of Syria (1840 C.E.)	165
9.3	The Mediterranean according to al-Iṣṭakhrī	167
9.4	The Western Mediterranean according to Ibn Ḥawqal (1086 C.E.)	168-9
9.5	Southern half of the St Sever Beatus map (mid eleventh century)	174-5
12.1	Penny of Godfrey of Bouillon with the inscription GODFRIDUS IEROSOLIMITANUS	240

Tables

1.1	Hours of daylight and marching time, Dyrrachion/Avlona to Thessalonikē	6
1.2	Modelled knight's troop	8
1.3	Maximum possible daily march	8
1.4	Provisioning medieval galley crews and Crusaders	10
1.5	Biscuit from grain according to Marino Sanudo and Angevin Registers	13
1.6	Provisions per day for horses transported by sea	16
1.7	Provision-carrying capacity of modelled knight's troop	19
1.8	Weights and measures	24
8.1	Feed requirements of animals using Roth's lower estimates	146
8.2	Carrying capacities of animals	146
8.3	Provisioning a cavalry force of 1,000 men and 250 remounts	151
8.4	Provisioning a cavalry force of 1,000 men and 250 remounts, using Roth's figures	152
8.5	Provisioning an army of 4,000 cavalry with 1,000 remounts	153
8.6	Provisioning an army of 4,000 cavalry with 1,000 remounts, using Roth's figures	154
8.7	Provisioning an army of 10,000, made up of 6,000 infantry and 4,000 cavalry, accompanied by 1,000 remounts	155
8.8	Provisioning an army of 10,000, made up of 6,000 infantry and 4,000 cavalry, accompanied by 1,000 remounts, using Roth's figures	156

Preface

From 30 September to 4 October 2002 a small group of scholars from around the world who had generously responded to my invitation met in Sydney for a week-long Workshop devoted to the logistics of Crusading and related military activities. They were Reuven Amitai, Bernard Bachrach, John Dotson, John France, Ruthy Gertwagen, John Haldon, Benjamin Kedar, Yaacov Lev, Tom Madden, Alan Murray, and Dick Unger. Two other scholars who learned about the workshop joined us: Charles Glasheen from Florida and Todd Mason from Melbourne. Todd's interests are in games theory and he made a valuable contribution to the Workshop, although he chose not to be included in its proceedings.

The Workshop was modelled on similar workshops which I had attended at the Institute for Advanced Study of the Hebrew University of Jerusalem in the past, with addition of use of the Web. It took the form of addresses by participants to drafts of their papers which had previously been posted on the Workshop's Website. Participants spoke for an hour or so, followed by a coffee break and then open discussion. We were gathered around a hollow square of tables in the centre of a large meeting room and there were chairs around the walls for other persons interested in attending. Quite a number in fact did so, especially on the Monday and Tuesday of the Workshop. These also could, and did, contribute to the round-table discussions after the presentations. The discussions were taped and later transcribed and posted on the Website with the drafts of the papers so that all participants could take advantage of them when finalizing their papers for publication, which had been negotiated in principle by Ashgate in advance.

The Workshop was an outstanding success, being lively and dynamic from John Haldon's opening session to John Dotson's concluding one. My own Public Lecture on the Monday evening was well attended and received. Throughout the Workshop participants contributed well to each other and, by and large, were receptive to what others had to say. That is not to say that they did not stick to their opinions when they considered them correct. However, given the disparate nature of the group, the varied nature of their fields of expertise, and the fact that very little has been written on the subject, a surprising amount of agreement was reached. John France's opening remarks reflected the fact that few scholars have ever considered matters concerning the logistics of warfare in the age of the Crusades and a glance through the Consolidated Bibliography will reveal just how little secondary literature there is on the subject. A considerable amount of attention has been devoted to military logistics in the classical and Early-Modern periods, but virtually none to military and naval logistics in the Middle Ages. I very deliberately extended the purview of the Workshop to include the Byzantine and Muslim worlds, but any glance at the existing historiography of Crusading expeditions such as the journey of the First Crusaders will show immediately that they have been considered in what is virtually a logistical vacuum. Frequently, it is as though Crusaders' human needs, the lands through which they passed, and the means by which they did so, did not exist.

Although agreement on specifics, for example required supplies of water and food for human beings, and even more so on that for animals, was never agreed upon by us all, a remarkable degree of agreement on parameters was reached, considering that we were virtually starting from scratch. There is no doubt, for example, that early Crusading expeditions by land and Byzantine campaigns by land required total resupply of participating forces many times over. The Workshop reached general agreement that such considerations affected profoundly relationships between Crusaders and Byzantines. There was also general agreement that parameters dictated by logistical considerations also affected politico-military outcomes in some circumstances, although everyone acknowledged that in the end logistics were not the sole determinant of human endeavour. None of us subscribe to logistical determinism any more than to geographic determinism or any other determinism.

Social events included a Reception hosted by the Centre for Medieval Studies on the Monday evening, an excursion on Sydney Harbour on the timber schooner *Boomerang*, originally a sailing ship built in 1903 but converted to power in the 1930s, and a concluding party at my home. Probably the highlight of the week for our overseas guests was the excursion on Sydney Harbour. A more enjoyable way to experience one of the world's great harbours is hard to imagine.

The Workshop was initially funded by a large grant from the University of Sydney, Pro Vice-Chancellor for Research's initiative in funding Areas of Research Excellence. Thanks to the energies of its Director, Professor Margaret Clunies Ross, the Centre for Medieval Studies was successful in applying for recognition of Medieval Studies as one such Area and the grant flowed from that. Additional funding was also obtained from the University of Sydney Mandelbaum Trust and from a donor who will remain unnamed. I am extremely grateful to all three for providing the funds which made this memorable gathering possible. I am also indebted to two of the participants who contributed towards their own expenses.

John Pryor
5 April, 2006

Abbreviations

AASS	*Acta Sanctorum quotquot tote orbe coluntur* ..., eds S. Coleti and J. B. A. Hieron, 51 vols (Venice, 1734-86).
AHES	*Annales d'histoire économique et sociale.*
AHR	*American historical review.*
ANRW	Aufstieg und Niedergang der römischen Welt.
AQDGM	Ausgewälte Quellen zur deutschen Geschichte des Mittelalters, 37 vols (Berlin 1955-82).
ASLSP	Atti della Società Ligure di Storia Patria.
BAR *IS*	British archaeological reports. International series.
BBA	Berliner byzantinische Arbeiten.
BEC	*Bibliothèque de l'Ecole des Chartes.*
BLVS	Bibliothek des literarischer Vereins in Stuttgart.
BMGS	*Byzantine and Modern Greek Studies.*
BSOAS	*Bulletin of the School of Oriental and African studies.*
CC *ContMed*	Corpus Christianorum Continuatio mediaevalis, 173 vols (Turnhout, 1971-2000).
CFHB	Corpus fontium historiae Byzantinae (Berlin et al., 1967-).
CIC	*Corpus iuris civilis*, eds T. Mommsen, P. Krueger, et al., 3 vols (Berlin, 1872-95) and re-editions.
CJ	*Codex Iustinianus*, ed. P. Krüger, in CIC, vol. 2, 15th ed. (Berlin, 1970).
CMH	*Cambridge medieval history*, eds J. B. Bury et al., 8 vols (Cambridge, 1911-36).
CSCO *ScriptAr*	Corpus Scriptorum Christianorum Orientalium Scriptores Arabici, ser. 3 (Paris, 1903-).
CSHB	Corpus Scriptorum Historiae Byzantinae, 34 vols (Bonn, 1828-53).
CTh	Theodosius II, *Theodosiani libri XVI cum Constitutionibus Sirmondianis, et leges novellae ad Theodosianum pertinentes*, eds T. Mommsen and P. Meyer, 2 vols (Berlin, 1905).

De Goeje, BGA	De Goeje, M. J., ed., *Bibliotheca geographorum Arabicorum*, 3rd ed. (Leiden, 1967).
DOP	*Dumbarton Oaks Papers.*
EHR	*English historical review.*
EI.2	*The encyclopedia of Islam*, new ed., eds H. A. R. Gibb et al., 11 vols (Leiden, 1960-2002).
FStI	Fonti per la Storia d'Italia.
HJAS	*Harvard journal of Asiatic studies.*
HR	*Historical research: the bulletin of the Institute of historical research.*
HSW	Hakluyt Society Works.
IJMES	*International journal of Middle East studies.*
IJMH	*International Journal of Maritime History.*
IJNA	*International journal of underwater exploration and nautical archaeology.*
JAH	*Journal of Asian history.*
JAOS	*Journal of the American Oriental Society.*
JESHO	*Journal of the social and economic history of the Orient.*
JHS	*Journal of Hellenic studies.*
JMS	*Journal of Mediterranean studies.*
JN	*Iustiniani Novellae*, ed. R. Schoell, in CIC, vol. 3, 9th ed. (Berlin 1968).
JRAS	*Journal of the Royal Asiatic society of Great Britain and Ireland, third series.*
JSAI	*Jerusalem studies in Arabic and Islam.*
MGH	Monumenta Germaniae historica.
— *DipReg*	Diplomata regum et imperatorum Germanie (Berlin, 1956-).
— *ScriptRerGerm*	Scriptores rerum Germanicarum in usum scholarum ex monumentis Germaniae historicis, Nova series, vols 1-7 (1922-30; rpt, Berlin, 1964); vols 8-13 (Berlin, 1955-67).
— *ScriptRerGerm.Cont*	Scriptores rerum Germanicarum in usum scholarum ex monumentis Germaniae historicis. Recusi or separatim editi (Berlin, 1871-2002).
— *SS*	Scriptores, 30 vols (1826-1934; rpt, Stuttgart and N.Y., 1963-4).
MHR	*Mediterranean historical review.*
MM	*The Mariner's Mirror.*

NCMH	*The New Cambridge Medieval History*, 7 vols (Cambridge, 1995-).
ODB	*Oxford dictionary of Byzantium*, ed. A. Kazhdan, 3 vols paginated continuously (Oxford and N.Y., 1991).
OED	Oxford English Dictionary, 2nd ed. (Oxford, 1992).
PL	Patrologiae cursus completus Series Latina, ed. J.-P. Migne, 217 vols (Paris, 1878-90).
PO	Patrologia orientalis, 47 vols [212 Nos] (Paris and Turnhout, 1903-98).
PPTSL	Palestine pilgrims' text society library, 13 vols (London, 1890-97).
RHC	Recueil des historiens des Croisades.
DArm	Documents Arméniens, 2 vols (1869, 1906; rpt, Farnborough, 1967).
HOcc	Historiens occidentaux, 5 vols (1844-95; rpt, N.Y., 1967).
HOr	Historiens orientaux, 5 vols (1872-1906; rpt, N.Y., 1967).
RHGF	Recueil des historiens des Gaules et de la France, new ed., ed. L. Deslisle, 19 vols (1869-80; rpt, Farnborough, 1967-8).
RISS *NS*	Rerum Italicarum scriptores Nova series, 34 tomes (Bologna, 1900-35).
ROL	*Revue de l'Orient latin.*
Rolls series	Rerum Britannicarum medii aevi scriptores or Chronicles and memorials of Great Britain and Ireland during the Middle Ages, 99 tomes (London, 1858-99).
SCH	*Studies in Church history.*
Setton, *HC*	Setton, K., ed., *A history of the Crusades*, 6 vols (Philadelphia and Madison, 1955-86).
SI	*Studia Islamica.*
SSCISAM	Settimane di studio del Centro italiano di studi sull'alto medioevo.
TAPS	Transactions of the American Philosophical Society.
TIB	Hunger, H., ed., *Tabula imperii Byzantini* (Vienna, 1976-).

Note on citations of Primary Sources, on Arabic, and on Geographical names

Citations of primary sources have been made throughout to the most recent, standard received edition of the works in their original languages. The only exceptions occur in a few cases where I considered that another edition may be more easily available to some readers and that the texts cited contained no differences of substance. In such cases both editions may be cited. Citations of Muslim authors have been made according to the *Agnomen* or *Kunya* or *Nisba* by which they are commonly known as listed alphabetically in the Consolidated Bibliography. The rest of their names then follow between commas. "al-" has been ignored for purposes of alphabetization.

Except in a few cases where contributors needed translations to be cited for particular reasons, they have not been. Where translations into English or French known to me exist, these have been included in the Bibliography for the assistance of non-specialist European-language readers; however, references to them have not been included in the notes.

For Arabic in transliteration, since there are many different systems in use and since I do not read Arabic myself, I chose to follow throughout the usage of the *Encyclopedia of Islam*, even though I appreciate that it uses a pedantic system followed by no one else. However, it was the only way in which I could regularize usage.

Citation of the very large number of place names used in this volume presented something of a problem since many places had a variety of names in different languages and at different times in the Middle Ages and these may or may not correspond to modern equivalents. To have adhered rigorously to the principle of using the name as used by the occupying power at the time of a particular citation in question would have involved too many absurdities; for example, Σιγγιδών or Σιγγιδόνον (Singidōn/Singidonon) for Beograd, or Belgrade, since the Byzantines occupied the town at the time of the First Crusade. Moreover, since places were frequently referred to by different names by occupying powers and by enemies who might wish to occupy them, more absurdities would have been created; for example to have referred to Edessa by the Arabic *al-Ruhā'* or Armenian *Ouṛha* on the grounds that it was in Muslim and then Armenian hands before the First Crusade or by the Latin *Rohas* by which it became known to the

Franks, rather by Edessa, derived from its Byzantine Greek name Ἔδεσσα, as is used commonly in all Crusading historiography, would have been absurd. Such a system would have been pedantic and would have created the need for a huge Gazetteer. On the other hand, not all medieval places still exist or even have modern equivalent names, so some use of medieval names was necessary. Moreover, some obscure place names cannot be found even in good quality atlases.

In the end I chose the principle that modern equivalent names would be used whenever the medieval place was actually on the same site as the modern place and the modern place name is also shown in an English language atlas which is commonly available: J. C. Bartholomew, *The world atlas*, 10th ed. (Edinburgh, 1975). For the rest medieval place names in commonly-used forms would be used and these would be listed in the Selective Gazetteer.

It is impossible to be absolutely consistent in such matters and in some cases where I considered that there were good reasons for doing so, usually for historical or cultural reasons, medieval names have been retained; for example, Constantinople rather than Istanbul, Dyrrachion rather than Durrës, Nicaea rather than Iznik, Al-Mahdīyya rather than Mahdia, Avarinos rather than Pilos or Navarino, and so on. No doubt some readers will find my system unsatisfactory, for example Edirne rather than Adrianople, but at least it is a system.

John Pryor

Map 1: Bohemond's march to Thessalonikē

Map 2: Greater Syria

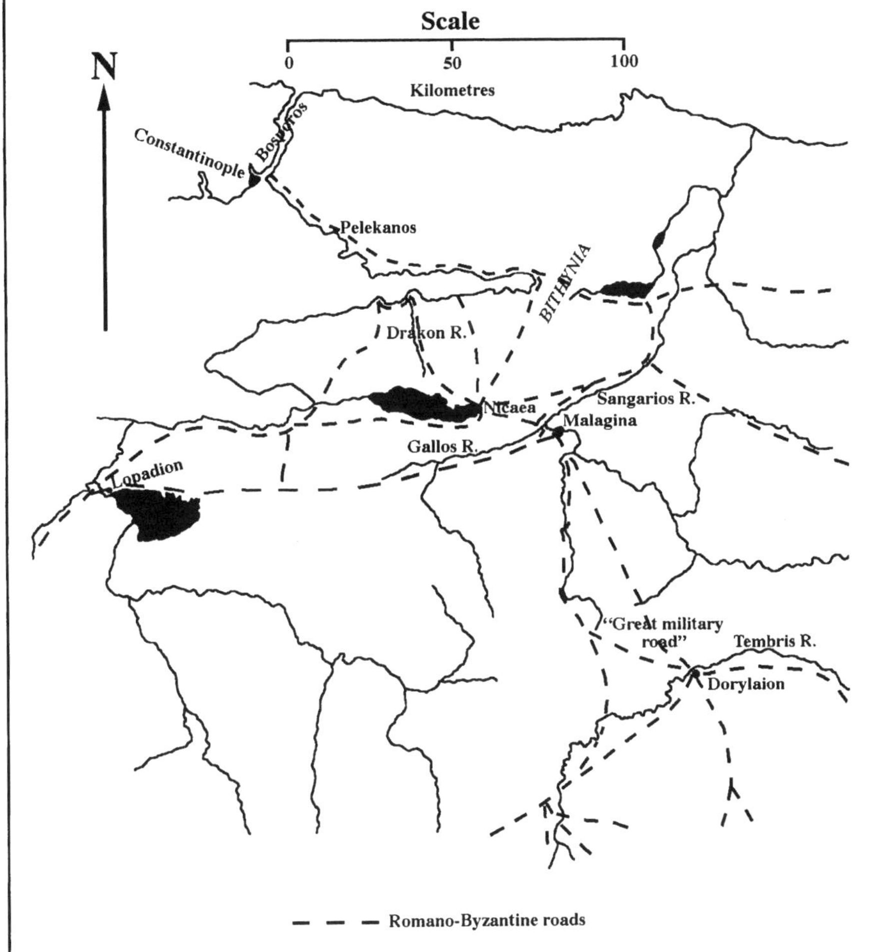

Map 3: From Constantinople to Dorylaion

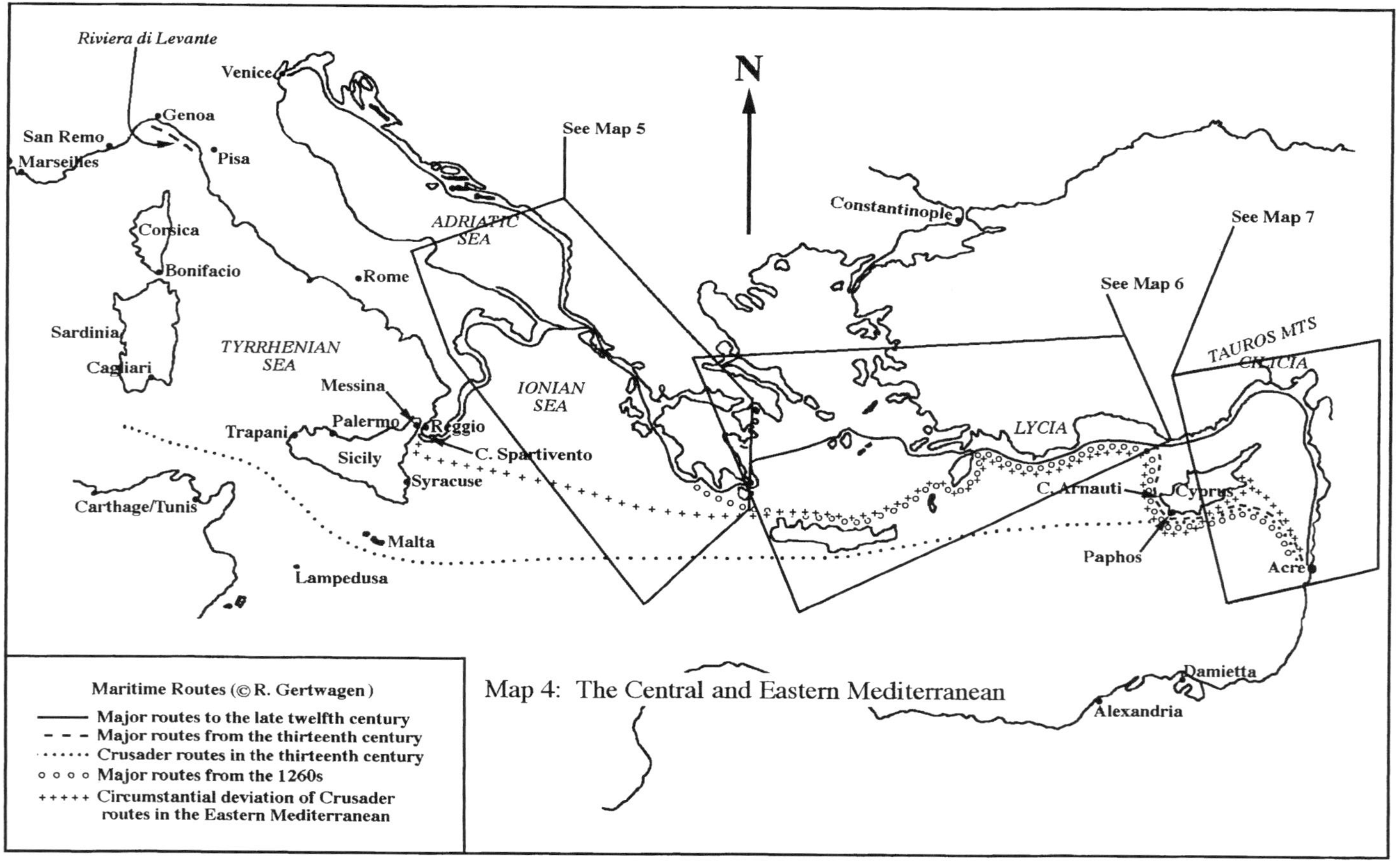

Map 4: The Central and Eastern Mediterranean

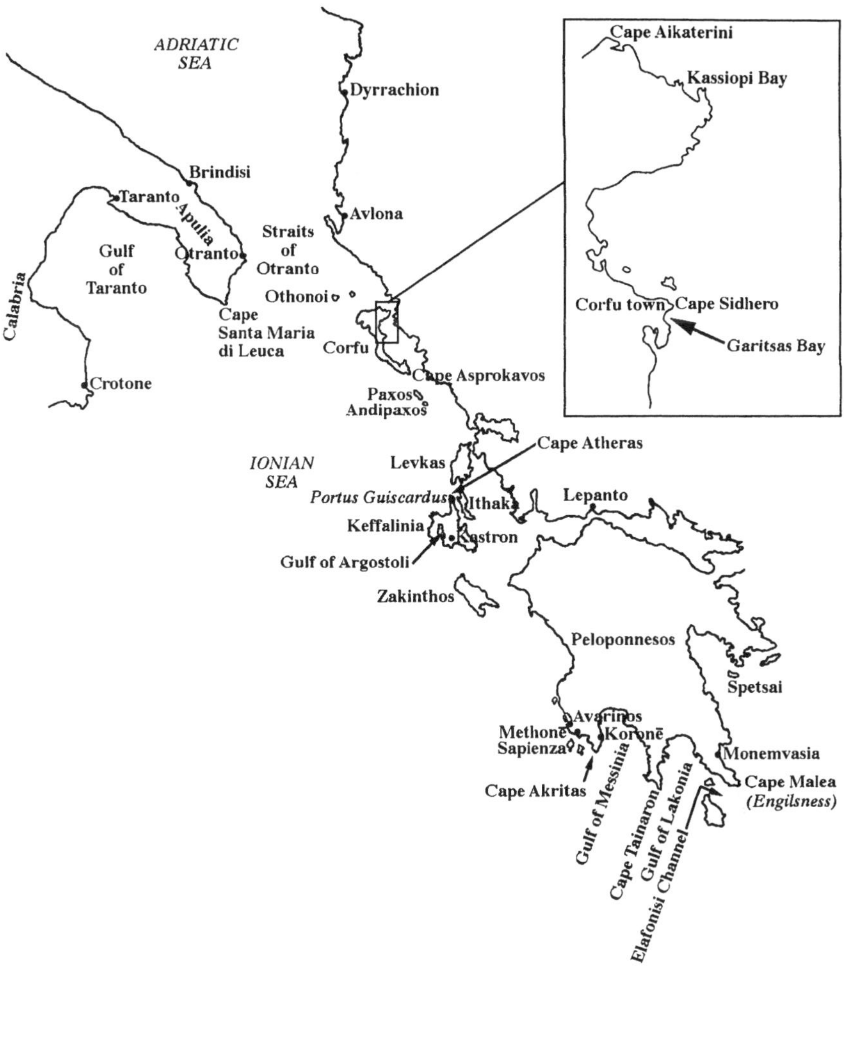

Map 5: The Ionian Sea

Map 6: The Southern Aegean and Lycia

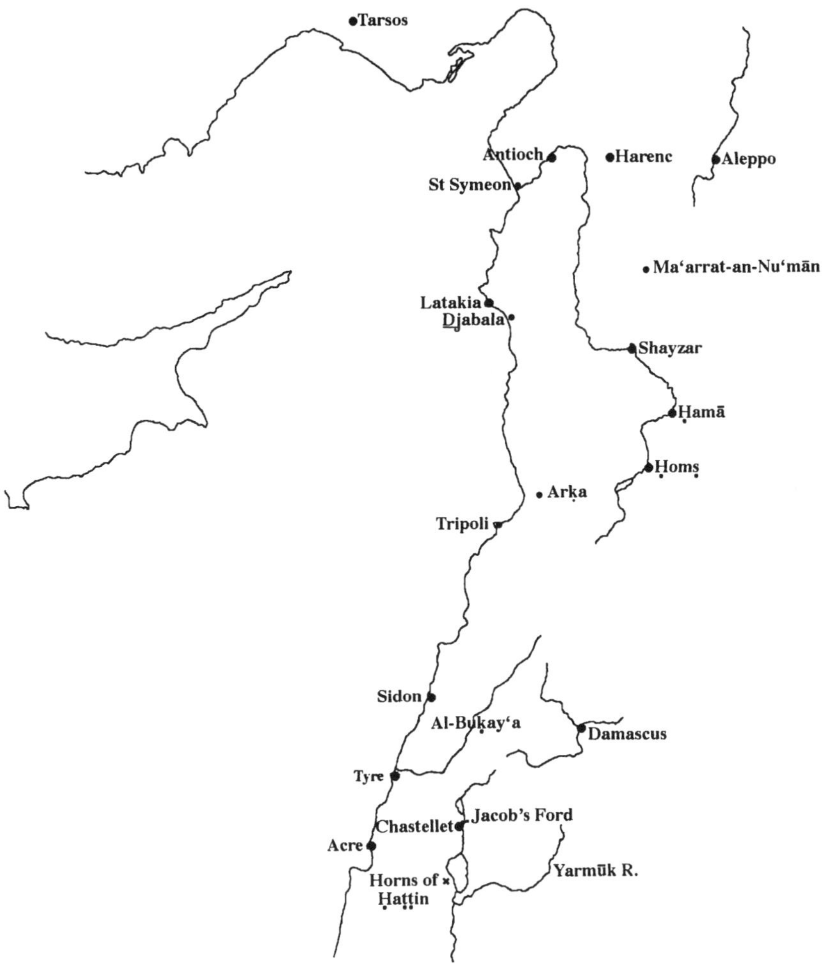

Map 7: Palestine and Syria

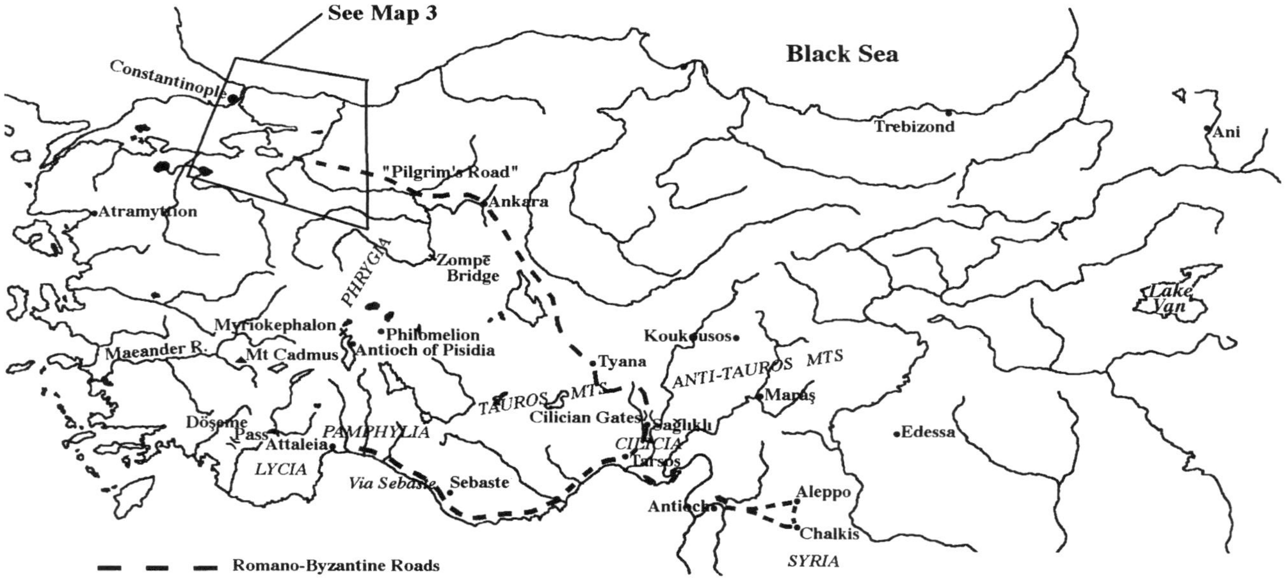

Map 8: Byzantine Asia Minor

Map 9: The Atlantic and Western Mediterranean

Map 10: From Clermont to Constantinople

Map 11: The Balkans

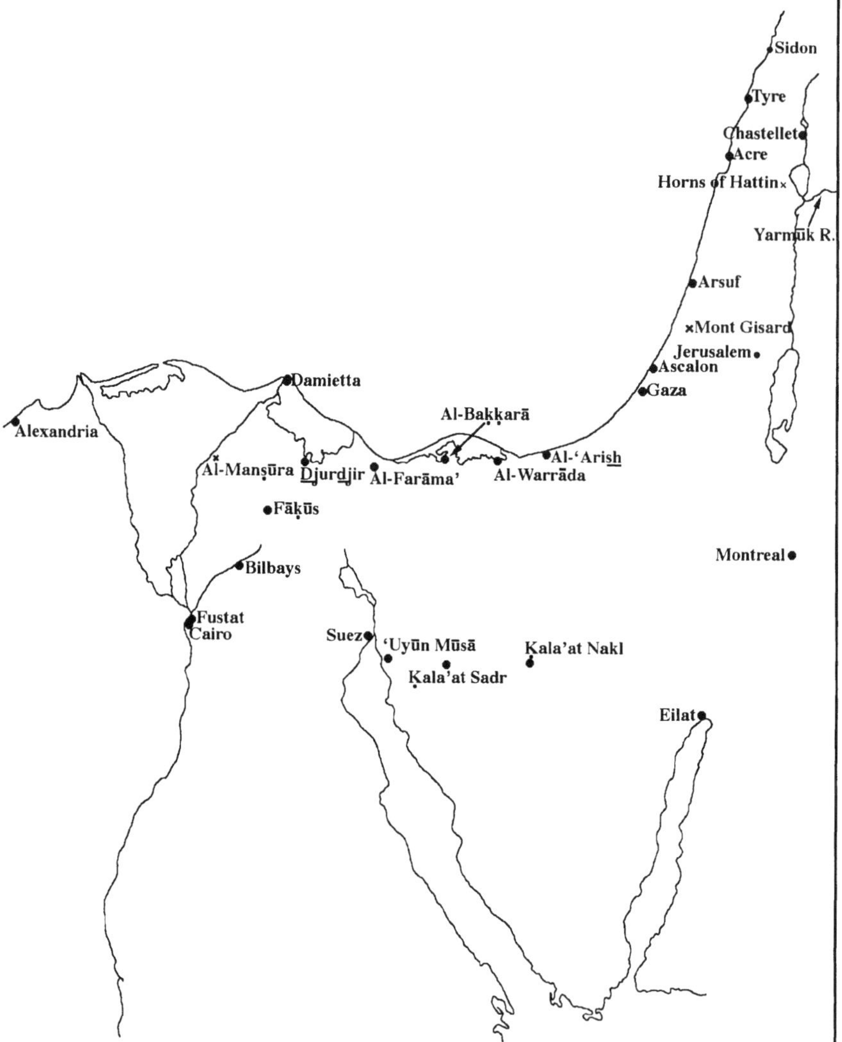

Map 12: Egypt and Palestine

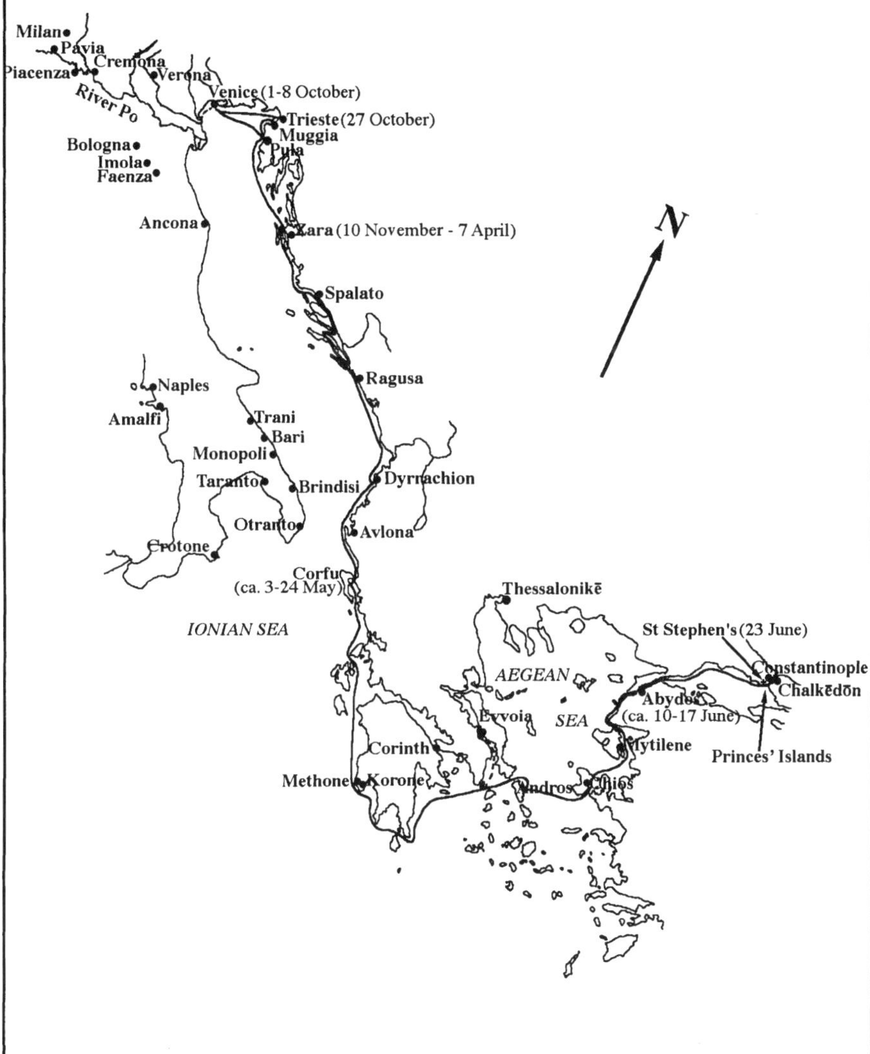

Map 13: The Fourth Crusade

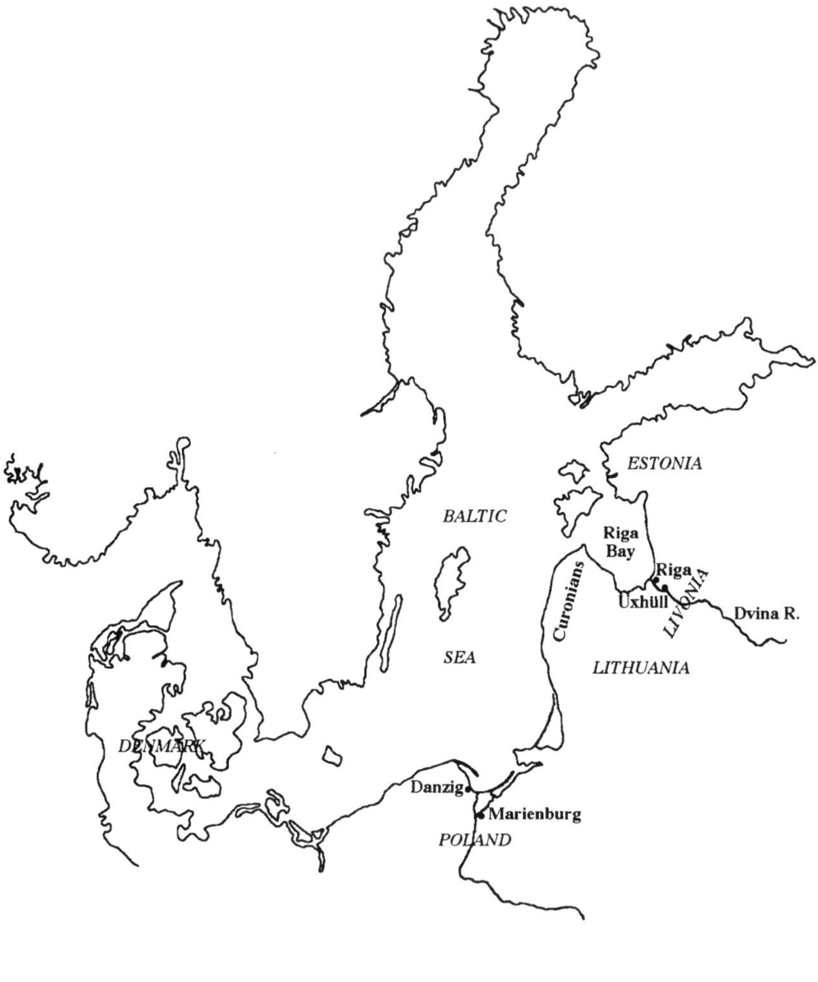

Map 14: The Baltic Sea

Chapter 1

Introduction: modelling Bohemond's march to Thessalonikē[1]

John H. Pryor

Bohemond of Taranto took the cross in September 1096 while besieging Amalfi with his uncle and half-brother. He and his forces then crossed the Adriatic in a piecemeal fashion in late October, which gives the lie to the story of the *Gesta Francorum* that he did not till then know about Clermont and the launching of the Crusade.[2] The anonymous author of the *Tudebodus imitatus et continuatus historia peregrinorum* wrote that he reached Avlona at All Saints, 1 November,[3] but it would have been impossible to have gathered the resources needed for such an expedition in such a short period of time. The forces landed either at Dyrrachion or at Avlona or just generally in "Bulgaria", and after landing their route took them to "Andrinopolis", which was Byzantine Drynopolis, on the river Drino near Gjirokaster.[4] Leaving there they marched through Albania and reached Kastoria at Christmas and crossed the Vardar outside Thessalonikē on 18 February. He himself reached Constantinople on 1 April and his forces on 26 April. This was a very slow march, and particularly slow from the landings to Kastoria, a distance of

1 Consult Map 1. Thanks to John France for editing my own chapter.

2 All of the "Crusade" accounts of Bohemond's taking of the cross are ultimately based on the *Gesta Francorum*. See *Gesta Francorum et aliorum Hierosolimitanorum*, ed. and trans. R. Hill, *The deeds of the Franks and the other pilgrims to Jerusalem* (Oxford, 1962), I[iii] (p. 7). But see also Geoffrey Malaterra, *De rebus gestis Rogerii Calabriae et Siciliae comitis et Roberti Guiscardi ducis fratris eius*, ed. E. Pontieri, in RISS *NS*, tome 5, parte 1 (Bologna, 1928), IV.24 (p. 102).

3 Anonymous, *Tudebodus imitatus et continuatus Historia peregrinorum*, in RHC *HOcc*, vol. 3, 165-229, §VIII (p. 177).

4 *Gesta Francorum*, I[iiii] (p. 8); Peter Tudebode, *Historia de Hierosolymitano itinere*, in *RHC HOcc*, vol. 3, 1-117, I.viii (p. 16); William of Tyre, *Willelmi Tyrensis archiepiscopi chronicon*, ed. R. B. C. Huygens, 2 vols, in CC *ContMed*, vols 63 & 63A [paginated continuously], 2.13 (p. 177); Albert of Aachen, *Historia Hierosolymitana*, in RHC *HOcc*, vol. 4, 265-713, II.xviii (p. 312); Guibert of Nogent, *Guibert de Nogent Dei gesta per Francos, et cinq autres textes*, ed. R. B. C. Huygens, in CC *ContMed*, vol. 127A, 76-352, III[ii] (p. 138); Baudri of Bourgeuil, *Baldrici, episcopi Dolensis, Historia Jerosolimitana*, in RHC *HOcc*, vol. 4, 1-111, I.xii-xvi (pp. 20-22); Robert of Rheims, *Historia Iherosolimitana*, in RHC *HOcc*, vol. 3, 721-882, II.xi (pp. 744-5); Anonymous, *Tudebodus imitatus*, §§VIII-X (p. 177); *Chronica monasterii Casinensis*, ed. H. Hoffmann, in MGH *SS*, vol. 34, IV.11 (pp. 476-7).

only around 350 kilometres. A little less than two months to cover 350 kilometres at an average of around six to seven kilometres a day was inordinately slow.

From reading sources we could only ever guess at how numerous were Bohemond's forces. Lupus Protospatharios stated that more than 500 knights followed him from Amalfi and Geoffrey Malaterra added that they were the warlike young men there. Albert of Aachen claimed that he was followed by 10,000 cavalry and very great numbers of infantry.[5] That was obviously ridiculous; however, he certainly was joined by other southern Norman lords as well as the French lords Bartholomew Boel, vice-lord of Chartres, and Count Geoffrey III of Roussillon.[6] He later gave Tancred 100 knights and 200 other men for his expedition through the Cilician Gates and then reinforced him at Tarsos with another 300 men.[7]

We will never be able to know from narrative sources, which are all that survive for Bohemond's march, the answers to questions such as how large were his forces and why his march was so slow. We must begin to approach problems in different ways, bringing to bear insights from collateral disciplines. Even if they do not provide definitive answers, these may at least facilitate understanding.

Bohemond had been in the Balkans fighting with his father against Alexios I Komnēnos in 1081-4.[8] He had been as far south as Avlona, Corfu, and Arta, along the *Via Egnatia* to Ohrid, Bitola, and the Vardar, and as far south-east as Trikkala and Larissa. He must have known the Balkan geography well. He had fought through the country between Kastoria and Ioannina and would have known that from the coast between Dyrrachion and Avlona there were really only three routes which could take him east to Thessalonikē because the stratigraphy of the middle

---

5 Lupus Protospatharios, *Annales*, ed. G. Pertz, in MGH *SS*, vol. 5 (Hanover, 1844), 51-63, p. 62; Malaterra, *De rebus gestis Rogerii*, IV.24 (p. 102); Albert of Aachen, *Historia Hierosolymitana*, II.xviii (p. 312).

6 The various sources mention his nephew Tancred and Tancred's brothers Robert and William; Richard of the Principate, Count of Salerno, and his brother Ranulf and the latter's son Richard; Robert of Anzi; Hermann of Canne; Robert of Sourdeval, the lord of Torosse; Robert of Molise, son of Tristan, the lord of Limosano; Humphrey, the son of Ralph; Richard of Caiazzo, the son of Count Ranulf; Geoffrey III, Count of Roussillon, with his brothers and the bishop of Roussillon; Bartholomew Boel, vice-lord of Chartres; Aubrey of Cagnano; Humphrey of Montescaglioso or his son Geoffrey; Gerard, the bishop of Ariano; Robert, the son of Gerard, Bohemond's constable and standard-bearer; Ralph the Red; and Bishop Peter of Anagni. See in addition to the sources in n. 3 above, H. Hagenmeyer, ed., *Epistulae et chartae ad historiam primi belli sacri spectantes* (1901; rpt, Hildesheim, 1973), ep. XIII (p. 156); *Vita de S. Petro episcopo Anagniae in Italia*, in AASS, *Augusti*, tome I (Venice, 1750), 230-41, p. 238.

7 Albert of Aachen, *Historia Hierosolymitana*, III.11 (p. 346); Ralph of Caen, *Gesta Tancredi in expeditione Hierosolymitana, auctore Radulfo Cadomensi, ejus familiari*, in RHC *HOcc*, vol. 3, 587-716, §33 (p. 630).

8 R. B. Yewdale, *Bohemond I, Prince of Antioch* (Princeton, 1924), pp. 13-24.

Balkans is characterized by river valleys running from north to south between mountain ranges. Passes allowing movement from west to east are few.[9] The first choice would have been to follow the main branch of the *Via Egnatia* from Dyrrachion up the Shkumbi river through Elbasan, Struga, Bitola, and Edhessa. That, however, would have meant passing through territory under Byzantine control and have involved supervision by Alexios's nephew John Komnēnos, governor of Dyrrachion. Given the earlier conflict between the Normans and the Empire, it is possible that he would have wished to avoid contact with Byzantine authorities until inevitable, although he may also have wished to do the opposite and announce his coming to placate Byzantine misgivings. The *Estoire de Jérusalem et d'Antioche* reported that a delegation was sent by Bohemond to Alexios to seek a safe conduct and returned with a positive reply.[10] As it stands, the report is a garbled impossibility but it may well reflect some foundation in fact. It would seem to have been eminently sensible and there is some evidence to suggest that Urban II and various Crusade leaders may have announced their coming to the emperor.[11]

A second choice would have been to head north on the branch of the *Via Egnatia* which ran from ancient Apollonia to connect with the main road at Rrogozhinë, on the Shkumbi. From where this road crossed the Seman at Kuch, another road mounted the valley of the Devol to Bilisht and Kastoria. From Kastoria a minor road ran further up the valley of the Aliakmon and then through the Lerinska gorge to rejoin the main *Via Egnatia* at Florina.

The third choice was to go south from Avlona down the valley of the Vjosa and Drino rivers to Drynopolis. From the head of the valley a road passed through the Metsovon pass and Trikkala to Larissa in Thessaly and from there another led via Veroia and the region of ancient Pella to Thessalonikē.

Bohemond chose none of these three routes *in toto*. From the coast he moved to Drynopolis, some 65 kilometres south of Avlona. His reasons for taking this route may have been related to logistics since the valley of the Vjosa and Drino rivers was extremely fertile and November came at the end of harvest. Many sources commented that provisions were abundant in the valley.[12] However, from the head of the valley some hard choices would have had to have been faced. He had to take either the road through the Metsovon pass to Trikkala or to cut north

---

9 On what follows see F. O'Sullivan, *The Egnatian Way* (Newton Abbot, 1972), esp. pp. 24-5; N. H. H. Sitwell, *Roman roads of Europe* (London, 1981), pp. 128-9, 138-9.

10 *Li estoire de Jérusalem et d'Antioche*, in RHC *HOcc*, vol. 5, 621-48, §V (pp. 627-8).

11 See P. E. D. Riant, "Inventaire critique des lettres historiques de croisades", *Archives de l'Orient Latin*, 1 (1881), 1-224, Nos 48, 51, 56, 60-62, 71 (pp. 112-13, 117, 120-21, 123).

12 *Gesta Francorum*, I[iiii] (p. 8); Peter Tudebode, *Historia de Hierosolymitano itinere*, II.viii (p. 16); Guibert of Nogent, *Dei gesta per Francos*, III[ii] (p. 138); Baudri of Bourgeuil, *Historia Jerosolimitana*, I.xvi (pp. 21-2); Robert of Rheims, *Historia Iherosolimitana*, II.xi (pp. 744-5); Anonymous, *Tudebodus imitatus*, §IX (p. 177).

across the ranges to Kastoria and from there to Florina. The first choice would have been easier but around twice as long. Moreover, Bohemond had been at Kastoria before and would have known it for the oasis that it was. Apparently, he continued south-east to the head of the Drino valley and from there followed the road east to Doliana.[13] East of there a left-hand branch turned north to the valley of the Voidhomatis until it joined the Sarandaporos. Roads led up its valley to Pevkofiton, from where tracks crossed the ranges for about 20 kilometres to pick up the valleys of either a western branch of the Aliakmon around Nestorion or the Stravopotamos, which fed into the Aliakmon at Argos Orestikon. From there one ascended the river to Kastoria. The only other sensible route to Kastoria would have been to head back north from Drynopolis to pick up the road which mounted the valleys of the Seman and Devol rivers to Bilisht and Kastoria; however, nothing in any of the sources suggests that Bohemond retraced his tracks back to the north.

From Kastoria he probably ascended the upper Aliakmon through the Lerinska gorge to pick up the *Via Egnatia* in the region of Bitola around Florina,[14] and from there he mostly followed the route of the Roman road to Constantinople.

There is no evidence that Bohemond's forces ever split into two or more divisions marching apart. In fact there is evidence that they were not split, at least when they crossed the Vardar. Count Geoffrey of Roussillon had not yet crossed the river when he was attacked by imperial forces but Tancred, who was presumably in the van with Bohemond, went back and rescued him. This could have occurred only if the army was travelling in one division.[15]

Depending upon where they landed, some forces may have followed the *Via Egnatia* south from Dyrrachion to Avlona and Bohemond must have picked it up again around Florina. This would have been the best of all the roads available to his forces and for the purposes of modelling his march the *Via Egnatia* may be considered to have offered a best-case scenario. Like all Roman roads it varied in width from place to place, generally broadening when approaching cities and

---

13 The following roads are marked on the oldest map available to me. They may not have existed as made roads in the Middle Ages, but there would have at least been tracks following their course. Al-Idrīsī enumerated not only the route of the *Via Egnatia* from Dyrrachion to Thessalonikē via Ohrid, Arnissa, and Edhessa, but also the route from Avlona to Drynopolis, and then either to Ioannina, or to Kastoria, said to be either two or one and a half days' journey from Drynopolis. The two versions differ. See Al-Idrīsī, Abū 'Abd Allāh Muḥammad ibn Allāh, al-Sharīf, *Kitāb Nuzhat al-mushtāk fi 'khtirāq al-āfāq*, trans. P.-A. Jaubert, *La géographie d'Edrisi*, 2 tomes (1836-40; rpt in 1 vol., Amsterdam, 1984), vol. 2, pp. 289-91; also trans. H. Bresc and A. Nef, *La première géographie de l'Occident* (Paris, 1999), pp. 405-6.

14 After leaving Kastoria Bohemond's forces sacked a fortress of heretics at a place named Pelagonia, medieval Latin for the region on the *Via Egnatia* between Bitola and Florina.

15 *Gesta Francorum*, I[iiii] (p. 9); William of Tyre, *Chronicon*, II.14 (p. 179).

around bends. However, in the countryside the normal width of the paved surface, was about ten Roman feet or 2.96 metres. In the mountains and forests and in defiles it could narrow to a mere six Roman feet: 1.77 metres.[16]

Fig. 1.1: Stretch of abandoned road probably covering the *Via Egnatia* near Struga at the northern end of the Lake of Ohrid
From F. O'Sullivan, *The Egnatian Way* (Newton Abbot, 1972), p. 88.
(All attempts to trace the author have failed)

The Romans had built well but by 1097 the road was over 1100 years old and there is no record of repairs after the fourth century, although Justinian probably made

16 O'Sullivan, *Egnatian Way*, pp. 28-30.

some in the sixth century.[17] It would have been in a state of considerable disrepair and would also have been silted over in places. Even where the pavement survived, it had been built with limestone slabs which wear well but are slippery when wet, and Bohemond made his march during the rains of winter and early spring. He would have followed the trace of the road rather than the road itself and where the ground was firm columns may have spread out on either side. But in defiles and through forests and along the beds of river valleys they would have been forced into the width of the ancient road. Where the ground was not firm, men attempting to spread out to the side of the road would have been hard pressed to maintain pace with those on it and that would have been undesirable. On a route march, all sections should move at the same pace. The road often ran through hilly country with scrub and forest to either side making it difficult to spread out. For modelling purposes, I assume that the column was confined to the trace of the ancient road and that that was only around three metres wide.

Table 1.1: Hours of daylight and marching time, Dyrrachion/Avlona to Thessalonikē[18]

	Civil twilight begins	Sunrise	Sun Transit	Sunset	Civil twilight ends	Maximum daily marching time
Dyrrachion, 31 October	0644	0712	1226	1739	1808	9.40 hours
Avlona, 31 October	0642	0711	1226	1740	1809	9.45 hours
Kastoria, 25 December	0727	0758	1236	1715	1746	8.33 hours
Edhessa, 25 January	0720	0750	1245	1740	1810	8.83 hours
Thessalonikē, 18 February	0651	0719	1242	1806	1834	9.72 hours
Average	0705	0735	1237	1740	1810	8.83 hours
Effective Average						9 hours
Effective duration		<——10.08 hours——>				
	<—————— 11.08 hours ——————>					

17 Ibid., pp. 140-49. See also Haldon below, p. 137.

18 United States Naval Observatory, Astronomical Applications Department, Sun and moon data for one day, 1800: http:/aa.usno.navy.mil.data/docs/RS_OneDay.html/, accessed 15 April, 2004. Days and months chosen to correspond to Bohemond's march. The year 1800 is chosen because it is the earliest date for which the US navy has compiled data, it eliminates the influence of mankind on climate in modern times, and because the climate at that time is considered to have been comparable to that at the turn of the eleventh and twelfth centuries.

From November to February, Bohemond's forces would have had around nine hours effective marching time per day. Sunrise to sunset would have been around ten hours and allowing for an hour after first light in order for the vanguard to eat, decamp, and tend to the animals before beginning the march to an hour before last light for the rearguard to end it in time to encamp, tend to the animals, and eat, then around nine hours would have been the average effective marching time.

What would have been the maximum numbers which could have followed Bohemond in one division along such roads given such marching times? To a degree the answer depends upon what may have been a minimal knight's "troop".[19] However, for the purposes of modelling it does not matter how large or small or variegated such troops may have been. Lords such as Bohemond and Geoffrey of Roussillon would have had many men and some poor knights may have had few. Some lords will have had several warhorses but poorer knights may have had only one. For the expeditions of the First Crusade, for which it would be stretching a point to even refer to the contingents as "armies" in any conventional meaning of that word, there was no doubt enormous variety, even amongst Bohemond's forces, which appear to have been amongst the most "professional" of the various contingents. Increasing or decreasing the size of the average troop, even changing the ratio between men and animals, would make little difference to the modelling.

For the purposes, I consider troops of one knight, one squire, one groom, and three footmen. The knight rides a riding horse. His warhorse carries its saddle and arms and feed, but no more. These were highly prized animals to be cared for for battle. In the later stages of the Crusade circumstances forced harsher treatment of them but that would not have been so at the outset. Knights would have had at least one attendant, a squire in popular speak, and at least one servant or groom. These may have walked when possible to rest the animals but they would have needed to be able to ride if necessary. Then there must have been pack animals for tents, camping equipment, clothing, arms, supplies, water if necessary, etc. The squire leads the warhorse from his own horse, with two pack animals behind it. The groom, also riding, leads two pack animals. It is possible to lead more than three horses from a lead horse, but the greater the number the more difficult it is. This would have been about as much as could have been managed handily over difficult terrain such as that to be found in the Balkans. Three footmen bring up the rear. Horses are about 2.5 metres long and another 2.5 metres must be allowed between each in case one rears or stumbles. Closed up more than this, one going down could bring the entire troop down.[20] The three footmen pace out at something

---

19 Consult J. France, *Western warfare in the age of the Crusades, 1000-1300* (Ithaca, 1999), p. 68; B. S. Bachrach, "William Rufus's plan for the invasion of the mainland in 1101", in R. P. Abels and B. S. Bachrach, eds, *The Normans and their adversaries at war: essays in memory of C. Warren Hollister* (Woodbridge, 2001), 31-63, pp. 47-50.

20 Cf. G. J. Wolseley [General Viscount], *The soldier's pocket-book for field service*, 5th ed. (London, 1886), p. 85, where he wrote that in hilly country on difficult roads animals

under a metre and about 2.5 metres needs to be allowed to the troop behind. The *Via Egnatia* allows two horses to be riddden abreast and the three footmen to march abreast. Standard marching rates for infantry have always been around three miles or 4.83 kilometres per hour and it is unlikely that that speed could have been bettered for an extended route march such as was undertaken by Bohemond's forces.

Table 1.2: Modelled knight's troop

			Metres		Metres
Knight		Squire	2.5	}	
			2.5	}	
Groom		War Horse	2.5	}	
			2.5	}	Total
Pack animal		Pack animal	2.5	}	22.5
			2.5	}	metres
Pack animal		Pack animal	2.5	}	
			2.5	}	
Footman	Footman	Footman	1	}	
			2.5	}	
			Next troop		

Each troop occupies 22.5 metres and the column moves at 4.83 kilometres, 4,830 metres per hour. So 4,830 ÷ 22.5 troops can pass any one point in one hour = 214.7 troops. In Table 1.3 each 214.7 troops is labelled a "company".

Table 1.3: Maximum possible daily march

				Head of company reaches camp			
		1					
		2	1				
		3	2	1			
		4	3	2	1		
		4.5	4	3	2	1	
Camp	0 hours 9 hours	0.5 hours 8.5 hours	1.5 hours 7.5 hours	2.5 hours 6.5 hours	3.5 hours 5.5 hours	4.5 hours	Camp
	4.5	4	3	2	1		
	3.5	3	2	1			
	2.5	2	1				
	1.5	1					
	0.5						
				Rear of company reaches camp			

should be disconnected from one another to avoid one bringing others down.

The vanguard would have had to encamp by sun transit around 1235 hours in order for the rearguard to encamp before dark, so a maximum of 4.5 companies by 214.7 troops could march in one division if maximum distance of march were to be achieved: 965 knights, squires, and grooms, 2,895 footmen, and 7,730 animals. In the conditions of the tracks and roads through the Pindos mountains, the best of which would have been the *Via Egnatia*, and within the parameters here, Bohemond's forces could have been no larger than this and his maximum march per day could have been only around 21-22 kilometres. If they were larger, the march would have slowed down accordingly. 5.5 companies would force the vanguard to encamp after only 3.5 hours and reduce the daily march to 17 kilometres. 6.5 companies would reduce it to 12 kilometres, 7.5 companies to 7 kilometres, and beyond that the division would effectively be unable to move at all.

At Kastoria Bohemond attempted to buy provisions from the inhabitants but was refused. Thereupon he took by force what he needed.[21] No doubt the inhabitants refused to sell not because they were afraid that the Crusaders were pillagers and murderers, as the *Gesta Francorum* claimed, but simply because whatever they had they needed to survive the winter. The grain that he would have needed above all would have been already sown as winter corn, leaving only sufficient to last to the next harvest. Kastoria did lie beside a lake in the valley of the upper reaches of the Aliakmon and was a natural place to reprovision after crossing the Pindos. But even in a fertile valley, reprovisioning a force such as that modelled here would have imposed a heavy strain on a subsistence economy.

Provisions which men may have needed to stay in good physical condition on campaign are much debated. For the Middle Ages little original evidence has been produced and historians have mostly relied upon the work of ancient historians and the evidence they have analysed. This is not the occasion on which to rehearse the evidence;[22] however, we do have four important pieces of medieval evidence, two

---

21 *Gesta Francorum*, I[iv] (p. 8); Anonymous, *Tudebodus imitatus*, §IX (p. 177).

22 John Haldon has reached a conclusion that daily rations weighed approximately 1.3 kilogrammes. See below and also his "The organisation and support of an expeditionary force: manpower and logistics in the Middle Byzantine period", in N. Oikonomides, ed., *Tó εμπόλεμο Βυζάντιο (9ος - 12ος αι.) / Byzantium at war (9th-12th c.)* (Athens, 1997), 111-51, pp. 124-6. For modelling purposes only, Bernard Bachrach works on a figure of one kilogramme of flour per man per day. Tom Madden calculates that the provisions to be provided by Venice for the Crusaders of the Fourth Crusade for up to a year would have amounted to some 500 litres, probably around 400 kilogrammes, sufficient for around 300 days at 1.3 kilogrammes per man per day. For antiquity see in particular D. W. Engels, *Alexander the Great and the logistics of the Macedonian army* (Berkeley 1978), pp. 123-9; J. P. Roth, *The logistics of the Roman army at war (264 B.C.-A.D. 235)* (Leiden, 1999), pp. 14-44. For the Early Modern period see C. Perjés, "Army provisioning, logistics and strategy in the second half of the 17th century", *Acta historica Academiae Scientiarum Hungaricae*, 16 (1970), 1-51, p. 5.

of which relate to provisions for galley oarsmen and the other two to provisions to be provided for Crusading expeditions by sea.

Table 1.4: Provisioning medieval galley crews and Crusaders[23]					
Provisions per man per day	Galleys of Charles I of Anjou, King of Sicily, ca. 1269-85[24]	*Contractus navigii domini regis cum Venetis*, Venice, 1268[25]	Marino Sanudo Torsello, *Secreta fidelium crucis*, ca. 1310[26]	*Informationes pro passagio transmarino*, Marseilles, 1318[27]	Average

[23] For metrology see Table 1.8 below.

[24] See J. H. Pryor, "The galleys of Charles I of Anjou, King of Sicily: ca. 1269-84", *Studies in medieval and Renaissance history*, n.s., 14 (1993), 33-103, pp. 84-8. The Angevin data in R. Filangieri, ed., *I registri della cancelleria angioina*, 33 vols (Naples, 1950-81) are problematic because the registers were destroyed in World War II and the reconstructions were made from transcriptions made before that, some of which disagree. It is unlikely that provisions differed by such dimensions. The transcriptions giving higher figures for salt "meat" or pork and cheese were probably incorrect.

[25] *Contractus navigii domini regis cum Venetis factus anno domini MCCLXVIII*, in C. du Chesne, *Historiae Francorum scriptores*, 5 vols (Paris, 1636-49), vol. 5, pp. 435-7: "Vobis domino regi ego Marchus Guirinus nuntius domini ducis Venetorum dico quod si transieritis per partes eius ad acquisitionem Terre Sancte, et passagium fuerit a festo beati Iohannis de mense iunii ad unum annum proximum venturum, quod ... Sequitur hic ea quae sunt necessaria ad ponendum in navibus pro sustentatione hominum et equorum quando passagium erit. Pro qualibet persona unum modium frumenti ad mensuram de Accor [Accon] liberatum [libratum] in pane et farina una quarta et dimidia vini ad mensuram Parisiis pro qualibet die, et tantum de aqua, carnes salsate, casei, oleum, alia legumina, armaturae militis pro se et equo suo, et duobus servientibus suis.".

There are five extant published versions of this text and one 16th-17th-century manuscript, the most recent being L. T. Belgrano, ed., *Documenti inediti riguardanti le due Crociate di San Ludovico IX* (Geneva, 1859), 378-82, pp. 380-81. That of Du Chesne is to be preferred for the most part. The emendation of *Accor*, to *Accon*, Acre, is his. That of *liberatum* to *libratum* is Belgrano's.

[26] Marino Sanudo Torsello, *Liber secretorum fidelium crucis super Terrae Sanctae recuperatione et conservatione*, ed. J. Bongars (1611; rpt Jerusalem, 1972), II.iv.x (60-64), esp. p. 60: "Igitur est notandum, quod qualibet die libra una et dimidia panis biscocti cuilibet homini impartitur. ... Datur etiam stipendiariorum cuilibet, omni die mensura una vini, quæ est quarta pars unius libræ, quæ libra in septuagenarium multiplicata numerum perficit mensuram quæbigontium vulgariter appellatur. ... Datur etiam cuilibet prædictorum, uncia una case omni die. ... Dantur similiter in xxx. diebus cuilibet præfatorum, de porcinis carnibus salitis, tres libræ, et totidem unciæ de prædictis. ... Distribuitur etiam in die cuilibet ex stipendiariis antedictis fabæ vel alicuius alterius leguminis, quadragesima particula unius quartarolæ, ad modum mensuræ Venetæ deputatæ.".

See F. C. Lane, "Diet and wages of seamen in the early fourteenth century", in his *Venice and history: the collected papers of Frederic C. Lane* (Baltimore, 1966), 263-8 [trans. from his "Salaires et régime alimentaire des marins au début du XIVe siècle",

Table 1.4 continued

Ship's biscuit	742.5 grammes[28]	365 grammes of grain both of *panis* and flour: total 730 grammes[29]	715 grammes	(1) half a sack of ship's biscuit for 60 days for the Crusaders[30] (2) 837 grammes of *panis* per day for galley crews	756 grammes

Annales: économies, sociétés, civilizations, 18 (1963), 133-8]. The figures here are Lane's, checked for accuracy.

27 A. de Boislisle, "Projet de Croisade du premièr Duc de Bourbon (1316-33)", *Annuaire-bulletin de la Société de l'histoire de France*, 9 (1872), 230-6, 246-55, p. 253: "Secuntur expense necessarie pro hominus et equis. Primo, levabit necessaria pro LX diebus ... Item, biscoitum: pro duabus personis, unum sacum appelatum *de moison*. Item, de vino, duas millayrolas pro qualibet persona. Item, de carnibus salsis, pro quatuor personis, unum quintale. Item, de caseis, pro XV personis, unum quintale. Item, de fabis, pro quatuor personis, unam eminam. Item, de ficeubus, pro XV personis, unam eminam. Item de lentigulis, pro XV personis, unam eminam. ... Secuntur arma necessaria pro galeis, videlicet: ... Item, sunt necessarii in qualibet galea CXXX homines, ... Item, habebunt isti homines, quolibet mense, de pane LXXX quintalia."

28 This figure differs slightly from that in Pryor, "The galleys of Charles I of Anjou", p. 84. Re-examination of the documents and elimination of some referring to galleys carrying passengers leads to the conclusion that the standard biscuit ration was 0.25 *cantarium* per man per month, 37 *cantaria* for galleys with 148 crew, 37.5 for those with 150, and 38 for those with 152. Only one document does not conform to the standard: Filangieri, *Registri della cancelleria angioina*, Reg. 70.444 (vol. 13, 150-63), p. 161. There the ration calculates out to a third of a *cantarium* per man per month. Either the transcription is faulty or some information is not revealed.

29 The problem is what the *modius* or *moggio* of Acre may have been since no nineteenth-century metrologists referred to it. However, two texts, both of which Jacoby suggests may have been based at this point on an official Venetian list of units of weights and measures, both equated the *modius/moggio/moça* of Acre to two Venetian *stai*: 166.6 litres. See D. Jacoby, "A Venetian manual of commercial practice from Crusader Acre", in G. Airaldi and B. Z. Kedar, eds, *I comuni Italiani nel Regno Crociato di Gerusalemme: atti del Colloquio "The Italian communes in the Crusading Kingdom of Jerusalem" (Jerusalem, May 24 – May 28, 1984)* [Collana storica di fonti e studi, 48] (Genoa, 1986), 401-28, p. 421.

The first is a Venetian merchant's manual compiled in Acre around 1270, Venice, Biblioteca Nazionale Marciana, MS. Ital. Class. XI, No. 87 (colloc. 7353), fols 1r-7r, an edition of which is in preparation by Jacoby. At fol. 4r, l. 16, the text reads: "Stera ij de grano de Venecia fase moço j in Acre.". I am grateful to David Jacoby for this information.

The second is the early fourteenth-century Venetian merchant's manual known as the *Zibaldone da Canal*. At fol. 42v, ll. 2-4, the text reads: "Formento e tuto ligume se vende a

Table 1.4 continued

Cheese	39.5 or 69 grammes		40 grammes	45 grammes	41.5 grammes
Salt "meat" or pork	49.5 or 99 grammes		52 grammes	170 grammes	90.5 grammes
Legumes	80 grammes[31]		0.13 litres = approx. 104 grammes	0.167 litres = approx. 133 grammes	106 grammes
Figs				0.037 litres	
Lentils				0.37 litres	
Wine	2.33 litres	2.8 litres	0.536 litres	2.13 litres	

There is remarkable consistency in these figures. Sanudo stated that the provisions were those customarily given to Venetian galley crews and provisions of this order were no doubt common across the Mediterranean around 1300 and probably had been so for centuries. They give a good insight into what men needed to remain in condition for moderate work: around 750 grammes of biscuit, 40 of cheese, a minimum of 50 of salt meat, and around 100 grammes of dried legumes. The meat and legumes for the Marseillese Crusaders of 1318 were either indulgent or reflected increasing difficulty in raising men for Crusades. The appalling performance of Angevin fleets in battle may be reflected in their wine allowance. Attempts to keep them quiet during weeks at sea may reflect those for the French Crusaders of 1268 and the Marseillese of 1318.

These results produced by applying values recorded by nineteenth-century metrologists are consistent and reasonable. One may doubt whether values recorded in the nineteenth century for traditional pre-metrification weights and measures were in fact the same as those of the thirteenth and fourteenth centuries; however, it does appear to have been so here. It is important to note that the biscuit allowance for Angevin crews calculated from the Neapolitan *cantarium* of approximately 89 kilogrammes, gives a biscuit ration consistent with other sources.

Satallia [Antalya] a mesura, è nome pinacho e geta inn Acre moço J e in Venexia stera IJ ...". See A. Stussi, ed., *Zibaldone da Canal: manoscritto mercantile del sec. XIV* (Venice, 1967), p. 70, and J. Dotson, *Merchant culture in fourteenth-century Venice: the Zibaldone da Canal* (Binghamton, 1994), p. 122.

30 This quantity is not quantifiable because a "sack" was not a measure.

31 In "The galleys of Charles I of Anjou", Table Four (p. 85), I erred at this point. The document reads one *salma* rather than one *cantarium*. The correction means that the weight should be approximately 80 grammes rather than 20 as stated in 1993 at a ratio of 0.80 kilogrammes per litre for dried legumes, which is approximately correct.

Around one kilogramme per man per day may be taken as that needing to be carried by men or animals when we take into account necessary extras. Crusaders could not have carried sufficient biscuit, or more probably grain, which would have lasted better, for more than a few weeks, however. They would have intended to acquire grain *en route*. Sanudo wrote that 90 *librae* of biscuit could be made from a Venetian *sextarium* of wheat and 105 *librae* from good wheat.[32] There is also Angevin data on grain required to produce biscuit.

Table 1.5: Biscuit from grain according to Marino Sanudo and Angevin Registers						
Marino Sanudo Torsello, *Secreta fidelium crucis*, IV.iv.x			Wheat One *Sextarium*	Biscuit *libre*	Wheat One litre	Biscuit Kgs
			1	90	1	0.515
And from good wheat:			1	105	1	0.601
Angevin Registers	Grain *Salme*	Biscuit *Cantaria*	Grain One *salma*	Biscuit *Cantaria*	Grain One litre	Biscuit Kgs
Reg. 14.1141 (vol. 4, pp. 173-5), here p. 175.	95	190	1	2	1	0.359
Reg. 21.79 (vol. 6, pp. 21-7), here p. 24.	50	75	1	1.5	1	0.269
Reg. 21.1881 (vol. 6, pp. 345-73), here p. 354	880.5	1320.75	1	1.5	1	0.269
here p. 365	975.5	1434.75	1	1.47	1	0.264
Reg. 121.30 (vol. 27, parte II, p. 461)[33]	1	1.75	1	1.75	1	0.314

32 Marino Sanudo Torsello, *Liber secretorum fidelium crucis*, II.iv.x (pp. 60-61): "Iterum est sciendum, quod sextarium panis biscocti, ex LXXXX. librarum numero computatur: sed ex boni frumenti sextario bene extrahuntur CV. libræ."

The two clauses appear to say different things. Checking the texts of the two manuscripts Bongars used at this point, Rome, Biblioteca Apostolica Vaticana, MSS. Reginae Cristinae 548, fol. 30r and Ottoboniani Lat. 906, lib. II, pars IV [unfoliated and the section in question not numbered], and those of five other manuscripts of the second and third redactions: Rome, Biblioteca Apostolica Vaticana, MS. Vat. Lat. 2003, fol. 34r; London, British Library, Add. MSS 27376, p. 46; Oxford, Bodleian Library, MS. Col. Tanner 190, fols 40v-41r; Brussels, Bib. Roy, Ms. 8347, fol. 32v; and Florence, Biblioteca Medicea Laurenziana, MS. Plut. XXI, 23, fol. 20r, the sentence should read: "Iterum/Item est sciendum quod sextarium panis biscoctati ex lxxxx. librarum numero computatur, sed ex boni frumenti sextario bene extrahuntur cv. libre." Biscuit weighed in *libre* had to be made from grain measured in *sextaria* and the second clause's sense outweighs that of the first.

33 This was an order to make biscuit for the entire fleet at the rate of 1.75 *cantaria* of

Some variation in the data may have reflected different grain qualities or baking processes and there is no need to doubt the data *per se*. Sanudo was very experienced and in the Angevin case several different historians, all experienced in the Neapolitan archives, were involved in the transcriptions.

Compare these figures to those compiled by Foxhall and Forbes from experimentation and classical sources.[34] Pliny the Elder gave an average weight for wheat from various parts of the Empire of 20.8 Roman *librae* per *modius*,[35] 780 grammes per litre, modified by Foxhall and Forbes to 772 grammes. Foxhall's experiments with milling and baking as done by troops in antiquity gave an extraction rate for flour after loss of bran of 94.6%,[36] meaning that 772 grammes of grain would yield 730 grammes of flour, and showed that bread weighed a half as much again as the flour. Extrapolating from his figures, one litre of wheat weighing 772 grammes would produce 730 grammes of flour and 1.075 kilogrammes of bread. Foxhall did not bake his bread again to make biscuit.

I weighed one litre of Australian wheat at 850 grammes. Wheat varies in weight according to species, variety, and water content and Australian wheat is very hard and dense, so I reduced the weight to 750 grammes. I had no way of milling it in a traditional way so I pulverized it in a processor and deducted 40 grammes for a bran allowance at Foxhall's figure of 94.6%: total 710 grammes. To this was added 10 grammes of salt, 14 grammes of yeast, and 500 grammes of water to make a very firm dough. The weight of the dough was 1,234 grammes. After first baking, the loaves weighed 1,110 grammes and after cooling and a second baking and cooling they weighed 1,025 grammes. The experiment yielded ratios as follows:

Volume of grain to weight of biscuit produced	1 litre	:	1.158 kgs
Weight of flour to weight of biscuit produced	1 kg.	:	1.444 kgs

Medieval biscuit may not have been leavened but it would have contained salt. I conclude that the ratio between weight of flour and weight of biscuit ought to be around 1:1.45. If biscuit was unleavened, if somewhat less water was used, or if it was further desiccated by drying, the figure may be reduced even further. Pliny the Elder wrote that *panis militaris* weighed one third more than the grain of which it

good and useful biscuit from each *salma* of grain "prout experientia compertum est".

34 L. Foxhall and H. A. Forbes, "Σιτομετρεία: the role of grain as a staple food in classical antiquity", *Chiron*, 12 (1982), 41-90.

35 Pliny, *Natural history*, Gaius Plinius Secundus, *Historia naturalis*, in H. Rackham, trans., Pliny, *Natural History*, 10 vols (London, 1938-62), XVIII.xii(66) (vol. 5, p. 232).

36 Note that the *Zibaldone da Canal* said that in the calculation of 100 *some* (*salme*) of wheat an allowance was made for 3 *some* and 3 *tonboli* (*thumuli*) of bran (*çemola*), i.e. 3.375%. See Stussi, *Zibaldone da Canal*, p. 52 and John Dotson's notes in *Merchant culture in fourteenth-century Venice*, p. 97.

was composed,[37] and in the eighteenth- to ninteenth-century British navy biscuit was still considered good if it weighed one third more than the flour.[38] I will use a weight ratio of 1:1.33. Remembering that analysis of men's rations resulted in confidence in the metrology, on this basis the 515 or 601 grammes of biscuit that Sanudo wrote could be produced from a litre of grain should have required only around 410-475 grammes of grain. Angevin biscuit, ranging from 269 grammes per litre of grain to 359 grammes, should have required as little as 210 to as much as 285 grammes of grain. These figures are far too low, even allowing for soft medieval wheat strains and ineficient threshing and winnowing techniques. There is something wrong, but all attempts to resolve the conundrum have failed.[39]

As important as provisions for men were those for animals. European horses were normally stall fed from antiquity to the nineteenth century. They were not the grass-grazing ponies of the Huns or Mongols and needed some grain, usually barley in the Middle Ages, in their diet. There is considerable debate about their grain, and also hay or grass, needs when in moderate work such as during a route march.[40]

Accepting the dangers of translating modern figures back to medieval animals, it is nevertheless instructive to consider daily rations recommended for cavalry horses during sea voyages in the ninteenth and early twentieth centuries: around 10-12 pounds of hay and around 10 pounds of grain or carrots.[41] Inactive horses

---

37 Pliny, *Natural history*, XVIII.xii(67) (vol. 5, p. 232): "... lex certa naturae ut in quocumque genere pani militari tertio portio ad grani pondus accedat, ...".

38 P. Kemp, ed., *The Oxford companion to ships and the sea* (Oxford, 1976), p. 85. Cf. Perjés, "Army provisioning, logistics and strategy", p. 5.

39 It is relevant that one Angevin document, the only one to state requirements in terms of grain rather than biscuit, gave a figure of 1250 *salme* of wheat for 25 galleys for two months, an average of 25 *salme* per galley per month or 2.46 litres of grain per man per day for a galley of 150 men. The daily ration of biscuit on the Angevin galleys was 742.5 grammes, which at a ratio of 314 grammes of biscuit to one litre of grain "prout experientia compertum est" would require 2.38 litres of grain, almost precisely what this one document stated. See Filangieri, *Registri della cancelleria angioina*, Reg. 13.902 (vol. 3, p. 276) [13 Aug., 1278].

40 A. Hyland, *Equus: the horse in the Roman world* (New Haven, 1990), pp. 87-90, accepts figures as low as 1.5 kilogrammes of barley and 4.5 of hay per horse per day. Roth, *Logistics of the Roman army at war*, pp. 62-5, concludes that around 2.5 kilogrammes of grain and 7 of hay or dry fodder per day is a reasonable estimate. P. Erdkamp, *Hunger and the sword: warfare and food supply in Roman Republican wars (264-30 B.C.)* (Amsterdam, 1998), pp. 37-9 is inclined to accept a figure as high as 5 kilogrammes of grain per horse per day. Haldon, "Organisation and support of an expeditionary force", pp. 126-7 and n. 57 concludes that around 2.5 kilogrammes of barley and 5.5 of hay were necessary.

41 See A. Shirley, *Remarks on the transport of cavalry and artillery* (London, 1854), p. 19 [6 pounds of oats, 10 pounds of hay, and 0.5 of a peck (1 gallon = 4.545 litres) of bran]; E. E. Martin, *The transport of horses by sea* (Calcutta, 1901), pp. 33-6 [two parts bran to one of oats plus some hay or chaff (no actual figures)]; M. H. Hayes, *Horses on board ship: a*

aboard ship would need only around two thirds the food when engaged in moderate work and that may be estimated at around 25 pounds, 11.34 kilogrammes, a half being grain and a half hay. General Wolseley gave the daily campaign ration of horses as 12 pounds each of oats and hay.[42] On a route march hay could be replaced by grazing, or more probably by cut grass, but that required two to three times the weight of grass as of hay because of its higher water component.

As far as I know, there are no medieval figures for the feed requirements of horses on campaign; however, there are some data for provisioning horses at sea.

Table 1.6: Provisions per day for horses transported by sea

	Treaty of Venice, Venice, April 1201[43]	*Contractus navigii domini regis cum Venetis*, Venice, 1268[44]	Registers of the Angevin archives, Naples 1278-80[45]	*Informationes pro passagio transmarino*, Marseilles, 1318[46]

guide to their management (London, 1902), pp. 176-7 [6 pounds of bran or 3 each of bran and oats, 17 pounds of hay, and 5 pounds of carrots; the Government allowance being 4 pounds of oats, 6 pounds of bran, and 12 pounds of hay]; F. Smith, *A manual of veterinary hygiene* (London, 1905), pp. 118, 915-17 [an average of 5 pounds of corn (grain), 4 pounds of bran, 10-12 pounds of hay, and 5 pounds of carrots]; Wolseley, *Soldier's pocketbook*, p. 217 [5 pounds of oats, 5 pounds of bran, 10 pounds of hay]; United States War Department, *Manual for stable sergeants* (Washington, 1917), pp. 63-4 [14 pounds of hay and 12 pounds of oats, corn, or barley]. See also Roth, *Logistics of the Roman army at war*, pp. 64-5.

42 Wolseley, *Soldier's pocket-book*, p. 74. Cf. Perjés, "Army provisioning, logistics and strategy", p. 17.

43 G. L. Fr. Tafel and G. M. Thomas, eds, *Urkunden zur älteren Handels- und Staatsgeschichte der Republik Venedig*, 3 vols (1856-7; rpt, Amsterdam, 1964), no. 92 (vol. 1, p. 365), no. 93 (vol. 1, p. 371): "Pro uno quoque equo modia [blave] tria ad mensuram Venetie, et de aqua, quantum sufficiens erit.". Consult Madden below, pp. 211-14.

44 *Contractus navigii domini regis cum Venetis*, pp. 435-7: "Vobis domino regi ego Marcchus Guirinus nuntius domini ducis Venetorum dico quod si transieritis per partes eius ad acquisitionem Terre Sancte, et passagium fuerit a festo beati Iohannis de mense iunii ad unum annum proximum venturum, quod ... Sequitur hic ea quae sunt necessaria ad ponendum in navibus pro sustentatione hominum et equorum quando passagium erit. ... Pro quolibet equo quatuor modia ordei ad mensuram de Accon currentem tempore, quo dominus rex Francorum erat ibi, una vata plena foeni, quae erit de tour novem pedum, et de longitudine quinque pedum, et quindecim quartas aque ad mensuram Parisiensis pro qualibet die."

45 See Filangieri, *Registri della cancelleria angioina*, Reg. 80.522 (vol. 18, pp. 245-6) [13 Aug., 1278]; Reg. 81.90, 267 (vol. 19, pp. 23, 66-7) [7 Sept, 1278; 14 Aug., 1278]; Reg. 95.213 (vol. 23, pp. 153-4) 8 Aug., 1280]; Reg. 101.43 (vol. 24, p. 116) [2 Nov., 1280].

46 De Boislisle, "Projet de Croisade du premièr Duc de Bourbon", pp. 253-4: "Item, si voluerint portare equos, portabit [navis] CXX cavallos. ... Secuntur expense necessarie pro hominibus et equis. Primo, levabit necessaria pro LX diebus pro equis, videlicet: quatuor eminas ordei et quatuor eminas finate [fenate?] pro quolibet equo. Item pro tribus equis,

Table 1.6 continued

Barley		1.83 litres	2.63 litres[47]	2.67 litres
Hay		3.7 litres-[48]		2.67 litres
Barley bran				0.67 litres = 0.16 kgs[49]
Grain	2.74 litres			
Water		27.94 litres		21.33 litres

The data vary somewhat, unsurprisingly. Louis IX's 1268 contract with Venice specified a hay ration twice that of barley while the Marseillese *Informationes pro passagio transmarino* of 1318 made the two equal. But, the totals were virtually identical. We must assume that in 1201 Venice was to provide a hay ration equal to the grain but not specified and that the same was true of Angevin horse transports. For the purposes of modelling, I average a barley ration for horses aboard medieval ships at 2.7 litres, in accordance with all of the data except Louis IX's contract with Venice. Barley is lighter by volume than wheat and I use Foxhall's figure of 587 grammes per litre.[50] At 587 grammes per litre 2.7 litres would have weighed 1.6 kilogrammes, which should be increased by a third for horses in work on a route march: 2.4 kilogrammes. This is low by comparison to nineteenth-century rations; however, that may be explained by logistics. On large nineteenth-century sailing ships and steamers stowage of provisions was no problem. On the smaller medieval *naves* it would have been more of a problem and on the very small horse-transport galleys it would have been critical. Rations on such ships would no doubt

---

unam basachiam de V cannis, plenam paleis. Item, formam [seonnum?] ordei, eminam unam pro qualibet bestia. ... Item aquam pro equis, M.M.CCCC. millayrolas.".

47 The Angevin documents are problematic. No less than five record that the standard allowance for feed for horses "*per noctem*" [or variants in Italian] was a third of a *thuminus*, a *tomolo*, according to the *salma generale*, giving the absurd figure of 18.44 litres per horse per night. However, for no other rations of any kind was the duration given in the registers in terms of "nights". Why would horses' rations alone be given in terms of nights and then be absurd? Because " *per octenam*" for a period of eight days, a seven-day week since days were counted inclusively, was mistranscribed as " *per noctem*". Anyone who has ever read a heavily abbreviated medieval chancery manuscript will understand that "*per octenam*" could very easily be transcribed as "*per noctem*". If it was not the modern historians who were responsible, then the medieval chancery scribes could well have been so themselves when translating whatever was the Franco-Italian vernacular for *octena* into Latin. This is the solution and the ration should be reduced to $18.4 \div 7 = 2.63$ litres, bringing it into alignment with other figures.

48 The external volume of the *vata* was approximately 1,355 litres, giving a daily hay ration under 3.7 litres over 365 days when the internal volume of the *vata* is considered.

49 Bran is light by volume. A litre compacted firmly weighs around 240 grammes.

50 Foxhall and Forbes, "Σιτομετρεία", p. 43.

have been kept to a bare minimum. In fact the ration accords roughly with those calculated for Romano-Byzantine cavalry horses. No doubt a warhorse would have received more, other horses less, and mules or asses even less again;[51] however, an average of 2.4 kilogrammes of hard fodder per horse per day is reasonable.

Horses, mules, and donkeys are nowhere near as efficient animals as men. If made to carry more than around 20% of their own weight they will suffer exhaustion and/or physical damage.[52] Medieval warhorses of 15 hands weighed around 550 kilogrammes and could be expected to carry up to 110 kilogrammes if necessary. Smaller pack animals and mules would be able to carry proportionately smaller loads. They would also consume proportionately less in feed. Terrain, track conditions, and required speed would also make a difference, of course. But in normal circumstances of non-forced route march over reasonable ground, a figure of around 200 lbs or 90 kgs seems to be an arguable figure for what the type of horses or mules found in medieval bagage trains might be expected to carry for an extended non-forced route march.[53] One text on imperial military expeditions produced for Emperor Constantine VII noted that cavalry horses carrying men could carry an additional 4 *modioi* of barley each for their fodder, unridden saddle horses could carry 8 *modioi* as well as their saddles, and pack animals 10 *modioi* each.[54] Arguably, the *modios* used here was the *basilikos modios* of 17.1 litres, which in terms of barley would have weighed around 10 kilogrammes.[55] A pack horse was apparently expected to carry a load of around 100 kilogrammes, a saddle horse around 80, and a ridden horse around 40. These figures are reasonable, if at the upper end of the range, and will be used in the modelling.[56]

---

51 One Angevin document put the ration for *asini*, asses or donkeys, at a sixth of a *tomolo*, half that for horses. See Filangieri, *Registri della cancelleria angioina*, Reg. 101.43 (vol. 24, p. 116) [2 Nov., 1280].

52 Smith, *Manual of veterinary hygiene*, p. 773.

53 Engels, *Alexander the Great and the logistics of the Macedonian army*, p. 129; B. S. Bachrach, "*Caballus et caballarius* in medieval warfare", in H. Chickering and T. H. Seiler, eds, *The study of chivalry: resources and approaches* (Kalamazoo, 1988), 173-211, pp. 177, 181; Haldon, "Organization and support of an expeditionary force", pp. 126-7.

54 Constantine VII Porphyrogennētos, *Constantine Porphyrogenitus: three treatises on imperial military expeditions*, ed. and trans. J. F. Haldon [CFHB, vol. 28] (Vienna, 1990), C.548-53 (pp. 128-9).

55 Cf. Haldon, "Organization and support of an expeditionary force", pp. 126-9. Haldon calculates the *basilikos modios* at 40 Roman pounds = 12.8 kilogrammes and for reasons then argued out arrives at the conclusion that the smaller *annonikos modios* was used and that the loads borne would have been approximately 85 kilogrammes, 68 kilogrammes, and 34 kilogrammes respectively. For my purposes it really does not matter a great deal and it is interesting that roughly the same conclusions were reached by two different methods.

56 General Wolseley considered that pack horses' maximum load should be 200 pounds, including the pack saddles, which averaged 30 pounds: load 77 kilogrammes. See Wolseley, *Soldier's pocket-book*, pp. 72, 77-8, 84.

In the modelled knight's troop the knight, squire, and groom would each carry 40 kilogrammes on their horses. The warhorse would carry its own saddle and the knight's accoutrements as well as its own hard fodder, and the pack horses would carry 100 kilogrammes each. Assuming that two of them had to carry equipment and only two could carry food and that the three footsoldiers could each carry 25 kilogrammes of food for the men,[57] we arrive at the following model.

Table 1.7: Provision-carrying capacity of modelled knight's troop

			Load
Knight (40 kgs)		Squire (40 kgs)	80 kgs
Groom (40 kgs)		Warhorse (proportionate)	40+ kgs
Pack Horse (100 kgs)		Pack Horse (100 kgs)	200 kgs
Pack Horse (0 kgs)		Pack Horse (0 kgs)	
Footman (25 kgs)	Footman (25 kgs)	Footman (25kgs)	75 kgs

Excluding the warhorse, the modelled troop has a total provision-carrying capability of 395 kilogrammes. Its daily consumption is 6 kilogrammes of human food and 16.8 of dry fodder for the horses: total, 22.8. It could not march more than 17-18 days without resupply. The warhorse would have carried around 42 kilogrammes for its own fodder. The inescapable conclusion is that Bohemond's forces would have required resupply around 6.5 times over between landing in Albania and crossing the Vardar. Playing around with the composition of the forces would make little difference. Increasing the number of animals in relation to men would merely exacerbate the problem since men consume less and can carry more proportionately than animals. Assuming that the forces brought with them supplies for the first 17-18 days, the 5,790 men in the modelled army would have required another 535 tonnes of wheat to reach the Vardar and the animals would have needed another 1,716 tonnes of barley. Somehow, that order of provisions had to be raised from the land. Is it any wonder that there were problems at Kastoria, and no doubt elsewhere? The valleys of the Vjosa and Drino rivers in November must truly have been lands of milk and honey for Bohemond's forces.

On a route march horses and men would require water and the horses would need grass. Horses are large animals but they have small stomachs, an average one containing only around a half a cubic foot of material. They cannot eat and drink much at a time but they need to do so often and take considerable time. General Wolseley insisted on giving small amounts of feed and water and allowing grazing whenever a break was taken on march.[58] Some water, and in some circumstances

[57] Twenty kilogrammes or 55 pounds is a reasonable load for a fit man. Cf. Haldon, "Organization and support of an expeditionary force", p. 124.

[58] See Wolseley, *Soldier's pocket-book*, pp. 73, 85.

cut grass, would have to have been carried in addition to the loads modelled above. Consumption on the march during the day may have alleviated to a degree the logistical considerations which follow; however, this cannot be quantified.

Finding water would not have been a problem through most of the Balkans but drinking it would have been. Horses require around eight gallons or 36.5 litres per day under moderate work and in mild climate.[59] If led to water in the most mathematically efficient manner, Bohemond's modelled 7,730 animals would have needed a lake or river with banks 8-10 kilometres long to water them simultaneously. Campgrounds like that would rarely have existed and watering processes must have been much more drawn out. Men would normally have brought water to animals in leather buckets rather than leading them to water. In places with no lake or running water and depending on wells, they would have needed around 14,000 well-lifts of 20-litre buckets to water the animals alone for a single day.[60] At an average of around eight litres per man per day, an accepted water requirement for troops in moderate work,[61] the 5,790 men would have required an additional 46,320 litres themselves, another 2,316 well-lifts. How could that be done in an overnight camp unless there were many wells? We are compelled to consider that watering such a force may well have taken the whole night in some circumstances and that that in itself may have influenced entire orders of march.

Finally, on a route march horses would have depended upon grazing rather than hay. The hay rations for medieval horse transports cannot be converted to weights because the compression and chopping of the hay are unknown; however, the nineteenth-century recommended hay ration was around 10-12 pounds.

---

59 Information supplied by the Department of Veterinary Anatomy and Pathology, University of Sydney. Haldon's estimates are similar: 5-8 gallons (22.75-36.4 litres) per day. See Haldon, "Expeditionary force", p. 127.

Lieutenant Martin reported that horses aboard ship required eight gallons per day and Captain Hayes that the Government allowance was 10 gallons per horse per day, which allowed for wastage, which could often be large. Colonel Shirley recommended 6 gallons per horse per day but his experience was on the cold North Atlantic run to Canada. See Martin, *Transport of horses*, p. 28; Hayes, *Horses on board ship*, p. 155; Shirley, *Remarks on the transport of cavalry and artillery*, p. 29; Smith, *Manual of veterinary hygiene*, pp. 747-9, 915-16.

Anne Hyland, an equestrian herself, considers a figure of five gallons, and more in hot weather, from practical experience. See Hyland, *Equus*, p. 90. General Wolseley wrote that horses could thrive on six gallons a day when not worked hard but would need 8-12 when at hard work, according to climate. See Wolseley, *Soldier's pocket-book*, p. 74. Eight gallons is about right for an extended route march through the Balkans in late autumn and winter.

60 This would, admittedly, be reduced substantially by animals grazing on grass. The water content of grass does reduce their need to drink water. See Hyland, *Equus*, p. 96. However, I know of no way to quantify this.

61 Cf. Wolseley, *Soldier's pocket-book*, p. 95.

Because of the higher water content of grass, horses accustomed to hay but turned over by necessity to grass will need two to three times the weight of grass as they would have had hay.[62] I opt for a medium of 11 pounds and 2.5 times and model a weight of 27.5 pounds or 12.5 kilogrammes of green grass per animal per day, which may be low. The Marquis de Puységur wrote that in the spring when meadow and corn was mowed horses' daily rations should be 50-60 pounds: 24.5-29 kilogrammes.[63]

Grazing green grass requires approximately 45 to 65 minutes per kilogramme, depending on the length of the grass and its distribution. A horse grazing 12.5 kilogrammes of grass would need most of the day and night not actually spent on the march to do so, and that does not include time needed to eat the dry fodder ration and drink.[64] The consequences of that for a route march are obvious but there is a fortunate mitigating factor. This is that horses do not require a great deal of sleep. By nature they will take a series of broken rests and feed in the intervals.[65]

What area would have produced that much grass? Haldon quite rightly points out that it would have depended upon "quality of pasturage, seasonal variations, and so forth" but estimates that 20 horses would have grazed an acre of medium quality pasture.[66] Hyland claims that an acre can produce around 1.5 tons, 1,525 kilogrammes, of "hay crop".[67] Even if by this she means green grass rather than dried hay, it should be capable of feeding not 20 but 120 animals. She realizes that ancient crops did not yield as much as modern ones, although one wonders whether that was true of natural grasses. Whatever the case, it is difficult to comprehend how one acre, 4,045 square metres, could possibly have grazed 120

---

62 The United States *Manual for stable sergeants* stated that the figure was three times. See *Manual for stable sergeants*, p. 64. Roth's calculation, based on Perjés, is that it should be two times. See Roth, *The logistics of the Roman army at war*, p. 64 based on Perjés, "Army provisioning, logistics, and strategy", p. 16.

63 J. F. de Chastenet, Marquis de Puységur, *Art de la guerre par principes et par règles*, 2 parts in 1 vol. (Paris, 1748), II.ii.11 (part 2, p. 30); 2 tomes (Paris, 1749), II.ii.11 (tome 2, p. 63). Puységur's rations for garrison horses were also high: 15 pounds of hay and 5 of straw or 18 pounds of hay alone, and a half a *boisseau* of oats, 13 litres. *Loc. cit.* in both cases. Wolseley recorded that when green forage was given in place of hay in Turkey during the Crimean War, 28 pounds of green fodder was given in place of 10 pounds of hay. See Wolseley, *Soldier's pocket-book*, p. 74.

64 *Manual for stable sergeants*, pp. 63-4. Cf. Bachrach, "Caballus et caballarius", p. 180; J. Haldon, *Warfare, state and society in the Byzantine world 565-1204* (London, 1999), p. 168 & n. 87; Idem, "Organization and support of an expeditionary force", pp. 126-7 & n. 58; Hyland, *Equus*, p. 90.

65 Smith, *Veterinary hygiene*, pp. 139, 143, 145.

66 Cf. Haldon, "Organization and support of an expeditionary force", pp. 126-7.

67 Hyland, *Equus*, p. 93. General Wolseley wrote that one acre of meadow gave from one to three tons of hay according to its quality. Even if he meant green hay grass, these are also very high figures. See Wolseley, *Soldier's pocket-book*, p. 87.

animals. The per animal quota of 33.71 square metres is a circle with a radius of only 3.275 metres. That a horse could graze 12.5 kilogrammes of green grass from such an area is difficult to accept. Van Creveld claims that an acre of green "fodder" could feed 50 horses, a figure accepted only with misgivings.[68] There is a need for more research in the sources and for experimentation. Haldon's figure may be low, Hyland's high. Fifty horses per acre would mean that at 12.5 kilogrammes of grass per animal per day, 50 horses would consume 625 kilogrammes, somewhat under a half of what Hyland claims good quality modern pasturage can produce. Each animal would have 80.9 square metres or a circle with a radius of 5.10 metres.

This seems to be a reasonable figure but it means that Bohemond's modelled 7,730 animals would have required grazing of 62.5 hectares, an area 790 metres square. If Haldon's lower figure of 20 horses per acre is used, the required area would have been 156.3 hectares, 1,250 metres square. Whatever the case it is apparent that finding campsites large enough to feed the horses every night would not have been an easy task along the tracks through the Pindos ranges.

---

[68] M. Van Creveld, *Supplying war: logistics from Wallenstein to Patton* (Cambridge, 1977), p. 34, supposedly on the authority of De Chastenet, *Art de la guerre* (Paris 1743), ii, 64.

Misgivings are aroused because Puységur did not use the measure, "acre". I have been unable to trace any 1743 edition of his *Art de guerre*; however, Van Creveld's calculation appears to be based on Perjés, "Army provisioning, logistics and strategy", pp. 16-17, where he says that he used a 1749 ed. of Puységur's *Art de guerre* and cites II, p. 64. In the 1748 ed. the information used by Perjés appears at II.ii.11 (part 2, p. 30) and I.xvi.5 (part 1, p. 190). In the 1749 edition used by Perjés the information appears at II.ii.11 (tome 2, p. 63) and I.xvi.5 (tome 1, p. 398).

Puységur wrote that fodder for a horse in the "green" season of May and June should be 50-60 *livres*, 24.5-29.5 kilogrammes per day, very high. Perjés made this 25 kilogrammes. Then Puységur wrote that each horse required a "*trousse*", a truss or bundle of fodder per day, which Perjés equated to the 25 kilogrammes. Puységur then suggested that commissaries needed to know how many square *toises* would be needed to produce a *trousse* if land was sown with wheat, rye, oats or other grain. He hypothesized that "*supposé*", putatively, 20 square *toises* of wheat (*froment*) produced one *trousse*, and he then posed a mathematical question of how many horses could be supported by 560,000 square *toises*, arriving at a figure of 28,000 and then halving that because of the indiscipline, trampling, and consumption of foraging troops. This was an exercise in mathematics, not a practical one in logistics. At this point Perjés wrote that a square *toise* was 76 square metres, whereas in fact it was approximately 3.8. He apparently meant 20 square *toises*. One hundred square metres would therefore feed 1.3 horses per day and a hectare 130 horses. Van Creveld apparently divided Perjés's 130 horses by 2.47105, the number of acres in a hectare, and came up with the figure of 52.5 horses, reducing it to 50 "even upon the most pessimistic assumption", and then made Puységur's "*froment*", which Perjés had made "corn", into "green fodder". It is difficult to comprehend Van Creveld's conclusions.

Green sown wheat will obviously produce more weight of fodder per unit of area than meadow and whether an acre of meadow could support 50 horses per day is arguable.

Moreover, horsess could not have been simply turned out to graze freely overnight because there would have been chaos next morning as units tried to round up their own. There would also have been fights.[69] Rather, they would have had to be picketed where troops were camped and at least some of the grass cut for them. Grooms and footmen attending to their needs would have been vital and sickles and leather buckets essential campaign equipment. Columella wrote that a good worker could mow a *iugerum* of meadow for hay in a day:[70] 240 by 120 Roman feet or 2,525 square metres. By this measure, the 0.625 square kilometres of grazing that Bohemond's modelled force would have required would have taken 247.5 man days to mow it. If the 965 grooms of the modelled knights' troops cut grass for the animals, it would have taken each of them 2.5 hours every day, assuming a 10-hour working day and ignoring what animals may have grazed themselves.

Many who may read this paper will conclude that it is hypothetical conjecture and some of my calculations may well be proved to be wrong. In most instances they are the first such to be applied to the First Crusade. Nevertheless, I believe that they do have some value. Almost without exception Crusade historians have not considered applying what can be known of the logistics of movement of columns of infantry and cavalry in later centuries. However, historians must begin to bring alternative methodologies to bear on questions for which conventional source analyses allow no answer. The intent here has not been to resolve the perplexing issues of Bohemond's march from Avlona to Thessalonikē, issues which will probably never be resolved beyond doubt, but rather to investigate parameters which may cast light on them. In the process I hope that ways of thinking about logistical problems may occur to those reading this volume and that they may bear them in mind when consulting my colleagues' papers.

---

69 See Hyland, *Equus*, p. 97, referring to Polybios, *The Histories*, in W. R. Paton, trans., *Polybius: the histories*, 6 vols (London, 1922-27), VI.33.11 (vol. 3, p. 344). Cf. Wolseley, *Soldier's pocket-book*, p. 72.

70 Columella, Lucius Junius Moderatus, *Res Rustica*, in H. B. Ash, ed. and trans., *Lucius Junius Moderatus Columella on agriculture*, 3 vols (London, 1941-55), XI.2.40 (vol. 3, p. 92). According to Perjés, "Army provisioning, logistics and strategy", p. 17 Chevennières claimed that one man could reap a "*fauche prée*", equated by Perjés to 559 "square fathoms", "toises carrées" one assumes, or 2,125 square metres in a day. I have stayed with Columella's figure.

Table 1.8: Weights and measures[71]			
		Ancient Rome	
Area	Iugerum		= 2,525 sq. metres
Dry goods	Modius	= 16 sextarii	= 8.62 litres
Length	Pes	= 4 palmi	= 0.296 metres
Weights	Libra	= 12 uncie	= 0.32745 kgs
		Byzantium	
Dry goods	Basilikos modios		= 17.1 litres
		Acre	
Dry goods	Modius/moggio/moça		= 166.6 litres
		Marseilles	
Dry goods	Emina	= 2 *Panaux*	= 40 litres
Liquids	Millayrola/millerole		= 64 litres
Weights	*Charge*	= 3 quintale	= 122.37 kgs
	Quintal	= 100 libre	= 40.79 kgs
		Naples	
Dry goods	Salma	= 8 tomoli	= 442.56 litres
	Thuminus/Tomolo	= 24 *misure*	= 55.32 litres
Length	Palmus	= 12 uncie	= 26.367 cms
Weights	Cantarium/cantaro	= 100 rotuli	= 89.10 kgs
	Rotulus	= 33.33 uncie	= 0.891 kgs
		Paris	
Area	Toise carrée	= 36 pieds carrés	= 3.799 sq. metres
Dry goods	Boisseau (of oats)	= 4 *picotins*	= 13.01 litres
Length	Toise	= 6 pieds	= 1.949 metres
Liquids	Quarta	= 2 *pintes*	= 1.86 litres
Weights	Livre	= 2 *marcs*	= 0.4895 kgs
		Venice	
Dry goods	Modius/moggio	= 4 sextaria/stai	= 333.27 litres
Length	Pes/piede	= 12 uncie/once	= 34.75 cms
Gross weights	Centinaio grosso	= 100 libre grosse	= 47.70 kgs
	Libra grossa	= 12 uncia grossa	= 0.48 kgs

[71] Based on the following sources: H. Doursther, *Dictionnaire universel des poids et mesures anciens et modernes contenant des tables des monnaies de tous les pays* (1840; rpt, Amsterdam, 1976); L. Martini, *Manuale di metrologia ossia misure, pesi e monete* (Turin, 1883); C. Salvati, *Misure e pesi nella documentazione storica dell'Italia del mezzogiorno* (Naples, 1970); R. Duncan-Jones, "The choenix, the artaba and the modius", *Zeitschrift für Papyrologie und Epigraphik*, 21 (1976), 43-52; *The Oxford dictionary of Byzantium*, ed. A. Kazhdan, 3 vols (New York and Oxford, 1991), vol. 2, p. 1388.

Chapter 2

The logistics of the Mongol-Mamlūk war, with special reference to the battle of Wādī 'l-Khaznadār, 1299 C. E.[1]

Reuven Amitai

It would not be amiss to suggest that the study of the military history of the late medieval Muslim world was first put on a sound scholarly footing by the late David Ayalon. In a career stretching some sixty years, Ayalon laid the groundwork for the study of military slavery in the Muslim world in general and the Mamlūk sultanate of Egypt and Syria in particular. Ayalon's *oeuvre* comprises many highly detailed technical studies of aspects of the military society of the Sultanate together with far-ranging interpretive essays dealing with important aspects of military life in the pre-modern Muslim world.[2] Of particular importance were his papers on the important, even decisive, role of the Turks and Mongols of the Eurasian steppes in the military and political life of Muslim countries in this era. While recent scholarship has called into question some of his conclusions,[3] on the whole Ayalon's model of the *mamlūk* institution has stood the test of time and remains the bedrock for the study of the Sultanate and much more.

I would like to emphasize one aspect of Ayalon's work which is relevant to the theme of the workshop, and then two reservations. The former is the extremely significant and useful studies on matters of military organization and logistics. These include, of course, the massive three-part study "Studies on the structure of the Mamluk army",[4] as well as "The system of payment in Mamluk military

---

1 Consult Map 2.

2 Most of Ayalon's papers have been collected conveniently in four Variorum volumes: D. Ayalon, *Islam and the Abode of War: military slaves and Islamic adversaries* (Aldershot, 1994); idem, *The Mamlūk military society: collected studies* (London, 1979); idem, *Outsiders in the lands of Islam: mamluks, Mongols and eunuchs* (London, 1988); idem, *Studies on the Mamlūks of Egypt (1250-1517)* (London, 1977). Mention should also be made of the monographs: *Gunpowder and firearms in the Mamluk kingdom: a challenge to a medieval society* (1956; 2nd ed., London, 1988) and the posthumously published *Eunuchs, Caliphs and sultans: a study in power relationships* (Jerusalem, 1999).

3 Note R. Irwin, "Gunpowder and firearms in the Mamluk sultanate reconsidered", in A. Levanoni and M. Winter, eds, *The Mamluks in Egyptian and Syrian politics and society* (Leiden, 2004), 117-39.

4 D. Ayalon, "Studies on the structure of the Mamluk army", *BSOAS*, 15 (1953), 203-228, 448-476; 16 (1954), 57-90 [rpt in *Studies on the Mamlūks*, No. I].

society",[5] "Notes on the *Furūsiyya* excerises and games in the Mamlūk sultanate",[6] and "Discharges from service, banishments and imprisonments in Mamlūk society."[7] I would go so far as to say that without these and other studies we would be unable to write a reasonable military history of the Mamlūk sultanate.

This being said, there are two important *lacunae* in Ayalon's scheme. First, with two small exceptions, he never tackled systematically how the Mamlūks actually fought;[8] and second, his studies of the Mamlūk army were generally conducted without comparison to their opponents. In his discussion of the failure of the Mamlūks to adopt firearms and other gunpowder weapons, he did cast more than a glance at the Ottomans and their acceptance and development of these weapons. But he rarely took a close look at what was going on over the frontier during the time of the early Sultanate. In other words, we find little discussion of either Frankish or Mongol fighting methods and related matters and how this affected the Mamlūk military machine or, conversely, how they themselves were influenced by their ongoing military encounter with the Sultanate.[9]

Let there be no misunderstanding. I am in awe of the scope and depth of Ayalon's research, and one can not criticize what scholars decide not to study, only what they actually do. I am, however, pointing to areas into which Ayalon did not delve deeply and suggesting that only by dealing with these can we move ahead in the study of the military history of the Sultanate and the region as a whole.

A significant step in this field was taken with the publication of Smith's important essay: "'Ayn Jālūt: Mamlūk success or Mongol failure?".[10] This article is an impressive display of interdisciplinary historical research. I have, however, some disagreements with its conclusions. Smith's discussion will be the basis of much of the following; although I hope to also contribute something new here.

Smith presents two major and interrelated theses, which may be summarized as follows. First, on the one hand, in general the Mongols were mediocre troops, not well trained and haphazardly equipped. They were mounted on steppe ponies

---

5 D. Ayalon, "The system of payment in Mamluk military society", *JESHO*, 1 (1958), 37-65, 257-96 [rpt in *Studies on the Mamlūks*, No. VIII].

6 D. Ayalon, "Notes on the *Furūsiyya* excerises and games in the Mamluk sultanate", *Scripta Hierosolymitana*, 9 (1961), 31-62 [rpt in *Mamlūk military society*, No. II].

7 D. Ayalon, "Discharges from service, banishments and imprisonments in Mamluk society", *Israel Oriental studies*, 3 (1972), 25-50 [rpt in *Mamlūk military society*, No. V].

8 The exceptions are some paragraphs in his articles "Ḥarb, iii - The Mamlūk sultanate", in EI.2, vol. 3, 184-90, and "Ḥimṣ (the battle of)", in EI.2, vol. 3, 402-3.

9 This being said, there is some discussion of the salient features of Mamlūk-Mongol relations, as well as of possible Mongol influence on Mamlūk administrative law, in the important study "The Great *Yāsa* of Chingiz Khān. A re-examination", *Studia Islamica*, 33 (1971), 97-140; 34 (1971), 151-80; 36 (1972), 113-58; 38 (1973), 107-56 [rpt in *Outsiders in Islam*, Nos IV (a)-(d)].

10 J. M. Smith, jr, "'Ayn Jālūt: Mamlūk success or Mongol failure", *HJAS*, 44 (1984), 307-45.

fed almost exclusively by pasturage which, although hardy, tired quickly, particularly in the rapid movement necessary in battle. The Mongol "nation-in-arms" made up for its amateurish soldiers by large numbers and the limitations of its mounts were compensated for partially by each trooper bringing several (generally five) on campaign. On the other hand, the Mamlūk sultans fielded smaller armies but their troops were well-trained and equipped, their armour and weapons manufactured professionally. Mamlūk horses were bigger than their Mongol counterparts and fed largely on grain and hay rather than by pasturage. Because of their meagre armour, smaller mounts, and inferior weapons (including bows), the Mongols preferred to fight by launching wave after wave of mounted archers, each retiring before engaging directly. Only after the enemy had been worn down, fled, or perhaps drawn into an ambush by a feigned retreat, would they allow themselves to become involved in hand-to-hand combat. By contrast, the better trained and equipped Mamlūks preferred to remain stationary either on horseback or dismounted and to absorb Mongol attacks and then to launch a frontal attack when conditions were right. Where numbers were more-or-less even or the Mongols did not enjoy significant numerical superiority, then the heavier and more professional Mamlūks enjoyed a decisive edge and stood a better chance for victory. As Smith writes: "... the Mongols made excellent amateur soldiers. But when these excellent amateurs met an equal force of the fine Mamlūk professionals, the result was predictable."[11]

Secondly, the Mongols sought to compensate for the inferiority of their soldiers, equipment, and mounts by bringing large armies to gain numerical superiority. They were, however, stymied in this effort by the logistical limitations of Syria. The pasture and water resources of greater Syria south of the Tauros Mountains and west of the Euphrates were not sufficient to sustain large Mongol armies composed almost exclusively of cavalry for any length of time. This situation was particularly acute during the long, dry Syrian summer, when pasture dried up and water resources were reduced severely. The Mongols, therefore, were condemned by logistics to lose the war with their long-term adversaries. They could maintain large armies in Syria for no more than a few months due to the limited carrying capacity of the country, and then only primarily in the winter. At the same time, smaller armies whose logistical needs were appropriate to the country could not hope to defeat qualitatively superior Mamlūk forces.

In the following I concentrate not on the first thesis, although I will return to it briefly eventually. Rather, I relate in some detail Smith's argument for the logistical limitations of Syria and then provide some comments of my own. Before commencing, I note that Morgan's paper, "The Mongols in Syria, 1260-1300", also

---

[11] Smith, "'Ayn Jālūt", p. 326. I do not do justice to Smith's detailed argument by this short summary. See, for example, the comparison of Mamlūk and Mongol archery on pp. 314-24.

appeared around the same time as the publication of Smith's paper.[12] In general, Morgan reached conclusions similar to those of Smith with regard to the logistical problems that Syria posed for the Mongols. His method, however, was different. Instead of a detailed analysis of what horses or large ponies needed in terms of water and fodder or pasture, and then examining what greater Syria had to offer, Morgan's approach was to collect references which hinted at, or stated explicitly that, the Mongols had problems feeding and watering their horses in Syria, particularly in the summer, and that therefore they were unable to maintain a large presence in Syria over time due to climatic and geographic conditions. Again, logistics dictated the outcome of repeated Mongol campaigns.[13]

At this point it is useful to give some of the calculations provided by Smith to strengthen his case for the logistical limitations of Syria. For example, he suggests that in 1299 the Īlkhān Maḥmūd Ghāzān took with him some 65,000 troops, all cavalry, in the campaign which led to the Mongol victory over the Mamlūks at Wādī 'l-Khaznadār near Ḥomṣ at the end of the year.[14] Waṣṣāf, a Persian writer close to the Īlkhānid court, stated explicitly that for this campaign each trooper set out with five ponies,[15] which fits what is known about the Mongols' preparations for campaigns. Ghāzān's army appears to have commenced this campaign with 325,000 ponies, which for the sake of simplicity I will round off to 300,000. One does not have to be a scholar of veterinary science to realize that such an army required a tremendous amount of fodder or pasture and water every day.

There were two possible ways to alleviate the problem. The first was to transport hay and grain (barley). Had the Mongols adopted this recourse, they would have required about five United States pounds or 2.27 kilogrammes each of

---

12 D. O. Morgan, "The Mongols in Syria, 1260-1300", in P. W. Edbury, ed., *Crusade and settlement: papers read at the First Conference of the Society for the Study of the Crusades and the Latin East and presented to R. C. Smail* (Cardiff, 1985), 231-5.

13 In this, Morgan was explicitly adopting the approach of Sinor, who suggested that the Mongols withdrew from eastern Europe in 1241 due to a lack of pasture. See D. Sinor, "Horse and pasture in Inner Asian history", *Oriens extremus*, 19 (1972), 171-83 [rpt in his *Inner Asia and its contacts with medieval Europe* (London, 1977), No. II].

14 Smith, "'Ayn Jālūt", p. 329. Numbers may, however, have been somewhat higher. See R. Amitai, "Whither the Ilkhanid army? Ghazan's first campaign into Syria (1299-1300)", in N. di Cosmo, ed., *Warfare in inner Asian history (500-1800)* [Handbuch der Orientalistik: Abt. 8, Zentralasien, vol. 6] (Leiden, 2002), 221-64, p. 236. Elsewhere (pp. 328, 332), Smith suggests that the Mongols generally entered Syria with some six *tümens* theoretically of 10,000 men each. In reality, however, *tümens* were not always, perhaps not even usually, up to full strength. See H. Ch'i-ch'ing, *The military establishment of the Yuan dynasty* (Cambridge Mass., 1978), pp. 170-71, n. 21.

15 Waṣṣāf al-Ḥaḍrat, *Ta'rīkh-i Waṣṣāf* (1852-3; rpt Teheran, 1959-60), p. 373, ll. 20-23. Waṣṣāf added that at the start of this campaign some 50,000 camels carried fodder, showing that the Mongols did indeed envisage logistical difficulties.

hay and of barley per horse per day.[16] For a *tümen* with 50,000 horses, 113.5 metric tonnes each of hay and of barley would have been needed per day.[17] For the some 300,000 horses mentioned above, the Mongol command would have had to provide 1,362 metric tonnes of fodder per day. Assuming that a camel load was about 181.5 kilogrammes,[18] this would have needed 7,504 camels just to transport one day's supplies. Waṣṣāf noted that in this campaign 50,000 camels were indeed dispatched with the army to carry fodder (*'ulāfa*). If this was the only source of food for the Mongol mounts, it would have sufficed for only six to seven days. It appears, then, that this supply train of camels was envisioned only as a stop-gap measure for emergencies, or that the fodder was destined only to supplement the tried and true method of pasturage. In any event, this is evidence that fodder was used, at least in part, in this campaign. Waṣṣāf also mentioned that the Mongols were twice resupplied with fodder and that they were forbidden by the Īlkhān to forage in the fields at Aleppo at least,[19] thereby implying that transported fodder was indeed at times the main form of food for the horses.

If transported fodder was only a partial solution to the problem of feeding their mounts, the Mongols had no choice but to depend mainly on the traditional method of pasturage, which had its own constraints. In order to emphasize the problems of pasturage, I can do no better than to cite some of Smith's figures and calculations:

> The Mongols' horses, judging by their modern descendants, weighed about 272 kilogrammes on average and were in fact only ponies, standing about $12^1/_2$ hands, or 1.27 metres (the horse-pony divide is 14 or $14^1/_2$ hands, 1.42-1.47 metres). The 'maintenance ration' for such an animal should provide 10.35 million calories (Mcal), according to modern calculations, plus 0.1365 Mcal for each hour of walking (at 6.45 kilometres per hour); consequently, on an 'average day' of Mongol campaigning, the average pony would need 10.9 Mcal. To obtain these calories from the common, good-quality steppe grass smooth brome (*bromus inermis*), for instance, which provides 2.36 Mcal/kilogramme of dry matter, and has 32.5% of dry matter in its fresh form, the pony would have to consume (10.9 ÷ 2.36 = 4.62 ÷ 32.5 x 100 = 14.21 kilogrammes

16 Smith, "'Ayn Jālūt", p. 332. Smith's metrology is in United States avoirdupois units of weights and measures. To be consistent with this volume his figures have been converted tacitly to metric equivalents. It should be noted that a wide range of figures are given for the calorific and other needs of various animals and figures provided here should be seen only as a rough guide to the logistical matters with which the Mongols and Mamlūks had to deal.

17 Smith, "'Ayn Jālūt", p. 332, citing J. W. Evans, et al., *The horse* (San Francisco, 1977), pp. 299-300, 324.

18 Smith, "'Ayn Jālūt", p. 322, n. 47 (400 United States pounds).

19 Waṣṣāf al-Ḥaḍrat, *Ta'rīkh*, p. 374. Whatever its veracity, this report may have the additional intent of portraying Maḥmūd Ghāzān as a pious Muslim ruler.

> =) 31.3 pounds of fresh bromegrass. And horses – or ponies – graze slowly. One authority calculates that a horse eats cut hay at a rate of 1.36 kilogrammes per hour and that grazing would go slower still. Even at 1.36 kilogrammes per hour, however, consumption of 14.21 kilogrammes would take almost 10½ hours. This lengthy grazing time accounts for the apparently leisurely pace of Mongol campaigns.[20]

How does this translate into real pasture? Again I paraphrase Smith's calculations. Here, the figure of 4.23 kilogrammes (dry weight) of bromegrass is used as the daily grazing need of a steppe pony.[21] Some 300,000 horses need about 1.27 million kilogrammes per day. Inner Asian pasturage at its peak provides 598.5 kilogrammes per hectare, although the exact amount varies. Assuming that Syrian grasslands were similar, which remains my working hypothesis even though it is admittedly a gratuitous assumption, a force of this size would need 2,122 hectares, around 21 square kilometres, of pasture per day.[22]

A second, no less important, matter raised by Smith is water supply. A steppe pony needed about 19 litres of water a day.[23] The 300,000 mounts which we have been using as the basis for our calculations would have needed around 5.7 million litres per day. In the winter months there should not have been a problem attaining these quantities. Firstly, about 9.58 litres could be obtained by eating the daily intake of fresh bromegrass (14.21 kilogrammes x 67.5% water). In addition, the main rivers of northern and central Syria would have been flowing at their peak in the winter. Near Aleppo the Quwayq has a maximum rate of 632 million litres per day. Near Ḥamā the Orontes reaches 337 million litres per day and the Baradā at Damascus and the al-A'waj south of there flow at 33.7 million litres per day. As would be expected, in summer the situation changes dramatically: the Quwayq shrinks to a flow of 6.8 million litres per day, the Orontes flows at 26.9 million

---

20 Smith, "'Ayn Jālūt", pp. 336-7 [Smith's measures converted tacitly to metric]. See also H. Epstein, *Domestic animals of China* (Farnham Royal, 1969), p. 100; Evans, *The horse*, pp. 259-62, 286, 288, 298-300, 324-9; C. Chenevix-Trench, *A history of horsemanship* (Harlow, 1970), p. 310.

21 I am not sure how this figure was reached. 32.5% of 31.3 United States pounds, the figures given as the percentage of dry matter in fresh grass and daily fresh grass needs of a steppe pony, yields a figure of around 10.2 pounds or 4.6 kilogrammes. I will, however, continue to use the number which Smith provides.

22 Smith, "'Ayn Jālūt", p. 338; I. V. Larin, *Pasture economy and meadow cultivation*, trans. from the Russian ed. of 1956 (Jerusalem, 1962), pp. 470, 539.

598.5 kilogrammes dry weight per hectare equals 242 kilogrammes per acre or 745 kilogrammes per acre of green grass at 32.5% dry weight to green as per n. 21 above. This figure is low by comparison to calculations for European meadow. Cf. pp. 21-3 above. [Ed.]

23 Smith, "'Ayn Jālūt", p. 339 (5 United States gallons), citing Evans, *The horse*, pp. 280-81. Again this is a low figure by comparison to European horses' requirements. Cf above p. 20 & n. 59. [Ed.]

litres per day, and the Baradā-A'waj at 8.3 million litres a day. These are average flows and in actuality they could be significantly less. Little solace is to be found in the grass. The moisture content of dry bromegrass is about 10% by weight.[24]

According to Smith, the dry conditions of summer and reduced pasturage then available, and the ultimately limited pasturelands of Syria even under more favourable circumstances, forced the Mongols to enter Syria with only part of their armies, in the neighbourhood of some 60,000 men, and prevented them from remaining in force once summer approached. Since the Mongols were on the whole inferior soldiers but could not bring to Syria a significantly larger army and maintain it there over time, they were therefore doomed to failure in their sixty-year war with the Mamlūks.

This does less than justice to Smith's impressive combination of textual work with consultation of veterinary, agricultural, geographical, and hydrological sciences. His detailed comparison of Mamlūk and Mongol archery is also an important and original exposition.[25] Smith has raised the whole subject of the Mongol-Mamlūk war to a new level, incidentally adding much to our understanding of the Mamlūk army. That being said, I register some disagreement with his arguments and conclusion. In the following I concentrate on logistics but permit myself some remarks about the relative quality of Mongol and Mamlūk cavalry. As a whole, the Mamlūks were not the supermen and the Mongols such wretched amateurs as claimed by Smith.

Many of the battles between the armies of the Īlkhāns and the Sultans were fought when the latter were far from the peak of their powers. In 1260, the Mamlūks, still ruling only in Egypt, were emerging from a decade of political disarray, hardly the circumstances for recruiting and training an army of choice cavalry.[26] At the battle of Ḥomṣ in 1281, the Mamlūks had just been through several years of political troubles and instability following the death of Baybars in 1277. Many experienced officers and troops had been arrested or removed from service and the army was just beginning to experience the stability and growth which characterized it through most of the 1280s.[27] The pitiful performance of the

---

24 Smith, "'Ayn Jālūt", pp. 339-40, citing the Syrian Ministry of National Economy, Dept. of Statistics, *Statistical Abstract of Syria 1950*, p. 4.

25 I hope to take up this matter in the future.

26 See R. Irwin, *The Middle East in the Middle Ages: the early Mamluk sultanate 1250-1382* (London, 1986), ch. 2, which he calls "The turbulent decade". For the somewhat motley composition of the Mamlūk army at 'Ayn Jālūt, see R. Amitai-Preiss, "'Ayn Jālūt revisited", *Tārīḫ: papers in Near Eastern studies*, 2 (1991), 119-50, pp. 126-7.

27 Ayalon was the first to note the somewhat sorry state of the Mamlūk army at this time. See his "Ḥimṣ (the battle of)", in EI.2, vol. 3, p. 402. For the battle see R. Amitai-Preiss, *Mongols and Mamluks: the Mamluk-Īlkhānid war, 1260-1281* (Cambridge, 1995), ch. 8: "Baybars's posthumous victory: the second battle of Homs (680/1281)".

Mamlūks at Wādī 'l-Khaznadār in 1299 may be attributed at least partially to problems of lack of unity, command, and overconfidence plaguing the army.[28]

The Mamlūks had difficulties defeating the Mongols even when in top form. In 1276, after over fifteen years of expansion, training, manœuvres, and operations, during which it had been honed to a high level of preparedness, almost the entire Mamlūk army under Baybars defeated only narrowly a much smaller Mongol and Georgian force at Elbistan. Returning to Syria the sultan was reported to have said:

> How can I rejoice? I had believed that if 10,000 horsemen of my army were to meet 30,000 Mongols, I would defeat them. But I met 7,000 [Mongols] with all my army. [The Mongol army] aroused panic and [my] army lost heart. [The Mongols] defeated the [Mamlūk] left. Without Allāh's grace, they would have defeated us. If I met them, and they were equal to the [Mamlūks in size], or larger than them, then [the matter] would not have turned out well.[29]

At Elbistan the outnumbered Mongols found themselves facing defeat.[30] Rather than flee or retreat, these "amateurs" fought to the death. This can be compared to the "professional" Mamlūks, who fled in complete disarray at Wādī 'l-Khaznadār when matters did not go as hoped, leading to a complete and embarrassing defeat.[31]

Here I concentrate more on the logistical aspects of Smith's argument and some points raised by Morgan.[32] The studies of both have certainly drawn attention to the importance of logistics and no subsequent discussion of the Mamlūk-Īlkhānid war can ignore them. However, were they as important and decisive as they have been argued to be so forcefully?

The logistical explanation may be too deterministic and reductionist since fundamentally it puts aside other factors such as leadership, morale, preparedness, ideology, internal politics, and international relations, including relations between the various Mongol khānates. These were all important matters which certainly had

---

28 See Amitai, "Whither the Ilkhanid army", pp. 227-9, 256.

29 See 'Abd al-Rahīm, continuation of Ibn Wāṣil, Abu 'Abd Allāh Jamāl al-Dīn Muḥammad, *Mufarrij al-kurūb fī akhbār Banī Ayyūb*, Paris, Bibliothèque Nationale, MS. Ar. 1702, fol. 439v. For the Anatolian campaign of 1276 and the battle of Elbistan, see Amitai-Preiss, *Mongols and Mamluks*, ch. 7: "Baybar's intervention in Seljuq Rūm".

30 Amitai-Preiss, *Mongols and Mamluks*, p 173, and Ibn Shaddād, 'Izz al-Dīn Abū 'Abd Allāh Muḥammad ibn 'Alī al-Ḥalabī, *Ta'rīkh al-Malik al-Ẓāhir*, ed. A. Khuṭayṭ (Wiesbaden, 1983), p. 172.

31 Amitai, "Whither the Ilkhanid army?", pp. 249-51; al-Yūnūnī, Quṭb al-Dīn Mūsā ibn Muḥammad, *Dhayl mir'āt al-zamān fī ta'rīkh al-a'yān*, in Li Guo, *Early Mamluk Syrian historiography: al-Yūnūnī's Dhayl Mir'āt al-Zamān*, 2 vols (Leiden, 1998), vol. 2, p. 98 [trans. vol. 1, pp. 133-4].

32 Amitai-Preiss, *Mongols and Mamluks*, pp. 225-9.

a bearing on the conduct and outcome of the Mamlūk-Īlkhānid war. For the Mamlūks this was a life or death struggle and consequently they devoted to it all available resources. For the Īlkhānids, the Mamlūk war was just one of several fronts, both internal as well as external, and thus was not always at the centre of their attention. This lack of attention to the Syrian front was one major reason for the Mongols' lack of success there, perhaps no less important than logistics.

Other reasons suggest that perhaps the logistical argument has been taken too far. First, the Mongols never controlled all of Syria and thus never exploited all of its pastureland. Second, they had no compunction seizing stored grain or pasturing their horses on cultivated fields, at least during campaigns. One thinks of the cultivated regions of Palestine, Transjordan (e.g. al-Karak), and Lebanon, which never hosted large Mongol forces for any length of time. Third, there were the water resources of these areas. Smith does not mention the Jordan and Litani, let alone other rivers in Lebanon. I suggest that, in the short run, Syria from Gaza in the south to the Euphrates in the north-east could have supported for some length of time, perhaps even into the summer if properly dispersed, Mongol forces of say 60,000 troops, each with his string of horses but without family and livestock, which would have been left behind during a campaign. If such a large force had been successful in delivering a decisive blow to the Mamlūks, a much smaller force, with families and herds, would have sufficed to hold the country.

Two additional points may be made. First, in 1950, there was enough livestock to support 80,000 nomadic families in modern-day Syria.[33] While some of the lands used to support this livestock may not perhaps have been appropriate for the type of pastoral nomadism practised by the Mongols, certainly some of them would have been. In any event, the Mongols would have had no problems of conscience in expelling indigenous nomads from these lands, be they Bedouin, Turkmen, or Kurds. Second, one contemporary source close to the Mamlūk court noted that during the reign of Baybars, some 40,000 Turkmen families fled Mongol-controlled areas for Syria, and found a home there.[34] The figure may well have been an exaggeration but it reveals that a not insignficant portion of Syria's pasturelands could have been used by nomads of Eurasian Steppe provenance.

In short, logistics may have been an important factor in the outcome of the Mongol campaigns into Mamlūk Syria but they were not the only or even a decisive factor. Or so I thought. Smith's counterattack, delivered in a courteous but firm manner, appeared in 1998.[35] Much of it dealt with a comparison of the tactics,

---

33 Smith, "'Ayn Jālūt", p. 309, citing *Statistical Abstract of Syria, 1950*, pp. 158-9. This, of course, included that part of modern Syria to the east of the Euphrates, but also excluded Lebanon, Jordan, and Israel, as well as a slice of south-eastern Turkey which might be considered part of geographical Syria in the Middle Ages.

34 Ibn Shaddād, *Ta'rīkh al-Malik al-Ẓāhir*, p. 335.

35 J. M. Smith, jr, "Nomads on ponies vs. slaves on horses", *JAOS*, 118 (1998), 154-62. On pp. 54-7 he gives an excellent precis of the contents of *Mongols and Mamluks*.

equipment, and training of the two armies, a continuation of his thinking expressed in "'Ayn Jālūt" but taking into account my reservations. Fundamentally, he reaffirmed his original approach and in the final pages turned to my criticism of logistical reductionism, as I might now call it.

Smith opened with a short discussion on the importance of Mongol transhumance for understanding the dynamic of the Mamlūk-Īlkhānid war.[36] For example, in 1261 Baybars apparently misunderstood the absence of Mongol forces from the environs of Baghdad, thinking that they had permanently withdrawn, and thus sent the 'Abbāsid claimant to the Caliphate, al-Mustanṣir, with a small force, perhaps as few as 400 men but probably more, to reclaim it. The reappearance of the Mongol garrison, back from summer pasture, meant the end of the Caliph's pretensions, and his life.[37]

Similarly, need for summer pastures forced Mongol leaders to withdraw the majority of their troops from Syria after successful campaigns, such as those of the winters of 1260-61 and 1299-1300. The small remaining forces were unable to deal with the Mamlūks, who then moved or returned to Syria. In 1260 Kitbuga Noyon's corps suffered defeat at 'Ayn Jālūt and in 1300 a rump force under Qutlugh-Shāh Noyan withdrew on the approach of the reconstituted Mamlūk army. The problems of conflicts with other Mongol rulers are discounted and it is even suggested that the evidence that these pro-Īlkhānid sources provide was disinformation.

Smith concedes that with cultivated fields and stored grain Syria had the capacity to feed the 300,000 ponies of a typical Mongol expeditionary force. He suggests, however, that a Mongol commander would not have been able to concentrate them and thus to meet a Mamlūk army in an effective way. This would have been compounded by the lack of water in summer.[38] In fact, the only river with a flow great enough to support the entire Mongol pony train was the Orontes in the environs of Ḥamā and Ḥomṣ.[39] Reiterating his calculations for the amount of pastureland needed per day for an army with 300,000 ponies or more, Smith notes that the steppe-like pastureland around the Orontes between Ḥamā and Ḥomṣ is

---

[36] This is a subject discussed in detail in J. M. Smith, jr, "Mongol nomadism and Middle Eastern geography: Qīshlāqs and Tümens", in R. Amitai-Preiss and D. O. Morgan, eds, *The Mongol Empire and its legacy* (Leiden, 1999), 39-56.

[37] See Amitai-Preiss, *Mongols and Mamluks*, pp. 56-9.

[38] I accept Smith's comment that I did not take into consideration fully that the flow of Syrian rivers can be much lower than the summer average.

[39] Smith, "Nomads on Ponies", p. 62, writes that this "is probably why the Mamluks chose to fight for Homs in 1281 and 1299". If I understand this correctly, the Mamlūk commanders thus wanted to deny the Mongols the opportunity to exploit the Orontes. On the other hand, it seems to me that had the Mamlūks held back, the Mongols might have seen that they had no alternative but to advance into the less watered southern part of Syria, as happened in the campaign of 1303 which ended with the Mamlūk victory of Marj al-Ṣuffār or Shaqḥab.

about 4,144 square kilometres, good for about 200 days or just over half a year. In other words, the Mongols could maintain in central Syria for about six months a large force which could take on the best that the Mamlūks could offer. This explains the withdrawal from Syria of the Mongol army as summer approached.

Two further arguments are made. The first, alluded to above, concerns the Turkmen who fled to Syria. Smith suggests that the fact that they played such a minor role in the continuing Mamlūk-Īlkhānid war in spite of their large numbers indicates that the conditions of the country led to "pastoral failure". Secondly, he notes that I suggested that the repeated Mongol attempts to take Syria and the efforts made to interest the West in joint campaigns show that the Mongol leadership, supposedly cognizant of logistical matters, thought that they were capable of taking and holding the country. He writes:

> Why the Mongols kept coming back, knowing that logistics prevented occupation, may be explained by their expectation that, under proper circumstances, they could destroy the Mamluk army or damage it so badly that even a small Mongol force could hold Syria; that there was something to this may be seen from their victory in 1299, and the defensive tactics of the weakened Mamluks in 1300.[40] In any case, if Syria could have supported the Mongol army, as Amitai-Preiss believes, it should not have needed to withdraw after winning in 1299. In sum, the Mongols could invade in great strength, they could defeat the Mamluks, but they could not stay in Syria long enough to exploit their success.[41]

I am still not convinced completely, although I begin to be more flexible. Let us begin with the Turkmen. They did play a role in the Mamlūk-Īlkhānid war. Unspecified numbers of them fought with the Mamlūks at Ḥomṣ in 1281, taking up an important position on the left flank.[42] They were not mentioned at Wādī 'l-Khaznadār in 1299 but they appeared in skirmishing against the Mongols in north Syria in 1271.[43] At least partially, why they were not mentioned frequently in the war with the Mongols is that many were settled in the coastal region of Palestine in areas reconquered from the Franks. There they played a not insignificant role and their descendants, at some point completely Arabized, remained in the area as a distinctive body through the Ottoman period and even into the twentieth century.[44] It should be noted that a recurring phenomenon in states using armies based on *mamlūks* was that Turkmen became marginalized. Such was the case in the

---

40 These "defensive tactics" are not completely clear to me.

41 Smith, "Nomads on ponies", p. 62.

42 Amitai-Preiss, *Mongols and Mamluks*, pp. 191-3.

43 Ibid., p. 125.

44 Ibid., 69-71. For the twentieth century, see T. B. Ashkenazi, *The Turkmans in Palestine (Ha-Turkmanim be-ereṣ-yisra'el)* (Jerusalem, 1931).

Sultanate of the Great Saljūqids, in that of the Saljūqids of Rum, and it was also the case in the Mamlūk sultanate. This does not indicate pastoral failure.

As to why the Mongols pulled out most of their troops in 1260 when things were going well, in spite of the important evidence in Hūlāgū's letter to Louis IX of France,[45] there were other, mainly political explanations: most importantly, the growing danger of an attack from the Golden Horde via the Caucasus. In any event, even if we accept the logistical explanation for the Īlkhān's withdrawal with the majority of his army, that does not really explain why Hūlāgū left such a small force for all of Syria. The entire country certainly could have held more than 10,000 men, even with families and herds. The decision to leave such a small force must have been a military one, based on needs elsewhere and faulty intelligence.[46] As for the campaign of 1299-1300, Heṭoum's evidence on the Chaghatayid incursion may have had more than post facto "spin".[47] There certainly was hostile Mongol activity in the eastern part of the Īlkhānate which might well have compelled Ghāzān to turn his attention to the east, let alone leave for the time being the campaign to the west.[48] In addition, one can wonder how long the ruler of a large and far-flung realm such as Ghāzān, who suffered his share of rebellions, could afford to dither in a distant province far from his centre.

The Mongols could bring large forces into central Syria and maintain them there for several months. In 1299-1300 they were large enough to lead to a complete Mongol victory, especially when taken together with other factors, many of them problems plaguing the Mamlūk army. Logistics have their place in understanding the way Mongol forces behaved in Syria but they should not be seen as the only or constant determining reasons.

Examining the campaign which led to the Mongol victory at Wādī 'l-Khaznadār in 1299, it can be seen that problems of feeding and watering the horses were of great importance for the conduct of the battle. Waṣṣāf reported that before the battle many Mongol ponies had died and others had become emaciated and incapacitated due to the long march. Because of this, Ghāzān ordered the whole army to fight on foot in order to encourage those troops who had lost their mounts,

---

[45] He wrote that he withdrew most of his men due to the lack of fodder and grazing in Syria. See P. Meyvaert, "An unknown letter of Hulagu, Il-Khan of Persia, to king Louis IX of France", *Viator*, 11 (1980), 245-60, p. 258. The possibility of post facto justification of a mistaken strategy cannot be discounted. Cf. Morgan, "Mongols in Syria", p. 233.

[46] See Amitai-Preiss, *Mongols and Mamluks*, pp. 28-9.

[47] Heṭoum, *La Flor des estoires de la Terre d'Orient*, in RHC *DArm*, vol. 2, 111-363, ch. XLI (p. 196).

[48] H. H. Howorth, *History of the Mongols from the 9th to the 19th century*, 4 vols (London, 1876-1927), vol. 3, p. 446; R. Grousset, *The Empire of the steppes: a history of central Asia*, trans. N. Walford (New Brunswick, 1970), p. 382; J. A. Boyle, "The dynastic and political history of the Īl-Khāns", in *The Cambridge history of Iran*, vol. 5, ed. J. A. Boyle (Cambridge, 1968), 303-421, p. 388.

to trick the Mamlūks, and to improve archery.[49] Waṣṣāf, however, may be misleading. While some of the Mongols dismounted during the battle, it remained basically a confrontation between cavalry. In any event, they remounted when the time was right and attacked the Mamlūks.[50] Before the battle itself Ghāzān had ordered his troops to disperse to forage and water the horses and he was thus caught unprepared for the Mamlūk assault on 22 December, having only 9,000 troops with him in the centre.[51] In spite of this inauspicious beginning, he was successful in repulsing the initial attack and then leading his troops to victory.

The most interesting evidence for the logistical difficulties faced by the Mongols occurred later in the campaign, after their victory and occupation of Damascus. Rashīd al-Dīn, the Persian historian and *wazīr* to the Īlkhāns was present in Syria during Ghāzān's campaign of 1299-1300. After describing the Mongol occupation of Damascus and unsuccessful efforts to take its citadel, he wrote: "Since the weather was turning hot, the Pādishāh [the Īlkhān Ghāzān] paid no attention to the people of the citadel [of Damascus] ... On Saturday, the 13th of Jumādā I [= 5 February 1300] he withdrew from Damascus."[52]

The passage did not say explicitly that there might be a problem feeding and watering the Mongol ponies and some have therefore misunderstood it.[53] But it did do more than hint that even in late winter logistical problems to come were giving the Mongol leadership some concern. This, of course, does not negate other matters which may have weighed on the Īlkhān's mind, such as problems on his eastern borders. At the same time, given the apparent massive defeat suffered by the Mamlūks less than two months before at Wādī 'l-Khaznadār, there was no reason to think that there would be danger from their direction in the near future. Ghāzān departed with most of his army but left a substantial force, whose exact size is unclear, under Qutlugh-Shāh Noyan in Damascus. Probably to the surprise of Mongol commanders and local notables alike, the Mamlūks managed to reorganize and were soon on their way back to Syria with a large army. Qutlugh-Shāh thought it prudent to withdraw. Logistical problems or not, even before his

---

49 Waṣṣāf al-Ḥaḍrat, *Ta'rīkh*, p. 233. See also Amitai, "Whither the Ilkhanid army?", p. 233; A. P. Martinez, "Some notes on the Īl-Xānid army", *Archivum Eurasiae medii aevi*, 6 (1986), 129-242, pp. 176-7.

50 Amitai, "Whither the Ilkhanid army?", pp. 246-50.

51 Waṣṣāf al-Ḥaḍrat, *Ta'rīkh*, p. 378.

52 Rashīd al-Dīn Abū 'l-Khayr Faḍl Allāh ibn 'Imād al-Dawla, *Jāmi' al-tawārīkh*, in K. Jahn, ed., *Geschichte Gāzān Ḫān's aus dem Ta'rīḫ-i-Mubārak-i-Gāzāni des Rasīd al-Dīn Faḍlallāh b. 'Imād al-Daula Abū 'l-Ḫair* (London, 1940), p. 130 [trans. based on W. M. Thackston, trans., *Rashiduddin Fazlullah's Jami'u 't-tawarikh: Compendium of chronicles: a history of the Mongols*, 3 vols (Cambridge Mass., 1998-9), vol. 3, p. 647].

53 Boyle, "History of the Īl-Khāns", p. 389; Howorth, *History of the Mongols*, vol. 3, p. 446.

departure Ghāzān had made it clear that he planned to return the following year.[54] The Īlkhān, at least, was not to be turned from Syria by logistical concerns.

The campaign of winter 1300-1301 was cut short by inclement weather. Both Mamlūks and Mongols had to return to their home bases.[55] There is prima facie evidence that a next Mongol invasion of Syria was postponed because of anticipated pasturage problems. Heṭoum wrote that Ghāzān had to postpone his attack on Syria in 1303 because the enemy had fired the pastures on the projected invasion route: "When Ghāzān had inquired into what the Saracens had done, considering that his horses would not be able to find anything on which they could live, he decided to remain that winter on the river Euphrates, and coming Easter when the the corn would begin to shoot, then he would take up his route."[56] Here we have the well-known Mamlūk "burners" (*munawwirūn*), about whom al-'Umarī provides detailed information, at work.[57] Heṭoum's chronology was imprecise since the Mongols, without Ghāzān, did invade Syria in early 1303. These actions are not confirmed by either Arabic or Persian sources and, such as they were, appear to have only slowed down the Mongols. The campaign led to their defeat at the battle of Marj al-Ṣuffār or Shaqhab on 20 April 1303.[58] It was to be almost another decade before the Mongols would try again to invade Syria.

Thus far we have followed the Mongol side of the logistical problems. However, we should consider the Mamlūk side also, concentrating on the battle of Wādī 'l-Khaznadār. Preliminary research on Mamlūk logistics had already been carried out by Ayalon, who wrote:

> Shortly before the army set out, various supplies were prepared in the stations lying along its route. These stocks (*iḳāmāt*, sing. *iḳāma*, or *iḳāmāt wa-anzāl*) consisted of barley, wheat, chickens, pigeons, geese, sweets, melons and various other kinds of food, as well as fire-wood, horses, riding-camels and camels of burden ... The military expedition was accompanied by a very large camel caravan, which carried its baggage (*thaḳal*, pl. *athḳāl*). Each Mamlūk participating in a campaign received at least one camel. Sometimes the Mamlūk received two camels, while the

---

54 Amitai, "Whither the Ilkhanid army?", pp. 253, 260.

55 Boyle, "History of the Īl-Khāns", p. 389; Irwin, *Middle East*, p. 101.

56 Heṭoum, *Flor des estoires*, ch. XLII (pp. 199-200): "Quant Casan ot entendu ce que les Sarazins avoient fait, regardant que ses chevaus ne porroient rien trover dont il peüssent vivre, prist conseil de demorer celui iver sur le flum Eufrates, e venant le pascor, quant lo ble comenceroit à borgioner, adonques prendroit son chemin."

57 Al-'Umarī, Aḥmad ibn Yaḥyā ibn Faḍl Allāh, *Al-Ta'rīf fī 'l-muṣṭalaḥ al-sharīf* (Cairo, 1312 A.H. [1894-5 C.E.]), pp. 201-3. See the discussion in Amitai-Preiss, *Mongols and Mamluks*, pp. 107, 126, 205-6.

58 On this campaign, see Boyle, "History of the Īl-Khāns", pp. 392-4; Smith, "'Ayn Jālūt", pp. 340-44, who suggests that "logistic problems led again to disaster".

> non-Mamlūk soldiers of the ḥalḳa received 3 camels per two men ... Mules were very rarely employed for carrying the baggage. They were used in 691/1292 by the Sultan's army in the region of Aleppo because most of the camels died in an epidemic ... The employment of wheeled vehicles (*'adjal*), mainly for carrying siege machines, was also extremely rare.[59]

Early Mamlūk armies were organized differently to those of the Mongols. Soldiers in Mamlūk armies, be they royal *mamlūks*, *mamlūks* of *amīrs*, or troopers in non-*mamlūk ḥalqa* formations, whether from Egypt or Syria, were invariably cavalry. The majority seem to have had only one mount, although some spare mounts (*janā'ib*) are noted. Their horses were larger than those of the Mongols, although there is no reason to assume that they were Arabians. One mid-fourteenth-century author noted that a major source of Mamlūk mounts was Cyrenaica. Horses from there were very strong and were a cross between Arabians and pack-horses. With the latter's sturdy legs, they were well suited to rough terrain.[60] According to Fidenzio of Padua, who wrote his *Book of the recovery of the Holy Land* shortly before the fall of Acre in 1291, Mamlūk horses were smaller and less vigorous than Frankish ones. They did not have harnesses like Frankish ones, which was an advantage since it permitted them to gallop with ease.[61] One might then construct a continuum of horse sizes from large to small in order: Frankish, Mamlūk, Mongol.

Given the dearth of pasturage in Egypt, horses there probably received most of their food from fodder. There is, however, some evidence that at least once a year horses were put out to pasture. In 1265 Baybars was somewhat stymied in his efforts to get out the Egyptian army in the face of a Mongol attack on the fortress of al-Bīra on the Euphrates, since the horses were out to pasture in the countryside.[62] There is no information how horses of Mamlūk armies in Syria were normally maintained but it is likely that the close proximity of pasturage near major cities permitted more pasturing of them than in Egypt.

Before campaigns Mamlūk officers and soldiers were normally given a *nafaqat al-safar*, a special campaign allowance distinct from their normal salary, to

---

59 Ayalon, "Ḥarb, iii - The Mamlūk sultanate", in EI.2, vol. 3, pp. 184-5.

60 Al-'Umarī, Aḥmad ibn Yaḥyā ibn Faḍl Allāh, *Masālik al-abṣār fī 'l-mamālik al-amṣār: mamālik Miṣr wa 'l-Shām wa 'l-Ḥijāz wa 'l-Yaman*, ed. A. F. Sayyid (Cairo, 1985), p. 101. See also Ayalon, "System of payment", pp. 263-4; Amitai-Preiss, *Mongols and Mamluks*, p. 218.

61 J. Paviot, "Comment reconquérir la Terre sainte et vaincre les Sarrasins?", in M. Balard, B. Z. Kedar and J. Riley-Smith, eds, *Dei gesta per Francos: études sur les croisades dédiées à Jean Richard / Crusade studies in honour of Jean Richard* (Aldershot, 2001), 79-85, p. 84.

62 Ibn 'Abd al-Ẓāhir, Muḥyī 'l-Dīn 'Abd Allāh, *Al-Rawḍ al-zāhir fī sīrat al-malik al-ẓāhir*, ed. 'A.-'A. al-Khuwayṭir (Riyadh, 1976), pp. 221-5, who is cited by later authors.

bring their equipment up to readiness and purchase supplies.[63] Whether this included fodder is unclear. Use of camels certainly gave them the capacity to carry some fodder and to do so was definitely necessary on some legs of the trip from Egypt, where the most important and largest sections of Mamlūk forces normally resided, to Syria, where most fighting took place. The fact that at times sultans forbad pasturing in fields in Syria indicates that some fodder was transported with armies.[64] It appears that local markets where fodder could also perhaps be purchased were set up along the way.[65] The army administration itself, or rather the *wazīr*, was also responsible in part for regular provisions for the army, probably including fodder.[66] Finally, there were large areas *en route* which could be used for pasturing horses, such as Shaqhab to the south of Damascus.[67] In short, it appears that feeding the horses was a combined affair which included private responsibility in the use of campaign allowances, some central provisioning, and the use of pasturage, perhaps sanctioned by commanding officers.

Amidst confusion and pressure before a battle, let alone the rough and tumble of combat, any system could well break down or simply not suffice. Let us return to Wādī 'l-Khaznadār to see how the Mamlūks had their own share of logistical problems. While waiting in southern Palestine for more exact intelligence on the Mongols, floods swept away much of their baggage and transport camels. Later, locusts probably wreaked havoc on surrounding pastures. The soldiers received their campaign allowance only when they reached Damascus because the command had waited until then to distribute it for fear that it would be misused. The soldiers, however, were reported to have been stingy with it in buying supplies because prices were high.[68] Having received a flow of intelligence about the approaching Mongol army, the sultan set off north from Damascus on 12 December 1299, taking up position in the plains north of Ḥomṣ. A group of sources, some contemporary, relate that the Mamlūks wore full battle gear for three days, wearing out men and mounts alike.[69] Baybars al-Manṣūrī, who during this

---

63 See Ayalon, "System of payment".

64 For example, Ibn al-Furāt, Nāsir al-Dīn Muḥammad ibn 'Abd al-Raḥīm, *Ta'rīkh al-duwal wa 'l-mulūk*, Vienna, Staatsbibliothek, MS. 814, fol. 65r.

65 This is hinted at in the sources cited in nn. 68-9 below.

66 L. Northrup, *From slave to sultan: the career of al-Manṣūr Qalāwūn and the consolidation of Mamluk rule in Egypt and Syria (678-689 A.H./1279-1290 A.D.)* (Stuttgart, 1998), p. 218.

67 W. J. Fischel, *Ibn Khaldūn and Tamerlane. Their historic meeting in Damascus, 1401 A.D. - 803 A.H.* (Berkeley, 1952), p. 43; idem, *Ibn Khaldūn in Egypt. His public functions and his historical research, 1382-1406* (Berkeley, 1967), pp. 44, 66.

68 Al-Maqrīzī, Taqī al-Dīn Abū 'l-'Abbās Aḥmad ibn 'Alī, *Al-Sulūk li-ma'rifat al-mulūk*, ed. M. M. Ziyādah and S. 'A. 'Ashūr, 10 vols (Cairo, 1934-73), vol. 1, p. 885.

69 Al-Jazarī, Shams al-Dīn Muḥammad, *Ḥawādith al-zamān*, Paris, Bibliothèque Nationale, MS. Ar. 6739, fol. 297b; al-Yūnīnī, *Dhayl Mir'āt al-zamān*, vol. 2, p. 97 [trans. vol. 1, p. 133]; Ibn al-Dawādārī, Abū Bakr ibn 'Abd Allāh, *Kanz al-durar wa-jāmi' al*

campaign was commander of the small Mamlūk force which remained in Cairo, wrote that the army rode the equivalent of three days' march in one.[70] This surely led to the exhaustion of men and horses. In another chronicle, the same author added that the Mamlūk horses were exhausted by the distance travelled and burden carried.[71] The Mamlūks entered the battle under less than auspicious circumstances, and this is not to take into consideration other problems such as the lack of a proper command structure and low morale.

To make matters worse, there was a supply problem. Prices had become expensive and there was a lack of fodder. Apparently soldiers and officers were expected to purchase some supplies, including fodder, on the private market and the merchants of Ḥomṣ and its vicinity plied a brisk trade just before the battle.

In short, the battle of Wādī 'l-Khaznadār was won by the Mongols in spite of logistical difficulties. The battle was lost by the Mamlūks because, *inter alia*, they faced logistical problems with which their normal system was unable to cope. Perhaps they were indeed exceptional soldiers; however, their logistical infrastructure still left much to be desired and let them down in this case. In addition, their large and generally well-cared-for horses also had their limitations, compounded it would seem by logistical failure. Grass-fed ponies, and the so-called amateurs who rode them, showed their mettle that day.

It may be instructive to compare Tīmūr Lang's invasion of Syria in the winter 1400-1401, during which he occupied Damascus for several months, taking Aleppo and Ḥomṣ on the way.[72] In principle, we should expect a great resemblance between the armies of Tīmūr and Ghāzān a century earlier and Tīmūr should have faced logistical difficulties similar to those which the Īlkhāns supposedly met.

A preliminary reading of some major Mamlūk and Tīmūrid sources,[73] and some outstanding modern studies of the period, does not indicate that Tīmūr

ghurar. Vol. 9: Al-Durr al-fākhir fī sīrat al-Malik al-Nāṣir (Der Bericht über den Sultan al-Malik an-Nāṣir Muḥammad ibn Qalā'ūn), ed. H. R. Roemer (Freiburg and Cairo, 1960), p. 16; Ibn Taghrī Birdī, Abū 'l-Maḥāsin, *Al-Nujūm al-zāhira fī mulūk miṣr wa'l-qāhira*, 16 vols (Cairo, 1930-56), vol. 8, p. 121; Al-Maqrīzī, *Sulūk*, vol. 1, p. 885. See also K. V. Zetterstéen, *Beiträge zur Geschichte der Mamlükensultane in den Jahren 690-741 der Higra, nach arabischen Handschriften* (Leiden, 1919), p. 58.

70 Baybars al-Manṣūrī al-Dawādār, Rukn al-Dīn, *Zubdat al-fikra fī ta'rīkh al-hijra: History of the early Mamluk period*, ed. D.S. Richards (Beirut, 1998), p. 331.

71 Baybars al-Manṣūrī al-Dawādār, Rukn al-Dīn, *Kitāb al-tuḥfa al-mulūkiyya fī 'l-dawla al-mamlūkiyya*, ed. 'Abd al-Ḥamīd S. Ḥamdān (Cairo, 1987), p. 157.

72 I owe the suggestion to make this comparison to Benjamin Kedar.

73 For the Mamlūks, Ibn Taghrī Birdī, *Al-Nujūm*, vol. 12, pp. 221-46; Al-Maqrīzī, *Sulūk*, vol. 3, pp. 1026-52. For Tīmūr, Shāmī, Niẓām al-Dīn, *Ẓafar-nāmah*, ed. F. Tauer, in *Histoire des conquêtes de Tamerlan intitulée Ẓafarnāmah*, 2 vols (Prague, 1937, 1956), vol. 1, pp. 230-36; Yazdī, Sharaf al-Dīn 'Alī, *Ẓafar-nāma*, in F. Petis de La Croix, the younger, trans., *Histoire de Timur-Bec connu sous le nom du Grand Tamerlan, empereur des Mongols et Tartares* (Paris, 1723), vol. 3, pp. 313-58; Ibn 'Arabshāh, Aḥmad ibn

encountered any problems feeding horses or that such problems prompted his withdrawal from Syria. For the latter, I suggest that his peripatetic nature had much to do with it. Having extracted as much wealth as could be expected from the region, he was ready to move on. Perhaps further research will uncover information which will shed more light on the problem.

There is, however, interesting evidence that some of the Mamlūk elite thought that Tīmūr might encounter logistical problems and that this could be turned to their advantage. Ibn Taghrī Birdī reported that at a war council in the sultan's presence before leaving Cairo, a senior *amīr* advised the sultan to take up position with his army at Gaza and that he, the *amīr*, should organize the defences of Damascus, which Tīmūr would not be able to take since it was so well defended and supplied. Since he had a large army and could not stay in one place for any long time, he would be forced either to move south to Gaza to meet the sultan's army, becoming caught between it and the Syrian army, or to leave and return to his own country. This would be because his army did not know the country and would suffer from lack of supplies. Since so much of the country had been destroyed, the *amīr* suggested that the latter possibility was greater. The suggestion was well received, although it was promptly ignored in spite of its logic.[74]

The *amīr* in question was Taghrī Birdī, the historian's father. Even if the story was fabricated to present the father as a paradigm of good advice, it has a certain credibility and importance as historical evidence. An invented story would only work if the contents made some sense. Taghrī Birdī's analysis as presented by his son was cogent because the Mamlūk elite understood that a Mongol-type army might encounter logistical difficulties in Syria. Whether Tīmūr really encountered such problems and how they influenced his actions has yet to be shown.

Muḥammad ibn 'Abd Allāh ibn Ibrahīm Shihāb, *Kitāb 'Ajā'ib al-maqdūr fī nawā'ib Tīmūr* (Calcutta, 1841-2), pp. 197-242 [trans. J. H. Sanders, *Tamerlane or Timur, the Great Amir* (London, 1936), pp. 133-66]. On p. 241 [trans. p. 165], there is mention of a plague of locusts which decimated the countryside of Syria as Tīmūr was leaving the country. There is, however, no indication that this affected the conqueror or his troops, who seem to have been well supplied with booty.

[74] Ibn Taghrī Birdī, *Al-Nujūm*, vol. 12, pp. 231-2.

Chapter 3

Crusader logistics: from victory at Nicaea to resupply at Dorylaion[1]

Bernard S. Bachrach

It took almost four months, 26 June-20 October, for the forces of the First Crusade to travel approximately 1,200 kilometres from Nicaea to Antioch.[2] This averages out at a rate of march of about ten kilometres per day, including all stops.[3] In part, the pace of the march must be seen in terms of conditions for the line of march traversed much very difficult terrain, although this was mitigated to some extent by the fact that the Crusaders mostly followed old Roman roads and, for the most part, the Byzantines had diligently kept the key arteries of military and commercial transportation in repair for centuries.[4] Normal difficulties encountered by large forces moving through country that held many opportunites for hostile action also undoubtedly affected the pace at which they moved.[5] In addition, the Crusaders faced terrible climatic conditions which saw temperatures probably average in excess of thirty degrees centigrade during the day throughout a good part of the march.[6] Finally, due to the fact that they were travelling mostly in the dry season, there were from time to time problems with obtaining adequate water.[7]

---

1 Consult Map 3.

2 See, for example, S. Runciman, *A history of the Crusades*, 3 vols (1951-54; rpt, Harmondsworth, 1965), vol. 1, pp. 183-8; J. France, *Victory in the East: a military history of the First Crusade* (Cambridge, 1994), pp. 171-2. The basic work was done by H. Hagenmeyer, *Chronologie de la première Croisade* (1898-1901; rpt, Hildesheim, 1973), pp. 81-105, but much detail needs to be revised.

3 From the departure from Dorylaion on 4 July and 20 October there were 108 days and thus fewer than 100 marching days.

4 See, for example, Runciman, *History*, vol. 1, pp. 183-8; France, *Victory in the East*, pp. 185-96.

5 For engagements fought during the march, see Runciman, *History*, vol. 1, pp. 188-93; France, *Victory in the East*, pp. 185-96; Hagenmeyer, *Chronologie*, pp. 86-105.

6 France, *Victory in the East*, pp. 137, 139 provides lower figures than those of Anonymous, *Baedeker's Turkey*, trans. W. Bell et al. (Basingstoke, 1994), pp. 28-9. However, in light of the medieval warming trend which made temperatures hotter that those today, it is likely that the situation was even worse than modern figures would suggest.

7 A. T. Hodge, *Roman aqueducts and water supply*, 2nd ed. (London, 2002), pp. 61-3; France, *Victory in the East*, pp. 189-90.

Evaluation of this rate of march must consider not only how the above factors impinged upon the pace, but also the frequency and lengths of stops.[8] For example, the distance between Nicaea and Dorylaion, as the crow flies, is something short of 95 kilometres.[9] The lead units left the environs of Nicaea on 26 June and, after two days encamped, the remaining forces left for Dorylaion on 29 June.[10] Early on 1 July, the third day of the second phase of the march, they fought a lengthy battle near Dorylaion.[11] Finally, on 4 July, the army left Dorylaion, apparently for Antioch.[12] Counting 26 June as a travel day, the forces spent eight days until 3 July traversing approximately 95 kilometres between Nicaea and Dorylaion. However, the actual time on the march was three days and a very small part of a fourth day.

---

8 Concerning the pace of columns, see J. W. Nesbitt, "The rate of march of Crusading armies in Europe", *Traditio*, 19 (1963), 167-82. France's view that a march of almost four months could be sustained with only 15 days of rest is, in my opinion, highly unrealistic. See *Victory in the East*, pp. 171-2; B. S. Bachrach, "Animals and warfare in early medieval Europe", in *L'uomo di fronte al mondo animale nell'alto medioevo* [SSCISAM, XXXI] (Spoleto, 1985), 707-64.

9 The old Byzantine fortress city of Dorylaion lay almost four kilometres northwest of the modern town of Eskişehir. The centre of modern Iznik is well within the walls of Nicaea and perhaps seven or eight hundred metres from the gate that led to the Roman road to Dorylaion.

Reconstructions of the course of the road from Nicaea to Dorylaion are controversial and the exact distance between the two cities has yet to be firmly established. However, like most other roads constructed during the first century C.E., it was probably as straight as Roman engineers could make it. It passed through modern Sögüt, which is directly between Nicaea and Dorylaion. The nature of travel suggests that commercial traffic could move comfortably from Nicaea to Dorylaion in three days and the empire's need for a classic road for military purposes between these two important population centres sustains the notion that it was designed for a three-day march.

See W. M. Ramsay, *The historical geography of Asia Minor* (London, 1890), pp. 197-221, esp. 212-13; H. Belke and N. Mersich, eds, *Phrygien und Pisidien* (TIB, vol. 7) (Vienna, 1990), pp. 141-3; W. M. Calder and G. E. Bean, eds, *A classical map of Asia Minor* (London, 1958); Runciman, *History*, vol. 1, p. 186, n. 1; S. Vryonis, jr, *The decline of medieval Hellenism in Asia Minor and the process of Islamization from the eleventh through the fifteenth century* (Berkeley, 1971), p. 31 and map facing p. 14.

10 For the chronology see Hagenmeyer, *Chronologie*, pp. 83-85; France, *Victory in the East*, pp. 170-71.

11 The date and time of the battle are generally agreed upon. See Hagenmeyer, *Chronologie*, pp. 86-88. However, its location is rather controversial. France, *Victory in the East*, pp. 171-2, places the battle some distance from Dorylaion.

12 Hagenmeyer, *Chronologie*, p. 89; Runciman, *History*, vol. 1, pp. 187-9; France, *Victory in the East*, p. 188, n. 119.

In that short period, the forces averaged about 30 kilometres per marching day. Such a rate of march was good by any pre-modern standards for large forces.[13]

The difference between average daily marching rates and distances that could be negotiated in a single day for several days only were the result of various conditions. The central problem facing all large armies for lengthy campaigns, whether on the march, in camp, or while undertaking sieges, was the need to supply the troops.[14] However, the study of logistics in regard to medieval warfare is in its infancy.[15] At present, very little is known in material terms, much less agreed upon, regarding supply systems underpinning the First Crusade; indeed, some are sceptical that anything resembling a system of logistics existed during it.[16] In this paper only one small part of the supply problem will be examined: feeding the forces from the victory at Nicaea on 19 June 1097 to the completion of resupply at Dorylaion on 3 July 1097.

All study of logistics must begin with numbers. When a commander undertakes a campaign, he must know the order of magnitude of forces under his command for a wide variety of reasons, among which is providing sufficient quantities of materiel, especially food, to sustain the campaign.[17] Organizing an effective supply system is most pressing when campaigns are likely to last long or when the theatre of operations is far from home base. The further from home forces operate, the more complex are supply problems and these are exacerbated when the aim is territorial conquest.[18]

Sound logistics are fundamental because troops and, to a lesser extent, camp followers and animals must be fed and watered. Demands created by objective realities, inherent in physiology, *au fond*, can be altered neither by literary

---

13 See, for example, Nesbitt, "Rate of march"; France, *Victory in the East*, pp. 171-2; D. W. Engels, *Alexander the Great and the logistics of the Macedonian army* (Berkeley 1978), pp. 153-6; Bachrach, "Animals and warfare".

14 See B. S. Bachrach, "Medieval siege warfare: a reconnaisance", *The journal of military history*, 58 (1994), 119-33 [rpt in his *Warfare and military organization*, No. IV].

15 B. S. Bachrach, "William Rufus's plan for the invasion of the mainland in 1101", in R. P. Abels and B. S. Bachrach, eds, *The Normans and their adversaries at war: essays in memory of C. Warren Hollister* (Woodbridge, 2001), 31-63.

16 Riley-Smith suggests that in terms of supply the Crusaders operated on an ad hoc basis. See J. Riley-Smith, *The First Crusade and the idea of Crusading* (London, 1986), pp. 65-9, 71, 88-90, where fear of starvation is highlighted. Against this see B. S. Bachrach, "The siege of Antioch: a study in military demography', *War in History*, 6 (1999), 127-46; Idem, "Some observations on the role of the Byzantine navy in the success of the First Crusade", *Journal of medieval military history*, 1 (2002), 83-100; Idem, "Some observations on administration and logistics of the siege of Nicaea", forthcoming in *War in history*.

17 See Bachrach, "Siege of Antioch".

18 B. S. Bachrach, *The anatomy of a little war: a diplomatic and military history of the Gundovald affair (568-586)* (Boulder, 1994), pp. 130-32; Idem, *Early Carolingian warfare: prelude to empire* (Philadelphia, 2001), pp. 228-39.

construction nor deconstruction.[19] The matrix of objective reality, food, weather, and terrain in relation to variables such as numbers of people and animals who must be fed and protected from the enemy and the elements, forms the basis for *Sachkritik*.[20] From a cultural perspective, it has long been understood that both the actual availability of supplies and also troops' perception of availability of adequate, if not satisfactory, supplies are of great importance in maintaining morale.[21]

The cliché that an army travels on its stomach is fundamentally true. It is a well known fact of military life that a hungry army is generally a dispirited army. Furthermore, serious lack of nourishment will over time undermine the physical and mental abilities of soldiers to fight effectively. Combat, especially the hand-to-hand combat of the Middle Ages, is exceptionally energy intensive and calorie expensive. Forces that lack confidence that they will be supplied effectively are more likely than not to be unable to sustain their mission.[22] However, religious fasts such as those reported in sources for the First Crusade may be assumed to have not damaged materially the strength of men who abstained from food for short periods.[23]

For most of the twentieth century historians regarded medieval armies as having been small, a doctrine for which the German military historian Hans Delbrück was effectively responsible. He argued that medieval chroniclers, or at least those upon whom he focused, often exaggerated the numbers of medieval armies to impossible orders of magnitude.[24] From a selection of accounts, written largely by chroniclers whose theological and other agendas he ignored, Delbrück concluded that medieval armies were poorly organized and lacked discipline. He

---

19 For a discussion of "brute fact" or objective reality from an epistemological perspective, which works as a fundamental corrective to excesses inherent in literary "theory" in terms of construction/deconstruction, see J. Searle, *The construction of social reality* (New York, 1995), pp. 34-5, 55-6.

20 For an explication of *Sachkritik* see G. A. Craig, "Delbrück: the military historian", in P. Paret, ed., *Makers of modern strategy: from Machiavelli to the nuclear age* (Princeton, 1986), 326-53, pp. 332-7.

21 See Bachrach, *Early Carolingian warfare*, pp. 132-8.

22 Bachrach, *Early Carolingian warfare*, p. 134. See also Riley-Smith, *First Crusade*, pp. 88-90, discussing the impact of starvation and fear of starvation.

23 See D. S. Bachrach, *Religion and the conduct of war c. 300- c. 1215* (Woodbridge, 2003), pp. 18, 25, 33-4, 42, 62, 96, 113, 117, 136, 151, 155, 159, 183.

24 H. Delbrück, *History of the art of war within the framework of political history*, trans. W. J. Renfroe, jr, 4 vols (Westport, 1975-85). Also of great importance is H. Delbrück, *Numbers in history: how the Greeks defeated the Persians, the Romans conquered the world, the Teutons overthrew the Roman Empire, and William the Norman took possession of England. Two lectures delivered before the University of London on October 6 and 7, 1913* (London, 1913). M. I. Finley, *Ancient history: evidence and models* (London, 1986), p. 71, draws attention to Delbrück's "obsession with numbers".

argued that the lack of productivity that he believed to be inherent in the medieval economy made it impossible to have large armies.[25]

Although Delbrück found the inflation of numbers by medieval ecclesiastical chroniclers worthy of detailed examination, and ultimately of drastic downward correction, it did not occur to him that they may also have exaggerated the poor organization and lack of discipline of the forces they described. He ignored much information from pre-Crusade times that indicated, for example, that the Carolingians and Anglo-Saxons had very sophisticated administrative systems.[26] When he did not ignore it, he dismissed surviving documents as evidencing the wishful thinking of men besotted with the image of the Roman world that they tried to imitate without success.[27]

Delbrück's influence was such that for much of the twentieth century medievalists reflexively rejected sources that gave large numbers for military forces.[28] As Werner, who began the process of criticizing Delbrück's methods and the work of his followers, put it: "In contrast to the extraordinary mistrust of critical efforts which accompany every attempt to establish a greater troop strength, it must be emphasized that the effective scholar does not gain distinction simply because he estimates the smallest possible number for an army but because his methods bring him closer to the truth and, in addition, he can prove the point."[29]

---

[25] Delbrück, *History of the art of war*, vol. 2, pp. 387-426; vol. 3, pp. 54-5, 97-8, 169. The *parti pris* of clerical historians is discussed by Goffart, who provides a model for how to deal with the distortion of military matters in such sources. See, for example, W. Goffart, *The narrators of barbarian history (A.D. 550-800): Jordanes, Gregory of Tours, Bede, and Paul the Deacon* (Princeton, 1988), pp. 172-83, dealing with slaughter and miracles.

[26] See J. Campbell, "The significance of the Anglo-Norman state in the administrative history of Western Europe" and K. F. Werner, "*Missus-Marchio-Comes*: entre l'administration centrale et l'administration locale de l'Empire Carolingien", both in W. Paravicini and K. F. Werner, eds, *Histoire comparée de l'administration (IVe-XVIIIe siècles): actes du XIVe colloque historique franco-allemand, Tours, 27 mars – 1er avril 1977 organisé en collaboration avec le Centre d'études supérieures de la Renaissance par l'Institut historique allemand de Paris* (Munich, 1980), 117-34, 191-239.

[27] Delbrück regarded the document trail as a sham. See *History of the Art of War*, vol. 3, pp. 17-18, 30-42. In the same vein T. Reuter, *Germany in the early Middle Ages, 800-1056* (London and New York, 1991), pp. 94-5.

[28] See, for example, discussion of figures for the armies of Harald Godwinson and William of Normandy at the battle of Hastings in B. S. Bachrach, "On the origins of William the Conqueror's horse transports", *Technology and culture*, 26 (1985), 505-31 [rpt in his *Warfare and military organization*, No. XIII].

[29] K. F. Werner, "Heeresorganization und Kriegsführung im deutschen Königreich des 10. und 11. Jahrhunderts", SSCISAM, 15 (Spoleto, 1968), 791-856, pp. 813-14, trans. Bachrach.

Werner's critique of earlier studies of early medieval military demography, paying particular attention to Delbrück and Ferdinand Lot and arguing cogently that the sources had been distorted ruthlessly to produce small numbers, was devastating.[30] Some minimalists still follow the pattern of distortion;[31] however, most medieval military historians have accepted the validity of the critique of Delbrück and his followers and Werner's view that military forces were much larger than Delbrück thought has been embraced.[32]

Subsequent studies have largely discredited major aspects of Delbrück's thesis and as a result several methodological imperatives have emerged.[33] Some medieval authors sometimes inflated the size of armies that they described. It is equally true, however, that some chroniclers on occasion provided numbers that were impossibly small. Thus, simply because a medieval figure for the size of an army may strike a modern scholar as being too large is no reason to reject it. Similarly, if a figure is thought to be too small, it must be investigated. All figures provided in sources must be scrutinized rigorously.

---

30 Werner, "Heeresorganization", pp. 813-15. See also Werner, "*Missus-Marchio-Comes*".

31 See, for example, M. G. Kellner, *Die Ungarneinfälle im Bild der Quellen bis 1150: von der "gens detestanda" zur "gens ad fidem Christi conversa"* (Munich, 1997). However contrast B. S. Bachrach, "Magyar-Ottonian warfare: à-propos a new minimalist interpretation", *Francia*, 27 (2001), 211-30. Reuter is led to assert that if Charlemagne's armies were more than 3-4,000 in number, a tenth or less of what medieval military historians maintain, they would have starved due to the lack of a satisfactory logistical infrastructure. See T. Reuter, "Carolingian and Ottonian warfare", in M. Keen, ed., *Medieval warfare: a history* (Oxford, 1999), 13-35. However, Reuter omits information in the very sources he uses which undermines his conclusions. When discussing the battle of Firenzuola, he claims that 50 men were killed while his source, Flodoard, wrote that 1,500 were killed. See Flodoard, *Annales*, ed. G. Pertz, in MGH *SS*, vol. 3, 363-408, Annus 923 (p. 373). See also *Annales Fuldenses*, ed. F. Kurze, in MGH *ScriptRerGerm.Cont* (Hanover, 1891), Annus 880 (pp. 94-6).

32 See, for example, P. Contamine, *War in the Middle Ages*, trans. M. Jones (Oxford, 1984), pp. 25-6; J. F. Verbruggen, *The art of warfare in Western Europe during the Middle Ages: from the eighth century to 1340*, trans. S. Willard and Mrs R. W. Southern, 2nd ed. rev. (Woodbridge, 1997), pp. 283-4. See also B. S. Bachrach, "The Hun army at the battle of Chalons (451): an essay in military demography", in K. Brunner and B. Merta, eds, *Ethnogenese und Überlieferung: Angewandte Methoden der Frühmittelalterforschung* (Vienna and Munich, 1994), 59-67; Idem, "Siege of Antioch"; C. R. Bowlus, *Franks, Moravians, and Magyars: the struggle for the middle Danube, 788-907* (Philadelphia, 1995), p. 19; France, *Victory in the East*, pp. 122-41; J. France, *Western warfare in the age of the Crusades, 1000-1300* (Ithaca, 1999), pp. 128-33.

33 See B. S. Bachrach, "Early medieval military demography: some observations on the methods of Hans Delbrück", in D. Kagay and L. J. Andrew Villalon, eds, *The circle of war in the Middle Ages* (Woodbridge, 1999), 3-20.

In the context of much recent writing on medieval military demography, scholars generally have been reasonable with regard to the size of First Crusade forces and figures in the 60-70,000 range are supported in some quarters; although, there are minimalists who look to half that figure.[34] France, concluded that "... a likely figure for the army at its greatest would be around the 50,000-60,000 mark including non-combatants." just after the surrender of Nicaea.[35] Since I focus here on the period immediately following the surrender, I will use France's figures despite the fact that I consider them to be somewhat low.[36]

Those unfamiliar with medieval military history may find problematic the size of the forces argued to have functioned during the First Crusade. However, it should be remembered that Alexander the Great kept field armies larger than the estimates for the Crusader forces in effective fighting condition for much longer than the operations undertaken during the Crusade and that Rome and Carthage also maintained exceptionally large armies for long periods of time under adverse conditions.[37] Furthermore, the primitive technology of early modern Europe did not keep armies of the sixteenth to eighteenth centuries at absurdly small levels. Indeed, Vienna was besieged in 1529 by 125,000 men with far fewer logistical assets available to them than the Crusaders had enjoyed at Antioch.[38]

The siege of Nicaea lasted for more than seven weeks and was marked by the building of extensive fortified encampments, perhaps some 10,000 metres of redoubts and ditches. These had to provide protection for 50,000 or more effectives and their non-combatant camp followers as well as for at least 10,000 horses. These defences comprised both a vallation of the city and a contravallation of the Crusader camp to protect the army from attack by either a relief force or local raiders. The Crusaders also constructed numerous siege engines of various types to sap towers and overreach defences. In addition, the besiegers made several efforts

34 Runciman put the figure at somewhere between 60,000 and 100,000 effectives and Duncalf accepted his estimates. Waas estimated a maximum of around 70,000. Lilie opted for a minimalist figure of not much more than 30,000. Riley-Smith is also a minimalist. See Runciman, *History*, vol. 1, pp. 169, 336-41; F. Duncalf, "The First Crusade: Clermont to Constantinople", in Setton, *HC*, vol. 1, 253-79; A. Waas, *Geschichte der Kreuzzüge*, 2 vols (Freiburg, 1956); R.-J. Lilie, *Byzantium and the Crusader states 1096-1204*, trans. J. C. Morris and J. E. Ridings (Oxford, 1993); Riley-Smith, *First Crusade*, p. 63.

35 France, *Victory in the East*, pp. 123-42, esp. pp. 124, 142.

36 See Bachrach, "Siege of Antioch".

37 Engels, *Alexander the Great*, pp. 146-52; P. A. Brunt, *Italian manpower, 225 B.C. – A.D. 14* (Oxford, 1971), pp. 417-512.

38 See J. Lynn, "Food, funds, and fortresses: resource mobilization and positional warfare in the campaigns of Louis XIV", in idem, *Feeding Mars*, 137-59; C. Duffy, *Siege warfare: the fortress in the Early Modern world, 1494-1660* (London, 1979), p. 201. Cf. Bachrach, "Siege of Antioch".

to storm the walls and they won a victory in pitched battle over the relieving army of Qïlïj Arslan, a force of some 10,000 troops, mostly mounted archers.[39]

Even taking into account the important contribution of the Byzantines, the operations of the Crusader forces at Nicaea were a formidable effort that certainly took a toll on both men and equipment.[40] However, by 26 June most forces, apparently apart from those wounded and unable to travel, were either on the march or preparing to depart. The last contingent to leave, the Aquitanian forces of Count Raymond of St Gilles, was on the road by 28 June.[41]

How was this exceptionally rapid transition carried out? After spending over seven weeks in a siege camp from which they had invested closely a great fortress city, how did the Crusader forces of 50,000 or more soldiers and non-combatants get in order in the course of a week all the equipment needed for a lengthy expedition? Who saw to the replacement of lost and damaged gear? When, where, and how was sufficient food and fodder gathered to initiate this new phase of the campaign? Who provided the wagons, carts, and pack horses to sustain the march? Who had been exercising the thousands of war horses and taking care of the even greater number of riding horses while the Crusaders were engaged in operations against Nicaea that required comparatively few horses?

When considering such questions, we must take into account a spectrum of variables. There is, however, one clear benchmark. Any conclusions regarding the condition of the forces, their animals, and equipment when they left Nicaea must take into account the fact that they were combat ready on the morning of 1 July when they won an overwhelming victory in the field against a large enemy force that launched a surprise attack against the forward column west of Dorylaion.[42]

Such fundamental questions permit some preliminary observations concerning the strategy that undergirded not only this phase of the Crusade but perhaps even the entire campaign. First, it must have been decided even before the siege of Nicaea that the forces would advance overland on Antioch and Jerusalem. It is likely that the leaders were informed about details of the post-Nicaea march at a conference convened by Alexios at Pelekanos.[43] The logistics required that between 26 June and 1 July, the planned date for the Crusaders to reach Dorylaion, a minimum of 600 metric tonnes of grain be made available to the forces. Supplying the Crusaders was the task, by and large, of the Byzantines and the

---

39 Bachrach, "Siege of Nicaea".

40 See Bachrach, "Siege of Nicaea"; France, *Victory in the East*, pp. 157-9.

41 Hagenmeyer, *Chronologie*, pp. 83-5; France, *Victory in the East*, p. 169.

42 See France, *Victory in the East*, pp. 175-84.

43 See Lilie, *Byzantium and the Crusader states*, pp. 26, 28-9; France, *Victory in the East*, pp. 165-7.

report of the author of the *Gesta Francorum* that the emperor promised to supply them with provisions and restore whatever they had lost is to be accepted.[44]

The Crusader strategy to fight overland through hostile territory for many months raises fundamental questions regarding their primary aims. Accounts of Urban II's views expressed at Clermont and later make it clear that the recovery of Jerusalem was his fundamental aim. There is no doubt that the extraordinary response to the preaching of the Crusade was a product of the desire of large numbers to participate in the recapture of Jerusalem by a pilgrimage under arms. Providing assistance to the Byzantines and relieving the plight of Christians under Muslim rule, two other motifs that emerge from accounts of Pope Urban's views, were secondary or even lesser goals. These latter would not have been much of an incentive to the thousands who joined the "People's Crusade" and the tens of thousands who followed them.

If the capture of Jerusalem was the primary aim, why did the leaders adopt such an inefficient means as an overland campaign to achieve it?[45] For very large forces to undertake a march of many months across some of the most difficult terrain in the Middle East was not the most effective way to execute a campaign aimed at capturing Jerusalem. Added to these liabilities was the fact that the weather would have been fit for neither man nor beast at the time of year of the march. The most efficacious way to deliver large Western forces mustered around Constantinople and intent upon taking Jerusalem would have been by a seaborne expedition.[46] It surely escaped the attention of none of the principals in 1097 that Byzantine fleets were largely in control of this part of the Mediterranean Sea.[47] If

---

44 Anonymous, *Gesta Francorum et aliorum Hierosolimitanorum*, ed. and trans. R. Hill, *The deeds of the Franks and the other pilgrims to Jerusalem* (Oxford, 1962), II.vi. (p. 12). See also Lilie, *Byzantium and the Crusader states*, p. 28; France, *Victory in the East*, p. 118.

45 See J. Shepard, "Cross-purposes: Alexius Comnenus and the First Crusade", in J. Phillips, ed., *The First Crusade: origins and impact* (Manchester, 1997), 107-129.

46 France preceptively recognizes the role of naval power during the Crusade and points out that Western chroniclers were not much interested in naval matters. See J. France, "The First Crusade as a naval enterprise", *MM*, 83 (1997), 389-97.

47 France, "First Crusade as naval enterprise"; Bachrach, "Byzantine navy".

France is rightly skeptical of Lewis's argument that Byzantine naval power was in decline at this time. In fact it was probably increasing in an absolute sense. However, Lewis was correct in regard to the relative position of Byzantium because of the growth of the naval power of the Italian maritime republics. See France, *Victory in the East*, p. 210, n. 46; A. R. Lewis, *Naval power and trade in the Mediterranean, 500-1100* (Princeton, 1951), pp. 225-49.

With regard to a Fāṭimid naval threat, France is probably correct in pointing out that Egyptian naval forces had sufficient access to water supplies because they controlled ports along the Syro-Palestinian coast as far north as Tripoli. See his "First Crusade as naval enterprise", p. 396, n. 8 and also J. H. Pryor, " 'Water, water everywhere, Nor any drop to drink.': water supplies for the fleets of the First Crusade", in M. Balard, B. Z. Kedar and J.

necessary, these fleets could be given support by the blue water assets of Pisa, Genoa, and Venice.[48] In addition, some northern European realms such as England appear to have had naval forces operating in the eastern Mediterranean during the Crusade. Norwegians, Danes, "Belgians", and perhaps even others from the north seem to have prepared ships for operations in the Mediterranean.[49]

These combined naval assets make it clear that the option was available, in material terms, for 10-15,000 of the 50,000 or so Crusaders who reached Constantinople in 1097 to be ferried rapidly with their horses to the Byzantine naval base on Cyprus, about 1,600 kilometres southeast.[50] From there, such a force of 10-15,000 could have been moved in an effective manner to the port at Jaffa, another 320 kilometres south-southeast. The fortifications at Jaffa were deemed incapable of being defended by the Muslims.[51] In addition, it is highly unlikely that both Western Christians and Byzantine military experts were unaware of the situation at Jaffa. The final passage would have needed only three or four days sailing and could not have been detected by the defenders of Jerusalem in a timely

---

Riley-Smith, eds, *Dei gesta per Francos: études sur les croisades dédiées à Jean Richard / Crusade studies in honour of Jean Richard* (Aldershot, 2001), 21-8. Why these assets were not utilized requires further study. The likelihood that the Egyptians could be co-opted was well understood by Alexios and he even informed the Crusaders of the need to develop good diplomatic relations with them. See M. A. Köhler, *Allianzen und Verträge zwischen frankischen und islamischen Herrschern im Vorderen Orient* (Berlin, 1991), pp. 1-72; France, *Victory in the East*, p. 166.

48 Concerning Italian fleets see France, *Victory in the East*, pp. 209, 219. Shepard is certainly correct in suggesting that Italian city fleets were no match for Byzantine fleets. See "Cross-purposes", p. 112. However, he assumes an adversarial relationship rather than a co-operative arrangement in support of the Crusade.

49 The "English" fleets require further study. On the one hand, Shepard argues that Anglo-Saxons in Byzantine service operated fleets in imperial service. On the other hand, Tyerman argues for fleets from England itself. France is probably correct in seeing at least two English fleets, one under imperial command and another coming from England. See Shephard, "English and Byzantium"; C. Tyerman, *England and the Crusades 1095-1588* (Chicago: 1988), pp. 19-21; France, "First Crusade as naval enterprise", p. 397, n. 24. See also N. Hooper, "Edgar the Ætheling: Anglo-Saxon prince, rebel and Crusader", *Anglo-Saxon England*, 14 (1985), 197-214; France, *Victory in the East*, pp. 215-20; C. W. David, *Robert Curthose, Duke of Normandy* (Cambridge Mass., 1920), pp. 230-44.

For a fleet from Norway see R. W. Unger, *The ship in the medieval economy 600-1600* (London and Montreal, 1980), p.129 and for others see France, *Victory in the East*, pp. 216-20.

50 Under the command of its governor Eumathios Philokalēs Cyprus was a major naval and military base. Eumathios had sufficient troops and ships available to take over the port at Laodicea from Robert of Normandy and to place a Byzantine garrison there. By mid-October 1097 Patriarch Symeon of Jerusalem was in exile on Cyprus and he seems not to have lacked for anything. The island apparently served as an off-shore headquarters of sorts and as an entrepôt for supplies for the Crusade. See Runciman, *History*, vol. 1, p. 222.

51 France, *Victory in the East*, pp. 326-7.

manner. Western or Byzantine strategists may have considered it possible to take the defenders of Jerusalem by surprise and the city would not have been able to be reinforced to the massive degree required to prevent its capture.[52]

Such a seaborne operation could also have aided the Byzantines in conquering Muslim territory and liberating Christians under Muslim rule. An army of 10-15,000 Crusaders supported by appropriate naval forces would have assured the Byzantines control of the Syro-Palestinian coastal cities from Tripoli to Ascalon. Indeed, the entire eastern Mediterranean littoral would have been at the mercy of such a force and the fleet that supported it. The half-dozen or so Muslim governments controlling various parts of the Mediterranean coast between Ascalon and Constantinople often did not co-operate with each other and sometimes were in open armed conflict.[53]

The attractiveness of a strategy of conquest based on a seaborne operation is obvious from a logistical perspective. When contrasted to the immense difficulties facing the land-based option that was in fact chosen, questions are raised that ultimately cause reflection on the nature of the Crusade itself. We must assume that the leaders at Constantinople discussed with Alexios and his military advisers how the campaign should be executed. However, the option of pursuing a seaborne strategy apparently did not find favour either with Alexios or with the leaders or perhaps even with both.

When the decision to undertake a land campaign was made, several crucial facts of military life were no doubt very clear. The Byzantines had made supplies available when the Crusaders arrived at Constantinople; indeed, they depended on Alexios for food. In many cases they had been able to maintain their march to the Byzantine capital only because logistical support had been provided. In addition, Alexios had provided key logistical and military support for the siege of Nicaea which had given the Christians the edge. Finally, whatever was required for the march from Nicaea even as far as Dorylaion had to have been provided by the Byzantines.[54]

Fulcher of Chartres made it clear that Alexios controlled the logistics in the initial stages of the Crusade in the East.[55] The Byzantines were able to determine

---

52 Even though the Crusaders had been operating in Muslim territory for more two years, the garrison at Jerusalem had still not been raised to a level that could defy them. See France, *Victory in the East*, pp. 327-66, esp. 330-31.

53 See France, *Victory in the East*, Fig. 3 (p. 109).

54 Bachrach, "Siege of Nicaea". See also Lilie, *Byzantium and the Crusader states*, p. 7.

55 Fulcher of Chartres, *Historia Hierosolymitana (1095-1127)*, ed. H. Hagenmeyer (Heidelberg, 1913), I.ix.3 (pp. 178-9): "erat enim omnibus hoc necesse, ut sic cum imperatore amicitiam consolidarent, sine cuius consilio et auxilio nostrum iter nequivimus expedire, neque illi, qui nos erant subsecuturi eodem tramite. Quibus ideo praebuit ipse imperator de nummismatibus suis et de pannis sericis quantum placuit; et de equis et de pecunia, qua nimis indigebant ad tantum iter explendum."

the strategy to be undertaken by the Crusaders and a land-based campaign was decided on. At least one of the major purposes of the oaths sworn by the leaders to the emperor was that they recognized him as a superior whom they would obey. Without Byzantine support there could be no Crusade. Indeed, it is probable that without major logistical support from the Byzantines, the main forces would sooner or later have suffered a fate similar to that of the People's Crusade or perhaps a disaster not unlike that which effectively ended the Crusade of 1101.

As a working hypothesis, Alexios made the key strategic decisions regarding the route that the Crusade would follow, his strategy adumbrated largely by what he believed would best serve his own and Byzantine interests.[56] As a result, he sent the Crusaders on a hellish campaign overland through very difficult terrain to retake great fortress cities such as Antioch and no less importantly to destroy any Muslim armies encountered. The reconquest of Jerusalem was not Alexios's primary aim and if the Crusaders were destroyed en route to Jerusalem, it is unlikely that the emperor would have shed many tears.[57]

The assertion that the Byzantines controlled the Crusaders' access to supplies and thus were able to dictate their course of action can be demonstrated in several ways.[58] A careful, though perhaps not uncontroversial, reading of the chronicles provides some evidence for this view. More importantly, it can be shown that between Nicaea and Jerusalem there was no way in which they could forage sufficiently on a regular basis in order to sustain themselves. This can be demonstrated by methods of *Sachkritik* focussing upon objective realities required to feed men and animals.

For arithmetical convenience France's minimal estimates will be used and it will be assumed that approximately 50,000 mouths had to be fed. Approximately 50,000 kilogrammes, 50 metric tonnes, of milled wheat were required each day simply to meet basic calorific needs. While such a daily ration would have provided sufficient calories, it would have been neither healthy in the long term nor conducive to high morale and is thus used here only for the purpose of a minimalist model.[59] However, even using this figure, the Crusaders would have needed each day grain that would have required 100 horse- or mule-drawn carts or

---

Shepard writes of Alexios's "virtual stranglehold on communications between the Levant and western Europe", which seems to have logistical implications; however, this is merely an inference. See his "Cross-purposes", p. 128.

56 Shepard, "Cross-purposes"; Lilie, *Byzantium and the Crusader states*, pp. 7-28; Runciman, *History*, vol. 1, pp. 193-4.

57 Cf. Lilie, *Byzantium and the Crusader states*, pp. 9-18.

58 See also Shepard, "Cross-purposes", pp. 112-13.

59 See Bachrach, *Early Carolingian warfare*, pp. 136-8.

wagons to carry it, each having a maximum load capacity of approximately 500-600 kilogrammes.[60]

In addition, they are estimated here, as a minimalist position, to have had at least 5,000 cavalry.[61] If each of these had on average only two horses, and the usual number was three, one for riding, one for equipment and supplies, and the third a war horse, an additional 50,000 kilogrammes of grain, either oats or spelt or barley, would have been needed daily.[62] Thus, in addition to 100 vehicles to carry grain for the forces each day, another 100 would have been needed for the animals. For a five-day march 1,000 vehicles would be required.

The number of draft animals and/or pack horses needed during any particular segment of the march depended upon three variables: first, the nature of the supply train; secondly what kind of foraging was intended; and thirdly, what amount of logistical support was provided either from imperial magazines or by having markets arranged.

A supply train of vehicles drawn by two horses or mules each and used solely to haul the minimum needs for one day would have needed approximately 450 animals because ten per cent or so over the bare minimum is needed to replace injured or dead animals.[63] Therefore, 2,000 or so kilogrammes of grain would be consumed each day by the draft animals. Like the war horses and riding horses, of course, these could not survive on only five kilogrammes of grain but required another five of grass or hay. This is assumed to have been available as forage along

---

60 See R. Lefebvre des Noëttes, *L'attelage: le cheval de selle à travers les âges*, 2 vols (Paris, 1931); A. C. Leighton, *Transport and communications in early medieval Europe AD 500-1100* (Newton Abbot, 1972); J. Langdon, *Horses, oxen and technological innovation: the use of draught animals in English Farming from 1066-1500* (Cambridge, 1986).

Recent efforts to revise the state of the question on the basis of archaeological evidence include G. Raepsaet, "La faiblesse de l'attelage antique: la fin d'un mythe?", *L'antiquité classique*, 48 (1979), 171-6; idem, "Attelage antiques dans le nord de la Gaule: les systèmes de traction par équides", *Trierer Zeitschrift*, 42 (1982), 215-73; J. Spruytte, *Early harness systems. Experimental studies: a contribution to the history of the horse, trans. M. Littauer* (London, 1983). These studies argue how inefficient harness could be positioned to become less inefficient but are unconvincing. C. W. Röring, *Untersuchungen zu römischen Reisewagen* (Koblenz, 1983) adds nothing new since the existence of pivotable front axles is in doubt.

61 France argues that there were some 7,000 "knights or lords" and may be right. I have used a lower figure so as not to prejudice the argument. See *Victory in the East*, p. 142.

62 For the number of horses used by mounted troops in the West during the later eleventh and early twelfth centuries, see Bachrach, "William Rufus", pp. 47-50. France adopts this traditional three-horse pattern as a minimum and concludes that there were at least 20,000 riding horses, pack horses, and war horses in the forces. See *Victory in the East*, p. 128.

63 France rightly focuses on loss of animals and problems of finding replacements. See *Victory in the East*, pp. 122-42, 280-82.

the route.[64] However, when it, and water as well, was not available, special measures had to be taken.[65] The grain needed for the draft animals, however, would have required each day an additional four vehicles and the animals to haul them. A ten-day march would require 40 additional vehicles and perhaps 90-100 additional animals just to feed the latter.

Using only the grain estimates for forces of 50,000 men and 10,000 horses, a ten-day march would require 2,000 vehicles. The size of the supply train has a material impact on the pace of the march. A horse-drawn vehicle capable of hauling 500 kilogrammes of grain occupies a rectangle on the ground of approximately 25 square metres (10 by 2.5).[66] A single line of 2,000 vehicles would need approximately 20 kilometres of road and if such a column could move at only 20 kilometres a day, the head would reach its destination at the same time that the tail would be leaving camp.

Obviously, the situation would not have been as simple as this. Under best conditions, and Roman roads were good, a column of horse-drawn vehicles could average around 30 kilometres a day for six days or less.[67] In addition, over much of the route traversed by the Crusaders the major Roman roads were about six to seven metres wide.[68] Such roads could accommodate side by side two vehicles of the type discussed above. Thus, 2,000 vehicles, a supply train for grain for the troops and cavalry horses, would string out for only 10 kilometres.

However, columns were composed not only of vehicles hauling grain for food. Both cavalry and foot soldiers were part of them and vehicles or pack animals carrying equipment such as tents, tools, rope, and metal parts for siege machines also required space.[69] We can, however, omit foot soldiery here because they could carry their own food in packs. It would have been possible for a double cavalry column to form up on roads approximately six metres wide and therefore the 5,000 cavalry would take up some 10,000 metres, while the 5,000 extra horses probably could be closed up somewhat. Perhaps a total of 8,000 metres for these. The forces of 50,000 could probably move in a more or less disciplined march formation at five abreast. Thus, the 50,000 would take up another 20,000 metres along the road. In all, the line of march with 2,000 wagons, 50,000 men, and 10,000 horses would string out some 40 kilometres.

---

64 See B. S. Bachrach, "*Caballus et caballarius* in medieval warfare", in H. Chickering and T. H. Seiler, eds, *The study of chivalry: resources and approaches* (Kalamazoo, 1988), 173-211 [Rpt in his *Warfare and military organization*, No. II], pp. 179-80.

65 The *Gesta Francorum* reported carrying water when it was believed that it would be in short supply or unavailable. See *Gesta Francorum*, IV[x] (p. 23).

66 Leighton, *Transport and communications*, pp. 77, 161.

67 See Bachrach, "Animals and warfare".

68 See V. W. Von Hagen, *The roads that led to Rome* (London, 1967), p. 134.

69 Cf. Bachrach, *Early Carolingian warfare*, pp. 107-18.

These estimates are consistent with comparable pre-modern historical situations. Engels has shown that, when operating in Asia Minor at a ratio of six infantry to one cavalry, the armies of Alexander the Great needed one baggage animal for 27 men and their horses for each day. Thus, Alexander's 65,000 men who crossed the Hellespont needed approximately 2,400 animals to carry one day's supplies. For five days 12,000 animals were needed. Engels concludes that "... it is highly improbable that the Macedonians ever employed more than 20,000 pack animals anywhere along their route." Thus, 65,000 men, of whom some 10,000 were cavalry, could march for only eight days before the required baggage train became too unwieldy and a breakdown in mobility would occur.[70]

Engels proposes that approximately each three and one third men needed one baggage animal for an eight-day march. However, most of these animals carried food for neither themselves nor the cavalry horses and men. The majority carried supplies such as tents, artillery, and siege equipment. After eight days, however, the food carried by those that were delegated to that task had to be replenished. Beyond the traffic problems caused by a train of 20,000 animals in addition to the 65,000 men, the animals in the supply train of a force such as this would after a finite period merely be carrying food to feed themselves.[71]

Turning to the Roman empire, figures are not dissimilar. Roman legions of 4,800 men were regularly accompanied on the march by approximately 1,400 mules carrying both supplies and baggage, on average one mule for approximately four men. The mules could carry 140 metric tonnes. They could be substituted for by 280 two-horse vehicles; however, these would need some 600 animals to pull them.[72] In addition, legions had artillery trains estimated at some 70 wagons and 160 draft animals. To these must be added pack animals accompanying auxiliaries, another 400 animals. A legion of around 5,000 men, horse and foot, might have as many as 2,000 pack animals and 70 vehicles or as few as 1,000 animals but more than 200 vehicles.[73]

Officers of a legion also had their own special baggage train and when several legions were brought together as armies, special baggage trains were required: command baggage trains and most importantly siege trains, which could be very large. That of the army of Marcus Antonius during the Parthian war has been estimated at 340 wagons requiring at least 800 draft animals.[74]

Such variables would seem to make it virtually certain that the Crusaders relied completely upon the Byzantines for logistical support during the two weeks

---

70 Engels, *Alexander the Great*, pp. 20, 144.

71 Engels, *Alexander the Great*, pp. 19-22.

72 J. P. Roth, *The logistics of the Roman army at war (264 B.C.-A.D. 235)* (Leiden, 1999), p. 83.

73 A. K. Goldsworthy, *The Roman army at war, 100 BC – AD 200* (Oxford, 1996), p. 290; Roth, *Logistics of the Roman army*, pp. 79-83.

74 Roth, *Logistics of the Roman army*, pp. 88, 91.

following the victory at Nicaea. The Muslim commander surrendered on very good terms on 19 June,[75] and most reports indicate that both the leaders of the Crusade and the rank and file were angered and disappointed by the manner in which the siege had been ended. As they saw it, they had been deprived of substantial booty which they had expected in consequence of promises made by Alexios. The emperor tried to mollify them by providing expensive gifts to the leaders and donatives of copper coins to the rank and file.[76] There is no report, however, that as a result of their disappointment, they took hostile action against the Byzantines, although Raymond of Aguilers suggested that there was some unrest in the ranks of the Aquitanian units.[77] Raymond of St Gilles's failure to meet Alexios at Pelekanos also may have been a manifestation of hostility toward him.[78] Fulcher of Chartres' report that some Crusaders deserted at this time may also be an index of displeasure at the course of events.[79]

The Crusaders proceeded with the campaign nevertheless, accompanied by a small Byzantine force under the command of general Tatikios. There is not a breath of rumour comparable to the earlier plot attributed to Bohemond by Albert of Aachen and Anna Komnēnē to use the Crusaders to capture the city of Constantinople.[80] Their docility following the surrender of Nicaea and the obvious command exercised by Alexios is highlighted by Fulcher of Chartres who wrote that the barons did not leave Nicaea until Alexios gave them permission.[81]

Latin sources make clear the Crusaders' negative feelings about their treatment by the emperor but curiously juxtapose their obedience to him. However, they do not report any efforts undertaken either by the leaders or by Byzantine officials to prepare for the forthcoming campaign. Nothing is said about the refurbishing of equipment that would have been necessary, nor is there any report of the extensive preparation of horses always required for a long march. Finally, not a word is mentioned concerning gathering foodstuffs. Indeed, the only information of a vaguely logistical nature is that some of the poor in the forces were provided with food by Alexios.[82]

---

75 Bachrach, "Siege of Nicaea".

76 Lilie, *Byzantium and the Crusader states*, p. 24; France, *Victory in the East*, pp. 165-6.

77 Raymond of Aguilers, *Historia Francorum qui ceperunt Iherusalem*, in RHC *HOcc*, vol. 3, 231-309, c. III (pp. 239-40); Raymond of Aguilers, *Liber Raimundi de Aguilers*, ed. J. H. and L. L. Hill (Paris, 1969), §4 (p. 44).

78 See Lilie, *Byzantium and the Crusader states*, pp. 28-9.

79 Fulcher of Chartres, *Historia Hierosolymitana*, I.x.5 (pp. 184-5).

80 Albert of Aachen, *Historia Hierosolymitana*, in RHC *HOcc*, vol. 4, 265-713, II.xiv (p. 309); Anna Komnēnē, *Alexiade*, ed. and Fr. trans. B. Leib, 3 vols (Paris, 1937-45), X.ix.1 (vol. 2, pp. 220-21).

81 Fulcher of Chartres, *Historia Hierosolymitana*, I.xi.1 (pp. 189-90).

82 Ibid., I.x (pp. 181-9).

It is necessary to reconstruct the course of events in logistical terms between 19 and 28 June when the last Crusader units began the march. The 50,000 men and 10,000 animals had to be fed until the march commenced and during the long week that many of the Crusaders remained encamped at Nicaea, a minimum of approximately 800 metric tonnes of milled grain, 1,600 cart loads, would have been needed.

It would be contra-intuitive to assume that the Crusaders, deprived of booty and glory after a bloody siege that had cost them dearly, would have tolerated a week-long diet of gruel or biscuit washed down by water. It may be suggested that they were provided with food and drink on a more lavish scale and that failure of Latin sources to mention this is consistent with their anti-Byzantine bias.[83] Indeed, they display a dual purpose of describing or implying the suffering of Crusaders on the one hand and the niggardly and often duplicitous behaviour of Byzantines on the other.

The distance between the Crusader encampments around Nicaea and Dorylaion was probably around 95 kilometres. However, they took not three days but more than double that time for the journey. In addition to supplying the army for a long week at Nicaea after the surrender, supplies for both men and animals had to be provided for the forthcoming march but the sources provide no information about the food carried when the forces left Nicaea. However, the Crusaders, who had scoured the region of Nicaea for supplies for more than seven weeks, could not have supplied themselves from local sources without active support from the Byzantines.

No matter what supplies were involved and from what other elements the Crusaders may have received some supplies, there can be no doubt that the regular logistical support was arranged by imperial officials from imperial land-based or naval assets, or by private businessmen. These latter were licensed by the administration or its agents, to whom had been delegated authority to issue the relevant documents. Terms such as *praeceptum*, *concessum*, and *iussio* were used in Latin sources to denote official *acta* giving permission to non-governmental entities to trade with the Crusaders.[84]

---

83 See S. B. Edgington, "The First Crusade: reviewing the evidence", in J. Phillips, ed., *The First Crusade: origins and impact* (Manchester, 1997), 57-77, pp. 64-5.

84 See Fulcher of Chartres, *Historia Hierosolymitana*, I.viii.9, x.6 (pp. 176, 185); Albert of Aachen, *Historia Hierosolymitana*, II.xxviii (p. 321).

It is possible, although seemingly unlikely, that the Crusaders could have obtained substantial suplies without imperial permission. The *Gesta Francorum* reported that Bohemond ordered that a great market, "maximum mercatum", be established by sea. However, the author's hostility to the Byzantines may have led to his omitting the emperor from the administrative process. See *Gesta Francorum*, II [vii] (p. 14).

The first day of the march on 26 June followed the Roman road south for some 30-35 kilometres.[85] The Crusaders probably camped around the great Byzantine *aplēkton* at Malagina on the left bank of the Sangarios river. That the Latin sources do not mention them being supplied at Nicaea for the march may well have been because Malagina, which traditionally had supplied Byzantine forces operating further south, was used to supply them.[86] If the *aplēkton* had been recovered by the Byzantines following the failed Muslim relief of Nicaea, it is probable that Alexios's officials had been busy after the fall of Nicaea, if not earlier, in storing supplies at Malagina. Consequently, the Crusaders need not be concerned with supplies for the first day.[87]

For this single day a minimum of 100 metric tonnes of grain was needed. Moreover, most of the forces remained at Malagina for two days.[88] Thus, rather than have the Crusaders depart from Nicaea with at least 600 wagon loads of grain for provisions for three days, 26-28 June, the Byzantines probably provided the necessary supplies at Malagina, thus enabling them to cover a greater distance and more quickly. In short, Alexios's officials provided for the Crusaders at Malagina in the same way that Roman and Byzantine forces had been provided for traditionally on the road to Dorylaion.

After encamping at Malagina, it was a march of two days to Dorylaion. When attacked by the Turks, Bohemond's vanguard fortified its camp, so clearly there must have been a supply train of some kind and it is most probable that it was the Byzantines who had provided supplies for this march as well, approximately 400 vehicles with grain for the men and horses and perhaps some extras from the magazine at Malagina.[89] It was known that there were enemy forces north of Dorylaion and, as a result, it is highly unlikely that the Crusaders would have foraged en route.[90] However, probably because of increased enemy presence, the

---

85 See Hagenmeyer, *Chronologie*, pp. 82-84; France, *Victory in the East*, pp. 169, 171-2. For the roads from Nicaea to Dorylaion see Belke and Mersich, *Phrygien und Pisidien*, pp. 141-3. There is no argument regarding the crossing of the bridge over the river Gallos at Levkas, some 25 kilometres from Nicaea. The question is how far beyond the bridge the lead units travelled before making camp. Since it would have been important not to cause traffic jams around the bridge, it must be assumed that the vanguard under Bohemond went 4-5 kilometres beyond the bridge before camping.

86 See Vryonis, *Medieval Hellenism*, p. 31.

87 Albert of Aachen explained the division of the forces into two divisions by the need to forage. However, other sources, including those written by participants, did not adduce this explanation. See France, *Victory in the East*, p. 137.

88 Hagenmeyer, *Chronologie*, pp. 83-5.

89 France, *Victory in the East*, p. 175.

90 France, *Victory in the East*, p. 172. Ramsay's assumption that Dorylaion was in Turkish hands and that the road as far the northern environs of Sögüt was also under control of the Turks has no basis in the sources. See his *Historical geography of Asia Minor*, pp. 212-13.

Crusaders did not reach Dorylaion on the second day.[91] Thus, they encamped several kilometres to the West in a well-watered area. Early next morning the march was resumed and the lead column was attacked by the army of Qïlïj Arslan. Their decisive victory allowed the Crusaders to reach Dorylaion late on the same day.

Following Dorylaion, Qïlïj Arslan was on the run. As a result, the Crusaders were well positioned to reach Dorylaion and obtain supplies from yet another Byzantine magazine in preparation for the next segment of their expedition. Yet, there was a significant problem. The Crusaders had captured a huge hoard of booty when they took the Turkish encampment after the battle and there was no way that it could be added to their supply train without significantly slowing it down. In addition, much of the Turkish supply train was hauled by camels and the Crusaders probably had little experience in handling these difficult beasts.[92] The resupply and refreshment of the Crusaders at Dorylaion required only two full days, 2 and 3 July, and booty taken from the Turks was probably sold or traded for additional supplies and cash.

Even examining a small segment of what proved to be a very lengthy and difficult march, it is clear that little could be done by the Crusaders without Byzantine support. Sending foraging parties into hinterlands beyond the Roman roads was a prescription for failure if not disaster, not only because enemy troops were scattered about in sufficient numbers to give even well-guarded wagon trains a very difficult time. Sending out foraging expeditions for several days without precise knowledge of where large caches of grain were to be found would more often than not be an exercise in futility. It must be remembered that something over 100 metric tonnes of grain per day had to be available within a day or two's journey of the line of march.

By contrast to such a haphazard hunt for resources, the Byzantines were in a position to inform locals with grain and other supplies to sell where the Crusader line of march was likely to be at any particular time. Indeed, making such arrangements may well have been one of Tatikios's primary tasks. In addition, the Byzantines had well established encampments and magazines along the major military roads in regions through much of which the Crusaders moved. The intelligence and good will of the Byzantines and the emperor's edict permitting markets to be made available were the operational essentials that made the march of the Crusaders both possible and successful.

---

91 Fulcher of Chartres, *Historia Hierosolymitana*, I.xi.1-2 (pp. 189-91); Ralph of Caen, *Gesta Tancredi in expeditione Hierosolymitana, auctore Radulfo Cadomensi, ejus familiari*, in RHC *HOcc*, vol. 3, 587-716, c. XXI (p. 621). France has a different set of distances because he neither regards the first encampment to have been at Malagina nor that the great military road was taken from Nicaea to Dorylaion. See *Victory in the East*,s pp. 171-2.

92 Fulcher of Chartres, *Historia Hierosolymitana*, I.xii.3 (p. 198).

They were at the mercy of the Byzantines, however. Many did not like this and the Latin chroniclers struggled to obscure the essential nature of the Byzantine contribution. Nevertheless, that they ultimately succeeded in reaching Jerusalem can be attributed in part to the exceptional effectiveness with which the Byzantines worked under very difficult conditions to keep the large forces supplied so that they could continue their march.

Chapter 4

Ship types and fleet composition at Genoa and Venice in the early thirteenth century[1]

John E. Dotson

The composition of Western war fleets changed considerably over the half-century between the establishment of the Latin Empire in 1204 and the outbreak in Acre of the War of St Sabas in 1256. In the wars between Genoa and Pisa in the Tyrrhenian Sea in the first half of the thirteenth century, fleets were often composed of a wide mix of sailing ships both large and small, galleys and smaller oared vessels of similar design, such as *sagitae/saette*, and intermediate types such as *taride*. Large sailing ships, or *navi*, were often mentioned as undertaking operations alone or as the primary warships in a fleet. However, by mid-century, when the first major war between Genoa and Venice broke out, galleys had clearly become the mainstay of war fleets.

In retrospect, Lane's view of Venetian, and by implication also of other Mediterranean navies', use of "low and fast long ships which were rowed" and of "high, wide, sailing ships" in fleet tactics over five centuries culminating in the Battle of Lepanto in 1571 was rather static. In his analysis, the slow, clumsy, but very defensible sailing ships, heavily armed with missile weapons, were sent out ahead of the main strike force of oared vessels to draw enemies in and disrupt them with missile fire. Then, the fast oared vessels would deliver the decisive blow.[2] While this assessment may have been broadly true, and may have reflected a kind of tactical ideal, contemporary descriptions of naval fights in the early thirteenth century indicate a greater degree of complexity and development than the model implies. The long history of oared fighting ships in the Mediterranean has tended to obscure some of the subtleties of this development. Just as the medieval Italian city-state was not a direct descendant of the ancient *polis*, neither did the medieval galley derive directly from the ancient trireme. Reliance on galley fleets as the mainstay of naval warfare in the late Middle Ages was incremental and influenced by technical, tactical, and economic considerations.

For this investigation I have relied mostly on the Genoese civic chronicle. It is continuous and detailed and, while not without its biases, was written by men who

---

1 Consult Maps 2, 4, 5, 6, 7, 9.

2 F. C. Lane, *Venice: a maritime republic* (Baltimore, 1973), p. 28 ff.

had access to information, who were familiar with the sea, and who were largely of a mind-set that valued accuracy in numbers and concrete facts. While the point of view is distinctly Genoese, and often of a particular party in that faction-ridden city, the authors of Genoa's civic chronicle were generally reliable when it came to such things as ship types and the numbers of vessels in fleets, both their own and those of their opponents. Caro compared fleet numbers in various actions as reported by Genoese and Venetian chroniclers and although there were differences they were always close. Venetian chroniclers often varied as much among themselves as they did from the Genoese reports.[3] Thus, the Genoese chronicle can be considered to be reliable as a guide to developing an understanding of early thirteenth-century war fleets.

Most of the forces of the First Crusade marched overland but it quickly became evident that any long-term Western presence on the eastern shores of the Mediterranean Sea would require reliable maritime resupply and reinforcement that could be ensured only by an effective projection of naval power into the area. The Italian maritime cities of Venice, Genoa, and Pisa had a religious commitment to Crusading and the strategic position, the technology, and also the economic incentive to establish a Western naval presence in eastern Mediterranean waters. The Genoese were famously the first to commit naval forces and to reap the rewards, followed quickly by the Pisans and the Venetians.[4] As a dividend for their participation, the Italian cities received quarters, rights, and trading privileges in the ports of Palestine and Syria. By the end of the twelfth century they had established their domination over the waters of the Mediterranean Sea and by that time they were one another's chief rivals and fiercest competitors; although, of course, there were other significant naval forces in the Byzantine Empire and Kingdom of Sicily. A kind of balance of power prevailed among them, punctuated by occasional violent encounters. In the waters of the Tyrrhenian Sea, Genoa and Pisa were at war for control of the great islands of Corsica and Sardinia as well for domination of Sicilian waters.[5]

After the conquest of Constantinople by the Fourth Crusade reestablished and strengthened Venice's dominant position in *Romania*, a new dynamic was created. Pisa, by then the smallest of the three maritime cities, also held the most precarious geopolitical strategic position. Locked in a struggle with Genoa in western waters,

---

3 G. Caro, *Genua und die Mächte am Mittelmeer, 1257-1311*, 2 vols (Halle, 1895, 1899); trans. O. Soardi, *Genova e la supremazia sul Mediterraneo (1257-1311)*, 2 vols [ASLSP, n.s., 14-15] (Genoa, 1974-5), vol. 1, p. 49, n. 71; p. 73, n. 9; p. 162, n. 19; etc.

4 See J. H. Pryor, " 'Water, water everywhere, Nor any drop to drink.': Water supplies for the fleets of the First Crusade", in M. Balard, B. Z. Kedar, and J. Riley-Smith, eds, *Dei gesta per Francos: études sur les croisades dédiées à Jean Richard / Crusade studies in honour of Jean Richard* (Aldershot, 2001), 21-8; J. France, "The First Crusade as a naval enterprise", *MM*, 83 (1997), 389-97.

5 See J. H. Pryor, "The maritime republics", in NCMH, vol. 5, 419-46.

she also faced powerful opposition in Tuscany from Florence and Lucca. Only an alliance with the Hohenstaufen emperors enabled her to maintain her position as a major power. Genoa, at war with Pisa in the west, increasingly came to view Venice as the enemy in the lucrative ports of the Eastern Mediterranean. In the twelfth century the Genoese had tried, with only limited success, to establish a commercial presence at Constantinople.[6] The Fourth Crusade conquest of Constantinople in 1204 made Genoese penetration of the markets of *Romania* only more difficult. The port of Acre in the Kingdom of Jerusalem became a flashpoint for rivalries between the three cities.

To consider properly the changes that took place in Italian fleets during the first half of the thirteenth century, it is necessary to form some idea of how contemporaries judged the fighting power of various types of ships. Those mentioned most commonly were *navi*, *galeae*, *sagitae/saette*, *buci*, *taride*, and *barche*. Another commonly encountered term, *legno*, literally meaning "plank", was analogous to the English word "vessel" in referring to water craft in a general sense. *Nave* corresponded to the English "ship" but was invariably used in documents of the late Middle Ages for capacious, broad-beamed cargo vessels carrying triangular lateen sails on two or three masts. The word has been adopted in its Latin (*navis*), Italian (*nave*), and French (*nef*) forms by modern scholars as a kind of technical term referring specifically to the lateen-rigged medieval ship. *Galeae* are understood by everyone to be long, narrow-beamed vessels capable of being sailed but designed for propulsion by oars. They are considered to have been the primary warships of the medieval and early modern Mediterranean. For most of the thirteenth century galleys had 25-28 benches per side with two oarsmen on each bench giving a range of 100-112 oarsmen. At the end of the thirteenth century, a third oarsman was added to each bench requiring the hull to be made slightly beamier but greatly increasing the available rowing and fighting power with 150 oarsmen on a 25-bench galley. The *sagitta/saetta* was much like a smaller galley but even narrower in proportion to its length and, usually, with one oarsman to a bench. The *bucio* and *tarida* were intermediate types that appear originally to have been narrow, oared vessels that evolved into purely sailing ships.[7]

As one would expect, any vessel was expected to match another of the same type and the larger of the two, or at least the one carrying more fighting men, was expected to prevail. In the instance of the fight between the *navi Carrocia* and *Leopardo* in 1204 described below, the *Carrocia* was sent in search of the *Leopardo* with a force superiority of approximately two to one. A larger, or more

---

6 S. A. Epstein, *Genoa and the Genoese, 958-1528* (Chapel Hill, 1996), p. 84 ff.

7 J. E. Dotson, "Merchant and naval influences on galley design at Venice and Genoa in the fourteenth century", in C. L. Symonds, ed., *New aspects of naval history: selected papers presented at the Fourth naval history Symposium, United States Naval Academy* (Annapolis, 1981), 20-32, p. 21.

heavily armed, *nave* was sent to deal with a weaker one. A little earlier in that same year a Pisan *saetta* of 100 oars,[8] accompanied by a *bucio* of 80 oars, cruised along the *Riviera di Levante* attacking and capturing local merchant shipping. The inhabitants of Portovenere equipped a single galley which they sent to wait near the mouth of the Arno for the Pisan raiders to return home. The *saetta* was captured and brought back to Genoa. Obviously, the Portoveneresi were confident that their galley would be a match for the Pisan *saetta*. On the surface of things, one would think that they were being foolhardy. Pryor has determined that war galleys of the Kingdom of Sicily under Charles I of Anjou between 1269 and 1284 had 108 oarsmen.[9] There is no reason to think that galleys were larger at the beginning of the century. To send such a vessel against two foes, one almost equal and the other at least three-quarters of the force of the Portovenerese galley, does not seem prudent. Of course, there is much that we do not know. Were the Pisan vessels encumbered with prizes and how badly had they been cut up in their raid? There is no mention of the *bucio* in the account of the capture of the *saetta* and perhaps it had been lost in an earlier fight. The distinguishing feature of a *saetta* was that it was even finer in its lines than a galley and was noted for speed, as its name suggests. Perhaps the slightly beamier galley would carry more marines or could provide a more stable fighting platform. At this distance in time it is impossible to grasp the very fine points of naval combat but we must assume that the Portoveneresi had reason to believe that their single galley disposed of sufficient force to overcome the Pisans, and events proved them right.

An estimate of the fighting power of large *navi* as well as of contemporary views of their proper use can be gained from the *Carrocia–Leopardo* fight alluded to above. The ambitious Genoese free-lance Alamanno da Costa set out in his *nave*, the *Carrocia*, with 500 fighting men aboard, in search of a Pisan corsair *nave*, the *Leopardo*. In the ensuing battle the *Leopardo* was captured and became an important part of Alamanno's force. The comparative size of the *Leopardo* can be estimated from the chronicler Ogerio Pane's report that 280 suits of armour were among the arms captured with the ship.[10] If this roughly represented the number of marines on board, *Leopardo* seems to have been heavily overmatched

---

8 This was a very large *saetta*. Jal quotes a number of references to *saette* of 30, 48, 58, 64, and 80 oars, most dating from the third quarter of the thirteenth century. See A. Jal, *Glossaire nautique: répertoire polyglotte de termes de marine anciens et modernes*, 1 vol. in 2 tomes (Paris, 1848), p. 1306.

9 J. H. Pryor, "The galleys of Charles I of Anjou, King of Sicily: ca. 1269-84", *Studies in medieval and Renaissance history*, n.s., 14 (1993), 33-103, pp. 81-4; idem, *Geography, technology, and war: studies in the maritime history of the Mediterranean, 649-1571* (Cambridge, 1988), p. 66.

10 *Annales Ianuenses*, eds L. T. Belgrano and C. Imperiale, *Annali genovesi di Caffaro e de' suoi continuatori dal MXCIX al MCCXCIII*, 5 vols [FStI] (Rome and Genoa, 1890-1929), vol. 2, p. 90; *Annales Ianuenses*, ed. G. H. Pertz, in MGH *SS*, vol. 18, 1-356, p. 121.

by *Carrocia*'s 500. Clearly, both these ships represented formidable fighting units. The further adventures of the *Leopardo* illustrate the relative merits of *navi* and galleys in naval combat. In 1205 Count Henry Pescatore of Malta equipped and manned the *Leopardo* and two galleys for a corsair expedition into the Aegean.[11] Shortly after entering the waters of *Romania* he encountered two Venetian *navi* returning from Constantinople: the *Falcone* and the *Rosa*. A classic sea chase worthy of C. S. Forester or Patrick O'Brien then ensued. The Venetian *navi* fled to the east pursued "day and night" by Count Enrico, who could not come up with them because of light winds. This would seem to have been ideal conditions for the galleys but apparently two galleys were an insufficient force to engage two *navi*. The Venetians abandoned the smaller *Rosa* and scuttled it, taking their cash and smaller goods with them onto the *Falcone*. As the *Rosa* was sinking, the galleys closed with it and offloaded two hundred bales of cloth onto the *Leopardo*. As the pursuit continued, the wind freshened and the *Falcone* was finally overhauled near Tyre. Ogerio Pane reported that there were 900 men on board with arms, armour, and a large amount of cash. This would have been the combined number of passengers and crew of the two *navi*.[12] The outcome of this encounter depended upon the offensive fighting power of the *Leopardo*. There was apparently something about this *nave* that suited it to the *guerre de course* since the Pisans had used it for that purpose and Alamanno da Costa had kept it and had lent it to Henry Pescatore for another corsair voyage. Perhaps it had outstanding sailing qualities. Certainly, once the wind had risen it overtook the *Falcone*.

The defensive power of a large *nave* has long been recognized by modern historians. The best-known example of this is probably found in the encounter between a Venetian merchant fleet and a Genoese galley fleet under Simone Grillo in 1264 in which the Genoese were able to capture all the Venetian ships except for the large *nave*, the *Roccafortis*, against whose bulk and towering sides the low-lying galleys were ineffective.[13] As late as the end of the fourteenth century the Venetian commander Carlo Zeno with more than a dozen galleys overhauled a large Genoese *cog*, the *Richigona* but was unable to take it until he had obtained the use of another *cog* to match its height.[14] The fight between the *Carrocia* and

---

[11] For Henry Pescatore see D. Abulafia, "Henry Count of Malta and his Mediterranean activities, 1203-1230", in A. T. Luttrell, ed., *Medieval Malta: studies on Malta before the Knights* (London, 1975), 104-25 [rpt in his *Italy, Sicily and the Mediterranean 1100-1400* (London, 1987), No. III].

[12] *Annales Ianuenses* (FStI), vol. 2, pp. 98-9; *Annales Ianuenses* (MGH *SS*), pp. 124-5.

[13] *Annales Ianuenses* (FStI), vol. 4, pp. 54-6; *Annales Ianuenses* (MGH *SS*), pp. 246-7. For a detailed account of this operation see J. E. Dotson, "Fleet operations in the first Genoese-Venetian War, 1264-1266", *Viator*, 30 (1999), 165-80.

[14] Daniele di Chinazzo, *Cronica de la guerra da veneciani a zenovesi*, ed. V. Lazzarini, in *Monumenti storici pubblicati dalla Deputazione di Storia Patria per le Venezie*, n.s., XI (Padua, 1958), pp. 218-20. See also Lane, *Venice*, p. 194.

the *Leopardo* and the later activities of the *Leopardo* in the East indicate clearly that 60 years before the *Roccafortis* battle, large *navi* were clearly thought capable of independent offensive action in *guerre de course*, the kind of warfare that is usually thought to place a premium on swiftness and manoeuvrability.

Even though a *nave* or a *nave* and a galley or two could operate as corsairs, wars were fought by major fleets, which in the early thirteenth century were still made up of a variety of ship types. The struggle that developed later in 1204 and in the following year between Pisa and Genoa for control of Syracuse demonstrated the extensive use of mixed fleets of various ship types. After the capture of the *Leopardo*, Alamanno da Costa sailed for Syracuse to seize that city from the "Pisan pirates" who, according to Ogerio Pane, lived there and preyed on the Genoese and everyone else.[15] He proceeded first to Malta where Henry Pescatore joined forces with him, bringing an unspecified number of galleys and men. They captured the city late in August 1204 and Alamanno was made Count of Syracuse by the commune of Genoa.[16] This blurring of the line between state and private activity was a common feature of Genoese public life and conduct of war. In 1205 hostilities between Pisa and Genoa continued and three Pisan *navi* captured in Provençal waters the Genoese *nave*, *Viola*, on its way to Al-Bijāya. They took their prize by way of Cagliari to Messina where they joined either four or ten more *navi* and 12 galleys and landed and attacked a small Genoese force there.[17] In this campaign galleys acted in the role that one associates with fast cruisers or frigates in later ages. While the main fleet remained in Messina harbour, the Pisan commander took two galleys west to Palermo. He was pursued and captured in a sharp action by two of Alamanno's galleys, one of them commanded by his son. Following the battle at Messina, a reinforced Pisan fleet of ten *navi*, 12 galleys, and "many other vessels" arrived at Syracuse with a substantial military force. For three-and-a-half months they laid siege to the city. All the while, Henry Pescatore gathered forces for a counterattack while the new Count of Syracuse saw to the defence of the city. Enrico, with four galleys and a number of *taride*, was joined at Messina by two Genoese *navi* returning from *Outremer*. At the urging of the Genoese, Pescatore quickly armed a further 16 *navi*, some galleys, and lesser vessels and sailed to Syracuse. There they destroyed a Pisan fleet whose numbers are reported as nine *navi*, 12 galleys, and 14 "*buciisque et barchis*".[18] Throughout the account of this campaign, *navi* figured prominently as fighting ships. They

15 *Annales Ianuenses* (FStI), vol. 2, p. 91; *Annales Ianuenses* (MGH *SS*), p. 121. The so-called *pirati* were in fact led by the Pisan count Ranieri Manente di Segalari, who was no more a *piratus* than Da Costa or Henry Pescatore. The word was a pejorative used commonly for enemies.

16 *Annales Ianuenses* (FStI), vol. 2, pp. 92; *Annales Ianuenses* (MGH *SS*), pp. 121-2.

17 Ibid. (FStI), vol. 2, p. 96; ibid. (MGH *SS*), p. 123.

18 Ibid. (FStI), vol. 2, pp. 97-8; ibid. (MGH *SS*), pp. 123-4.

were invariably listed before galleys in lists which always give the numbers of vessels in descending order of size and, presumably, of perceived importance.

In 1234 the Genoese launched their largest naval effort up until that time, and probably their largest until the great fleet of 1295. As early as 1231 a conflict had began between the *amīr* of Morocco and the Almohad Caliph that threatened the interests of Genoese resident merchants in Ceuta. Diplomacy by the Genoese and the *amīr* of Seville averted war at that time.[19] But on 24 August 1234, the Genoese residents of Ceuta learned that the Caliph Abū Muḥammad ‘Abd al-Wāḥid II al-Rashīd was about to attack Ceuta with a large force of Christian mercenaries. The anonymous Genoese chronicler called them *Calcurini* and described them as “Crusaders” when he wrote that “the Genoese who were in those parts with many *navi* and a very great quantity of bezants, merchandise, and goods were struck with fear. They feared the loss of their persons and property if that land were taken by the *Calcurini*. Equally, they feared to fight against Christians signed with the cross”.[20] In the end, they armed ten of their “largest and best” *navi* to try to prevent their opponents from crossing the Straits of Gibraltar. However, their fleet was scattered by a fireship and, abandoning their goods in Ceuta, they withdrew with the ten *navi* to Malaga. There they regrouped, and assembled 600 well-equipped men whom they sent in two ships to assist in the defence of Ceuta. Another four ships were sent to Genoa to seek help. The remainder went to Tunis where they were joined by the first two after they had landed the defence force. When the request for help, including a letter from the *amīr* of Ceuta, reached Genoa, the *podestà* and the Council authorized a fleet of 18 galleys to join with the four ships from Ceuta to sail to its relief. This appears to have required a great financial effort on the part of the commune since it sold the revenue from the salt tax for the next ten years to raise 18,000 *libbre* to pay for outfitting the fleet.[21] In the ensuing battle much of the city, including the Genoese *fondaco,* was burned. Except for the four *navi* that had come from Ceuta the entire fleet was composed of galleys.

The following year four galleys were sent to rescue those Genoese who remained in the city. In a strange passage the anonymous Genoese chronicler wrote that no *milites* were sent but, if they had been, the Genoese would have taken the city with an attack from the land. Since there were no cavalry, the assault was mounted from the sea. Two trebuchets were built on two *navi* and the fleet of 70 large *navi*, 30 smaller *navi*, 20 galleys and “many other small ships” kept up a continual bombardment of the city with the trebuchets and other machines.[22] This huge fleet seems not to have been sent all in a body from Genoa, but was made up

---

19 Ibid. (FStI), vol. 3, pp. 56-7; ibid. (MGH *SS*), p. 177.

20 Ibid. (FStI), vol. 3, p. 72; ibid. (MGH *SS*), p. 183.

21 Ibid. (FStI), vol. 3, p. 73; ibid. (MGH *SS*), p. 183. See also Epstein, *Genoa and the Genoese*, p. 122.

22 *Annales Ianuenses* (FStI), vol. 3, p. 75; *Annales Ianuenses* (MGH *SS*), p. 184.

of ships that had been sent earlier, some that had been in other ports in Spain and the Maghrib, as well as at least the four additional galleys mentioned above. Nonetheless, it represented a very large effort.[23] It succeeded in persuading Abū Muḥammad to come to terms. Again, this was a mixed fleet, with many more *navi* than galleys. In this case, it is easy to see that *navi* were probably better suited to mounting siege artillery for shore bombardment than galleys would have been. Nonetheless, the situation was very similar to the one in which the Genoese found themselves a little more than twenty years later at the other end of the Mediterranean when their compatriots came under attack by superior forces in Acre. In that situation, however, large *navi* were not a part of their force.

At Acre a complicated web of grievances developed, including a dispute over the monastery of St Sabas which would give the ensuing war its name. These quarrels finally led to open war in 1257 as the governments and people of the metropolises were dragged into the dispute. In that year, the Venetians sent a *nave* and 13 galleys to escort their merchant fleet to Acre.[24] When they arrived, fighting began in the city as, according to an anonymous Genoese chronicler, the Genoese in Acre took up arms and began a war with the Pisans and the Venetians.[25] From the Venetian and Pisan point of view this must have seemed a very good time to attack the Genoese in the East since they were distracted in the western Mediterranean and off balance at home. Pisa certainly began to find it advantageous to ally with Venice since she was heavily involved in a renewal of her war with Genoa for domination of Sardinia. Earlier in that year there had been a revolution in Genoa that had brought Guglielmo Boccanegra to power as *capitano del popolo* and had created a Council of Elders as a key element in the new government.[26] This new regime responded quickly to the developing situation. Shortly after taking power, the new Captain and Council equipped a fleet of 16 galleys to counter Pisan moves in Sardinia.[27] Equipping this fleet did not occasion the kind of extraordinary effort to finance it as did the 14 galleys some 23 years earlier, but it must have been a substantial expense for an effort which ended in failure when they were unable to break through the Pisan siege to resupply their forces.

---

23 R. S. Lopez, *Storia delle colonie genovesi nel Mediterraneo* (Bologna, 1938), pp. 173-6 estimated that this was probably the largest Genoese fleet that had ever been gathered up to that time.

24 Martino da Canale, *Les estoires de Venise: cronaca veneziana in lingua francese dalle origini al 1275*, ed. A. Limentani [Civiltà veneziana, fonti e testi, XII, serie terza, 3] (Florence, 1972), p. 159.

25 *Annales Ianuenses* (FStI), vol. 4, p. 32; *Annales Ianuenses* (MGH *SS*), p. 238.

26 Ibid. (FStI), vol. 4, pp. 25-6; ibid. (MGH *SS*), p. 236. See Epstein, *Genoa and the Genoese*, p. 137 ff.

27 Annales Ianuenses (FStI), vol. 4, p. 28; *Annales Ianuenses* (MGH *SS*), p. 237.

Learning of the galleys that the Venetians had sent to *Outremer* with their merchant caravan, the Genoese dispatched ten galleys to accompany their own merchant fleet to the East. *En route* they were scattered by a storm. Four galleys returned to Genoa and the remaining six limped on to Tyre with broken oars and other damage where they found "many galleys and *saettie* of the Genoese". From the context most of these must have been privately owned and operated. Nineteen galleys were quickly equipped for war. When a Venetian fleet of 23 galleys was sighted off Tyre, this scratch fleet tumbled out of the harbour in a completely undisciplined fashion and was easily defeated.[28] Late in June of 1258, the Genoese assembled 48 galleys and four *navi* under the command of the aged hero Rosso della Turca for the relief of their compatriots in Acre. The Venetian fleet of some 40 galleys as well as *taride* was disrupted by a contrary wind as it came out to engage the Genoese but Rosso failed to press the advantage that this gave him. In the end he was defeated with a loss of more than 20 galleys and 1,700 men.[29] It is plain from the accounts of the preparation of the fleets and in the accounts of the battles that by this time the galley had become the principal warship. Fleets were estimated in terms of the numbers of galleys that constituted them. *Navi*, *saettie*, and *taride* were mentioned but only in passing and, except for the four Genoese *navi*, they were not numbered. There is no indication that the *navi* played any significant role in the battles.

In other major battles of the first Genoese-Venetian War the galley emerged even more clearly as the primary, if not the only, ship of war. In 1263 Genoa sent 25 galleys, a *saetta*, and five *barche*, equipped at a cost of 36,000 *libbre genovesi*, to *Romania* to prosecute the war with the Venetians.[30] No *navi* were included in the fleet and the smaller vessels were certainly included as auxiliaries. The fleet left Genoa in late May and travelled to Constantinople. Later in the year, now augmented to 38 galleys, it was transporting supplies to Monemvasia in the Peloponnesos for the Byzantine Emperor when it encountered a fleet of 32 Venetian galleys near the island of Spetsai at the mouth of the Gulf of Argolis. In the fight that followed, a claim of immunity by the Venetians on the grounds that they were Crusaders threw the Genoese into confusion and only a part of them pressed the attack. At least this is the explanation offered by the Genoese chronicler. They lost four galleys and the remainder sought refuge in Monemvasia.[31]

---

28 Ibid. (FStI), vol. 4, pp. 33-4; ibid. (MGH *SS*), p. 239.

29 Ibid. (FStI), vol. 4, pp. 34-5; ibid. (MGH *SS*), pp. 239-40. See Caro, *Genova*, vol. 1, p. 74 ff.

30 This was exactly twice the cost of equipping the fleet of 1234 and an indication of rising costs in the intervening period. This time, the commune raised the money by a forced loan instead of selling future tax income. *Annales Ianuenses* (FStI), vol. 4, pp. 49-50; *Annales Ianuenses* (MGH *SS*), p. 245.

31 Ibid. (FStI), vol. 4, p. 51; ibid. (MGH *SS*), p. 245.

In 1264 Genoa equipped two large *navi*, described as being of "exceptional size", and 20 galleys with 3,300 men aboard, or an average of 175 men per galley.[32] This fleet would embark upon one of the most successful and well-executed operations carried out by any medieval naval commander. The admiral in command, Simone Grillo, convinced the Venetians that he was headed for *Romania* by an excellent use of misdirection and intelligence gathering but in fact hid his fleet at Malta. When the Venetian war fleet headed east in what they believed was pursuit of him, he detached the two *navi* and three galleys to the Maghrib to trade. Of the remaining galleys, he sent one to shadow the Venetian fleet and took the remaining 16 to the Adriatic to intercept a Venetian trading caravan. Near Dyrrachion, he intercepted the Venetian fleet of some 20 vessels. One of them was a huge *nave* known as the *Roccafortis*, one of the largest vessels afloat at the time.[33] Additionally, there were 18 or 19 smaller vessels. An anonymous Genoese chronicler identified two other *navi*, 13 *taride*, a *panzone*, two galleys, and a *saetta*. In a savage battle that lasted until midnight, one of the Venetian *navi* was sunk, a *tarida* was burned, and the rest were captured except for the *Roccafortis*. The large *nave* escaped after most of the men from the other vessels had managed to withdraw to the safety of its high sides.[34] One must wonder whether Grillo might have captured even the *Roccafortis* if he had had the two large *navi* with which he had begun his voyage.

In May of the following year a single galley departed from Genoa as a commerce raider. Cruising around the Peloponnesos, it captured, in succession, a galley, a large *tarida*, and a large *panzone*. In September it returned to Genoa with 82 Venetian prisoners and the loot from the prizes. None of these were particularly large vessels and the *panzone* with 45 men aboard put up a considerable fight before being taken.[35] The chroniclers Lanfranco Pignolo and his associates noted that the commune's share of the loot from the Venetian ships amounted to 2,000 *libbre genovese*. This was the kind of corsair raid that had been entrusted to *navi* or *navi* in conjunction with galleys early in the century.

In other fleets galleys also constituted the major fighting elements after the middle of the century. In the same year 1265, Charles of Anjou proceeded toward Rome with a fleet of 25 galleys and 13 small vessels transporting 500 cavalry and 1,000 crossbowmen. He was confined in the mouth of the Tiber by 60 galleys belonging to Manfred of Hohenstaufen. Forty of these were told off and raided

---

32 Ibid. (FStI), vol. 4, p. 54; ibid. (MGH *SS*), p. 246.

33 Lane, *Venice*, p. 77.

34 Martino da Canale, *Les estoires de Venise*, pp. 202-5; Annales *Ianuenses* (FStI), vol. 4, p. 55; *Annales Ianuenses* (MGH *SS*), p. 247. See Dotson, "Fleet operations", pp. 174-5; F. C. Lane, *Venetian ships and shipbuilders of the Renaissance* (Baltimore, 1934), pp. 4-6.

35 *Annales Ianuenses* (FStI), vol. 4, p. 70; *Annales Ianuenses* (MGH *SS*), p. 251.

toward Provence capturing a merchant *nave* and then scattering 12 Angevin galleys near San Remo.[36]

In 1266 Genoa equipped 18 galleys and a large *nave* belonging to the commune to sail against the Venetians under the command of Lanfranco Borborino. The fleet left Genoa in April, at which time the commune armed a further nine galleys. Borborino waited in Bonifacio in Corsica until the additional galleys arrived. He then abandoned the *nave* and distributed picked men from its crew among the galleys. In the meantime the Venetian Jacopo Dandolo had been harassing Genoese and other Ligurian shipping in the Sicilian Channel between Sicily and Tunis. Borborino, hearing that the Venetians were operating near Sicily with 30 galleys, sailed with his 27 reinforced galleys to seek them out. Dandolo had, in fact, been joined by a further 14 galleys and two *saette*. The two fleets groped for one another and met near Trapani on June 24. The Genoese fleet was handled poorly and its commanders were later convicted of incompetence by a board of inquiry. All of their galleys were captured by the Venetians, who burned three of them and towed away the remaining 24, each towed by one of the 24 Venetian galleys.[37]

A number of factors may account for the decline of the presence of *navi* in fleets from the mid-thirteenth century. Early in the century, especially in the navies of the western Mediterranean, fleets were often assembled from whatever vessels were at hand. Even the great Genoese fleet at Ceuta seems to have been of this *ad hoc* nature. Since *navi*, and smaller merchant vessels such as *bucii*, *taride*, and, as the sources often have it other, lesser *legni*, were the most common merchant vessels of the age, they naturally figured prominently in these scratch fleets for the simple reason that there were generally more of them available. In the *guerre de course* the *nave* would often be the vessel of choice because only a *nave* could attack another *nave* with a reasonable assurance of success. Secondly, the carrying capacity of a large *nave* could transport large amounts of loot. The Pisan *saetta* and *bucio-galera* that raided the *Riviera di Levante* are reported to have taken many *buci*, vessels that were smaller and slower and that in their merchant configuration carried smaller crews than the Pisan raiders. For longer range operations, a *nave* would be of great advantage. Alamanno da Costa's combination of the *Leopardo* and two galleys seems almost ideal for the kind of corsair expedition that he undertook. Yet, when communes outfitted fleets, they tended to rely heavily, if not exclusively, on galleys. When Genoa equipped a fleet in response to the situation in Ceuta in 1234, she dispatched 18 galleys. The four *navi* that accompanied them were those that had come to seek assistance from the home city. In the battles around Acre and Tyre in the early stages of the War of St Sabas there was

36 Ibid. (FStI), vol. 4, p. 74; ibid. (MGH *SS*), p. 252.

37 Ibid. (FStI), vol. 4, pp. 89-91; ibid. (MGH *SS*), pp. 256-7. See Dotson, "Fleet Operations", pp. 178-9.

occasional mention of *navi* and some other types, but clearly galleys were the mainstay of the fleets. In subsequent actions, galleys were even more prominent. In 1264, when Simone Grillo launched his brilliant raid on the Venetian caravan that was organized around the *Roccafortis*, he dismissed the two large *navi* that had originally accompanied his fleet and proceeded with galleys alone. In the event, his galleys were able to take the Venetian *taride* and other small vessels, but the *Roccafortis* was too powerful for them.

This analysis of some representative naval actions of the first half of the thirteenth century reveals several points of interest regarding the uses of various ship types. It is clear that, barring unusual circumstances, only a *nave* could fight another *nave* successfully. Smaller types could be useful as privateering or scouting vessels or in other supplementary roles. Galleys are often described in modern works as fast or quick and by other adjectives that imply speed but, while it is true that galleys under oars could probably reach relatively fast sprint speeds, they could not maintain those speeds for very long. Twenty minutes at maximum effort would leave a crew exhausted and one should remember that after making contact, galley crews were expected to take part in combat. Galleys could also reach relatively fast speeds under sail, but again only under very specific conditions. With a good breeze dead astern they could sail as quickly as they could be rowed and for a much longer time if the wind held. However, a quartering or beam wind could actually be dangerous because of their narrow beam, low freeboard, and protruding oars. Like other pre-modern sailing ships, they were unable to make much progress against contrary winds. For a brief time, or against a light breeze, they could row but aerodynamic and hydrodynamic resistance made that difficult work. Most indications are that over long distances they averaged about the same cruising speeds as large *navi*.[38] More importantly, oars gave galleys the ability to manoeuvre more precisely in formation with other galleys than *navi* could hope to do. By the seventeenth century, fully developed sailing ships with many sails, some of them devoted to the purpose of helping in the steering of the ship, could closely coordinate their movements in line of battle; however, this was beyond the capability of medieval lateen-rigged *navi* which carried only one sail on each mast. The value of manoeuvrability in battle is not inconsequential. When it was necessary to coordinate the movement of more than a handful of vessels in a fleet action, only those powered by oars could be depended upon. Among these, the galley was pre-eminent.

The factors that made the galley the warship of choice in the Mediterranean environment were multiple. In Mediterranean conditions they were relatively seaworthy and in bad weather they could find refuge in small ports or could be

---

38 J. E. Dotson, "Economics and logistics of galley warfare", in J. Morisson, ed., *The age of the galley: Mediterranean oared vessels since pre-classical times* (London, 1995), 217-23, p. 220; Pryor, *Geography, technology and war*, p. 86.

beached if necessary. Galleys carried large crews and therefore had considerable fighting force. The 175 men in each of Simone Grillo's galleys were considerably fewer than, for example, the 500 men in Alamanno da Costa's *Carrocia*, but Grillo had many galleys. That may be a crucial point in the discussion of the developments in fleet composition. Galleys could be expected to engage and defeat any types of ship except for very large *navi* and those were, in fact, few in number. Galley hulls were much less complex and expensive than a *nave*. Evidence for the cost of hulls is rare, but it appears that the average cost of a galley might be only about one-tenth that of a really large *nave*.[39] When the Genoese and Pisans were assembling fleets from largely private vessels it is not surprising that many types were used. The *nave* was a versatile ship, more useful to private individuals who would own shares in vessels intended primarily for mercantile activities. The balance of benefits was different for the communes as they increasingly became involved in long-term warfare. The economics of the situation is clearly evident in Simone Grillo's operation. He abandoned his two *navi*, described as very large, that might have given him a chance to capture the *Roccafortis* as well as the smaller Venetian vessels. But, the cost/benefit calculus of medieval naval warfare supported him in this decision. He was not able to capture the *Roccafortis* but he did seize or destroy all the rest of the Venetian caravan and he forced the *Roccafortis* to turn back to Venice. His two *navi* were able to carry out a profitable trading voyage while he denied the same opportunity to the Venetians. While large *navi* were effective fighting ships when they were manned and equipped for war, they were just too expensive and too valuable in other roles to use extensively in that capacity. As naval war was increasingly transformed into an activity under state direction, as more and more often fleet met fleet in a struggle for maritime domination, the galley became the preferred weapon system.

---

[39] H. C. Krueger, *Navi e proprietà navale a Genova, seconda metà del sec. XII* (Genoa, 1985), pp. 160-61, 164-5; E. H. Byrne, *Genoese shipping in the twelfth and thirteenth centuries* (Cambridge Mass., 1930), pp. 22-4.

Chapter 5

Logistics and the Second Crusade[1]

John France

For some the notion of a Workshop on Crusader logistics has seemed a narrow subject for which it is assumed that sources are lacking. The general assumption underlying such views is that it is possible to make something of logistics only occasionally, particularly for the later Crusades but not for the earlier ones. These are not unreasonable attitudes in view of the limited literature currently available on the subject and, indeed, on medieval military logistics in general. However, this is far from to justify such comments. Military logistics is a vital subject which was drawn firmly to my attention when I was writing about the First Crusade, whose journey was powerfully influenced by needs of supply.[2] On this occasion I have turned to the Second Crusade, which is of particular interest for several reasons.

First, the sources provide a substantial amount of information about logistics and it is worthwhile considering why this is so. Secondly, that information is consonant with what we already know about the nature of armies in general and Crusading armies in particular. Thirdly, consideration of the logistics of this event raises very important issues about the nature of the Second Crusade and Crusading in general and this has led me to examine critically the myth of the Second Crusade, what might be called the "Constable thesis". Almost everything that has been written about the Second Crusade in the last half-century has been dominated by ideas expressed by Giles Constable in an article written in 1953 which is distinguished but whose conclusions stand in need of revision.[3] Far from being a narrow and dead-end subject, consideration of the logistics of the Second Crusade has opened up the subject in the widest possible sense.

It may seem odd that some of the sources for the Second Crusade should be so lavish in their references to logistics. This is in part because some letters of the

---

1 Consult Maps 3, 8, 9, 10, 11.

2 J. France, "La stratégie arménienne de la première croisade", in C. Mutafian, ed., *Les Lusignans et l'Outremer: actes du colloque Poitiers-Lusignan, octobre 1993* (Poitiers, 1995), 141-9; idem, *Victory in the East: a military history of the First Crusade* (Cambridge, 1994).

3 G. Constable, "The Second Crusade as seen by Contemporaries", *Traditio*, 9 (1953), 213-79. This has somewhat overshadowed the account by V. G. Berry, "The Second Crusade", in Setton, *HC*, vol. 1, 463-512. A very significant dissenting voice is that of A. J. Forey, "The Second Crusade: scope and objectives", *Durham University journal*, 86 (1994), 165-75, which deserves to be better known.

major leaders, Conrad III and Louis VII, have come down to us, although these letters tend to mention logistics only briefly, as in Louis VII's letter to Suger of St Denis from the Hungarian frontier noting that the army was well-fed.[4] A letter of St Bernard seeking to recruit merchants may perhaps explain the presence of such people, who are particularly noted in the French army at Worms, but it is clear that others referred to at this time were simply local businessmen seeking to profit from the need of the army for supplies.[5] Pope Eugenius III tried to expedite the movement of Crusaders to *Outremer* by writing to Manuel I Komnēnos and a considerable diplomatic effort was made to persuade the king of Hungary to allow passage to the Crusaders.[6] It ought not to be a surprise that leaders were concerned with such matters because the movement of any major army raised the gravest logistical problems and there was every reason why the two leaders should make reference to them. Commonly, of course, the source material fails us and until recently historians therefore ignored such matters. However, only about a quarter of a century after the Second Crusade the English royal records provide an insight into the preparations needed to launch and sustain a military expedition, perhaps some 10,000 strong, to Ireland. Supplies were sent from all over England, including 60,000 nails and 1,000 shovels from the ironworks of the Forest of Dean.[7] The armies of the Second Crusade were certainly no smaller than this, so what we see in the royal letters can be no more than the tip of the iceberg.

Much more substantial information about problems of supply is provided by Odo of Deuil, who was much concerned with such matters. This has been little remarked upon by modern historians, who have found so much else of interest in Odo's account that they have rather ignored this aspect. As the chaplain of Louis VII, Odo was not just an eyewitness. He attended meetings of the French leadership, whose deliberations he reported, was very well informed, and probably wrote the king's own letters. It is not too much to say that he was a player rather than merely an observer and this makes the extent of his comments on logistics

---

4 For the letters of Conrad III of Germany, see F. Hausmann, ed., *Die Urkunden Konrads III. und seines Sohnes Heinrich*, in MGH *DipReg*, vol. 9. For those of Louis VII, see Suger of St Denis, *Epistolae Sugerii abbatis S. Dionysii*, in RHGF, vol. 15, 483-532, nos 11, 12 (pp. 487).

5 B. S. James, trans., *Letters of Bernard of Clairvaux*, 2nd ed. (Stroud, 1998), no. 391 (p. 462).

6 Eugenius III, Pope, *Epistolae Eugenii III Papae*, in RHGF, vol. 15, 426-83, no. 25 (pp. 440-1); Odo of Deuil, *De profectione Ludovici VII in orientem*, ed. and trans. V. G. Berry (New York, 1948), bks I-II (pp. 11, n. 24, 22-3). On tensions between Germany and Hungary and their bearing on the Crusade, see J. Phillips, "Papacy, Empire and the Second Crusade", in J. Phillips and M. Hoch, eds, *The Second Crusade: scope and consequences* (Manchester, 2001), 15-31, p. 20.

7 R. Bartlett, "Provisioning war in the twelfth century", in A. McKay with D. Ditchburn, eds, *Atlas of Medieval Europe* (London, 1997), 125-7.

very interesting. His was not a dispassionate and detached account and he clearly had his own views on how the Crusade should have been conducted and was not afraid to express them in a sour and critical tone. He was also highly manipulative of the facts in some respects. But his account, unlike those of the First Crusade, was not dominated by the shining vision of ultimate success, so his focus was quite different. He sought constantly to explain disaster and problems of supplying the army formed a natural theme, especially as to some extent they could be blamed upon the hated Byzantines. It is a great pity that the literary sources for the journey of Conrad III are so limited but Odo has something to say about his logistical achievements and difficulties also.[8]

Louis VII's decision to undertake a major expedition to assist the European settlements in the eastern Mediterranean was the first major land expedition of its kind since the First Crusade and it was inevitable that all concerned should look to the earlier journey as a precedent. Indeed, in *Quantum praedecessores*, Eugenius III was avowedly following Urban II's pattern.[9] Louis, it is often said, was apparently much attached to this precedent, and is alleged to have refused to agree to the idea of crossing the Dardanelles at St George of Sestos opposite Abydos on the grounds that "he had never heard that Franks" had used this route.[10] This may have meant that the First Crusade had not used this route. However, we should perhaps consider this more critically. Louis may have had other reasons for declining to take the proposed route. He would have known that he was in a stronger position to reach agreement with Manuel Komnēnos over the nature of relations between them if his army was close to Constantinople. Moreover, he had been willing to consider taking the route via South Italy which his great-uncle, Hugh of Vermandois, had taken during the First Crusade since he wrote to Roger II of Sicily suggesting that the French would need the support of his kingdom, to which Roger responded with the offer of a fleet and full logistical support.[11] To what this offer amounted is not entirely clear since, if the idea was to take the whole French force all the way to *Outremer*, it would have been a remarkably generous offer which clearly would have simplified all of Louis's logistical problems. It may be that the offer was to take only Louis VII's own troops, who actually formed the core of the whole French force. It is even possible that all that was implied was free transport and supplies across the Adriatic if Louis decided to follow the route of the northern French on the First Crusade. Although Louis was not firmly wedded to the Danube route, the South Italian option, in whatever form

---

8 Otto of Freising, *Ottonis et Rahewini Gesta Friderici I. imperatoris*, 3rd ed., ed. G. Waitz, in MGH *ScriptRerGerm.Cont.* (Hanover, 1912), I.xliiii-xlvii (pp. 63-7) gives only the most brief account of Conrad's journey.

9 Eugenius III, *Epistolae*, RHGF, no. 8 (pp. 429-30) [ed. Migne, *Eugenii III Pontificis Romani epistolae et privilegia*, in PL, vol. 180, coll. 1013-1614, no. 48 (coll. 1064-6)].

10 Odo of Deuil, *De profectione Ludovici VII*, bk III (pp. 58-9).

11 Ibid., bk I (pp. 10-15).

it came, was rejected for reasons unknown. The need for co-operation with the Byzantines, who were deeply suspicious of Roger, may have persuaded Louis. This left only the route down the Danube valley, which the Germans had always been determined to take. The fact that this had been the route taken by Godfrey of Bouillon, first ruler of the Holy Sepulchre, may have made it highly acceptable. However, the choice clearly presented closely interconnected political and logistical challenges.

The army could not be permitted to forage while it crossed the friendly territories of the German Empire, the kingdom of Hungary, and the Byzantine Empire. This posed a very major problem and the example of the First Crusade makes clear how difficult it was. Although the first elements of the People's Crusades under Walter SansAvoir and Peter the Hermit seem to have been peaceful and well-supplied, other groups got out of hand badly while crossing Hungary, whose king dealt with them harshly. Shortly after, Godfrey had been obliged to give hostages for the conduct of his forces before he was allowed into the kingdom. Raymond of St Gilles had taken his army down the east coast of the Adriatic to reach Byzantine territory at Dyrrachion. By the time he arrived, his forces were short of food and fell rapidly to pillaging, with disastrous results when they were scattered by imperial troops.[12]

Moreover, Louis would have known that his army would be following the Germans down the Danube valley. As Odo wrote, on the one hand, this had the enormous advantage that the French could use bridges built to cross rivers by Conrad's army. On the other hand, however, it had the distinct disadvantage that the Germans would have the first opportunity to buy food, perhaps leaving the French short. Rapid progress across Central and Eastern Europe and adequate logistical support were made possible by careful diplomatic and political preparations. These were begun by Eugenius III, who wrote to the Emperor Manuel, while Louis approached both Geza II of Hungary and Roger of Sicily, asking them to grant free passage through their lands and to provide food supplies. Manuel immediately linked such agreement to conditions, but the other monarchs replied quickly and positively.[13] The Byzantine attitude led to a complex series of negotiations when Manuel demanded oaths of homage from the western kings and other leaders on the model of the First Crusade, negotiations which were not resolved until after the arrival of the German and French armies at Constantinople. The connection between successful diplomacy and good supply was graphically illustrated by their troubled course. A French group which had travelled ahead of

---

[12] Raymond of Aguilers, *Liber Raimundi de Aguilers*, ed. J. H. and L. L. Hill (Paris, 1969), §§1-2 (pp. 37-41) [Raymond of Aguilers, *Historia Francorum qui ceperunt Iherusalem*, in RHC *HOcc*, vol. 3, 231-309, §§1-2 (pp. 235-8)]. See France, *Victory in the East*, pp. 88-93, 104-6.

[13] Eugenius III, *Epistolae*, RHGF, no. 25 (pp. 440-1); Berry, "Second Crusade", pp. 469-70.

Louis VII was forced to cross the Bosporos with the Germans when the Byzantines threatened to withdraw their market privileges. Subsequently the course of negotiations between Manuel and Louis was punctuated by the withdrawal and then restoration of market privileges. Logistical support was clearly a key weapon in the Byzantine armoury.[14]

The other major power along the route was Hungary. Both Conrad and Louis needed Geza's consent to cross and a guarantee of supplies, and Conrad's need raised especially delicate issues since a year earlier he had unsuccessfully supported German intervention in a succession struggle in Hungary. Geza was fearful that the large German force might take its revenge but, after a delay, agreement was reached. Louis VII had at a very early date written to Hungary and received consent. However, a difficulty did arise. Geza's rival Boris wanted to reach Byzantine territory by hiding among the German army, but Conrad, in his anxiety to keep the Hungarians happy, refused to countenance this. Louis VII discovered Boris among the ranks of the French but refused to surrender him to Geza, who was deeply offended but lacked the military power to enforce his will. It seems likely that this story became known only shortly before the army reached imperial territory, so the "supply card" was of very little value to Geza.[15]

In fact both the Germans and French managed their journeys to Constantinople very well indeed. Odo of Deuil commented on the good markets provided by the Hungarians and the ease of the passage, which seems to have been quite rapid. There are always difficulties in working out how fast armies travelled because precise dates of departure and arrival are not always known. However, the indications are that the two monarchs made much faster time than their First-Crusade predecessors. Godfrey of Bouillon, who left in mid-August 1096, arrived at Constantinople only on 23 December, a march of approximately 130 days. Conrad gathered his army at Regensburg in late May and arrived at Constantinople on 10 September, a march of just over 112 days; however, he had a shorter distance to cover. Louis VII concentrated the French at Metz in mid-June and arrived at Constantinople on 4 October, a march of 110 days; although he had a rather longer distance to cover than had Godfrey and a very substantially longer one than Conrad, whose route he followed from Regensburg. The armies of both Godfrey and Conrad faced political complications and had to spend time negotiating for their passage of Hungary: Godfrey, because the disorders of the People's Crusades had upset the Hungarian king Coloman I and Conrad because of long-standing political tensions between himself and Geza. However, overall Conrad was much faster than Godfrey and presumably this was a result of his careful preparations: provision of a major fleet on the Danube to carry his baggage to Byzantine territory and the carrying of prefabricated bridges to facilitate river

---

14 Odo of Deuil, *De profectione Ludovici VII*, bk III (pp. 50-51), bk IV (pp. 72-5, 82-3).

15 Ibid., bk II (pp. 32-8). See Berry, "Second Crusade", p. 470.

crossings. This is particularly impressive because the armies of Conrad and Louis both seem to have been considerably larger than that of Godfrey.[16]

Now, I have referred, as almost all writing on the subject does, to the armies of Louis VII and Conrad III but if we are to understand their logistics we need to understand what an army was like in the twelfth century. And to understand that, it is necessary to grasp something of the reality of a kingdom. Twelfth-century royal government was mediated by great lords. They controlled great resources and a king's control over them was fundamentally a matter of circumstance and personality. Odo of Deuil carefully pointed to this reality in his account of the assembly of Louis's army at Metz: "Although the king found nothing there which was subject to him by sovereign power, he nevertheless discovered that all were voluntarily willing to behave as subjects, as had already happened at Verdun."[17]

The implications are clear. Outside his own lands, Louis could not command. He could only cajole, persuade, occasionally coerce. In a military sense this meant that apart from those from his own demesne lands who joined the Crusade and those great lords in his immediate entourage, the king could not command unless men chose to obey. This situation was, of course, complicated by the conditions of a Crusade on which all were pilgrims equally sworn to God's service. The same problem had plagued the First Crusade. Then men and women had grouped themselves around major magnates. However, when circumstances were appropriate many of them had felt themselves free to change allegiances. One such was Gaston of Béarn, who began the Crusade with Raymond of St Gilles but ended it in association with Tancred, Raymond's bitter enemy.[18] Moreover there was no obvious leader and the whole expedition was run by an uneasy, bickering committee of magnates. Holding these disparate elements together proved very difficult and ultimately sizeable forces did not continue on to Jerusalem. Louis enjoyed the prestige of kingship, but relatively little direct authority.

It is conventional to contrast Conrad's large but ill-disciplined forces with Louis's smaller but more controlled ones and there is plenty of evidence for the disorganization and ill-discipline of the Germans. Odo wrote that although supplies were plentiful in the Byzantine lands, the Germans pillaged savagely and even

---

16 J. W. Nesbitt, "The rate of march of Crusading armies in Europe", *Traditio*, 19 (1963), 167-82; France, *Victory in the East*, pp. 90-107. It is very difficult to estimate the size of Crusading armies. The First Crusade probably mobilized about 100,000 people, but many less, perhaps 60,000, gathered at Nicaea. See France, *Victory in the East*, pp. 122-42. It seems unlikely that Godfrey's army was at any stage larger than 8,000. All one can say about the armies of the Second Crusade is that they give the impression of being substantially larger than this.

17 Odo of Deuil, *De profectione Ludovici VII*, bk II (p. 21): "Ubi rex cum iure dominii nihil suum invenerit, omnes tamen invenit ex gratia, sicut Verduno iam fecerat, quasi servos."

18 France, *Victory in the East*, pp. 331, 344.

treated quite badly French Crusaders searching for food. A little later he recorded that the Germans showed "no moderation" and constantly fought with local people. Undoubtedly the Germans were unlucky to suffer losses of men and supplies in a flash flood as they were approaching Constantinople;[19] however, when they reached Constantinople itself they sacked the pleasure-park of the *Philopation*. Odo wrote that they were divided and uncertain even at the moment of crisis when they faced the Turks and were defeated at Dorylaion. In a letter to Wibald of Corvey, Conrad III himself explained the defeat in part by saying that his army had been gravely short of food at the time.[20]

We only have Odo's word for the indiscipline of the Germans but certain factors suggest that he was telling the truth. In the first place, tension between Manuel and his former ally, Conrad, over the disorderly conduct of the German army at Constantinople appears to bear this out. In the second, Odo gave great prominence to the misdeeds of the Germans but did not hide the loose organization of his own forces, though historians have tended to ignore this. At Metz, where Louis gathered his army in mid-June 1147, he proclaimed the "Laws of the camp" to which all the leaders swore obedience. "But because they did not observe them well, I have not preserved them either", commented Odo.[21] The French paused at Worms in late June to await the arrival of Normans and English under bishop Arnulf of Lisieux, and merchants brought food there by river. "Here", wrote Odo, "... we first perceived the foolish arrogance of our people."[22] The French and their suppliers quarrelled and a riot ensued. At about this time a sizeable contingent of the army, worried by the cost of food, broke away and travelled on via Italy. As the main force entered the Byzantine empire many fell to pillaging and at least one group went on ahead in the hope of finding cheaper food. Its reputation became so bad that cities on its route became terrorized and provided only poor supply, thus increasing the need for pillaging. Odo admitted that at Constantinople very fair arrangements were made by Manuel for provisioning the army but, despite this, the French still pillaged and looted. The most revealing passage about the nature of provisioning, however, concerns their journey through Byzantine territory when Michael Branas, the governor of Niš attended to the king's needs and saw that his troops were fed. Louis passed on this food even at his own expense: "... but there preceded and followed him [Louis VII] many divisions who gained plenty for themselves, either from the market, whenever that was possible, or from plunder,

---

19 Odo of Deuil, *De profectione Ludovici VII*, bk III (pp. 46-49); Otto of Freising, *Gesta Friderici I*, I.xlvii (pp. 65-7).

20 Odo of Deuil, *De profectione Ludovici VII*, bk III (pp. 40-43, 46-9), bk V (pp 90-93); Conrad III in Hausmann, *Konrads III*, no. 195 (pp. 354-5).

21 Odo of Deuil, *De profectione Ludovici VII*, bk II (pp. 20-21): "...; statuitque leges paci ceterisque utilitatibus in via necessarias, quas principes sacramentis et fide firmaverunt. Sed quia ipsi non bene tenuerunt eas, nec ego retinui."

22 Ibid., bk II (p. 22): "Hinc primam nostri populi stultam superbiam sensimus."

because they had the power to do that."[23] Louis saw to those who were in a special sense *his* men, but not to others. This should not be a surprise, for his division was all that he could completely command and therefore all that he could supply. The rest had to make shift for what supplies they could and their obedience was, accordingly, qualified. What is known in detail of the logistics sustains this picture amply.

At Regensburg the French leaders met a Byzantine delegation which demanded oaths of security from Louis and threatened that if he did not comply Manuel would order the burning of supplies prepared for him. In the event a decision on the question of the oaths was delayed. At the same city a supply fleet was assembled, perhaps raised by Alvisus of Arras and Leo of St Bertin who had been sent ahead to Regensburg to open discussions with the Byzantines. This fleet was described at first as ample but by the time they crossed the Drava river in Hungary it was described as small, which suggests that only a few ships accompanied them through Hungary. Here provisions were plentiful but some in the army, concerned about supply in Byzantine lands ahead, embarked carts with two or four horses to carry their supplies. These apparently broke down on the roads frequently, blocking the passage of pack-horses and holding up the army badly. The clear picture is of private provision, nobody in overall charge, and resulting chaos.[24]

This pattern of individual provision meant that each Crusader had to carry money, and Louis was much exercised by this need. In both his letters to Suger of St Denis, from the "Gates of Hungary" and Constantinople, he reminded his regent of his financial needs.[25] A consequence of this forms a recurrent theme in Odo's account of the march: the presence in and near the army of money-changers and merchants.[26] Odo commented that when the army entered Byzantine territory, although food was provided, a very poor rate of exchange was given, making it dear. Presumably merchants and money-changers in the army were there to arrange such matters with their Byzantine counterparts on behalf of the French. At Constantinople the rate of exchange was very good, but this changed when the army crossed the Bosporos and remained waiting for elements of their forces which had lagged behind. Odo provided a fascinating insight into the mechanics of supply. The tables of the money-changers, set out on the shore by their ships,

---

23 Ibid., bk II (p. 24), bk III (pp. 40-42, 44): "...; multae vero illum praecedebant acies et sequebantur, vel de foro si poterant, vel de praedis quia hoc poterant sibi abundantiam conquirentes."

24 Ibid., bk II (pp. 24-5, 28-32).

25 Suger, *Epistolae*, nos 12, 13 (pp. 487-8).

26 The issue of exchange-rates has been explored in A. E. Laiou, "Byzantine trade with Christians and Muslims and the Crusades", in A. E. Laiou and R. P. Mottahedeh, eds, *The Crusades from the perspectives of Byzantium and the Muslim world* (Washington, 2001), 157-96, esp. pp. 168-79.

groaned with money that the Crusaders had changed and "... the silver vessels which they [the money-changers] had bought from us."[27] Discontent finally exploded and a Fleming led an assault on the money-changers who fled in such haste that innocent Crusaders trapped aboard their ships were beaten by the locals in revenge. Louis intervened immediately because disruption of money-changing facilities would have caused a major hiatus in the whole supply system. He demanded the surrender of the Fleming from the count of Flanders, hanged the man, and arranged restitution for the merchants and money-changers through the bishop of Langres.[28] This all took time and in the absence of provisioning many ate what they had been saving for their journey into Anatolia. The implication of this concern with exchange-rates is that individuals and groups provided for themselves according to their means, using money and any other assets brought with them. It explains the paradox that there was pillaging and rioting in the midst of admitted plenty. Perhaps the king and other great men made some provision for the lesser people but this was a matter of charity rather than of central control.

All this time Louis VII was bickering over the terms of the oaths demanded by Manuel and also awaiting various French contingents which arrived late at Constantinople. There was further delay when news arrived of the defeat of the German army at Dorylaion on 25 October. When agreement was finally reached, it included arrangements for provisioning. The French took imperial advice and sensibly set off through Byzantine lands in south-west Anatolia, the emperor guaranteeing that cities there would provide markets and fair rates of exchange. If this was not forthcoming they would have the right to pillage.[29] The Crusade was now moving into areas where different problems of supply awaited. In reality, the Byzantines controlled only the cities of the Anatolian coast, although they had a tenuous hold over some of the plateau but subject to Turkish raids. It is hardly surprising that the coastal cities were suspicious and conceded supplies only grudgingly.

The route inland up the Maeander valley passed through territory still nominally Byzantine but it was so vulnerable that the army very quickly came into violent conflict with the Turks. Attacks continued, with a major struggle occurring at a ford outside Antioch of Pisidia.[30] The French army certainly seems to have been better organized than the German one because it continued to advance despite these attacks. At Mount Cadmus, however, on 8 January 1148 indiscipline and problems of control finally made themselves felt. The vanguard went much too far ahead of the main force and the Turks were able to inflict a major defeat upon the

---

27 Odo of Deuil, *De profectione Ludovici VII*, bk II (pp. 22-3), bk III (pp. 40-41), bk IV (pp. 66-7, 74-5): "...; fulgent auro tabulae, vasisque argenteis quae a nostris emerant onustantur."

28 Ibid., bk IV (pp. 74-5).

29 Ibid., bk IV (pp. 74-83, 90-91).

30 Ibid., bk VI (pp. 106-7, 110-13).

French. Louis has often been criticized as a poor commander, but he did a remarkable job in holding his army together in the struggle that followed. He had, after all, to persuade rather than command. Immediately after this he placed the army under the guidance of the Templars, who imposed much tighter control, particularly persuading troops to take oaths of confraternity.

Odo of Deuil made it clear that during the grim winter march across the mountains the "sumpter horses", *summarii,* of the baggage train were all-important. As the vanguard raced ahead and the mass of the following army picked out the trail, these animals often fell, had to be pulled through river crossings, and when they died provided food for the starving army. Odo alleged that, unable to prevent the army advancing, the Greeks and Turks resorted to scorched earth tactics by driving off flocks and burning supplies.[31]

The immediate object of their march was Antalya, a major Byzantine outpost on the southern coast of Anatolia. Its people were able to provide supplies, but the problem now became lack of horses since many had perished and there was insufficient fodder for those that survived. When it became apparent that the Byzantines could not provide enough ships to lift the whole army to Antioch, Louis VII seems to have wanted to keep his army together and to march on. However, this provoked debate in the higher ranks of the army which again exposed the limitations of his authority, especially since he had lost his elite guard during the fighting on Mount Cadmus. The nobles were unwilling to march on because of the shortage of horses and because there were sufficient ships to take at least them to Antioch. Louis had to agree. The army divided, with the better off proceeding by sea and the rest left to march.[32]

In the end it was not simply supply which defeated the French, but their own indiscipline and misfortune. Despite the difficulties they reorganized, and that was a considerable achievement after Mount Cadmus. Even after a difficult winter march they still had a baggage-train of sorts when they arrived at Antalya, where food was available at least. But their leaders quailed before the prospect of continuing to fight on in those conditions as supplies dwindled and horses died. It has often been pointed out that had the Northern fleet reached them at Antalya, instead of wintering at Lisbon after the capture of that city, events might have been very different. It is to a consideration of that fleet and its supposed role that we must now turn.

The northern sea-borne expedition was almost certainly the product of St Bernard's preaching of the Crusade. Some of its major elements came from Cologne and Flanders, areas where he is known to have been active, although German sources show a partiality for the original preacher of the Crusade in those parts, a Cistercian monk named Radulf. It is possible that his combination of

---

31 Ibid., bk VI (pp. 118-21), bk VII (pp. 126-9).

32 Ibid., bk VI (pp. 128-33).

Crusading zeal and anti-Semitism appealed especially to the mercantile and sea-going people who were the backbone of the expedition.[33] However, it is interesting that they chose as their leaders members of the traditional ruling elite, Count Arnold of Aerschot, a nephew of Godfrey of Bouillon, and Christian of Ghistelles who was closely associated with Count Thierry of Flanders. The origins of the very substantial English participation in this expedition are more problematic. When he was active in Flanders in the summer of 1146, St Bernard wrote to England and Italy seeking support and Eugenius III had done likewise.[34] A number of English barons did join Louis VII but quite how the English merchants and shipowners who made up about half the fleet which gathered at Dartmouth on 19 May 1147 were recruited we do not know. They, like the Germans and Flemings, looked to members of the traditional elite for leadership in the persons of two noblemen, Hervey of Glanvill and Saher of Archelle.[35]

It is well known that St Bernard and Eugenius III agreed to Crusade status for expeditions by Saxon nobles against the Wends of north-eastern Germany and by Ramon Berenguer IV in eastern Spain.[36] Bernard saw these expeditions as opportunities for salvation on a par with those to Jerusalem, which may well have been sound theology. It is certainly true that some of these expeditions were seen as parts of a single sacred war against enemies of Christendom. But this should not delude us into thinking that this represented a real plan. Eugenius III referred to the Crusade to the Holy Land and the campaigns which he had sanctioned in eastern Spain in a letter extending Crusader privileges to the expedition against the Slavs.[37] However, this was no more than a gloss on reality as was Helmold of Bosau's comment that: "The initiators of the expedition, however, deemed it advisable to design one part of the army for the eastern regions, another for Spain, and a third against the Slavs who live hard by us." Helmold was writing some 20 years later, though this did not prevent him from exposing the political shenanigans of the

---

33 Berry, "Second Crusade", pp. 472-3. S. B. Edgington, "The Lisbon letter of the Second Crusade', *HR*, 69 (1996), 328-39, pp. 336-9 provides a complete Latin text of the German letters which describe the siege of Lisbon. There is an English translation in S. B. Edgington, "Albert of Aachen, St Bernard and the Second Crusade", in Phillips and Hoch, *The Second Crusade*, 54-70, pp. 61-7 where she comments on: " ... the tantalizing ambiguity of the original Latin with regard to Radulf and Bernard: ..."

34 Constable, "Second Crusade", p. 261.

35 The most important account of the Lisbon expedition, the *De expugnatione Lyxbonensi*, was written by an anonymous Anglo-French priest closely associated with Hervey of Glanvill, now identified as a certain Raol. He provided a list of leaders, noting also two more obscure figures: Simon of Dover, who led the men of Kent, and Andrew, who led those of London.

36 Berry, "Second Crusade", pp. 483, 493.

37 Eugenius III, *Epistolae*, PL, no. 166 (coll. 1203-4).

Saxon and Danish "Crusaders" of 1147. Ramon Berenguer similarly had his own axes to grind in eastern Spain.[38]

The unitarian view of the Crusade is no more than a rationalization for the fact that Bernard and Eugenius made the best of all offers received. This Crusade was planned only in the most limited sense, and certainly not as a broad assault on all non-Christians. Eugenius III proclaimed it in response to the fall of Edessa and addressed *Quantum Praedecessores* to Louis VII and the French. He seems to have been surprised by the participation of Conrad III, to whom he never addressed a Crusading bull and who he declined to visit even when clear opportunity offered.[39] The Pope never seems to have followed up Manuel Komnēnos's demand that he guarantee the behaviour of Crusaders in the Byzantine empire. He appointed Henry of Olmütz to negotiate with Manuel on relations between the Latin and Greek Churches but when this cleric joined the Wendish crusade, does not appear to have appointed a replacement.[40] Louis VII and Conrad certainly coordinated their movements down the Danube valley, but there is no evidence that they intended such cooperation to continue beyond this. Odo of Deuil was singularly unsurprised by the rapid departure of the Germans from Constantinople. Nor is there any indication that Rome or St Bernard tried to coordinate their movements with those of other expeditions.

There is no specific indication that the naval expedition which gathered at Dartmouth in May 1147 had any Papal endorsement for an attack on Muslim-held Lisbon. The only indication that this may have been the case is an alleged letter of St Bernard which suggests that the saint had received a letter from Afonso I Henriques of Portugal carried by the king's brother Peter with an unspecified request to which both Bernard and the Pope were pleased to accede. However, what was the nature of this request is not clear, the dating is unclear, and the letter may well be a forgery. A strong argument has been made by Livermore for its authenticity, but it seems to me that the whole thing founders on its vagueness: "What I have done in the matter will be evident from the outcome of it, as you will see for yourself. You will see with what promptitude I have complied with your request and with the exigencies of the affair. ... My son, Roland, is bringing you the documents which set forth the liberality of the Holy See." As for Afonso's

---

[38] Helmold of Bosau, *Helmoldi presbyteri Chronica Slavorum*, 3rd ed., ed. B. Schmeidler, in MGH *ScriptRerGerm.Cont.*, I.lix (p. 115): "Visum fuit auctoribus expedicionis partem exercitus unam destinari in partes orientis, alteram in Hyspaniam, terciam vero ad Slavos, qui iuxta nos habitant." See also Constable, "Second Crusade", p. 223.

[39] As noted by Hiestand, "Papacy and the Second Crusade", pp. 37, 41.

[40] Manuel to Eugenius III in Eugenius III, *Epistolae*, RHGF, no. 25 (pp. 440-1); Eugenius to Henry in Eugenius III, *Epistolae*, PL, nos 204 (coll. 1251-2), 214 (col. 1262). See also R. Hiestand, "The Papacy and the Second Crusade", in Phillips and Hoch, *Second Crusade*, 32-53, p. 44.

brother Peter, the letter said that: "... he is campaigning in Lotharingia and will soon be fighting in the hosts of the Lord." This last might refer to the Second Crusade or to fighting in the Iberian Peninsula, or indeed to something else.[41] If the reference is to Peter participating in the Second Crusade, the letter must date from before December 1146 because it was then that Bernard brokered a peace amongst bickering factions in Lotharingia.[42] This seems unlikely because the indications are that the attack on Lisbon was the product of circumstances which became clear only well into 1147.

Afonso of Portugal certainly realized that a Crusader fleet would put in to port in Portugal. According to the *De expugnatione Lyxbonensi*, when it put into Oporto the local bishop related that the king had written to tell him that if a Frankish fleet should arrive there he was to receive it well and urge those on board to make an agreement with Afonso. But there is no need to account for this by assuming that arrangements had been made via St Bernard. Indeed, that the bishop of Oporto had been informed by the king only days before the arrival of the fleet argues against it. He would have known of any major diplomatic initiative involving senior European clergy. Moreover, the author of the *De expugnatione Lyxbonensi* explained that the Crusaders themselves informed the king of their coming via five ships sent ahead of the main fleet.[43] Circumstances suggest that the siege was arranged at short notice and that the key event was Afonso's capture of the fortress of Santarém by surprise attack on 15 March 1147, which opened real prospects for an attack on Lisbon. On the one hand, the capture of Santarém was made possible only by a complex sequence of political and military events which fatally weakened its Muslim defenders, but until virtually the last moment the king could have found himself on the defensive. It must have taken some time for new possibilities to become apparent and this would barely have left time for a diplomatic mission to be mounted. On the other hand, Afonso had attacked Lisbon some five years earlier with the support of an Anglo-Norman fleet, so he would have known to where to address an appeal for help.[44] The message borne by the five ships must have been somewhat positive, but we do not know its content or from whom it came. It could hardly have been a full commitment in view of the later tensions amongst the Crusaders about accepting the king's invitation but it is

---

41 Bernard of Clairvaux, St, *Epistolae*, in PL, vol. 182, coll. 67-716, no. 308 (coll. 512-13): "Quid in hoc egimus, et exitus pro nobis, et vos per exitum comprobabitis: animi promptitudinem ex injuncta sollicitudine, vel saltem ex memorata necessitudine colligetis. ... Frater Rolandus, filius noster, apostolicae largitatis litteras defert." Trans. James, *Letters of Bernard of Clairvaux*, no. 397 (p. 469). See also H. Livermore, "The Conquest of Lisbon and its author", *Portuguese Studies*, 6 (1990), 1-16.

42 A. Lane Poole, "Germany, 1125-1152", in CMH, vol. 5, 334-59, pp. 350-51.

43 Raol, *De expugnatione Lyxbonensi / The conquest of Lisbon*, ed. and trans. C. W. David, rev. J. Phillips (1936; rpt, New York, 2001), pp. 68-9, 98-9.

44 H. Livermore, *A new history of Portugal* (Cambridge, 1966), pp. 54-6.

notable that Hervey of Glanvill emerges as a great ally of Afonso, and he was in command of at least a section of the Anglo-Norman forces.

Most of our detailed knowledge of the siege of Lisbon comes from a single source: the *De expugnatione Lyxbonensi*. This was written by a Norman priest who participated and fought in the expedition in close association with Hervey of Glanvill and it is generally accepted that he was the unnamed priest who delivered a stirring exhortatory sermon as the army prepared for the final assault which captured Lisbon on 24 October 1047. His account is cast in the form of a letter to Osbert of Bawdsey who was a close associate of the Glanvill family. It has been convincingly argued that the writer was a priest called Raol who cared for the Anglo-Norman cemetery at Lisbon and gave it to the Augustinian Priory of Holy Cross at Coimbra along with 200 silver marks for its care.[45] Its outline of events is clearly supported by another letter which Winand, a German priest and participant, wrote to the archbishop of Cologne. This German source appears to have been absolutely contemporary with the events it described.[46]

What is striking, however, is that the *De expugnatione*, far from being the work of "a simple Crusader" as Constable suggested, was actually a very sophisticated literary production. The work was dotted liberally with Biblical citations and appears to show acquaintance with much contemporary thinking about Crusading and penitential themes. The writer provided ample discursions on places visited by the fleet and certainly made heavy use of Solinus' *Collectanea rerum memorabilium* and perhaps other classical works as well. The sense is of a work that had been compiled very carefully. A recent analysis has shown that it is structured around four major speeches or exchanges of speeches: the address of Peter of Oporto to the Crusaders asking them to join King Afonso in an attack on Lisbon, a dialogue between (chiefly) Hervey of Glanvill and those dissenting from the king's proposals, a dialogue between archbishop John of Braga and a Muslim elder when the Crusaders attempted to obtain a voluntary surrender of the city, and a sermon given by Raol as the army prepared for the final assault on Lisbon. These passages of direct speech drew on a common body of ideas and even phraseology, all tending to justify the expedition, held it up as an example of what God favoured and, more particularly, excused the participants from the charge of having been concerned only with loot.[47]

The importance of these passages can be revealed by a simple statistic. In David's edition, they occupy 48 pages of a total of 132, some 36% of the text. The

---

45 Livermore, "Conquest of Lisbon", pp. 3-7. For the sermon see Raol, *De expugnatione Lyxbonensi*, pp. 146-59.

46 Edgington, "Lisbon letter", pp. 328-39; idem, "Albert of Aachen", pp. 54-70.

47 Constable, "Second Crusade", p. 221; Raol, *De expugnatione Lyxbonensi*, pp. 68-85, 104-11, 114-25, 146-59; J. Phillips, "Ideas of Crusade and Holy War in *De expugnatione Lyxbonensi (The Conquest of Lisbon)*", in R. N. Swanson, ed., *The Holy Land, holy lands and Christian history* [SCH, vol. 36] (Woodbridge, 2000), 123-41.

unity of their content suggests that they did not represent precisely what the speakers said, and indeed a marginal note by Hervey admitted that these were not his exact words.[48] The work was reflective rather than merely narrative, yet the narrative was vivid and alive. There is one hint of when it was written. In his description of La Coruña the author related that at the port there was a "bridge" of 24 arches extending out into the sea (we might rather call it a jetty), which, he wrote, "... are now visible which were not in sight two years ago."[49] Now this occurs in the context of one of the elaborate descriptions of places visited of which the author was so fond and which drew heavily on his classical learning. This suggests that the text as we have it was written just over two years after the events it described. The author was describing La Coruña as it was in June 1147 when the "bridge" was not in sight but he had evidently seen it again some two years later. As a pious priest he almost certainly had gone on to *Outremer* where the Crusade fizzled out in July 1148 before Damascus. It is entirely possible, therefore, that if he stayed until 1149, as did Louis VII, that he saw La Coruña on his way back some two years after his first visit. This would suggest that what we have is the original and full text of the work dating from a relatively short time after the events it described, rather than a revised and rewritten form of an originally brief text, which is suggested by the fact that the sole manuscript seems to belong to the 1170s.

Why in all this careful justification did the author not say simply that Eugenius III had sanctioned the expedition against Lisbon? Why, moreover, did at least some of the Crusaders feel the need to renew their vows by retaking the cross at Oporto and again on the eve of the final assault?[50] Above all, why was it that when the leaders came to persuade the army to accept Afonso's terms, Hervey of Glanvill's speech made no mention of any Papal approval? More generally we have to ask why such careful self-justification marks the *De expugnatione Lyxbonensi*? If there is one theme which predominates in it, it is rebuttal of the charge that the attack was undertaken for greed. Now, it has often been said that the Lisbon expedition attracted relatively little attention because it was not led by great men, and it may well be true that the social elite denigrated it by casting doubt on its motives. However, they could do this the more easily, I suggest, because no higher sanction had ever been given for the diversion.

---

48 Phillips, "Ideas of Crusade", pp. 125-6.

49 Raol, *De expugnatione Lyxbonensi*, pp. 64-5.

50 Raol, *De expugnatione Lyxbonensi*, pp. 104-5, 158-9. This raises the question of the applicability of Crusader vows. In 1217 a Rhenish and Dutch group on their way to the Fifth Crusade stopped at Lisbon and fought very successfully in Spain, but Pope Honorius III refused to grant them Crusading privileges and urged them on their way to Egypt. See D. W. Lomax, *The reconquest of Spain* (London, 1978), p. 132. However, by 1217 definitions were much closer. In 1147 many may have assumed that any act against Muslims was covered by their oath.

So what was the naval expedition which set off in the spring of 1147? Well, the first thing that must be said is that it was large. The author of the *De expugnatione Lyxbonensi* wrote that 164 ships gathered at Dartmouth and the German letter stated there were "about two hundred ships". There is evidence that a force of 13,000 besieged Lisbon, though this surely included forces of the Portuguese king.[51] It can be no coincidence that it set off at roughly the same time as the other major expeditions. Presumably, following the precedent of the First Crusade some date had been set. Quite simply these were men who had chosen to travel to *Outremer* by sea but in no sense did they constitute a coordinated supply and support fleet. We know that the fleet actually took a month to reach Lisbon where, allowing for delays, it would have been two to three weeks sailing from the Mediterranean which it would have entered in early July. A summer passage across the Mediterranean from west to east would typically have been 28-42 days, producing an arrival in *Outremer* in mid-August.[52] When King Afonso's terms for participation in the siege of Lisbon were put to the fleet as a whole, one of the Norman leaders, William Veil, objected vigorously, partly because he had participated in an earlier attack on Lisbon in 1142 which he felt the king had betrayed. More importantly, he argued that to delay would compromise their chances of making a good passage. He made it clear that he did not expect to sail directly to *Outremer* because once in the Mediterranean he wanted to pillage Muslim shipping. He was probably reckoning on an arrival in September 1147, which would still have been within the summer sailing season. However, any substantial delay, and the earlier attack on the city in 1142 had proved that Lisbon was capable of strong resistance, would mean missing the sailing season and awaiting the next one in the spring of 1148.

Ultimately, we know that some Crusaders went on from Lisbon on 1 February 1048, attacked Faro, and reached *Outremer* later in the year in time to join in the abortive Damascus expedition of July 1148. It is unlikely that anybody in early 1147 expected that the great armies of the kings would take so long to reach *Outremer*. Given the problems of wind and weather and the fact that these were quite obviously Crusaders, though with their own agendas, it seems unlikely that there could have been any thought of their supporting the march of the French and German armies. These men had chosen to go to *Outremer* by sea. They had chosen their own leaders and would have seen themselves as being free agents who might or might not join others as conscience and interest decided.

---

[51] Raol, *De expugnatione Lyxbonensi*, pp. 52-3 and n. 4; Edgington, "Albert of Aachen", p. 63. As in the case of the land armies it is difficult to estimate numbers. However, if there were 10,000 on the ships participating in the siege, this would represent an average per ship of less than 100. Ships, of course, varied in size. When they reached *Outremer* the men of the fleet seem to have joined Conrad III and made up a substantial force.

[52] J. H. Pryor, *Geography, technology, and war: studies in the maritime history of the Mediterranean, 649-1571* (Cambridge, 1988), pp. 3-4, 88.

Crusading in the twelfth century was neither a coherent nor an articulated doctrine, nor was it very distinct from "Holy War". Papal initiative, which we perceive as the distinctive feature of Crusade, was not yet defined and elaborated as such.[53] In 1146 Eugenius III, aided and abetted by St Bernard, proposed an expedition to the Holy Land. But Eugenius and Bernard could only propose and enthuse, not control. To maintain influence they were obliged to agree to accept other destinations, although at first Jerusalem had been their very evident concern. It is clear that many in the fleet had doubts about the attack on Lisbon and that the issue of authorization loomed large in their minds and that this produced the *De expugnatione Lyxbonensi* as we have it today. Others saw little reason why they should not do as they thought best within the framework of an ultimate objective. But none conceived of themselves as being part of a master plan in which theirs was a subordinate role. It is we who have tended to look at the fleet in that way because the much smaller naval forces of the First Crusade did serve that purpose. This, however, was largely a matter of circumstances.

The Second Crusade is an interesting study from the point of view of logistics. The sources testify to problems of supply and to the importance which leaders attached to solving them. At the same time an examination of logistical problems is inseparable from questions of overall organisation. The view put forward here is that Eugenius III and Bernard of Clairvaux were the inspirers, but not the organisers, of the expedition we know as the Second Crusade and that the separate forces acted according to their own agendas, which the inspirers were prepared to accept for the most part. These forces suffered from all the ills of medieval armies in general, compounded by the particular problems of Crusading armies. In short, what we see is ramshackle by modern standards but it approximated to what was possible in that age. We should not try to reorganize it retrospectively.

---

53 For current ideas about what constituted a Crusade see J. Riley-Smith, *What were the Crusades*, 3rd ed. (Houndmills, 2002). For a critique see C. Tyerman, *The invention of the Crusades* (London, 1998).

Chapter 6

Harbours and facilities along the eastern Mediterranean sea lanes to *Outremer*[1]

Ruthy Gertwagen

Varied in composition, many fleets moved from the West to the Levant and the Holy Land during the Crusades. Some, such as the Venetian fleets of 1099-1100 and of 1217, which transported Andrew of Hungary to Acre, Philip Augustus's of 1191, the fleet that arrived from Marseilles in 1239, and most of those from Northern Europe and England, consisted of sailing ships of the *naves* type only. On rare occasions only did fleets consist solely of galleys: the Genoese fleet of 1097, the Venetian fleet of 1220 for the Fifth Crusade, and Frederick II's of 1228. One of the Genoese galleys of 1097 was a transport galley, a *sandanum*, and Frederick's fleet included transport galleys, *taride*, for horses and materiel, as well as battle galleys. In 1107 Sigurd of Norway probably had both longships for battle and transports of the type of the Northern *knorr*.[2] Other fleets had both *naves* and galleys; for example, the Pisans in 1098-9.[3] *Naves* also transported warriors and their horses, as the Venetians did in 1122-3. Only from the Third Crusade, however, did fleets including horse transport galleys with hull ports at the stern,

1 Consult Maps 2, 4, 5, 6, 7, 12.

2 The composition is nowhere described. Snorri Sturluson, quoting the poet Thórarrin Stuttfeld, referred to the ships simply as *skip*. Albert of Aachen called them *buzae*. Snorri Sturluson, *Heimskringla*, ed. B. S. Kristjánsdóttir, et al., 3 vols (Reykjavík, 1991), *Magnússona Saga*, §3 (vol. 2, p. 720); Albert of Aachen, *Historia Hierosolymitana*, in RHC *HOcc*, vol. 4, 265-713, XI.xxvi (p. 675). However, since Sigurd fought naval engagements against Muslim forces *en route*, he presumably had some long ships.

3 Bernardo Maragone wrote that the fleet was 120 *naves*, ships in the generic sense. See Bernardo Maragone, *Annales Pisani*, ed. M. L. Gentile, *Gli Annales Pisani di Bernardo Maragone*, in RISS *NS*, tome 6, parte 2, 1-74, A.D. MXCVIII (p. 7). However, Anna Komnēnē wrote that the fleet included δρόμωνες, by which galleys must have been intended, and Andrea Dandolo that the Pisans had 50 galleys. Since they engaged both Byzantines and Venetians, they must have had some galleys. Anna Komnēnē, *Alexiade*, ed. and Fr. trans. B. Leib, 3 vols (Paris, 1937-45), XI.x.1-8 (vol. 3, pp. 41-5); Anonymous monk of the Lido, *Monachi anonymi Littorensis historia de translatione sanctorum Magni Nicolai, terra marique miraculis gloriosi, ejusdem avunculi, alterius Nicolai, Theodorique, martyris pretiosi, de civitate Mirea in monasterium S. Nicolai de Littore Venetiarum. 6 Dec. 1100*, in RHC *HOcc*, vol. 5, 253-81, §VI (pp. 257-8); Andrea Dandolo, *Chronica per extensum descripta, aa. 46-1280 d. C.*, ed. E. Pastorello [RISS *NS*, tome. 12, part 1] (Bologna, 1938), c. x(l) (p. 221).

known as *taride*, *chelandre*, or *salandria*, come from the West to *Outremer*. Philip Augustus, however, shipped his horses on *naves*.

The shape, size, and gearing of ships, and cargoes carried, all impacted on the logistics involved and consequently on landfalls made. Moreover, in the case of Crusader fleets, whose purpose was to deliver armies and their equipment and horses safely to *Outremer*, and sometimes directly to a battle zone, the logistical requirements must have been greater than for normal commercial traffic and the techniques necessitated thereby out of the ordinary. Re-provisioning with food and water was indispensable. Food was vital for energy to manœuvre heavy rudders and large lateen sails and even more so for galley oarsmen to maintain high levels of exertion. Ship's biscuit provided glucose and was the basis of seamen's diets. This also included vegetables or legumes, which provided B vitamins and minerals, and onions and garlic, which contained vitamins C and B complex, anti-oxidants, anti-dermatomycosis and anti-viral, and minerals. Fruit contained mainly vitamins. Cheese and salt meat or fish all contained proteins. Olive oil provided omega 9. Rice provided vitamins E, B complex, and minerals. Wine softened the hard biscuit and ameliorated the tastelessness of the food. Water, however, was the main "fuel".

Historical sources rarely mentioned logistical issues such as watering because authors regarded them as too commonplace to discuss unless relevant to other issues they were addressing. No ports of call were recorded for the Genoese expedition of 1097 which reached St Symeon in late November.[4] In the case of the Pisan fleet of 1098-9, lay-overs were related to its assault on the Ionian islands.[5] The Pisans intended to winter in the Ionian and the clashes there must have been a consequence of authorities refusing permission.[6] Fears of strain on local economies from heavy provisioning could lead to conflict with locals. Later, the Pisans also pillaged the Byzantine Dodecanese, most probably for similar reasons.[7]

Sigurd of Norway's wintering in Sicily was attributed in the sagas to a glorification of the king with respect to his Norman racial relatives rather than to any logistical considerations. Roger of Hoveden, however, recorded that Richard Cœur de Lion needed to repair shipworm damage when he wintered in Sicily in

---

4 *Annales Ianuenses*, eds L. T. Belgrano and C. Imperiale, *Annali genovesi di Caffaro e de' suoi continuatori dal MXCIX al MCCXCIII*, 5 vols [FStI] (Rome and Genoa, 1890-1929), vol. 1, pp. 5, 13 [*Annales Ianuenses*, ed. G. H. Pertz, in MGH *SS*, vol. 18, 1-356, pp. 11, 14]; Caffaro de Caschifelone, *De liberatione civitatum Orientis liber*, ed. L. T. Belgrano, in *Annales Ianuenses* (FStI), vol. 1, 95-124, p. 102.

5 Anna Komnēnē, *Alexiade*, XI.x.1 (vol. 3, pp. 41-2); Maragone, *Annales Pisani*, p. 7.

6 J. H. Pryor, " 'Water, water everywhere, Nor any drop to drink.': water supplies for the fleets of the First Crusade", in M. Balard, B. Z. Kedar, and J. Riley-Smith, eds, *Dei gesta per Francos: études sur les croisades dédiées à Jean Richard / Crusade studies in honour of Jean Richard* (Aldershot, 2001), 21-8, p. 25 and n. 19.

7 Anna Komnēnē, *Alexiade*, XI.x.1-8 (vol. 3, pp. 41-5).

1190-91.[8] The chroniclers referred to provisioning only on reaching Rhodes two weeks after leaving Messina, notwithstanding emphasizing that the fleet took refuge from a storm to recover and regroup somewhere on the north-west coast of Crete, most probably in the Gulf of Chandax, where there were six streams.[9] At Rhodes food alone was mentioned. However, the *Eracles* claimed that when the *buss* carrying Richard's sister and bride-to-be was driven to Cyprus by a storm but the women declined emperor Isaac Ducas Komnēnos's invitation to be his guests ashore, he cut off water supplies in an attempt to coerce them. When Richard himself arrived, Isaac's officials refused to allow his men to land and to provide food and water and this provoked him to attack Limassol and led to the conquest of the island. When leaving Cyprus, Richard loaded his ships with provisions needed at Acre.[10]

Frederick II made only two known lay-overs for provisioning, firstly in Keffalinia six days after leaving Brindisi. By now the Ionian islands belonged to the Kingdom of Sicily and provisions had been laid up by Frederick's count on the island. Even though the only place at which watering was specifically mentioned was at Phoinikous on the fourth day after leaving Crete, we may assume that Chandax was used also for watering.[11]

Naves had more space for cargo, provisions, and water. Since no irregular incidents were mentioned for the fleet of Philip Augustus, which took 22 days to reach Acre from Messina in 1191, we may assume that it carried sufficient provisions and water to reach its destination.[12] When sailing with galleys as a fleet, however, *naves* had to make as many landfalls as they did.

---

8 Roger of Hoveden, *Chronica*, ed. W. Stubbs (Rolls series, tome 51), 4 vols (London, 1868-71), vol. 3, pp. 71-2.

9 See Cristoforo Buondelmonti, *Descriptio insule Crete et liber insularum, cap. XI: Creta*, ed. M.-A. van Spitael (Iraklion, 1981), pp. 131-50. See also R. Gertwagen, "The Venetian port of Candia, Crete, 1299-1363: construction and maintenance", *MHR*, 3 (1988), 141-58, p. 143 and map p. 145.

10 Roger of Hoveden, *Chronica*, vol. 3, pp. 105-6; *Chronicles and memorials of the reign of Richard I. Vol. I: Itinerarium peregrinorum et gesta regis Ricardi*, ed. W. Stubbs (Rolls series, tome 38) vol. 1, II.27, 42 (pp. 177-80, 204-5); Ambroise, *L'estoire de la Guerre Sainte: histoire en vers de la Troisième Croisade (1190-1192)*, ed. G. Paris (Paris, 1897), ll. 1261-8, 1301-2, 1310-13, 2103-4; Anonymous, *L'estoire de Eracles empereur et la conqueste de la terre d'Outremer*, in RHC *HOcc*, vols 1.1, 1.2, 2, XXIV.xviii, xxii (vol. 2, pp. 160, 163).

11 *Breve chronicon de rebus Siculis a Roberti Guiscardi temporibus inde ad annum 1250*, ..., in J.-L.-A. Huillard Bréholles, *Historia diplomatica Friderici Secundi*, 6 vols (Paris, 1852-61), vol. 1, part 2, 887-908, pp. 898-900. See J. H. Pryor, "The Crusade of Emperor Frederick II, 1220-29: the implications of the maritime evidence", *The American Neptune*, 52 (1992), 113-32, pp. 125-7.

12 Roger of Hoveden, *Chronica*, vol. 3, p. 100; *Itinerarium peregrinorum* (Stubbs), II.26-8 (vol. 1, pp. 175, 181).

The Venetian fleet of 1122-3 was the first composite one of this kind. Wintering on Corfu and leaving in late March, it arrived off *Outremer* in early June after a very slow voyage of two months during what should have been a favourable sailing season, which is to be explained by Fulcher of Chartres' declaration that it sailed only by day whenever possible and made port to water the crews, passengers, and horses "*cotidie*", lit. "daily" or "every night".[13] This is the first occasion known on which warriors and their horses were transported directly to *Outremer* from the West and during which experience of water requirements would have been gained.

The Venetians kept their fleet together and prevented it scattering in veering winds by voyaging by day only, albeit that lanterns for night voyages had been used as early as the sixth century by Belisarios in the Byzantine fleet that sailed for Vandal Carthage. Belisarios had to cope with strong winds astern, calms, contrary winds, and night voyaging and Prokopios conveys the pilots' dexterity and the seamen's physical efforts to keep it together. At Zakynthos Belisarios took aboard sufficient water for the crossing to Sicily but much of it putrefied.[14]

In 1191 Richard Cœur de Lion also supposedly attempted to control the fleet's voyage as one unit. According to Richard of Devizes, the fleet was arranged in eight divisions, one behind the other. On leaving Messina, the sound of a trumpet could be heard between divisions and a man's shout between the ships in each division. The king and the galleys brought up the rear.[15] The fleet moved through the Straits into the Ionian but was then scattered by a storm. The voyage continued by night, using lanterns on Richard's ship.[16]

Galleys of all kinds were poor sailers because they were designed to be rowed and could use their sails only with moderate breezes astern. Nor was the upwind performance of *naves* much better than that of galleys. Because of their rounded hull configuration and lack of deep keel, they made much leeway and with winds abeam it was difficult to hold a real course at 90° to the wind. In autumn 1184 the Genoese ship on which Ibn Jubayr was travelling was struck by a Westerly gale

---

13 Fulcher of Chartres, *Historia Hierosolymitana (1095-1127)*, ed. H. Hagenmeyer (Heidelberg, 1913), III.xiv-xv (pp. 656-8), trans. in Pryor, "Water, water everywhere", pp. 22-3.

14 Prokopios of Caesarea, *History of the wars*, in *Procopius*, 7 vols, ed. and trans. H. B. Dewing (London, 1914-40), vols 1-6, III.xiii.1-8, 21-4 (vol. 2, pp. 116-18, 124-5).

15 Richard of Devizes, *Chronicon Richardi Divisensis de tempore regis Richardi Primi / The Chronicle of Richard of Devizes of the time of King Richard the First*, ed. and trans. J. T. Appleby (London, 1963), pp. 15, 35. It should be appreciated that as an ecclesiastical chronicler Richard of Devizes was not acquainted with such matters. He either delivered someone else's fantasy or a report of an eyewitness of an experience that was only momentary. No medieval fleet could ever have maintained station such as this for more than moments.

16 *Itinerarium peregrinorum* (Stubbs), II.xxvi-xxvii (vol. 1, pp. 174-80).

when leaving the south coast of Crete and was blow back past Andikithera onto the island's north coast.[17] Consequently, when winds veered or strengthened it became very difficult to keep composite fleets together. It was safer and easier to make land and rest crews and passengers.

Constant moderate following winds were in fact quite rare, even in spring and autumn.[18] Philip Augustus did enjoy them but Richard Cœur de Lion, who left Messina shortly after him, endured strong winds and storms all the way to Cyprus. Frederick II enjoyed moderate following winds most of the time in June-July 1228 and sailed by night to take advantage of them, reaching Rhodes after 16 days and Cyprus on the 24th day even though he made frequent stops, laying over for a night at Methone because of rising seas. The fleet made the 275 kilometres from Chandax to Rhodes island in 36 hours on 12-13 July at an average of 4.1 knots (7.6 kilometres per hour), which suggests that it was pushed by a fresh breeze. Handling the sails to remain at a safe angle of heel must have required great physical effort because at Rhodes the fleet anchored for 36 hours to rest the crews.[19] They went ashore only at Phoinikous. Thirty years earlier Richard also had encountered strong northerly winds between Crete and Rhodes which tossed the ships, billowing the sails, bending the masts, and driving the fleet towards southern Rhodes, where it found surf along the coast and no refuge. The chroniclers recorded that the fleet remained at Rhodes for ten days because Richard was sick, but also to wait for lost galleys and *naves* to come in.[20]

Eyewitnesses, the *Itinerarium peregrinorum* and Ambroise for Richard Cœur de Lion's and Pseudo-Brocardus and Joinville for Louis IX's expedition of 1249, conveyed graphically the poor physical and psychological condition of those aboard during heavy weather: losing sensibility leading to suicidal tendencies, sea sickness, lack of appetite causing weakness and eventually death, and fear and loud cries for salvation causing stress. According to Pseudo-Brocardus life aboard was unbearable for the French, who were not accustomed to sea voyaging, and for

---

17 Ibn Jubayr al-Kinānī, Abu 'l-Ḥusayn Muḥammad ibn Aḥmad, *Riḥla*, trans. R. J. C. Broadhurst, *The travels of Ibn Jubayr* (London, 1952), Sha'bān 1184 (pp. 331-2).

18 See the wind roses for April and July in Great Britain, Admiralty, Hydrographic Office, *Mediterranean Pilot. Volume I: the coasts of Spain* ..., 10th ed. (London, 1978), diagrams 12 and 13.

19 The average heading would have been East, then roughly North-East. In July the fleet would almost certainly have had quite strong winds on the port beam, from West-North-West around to North. See Great Britain, Admiralty, Hydrographic Office, *Mediterranean Pilot. Volume IV: the islands of the Grecian archipelago with the adjacent coasts* ..., 9th ed. (London, 1968), climatic tables for Khania, Iraklion, Thira, and Rhodos (pp. 23, 24, 26, 33) and Figure 6: Wind frequency distribution - July.

20 *Itinerarium peregrinorum* (Stubbs), II.27 (pp. 179-80); Ambroise, *Estoire de la Guerre Sainte*, ll. 1268-1315 (coll. 34-6). In the late fifteenth century Felix Fabri gave a realistic description of the difficult conditions aboard Venetian great galleys during strong contrary and astern winds. See Felix Fabri, *Evagatorium* (Hassler), vol. 1, pp. 155, 161-2.

whom tasteless fat food, stinking putrid water, and close conditions and other things that caused illness, were unbearable. He advised sea voyages only for seamen and those accustomed to them or not affected by tempests and perils and also making port, even in uninhabited havens, during heavy weather.[21]

Heavy weather worsened the condition of horses taken from their natural environment and stalled in dark, narrow holds with dense, suffocating air.[22] Horses are affected badly by sea travel and to minimize their suffering from the ships' rolling they were stabled fore and aft aboard *taride*, restrained by under-belly slings to keep them on their feet.[23] In foul weather pounding seas must have terrified the animals. Due to poor conditions aboard, they were not groomed and some became weak and eventually died. There is no evidence that they were disembarked during short stops and constant standing restrained against movement made those which survived extremely stiff.[24] However, mooring would at least have permitted removing the slings, cleaning stalls, pumping water from bilges, and feeding and watering them.

After long voyages men and horses lost almost completely the ability to fight. At Cyprus Richard's men did not dare land from the boats to fight until the king led the way, and he could not use his horses. Later he ordered them exercized, "... for they were all numb and lame and dazed from the month that they had been on the sea without ever being able to lie down."[25] The Crusade of 1239 that left from Marseilles was hit by an easterly storm two days out from Acre that "... scattered them so violently that they believed themselves to be totally lost." Supposedly, they were eventually driven to Cyprus, Sicily, Corsica, and Sardinia, but regrouped in Sicily and reached Acre later in September. The Crusaders "... rested and

---

21 *Itinerarium peregrinorum* (Stubbs), II.27-8 (pp. 178-81); Ambroise, *Estoire de la Guerre Sainte*, ll. 1204-1312 (coll. 34-6); Pseudo-Brocardus, *Directorium ad Passagium faciendum*, in RHC *DArm*, vol. 2, 365-517, III (p. 412), IV (p. 421); Jean de Joinville, *Vie de saint Louis*, ed. J. Monfrin (Paris, 1995), §§127-9 (pp. 62-5).

Pseudo-Brocardus was almost certainly William Adam, the Dominican missionary and archbishop of Sultaniyah.

22 Pseudo-Brocardus, *Directorium*, III (p. 412).

23 J. H. Pryor, "From dromōn to galea: Mediterranean bireme galleys AD 500-1300", in J. Morrison, ed., *The age of the galley: Mediterranean oared vessels since pre-classical times* (London, 1995), 101-116, pp. 115-16. How they may have been stabled aboard *naves* is unknown, although it was probably fore-and-aft also.

24 Pseudo-Brocardus, *Directorium*, III (p. 412).

Ibn Jubayr wrote that the pounding of the waves tossed his ship around, as large as it was, like a fragile branch of a tree. See Ibn Jubayr al-Kinānī, Abu 'l-Ḥusayn Muḥammad ibn Aḥmad, *Riḥla*, Fr. trans. M. Gaudefroy-Demombynes, *Ibn Jobair - Voyages*, 4 parts (Paris, 1949-65), 3rd part, p. 370; Ibn Jubayr, *Riḥla* (Broadhurst), p. 331.

25 *Itinerarium peregrinorum* (Stubbs), II.32 (pp. 190-91); Ambroise, *Estoire de la Guerre Sainte*, ll. 1565-78 (coll. 42-3): "... Car il erent tut engurdi / E deboistié e esturdi / D'un mois qu'orent en mer esté / E sanz jesir toz jorz esté."

refreshed themselves for a time because of the great distress and the great fear and great peril that they had had on the sea."[26] According to Joinville, following a storm immediately after leaving Limassol which dispersed and sank many vessels of Louis IX's fleet of 1249, mainly galleys, *taride*, and small boats, the regrouped fleet reached Damietta after four days.[27] According to the letter of John Sarrasin, the storm caused many difficulties and much distress. On arriving off Damietta Louis therefore ordered that only those who were by then still fit to do so should participate in the landing. They would board galleys and small boats to go ashore.[28] The king was aware of the logistical disadvantage of lack of a port in which to shelter and to enable men and animals to recover.[29]

Crusading maritime itineraries show inconsistency in sea lanes followed and landfalls and ports made and no navigation guides for Mediterranean sea lanes survive from the twelfth and early thirteenth centuries, although there were treatises discussing harbours, ports, port towns, and distances and directions between them. Al Idrīsī's geography and map of 1154 gave detailed anthropological, economic, and geographic descriptions, including information on water sources and markets; however, he was primarily interested in geography rather than sea lanes and did not indicate routes from one place to another. Traders, travellers, and captains or pilots provided him with information.[30] Nevertheless, his geography may be considered a portolan and *compasso da navigare*, a precursor of the *Lo compasso de navegare* compiled in the mid thirteenth century.

This latter included ports of call, using the same term, *portus*, for artificial harbours created by moles and/or closed by chains as for natural havens with adequate depth of water, not necessarily, however, providing protection against prevailing winds.[31] Directions and distances indicated, but did not necessarily

---

[26] *Continuation de Guillaume de Tyr de 1229 à 1261 dite du manuscrit de Rothelin*, in RHC *HOcc*, vol. 2, 483-639, §20 (p. 529): "... qui durement les esparpilla qu'il cuidierent tuit estre pardu. ... se reposerent et rafreschirent une piece pour les granz annuiz, et les granz poourz et les granz perilz qu'il avoient euz en la mer."

[27] Joinville, *Vie de saint Louis*, §§146-8 (pp. 72-3). See also Pryor, "Crusade of Emperor Frederick II", p. 117.

[28] *Rothelin continuation*, §§44, 59 (pp. 571, 589). See also the letter of Guy, a knight of the household of the viscount of Melun, to B. of Chartres on the capture of Damietta in Matthew Paris, *Mathaei Parisiensis monachi Sancti Albani, Chronica Majora*, ed. H. R. Luard (Rolls series, tome 57), 7 vols (London, 1872-83), Additamenta, no. 81 (vol. 6, pp. 155-62).

[29] Joinville, *Vie de saint Louis*, §149 (pp. 72-5).

[30] Al-Idrīsī, Abū 'Abd Allāh Muḥammad ibn Allāh, Al-Sharīf, *Kitāb Nuzhat al-mushtāk fi 'khtirāq al-āfāq*, part. Fr. trans. H. Bresc and A. Nef, *La première géographie de l'Occident* (Paris, 1999), pp. 19-22.

[31] For example, B. R. Motzo, ed., *Il compasso da navigare: opera italiana della metà di secolo XIII* (Cagliari, 1947), p. 28: "Trani è porto de molo"; p. 29: "Ancona è bom porto de molo ... E lo porto è facto de molo."

mean, favoured sea routes, nor why particular routes were followed from one point to another. Once a route was chosen, however, *Lo compasso de navegare* reported landfalls and access to them and topographic dangers to be avoided. To ascertain what routes indicated by *Lo compasso de navegare* were favoured sea lanes, one should consult different types of sources, including modern pilot books providing information on wind and current regimes, those dealing with the technological limitations of medieval ships, and medieval voyage reports.

Reporting on the Third Crusade, Roger of Hoveden wrote that with following winds *naves* could make a shorter high-seas voyage from Marseilles to Acre via the Sicilian Channel and south of Crete without ever sighting land. By "shorter" he referred to avoiding landfalls and manœuvring between islands. He also claimed that the southern route was safer, most probably meaning that it avoided the hazardous coastal topography of the Ionian and Aegean, shoals, reefs, capes, and off-shore islets, some of which he mentioned while reporting Philip Augustus's return voyage. He warned, however, against approaching the hostile North African coast, onto which ships might be driven by northerlies. He also advised galleys to avoid the high-seas route since they would risk being sunk in storms.[32] Roger was no seaman but he undoubtedly reported the experience of seamen or passengers in explaining why Richard took the northern route in spite of the tidal rips and whirlpools in the Straits of Messina and the storms encountered after having left them.

The voyages of the Genoese *naves* that carried Ibn Jubayr to Alexandria in March 1183 and James of Vitry to Acre in late September 1216, as well as those of the Crusades of 1239 and of Louis the IX in 1249, demonstrate that heavy seas were dangerous for *naves* also.[33] The southern coast of Crete, where bays were open to both southerly winds and strong katabatic squalls off the Cretan mountains,

---

32 Roger of Hoveden, *Chronica*, vol. 3, pp. 160-61: "Et est sciendum, quod si ventus prosper fuerit illis, qui a Marsilia ad Accon ire voluerint, dimittent insulam de Sardena, et insulam de Sicilia, et insulam de Creta, longe in sinistra parte navigii; et, si rectum cursum tenuerint, non videbunt terram donec videant terram Suliae; et via illa brevior est et securior sed cavendum est illis ne nimium declinent in dextris navigii, propter Barbariae et alias insulas multas, in quibus habitant pagani sub imperatore Africæ. Sed galeæ non possunt neque audent illam viam ire, quia si tempestas supervenisset, ex facili submergerentur; et ideo oportet eas semper ire prope terram."

For Philip's return voyage see vol. 3, pp. 155-60, 165-6.

33 Ibn Jubayr, *Riḥla*, Shawwal 1183 [(Gaudefroy-Demombynes), part 1, pp. 26-9; (Broadhurst), pp. 27-9]; Jacques de Vitry, *Lettres*, ed. R. B. C. Huygens, in CC *ContMed*, vol. 171, 491-652 [rpt of *Lettres de Jacques de Vitry (1160/1170-1240) évêque de Saint Jean-d'Acre* (Leiden, 1960)], Ep. 2, ll. 72-104 (pp. 560-62); Joinville, *Vie de saint Louis*, §127 (pp. 62-3).

provided no shelter. According to Buondelmonti, this coast could be navigated only by skilled local crews who knew its topography well.[34]

Calms could be just as damaging as storms wrote Pseudo-Brocardus.[35] In the late fifteenth century Felix Fabri was more graphic: "... everything that is on board becomes putrid, and foul, and mouldy; the water begins to stink, the wine becomes undrinkable, meat, even when dried and smoked, becomes full of maggots, and all of a sudden there spring into life innumerable flies, gnats, fleas, lice, worms, mice and rats. ... I have seen few men die on board ships during storms, but many I have seen sicken and die during these calms."[36] Galleys might survive calms better than *naves* because they could be rowed; however, rowing required plentiful food and water, especially in hot, calm weather.

There was no consistency in sea lanes followed and land falls made by various Crusades. Philip Augustus may have followed in reverse the same itinerary as on his way home: along the coasts of Cilicia and Lycia, through the Dodecanese, across the southern Aegean to the Peloponnesos, and into the Ionian, or he may have dropped down from Rhodes to Cyprus.[37] Richard Cœur de Lion in 1191 and Frederick II in 1228 both took the route from Crete to Rhodes and thence to Cyprus.

The position of Crete is curious. The further crossing to Rhodes required a passage exposed to strong winds and, in fact, Rhodes could be reached more easily from the Ionian via the Peloponnesos, Cyclades, and Dodecanese. Crete was marginal to the main Aegean trunk routes and the Venetians did not include it in the partition agreement of 1204. They eventually purchased the island solely to prevent it falling to the Genoese.

Crete became important from 1230 due to geopolitical shifts in the eastern Mediterranean which began with the Byzantines of Nicaea conquering the southern Dodecanese. Their control of the eastern Aegean islands from Lesbos to Karpathos forced Latins to abandon the main Aegean crossing via Naxos and Amorgos. Until

---

[34] R. Gertwagen, "The concept of ports in the medieval eastern Mediterranean: construction and maintenance on Crete to the end of the fifteenth Century", *IJMH*, 12 (2000), 177-241, p. 192 and n. 23; Buondelmonti, *Descriptio insule Crete*, pp. 24-6. Buondelmonti's ship was driven to southern Crete by a storm.

[35] Pseudo-Brocardus, *Directorium*, III (p. 412): "Interdum in alto mari deficiet eis [navibus] ventus, et tunc non possunt retrocedere vel procedere, nec declinare ad dexteram vel sinistram. Que quidem omnia sunt in magnum vie dispendium, in magnam materiam expensarum et in personarum et equorum non modicum detrimentum."

[36] Felix Fabri, *The book of the wanderings of brother Felix Fabri*, trans. A. Stewart, 2 vols, 4 parts, in PPTSL, vols 7-10 (London, 1893-6), vol. 1, part 1, p. 124, from Felix Fabri, *Evagatorium* (Hassler), vol. 1, p. 116.

[37] For the outward voyage see *Itinerarium peregrinorum* (Stubbs), II.26 (p. 175); Roger of Hoveden, *Chronica*, vol. 3, p. 100. For Philip's return voyage see Roger of Hoveden, *Chronica*, vol. 3, pp. 155-6, 160-61, 165-6.

the establishment in 1268 of amicable relationships with the revived Byzantine Empire, Crete became a port of call for all Western ships sailing between the northern Aegean and the Levant. To and from the Levant ships had to sail the high seas west from Cyprus, staying as far south of Byzantine-held Rhodes and Karpathos as possible. Buondelmonti's experience in the fifteenth century while sailing from Karpathos to north-eastern Crete shows that strong northerlies on the starboard beam could easily push ships onto its unfavoured southern coast. Crete became a key point for Venetian maritime traffic in the Eastern Mediterranean only after they gained control of the whole island and signed a final treaty with the Byzantines in 1302.[38]

Richard Cœur de Lion made Crete only unintentionally. After leaving Messina, the fleet lay overnight somewhere between Calabria and Mount Etna but the wind then veered to contrary, probably a north-easterly from the Straits of Otranto, and hit the fleet near Cape Spartivento, dispersing it. Otherwise Richard would have followed the coast to Apulia and the Ionian.[39] The reasons for Frederick II's landfall on Crete are unclear. The local Byzantine population had begun its long revolt against Venetian rule and Muslim Chandax was still in ruins and silted and during his tenure of the island between 1206 and 1211 Henry Pescatore had built Palaiokastron in the Bay of Fraschea on the north-western side of the Gulf of Chandax as his administration seat. Fraschea's bay would have provided shelter and adequate depth of water and Frederick's fleet must have moored there.[40]

Galley fleets that did not provision in Sicily would have put in to Reggio, where there was abundant water, attested to by the existence of bathhouses, and whose markets provided fruit and vegetables.[41] Entering the Ionian, fleets were exposed to prevailing northerlies as well as to the changeable and often inclement weather characteristic of it. Storms generated locally, either the N.E. *Tormenta* from the Straits of Otranto or the S.E. *Libecchi*, are quite common. Modern observations record stormy winds blowing from every direction and reaching the force of seven knots between November and mid March and, less frequently,

---

38 See Gertwagen, "Concept of ports", pp. 183-92, 199-202.

39 Ambroise, *Estoire de la Guerre Sainte*, ll. 1203-60; *Itinerarium peregrinorum* (Stubbs), II.27 (pp. 178-9). In addition to the prevailing winds there are locally created storms. See R. Heikell, *Italian waters Pilot: a yachtsman's guide to the west and south coasts of Italy with the islands of Sardinia, Sicily and Malta*, 6th ed. (London, 2002), pp. 375-6.

40 Gertwagen, "Concept of ports", pp. 189, 194-5. This is now contrary to what I wrote at pp. 186-7 on Frederick's itinerary.

41 Al-Idrīsī, Abū 'Abd Allāh Muḥammad ibn Allāh, al-Sharīf, *Kitāb Nuzhat al-mushtāk fi 'khtirāq al-āfāq*, trans. P.-A. Jaubert, *La géographie d'Edrisi*, 2 tomes (1836-40; rpt in 1 vol., Amsterdam, 1984), vol. 2, p. 116 [trans. Bresc and Nef, p. 341].

between May and September. The historic record also attests to calms.[42] The Gulf of Taranto, however, provided shelter and numerous small rivers along the east coast of Calabria provided water and moorings at their mouths. On one of them in the south and another near Taranto there were water mills and near Cape Santa Maria de Leuca ships could find shelter as well as water.[43] From there they turned east towards Corfu, the focal point of the three coastal routes to the Eastern Mediterranean.

A second originated in Brindisi or Otranto in Apulia. In 1126 the young Bohemond II left for Antioch from Otranto and in 1228 Frederick II headed south from Brindisi to Otranto and then south-east past Othonoi to Corfu. He lay overnight at Othonoi, either in Fiki bay in the north offering shelter against south and south-east winds or in Ammou bay in the south offering shelter against north-west winds.[44] The third route down the east coast of the Adriatic was invariably followed by Venetian fleets.

The three routes met at Kassiopi Bay on north-east Corfu, called *Santa Maria de Chaxopo* by a portolan of the 1490s after the tiny Gazapolis islands that close the bay.[45] Along the shore there was a church named after the Virgin of Cazopa, a patron of seamen, who used to take on water and frequent the church. They anchored at the secure eastern end of the bay.[46] Robert Guiscard built a castle which developed into a town here, and fought the Venetians and Byzantines in 1085.[47] As an almost obligatory port of call, it attracted pirates, including those who waylaid Richard Cœur de Lion when he stopped there on his way back from *Outremer*. After having recognized one of his crew, they treated him with respect

---

42 Great Britain, Admiralty, Hydrographic Office, *Mediterranean Pilot. Volume III: the Ionian Sea bordering the western coast of Greece ...*, 10th ed. (Taunton, 1988), §§1.182-1.192 (pp. 26-32), §1.210 (p. 38) [Methoni Climatic Table]. On calms see Prokopios, *History of the wars*, III.xiii.10-11 (vol. 2, pp. 120-21).

43 Al-Idrīsī, *Géographie*, IV.3 [(Jaubert), vol. 2, pp. 117-19; (Bresc and Nef), pp. 341-3]; Motzo, *Compasso da navigare*, pp. 24-6.

44 Fulcher of Chartres, *Historia Hierosolymitana*, III.lvii.2 (pp. 806-7); *Breve chronicon de rebus Siculis*, p. 898. See R. Heikell, *Greek waters Pilot: a yachtsman's guide to the coasts and islands of Greece*, 8th ed. (London, 2001), pp. 43-4.

45 "Portolan del colfo di Venetia dalla banda di greco", in G. B. Dosio, ed., *Ragioni antique spettanti all'arte del mare et fabriche de vasselli: manoscritto nautico del sec. XV* (Venice, 1987), 61-3, p. 63.

46 Cristoforo Buondelmonti, *Isolario*, Venice, Biblioteca Nazionale Marciana, MS. Ital. Class. VI, Cod. XIX (= 6087), fol. 2r; Felix Fabri, *Evagatorium* (Hassler), vol. 1, p. 43; Heikell, *Greek waters Pilot*, pp. 46-7.

47 Anna Komnēnē, *Alexiade*, VI.v.5 (vol. 2, p. 52). See also E. Malamut, *Les îles de l'Empire Byzantin VIIIe – XIIe siècles*, 2 vols paginated continuously (Paris, 1988), p. 185.

and he leased two ships from them to carry him to Zara.[48] Philip Augustus on the other hand, ignored this anchorage, most probably because the pirates were responsible for the fact that the town was deserted. Roger of Hoveden claimed that there was no one who dared to go there because of the many snakes and praised instead *Sancta Carenta*, a bay with a deserted town on the Albanian mainland.[49] On the morrow of Venice's taking over Corfu in 1386, the Senate ordered Kassiopi's walls and wells demolished for fear that Genoese visiting the church might land, conquer the place, and settle there. The resolution stressed retaining the church alone to avoid the large investment required to man the castle.[50] A Genoese squadron of eight galleys, eight *gorabii*, and a great *navis* may have stayed either here or at Corfu town at the tip of Cape Sidhero further south for two days in 1101.[51]

In the twelfth century Corfu town was a castle on the eastern hill of the cape with natural and artificial defences. Known as *Castello Maris*, or in later documents as *Castel Vecchio*, it was mentioned as early as the eleventh century. Re-fortified by the Normans when they occupied Corfu in 1147-9, it was prosperous according to al-Idrīsī. It was besieged by the Venetians in 1122 and, according to Nikētas Chōniatēs, by the Fourth Crusade in 1203. In the thirteenth century it was enlarged by building another fortification on the western hill of the cape, known later as the *Castel a Terra* or the *Castel Nuovo*, and a suburb, *burgus*, developed between the two fortified hills.[52] The medieval town had no stream or wells for water. Only in 1488 did a local inhabitant declare that there was a water source under the rock shelf on which the city was built and only in 1495 did the

---

48 *Itinerarium peregrinorum* (Stubbs), VI.xxxviii (p. 442); Ralph of Coggeshall, *Chronicon Anglicanum*, ed. J. Stevenson [Rolls Series, tome 66] (London, 1875), 1-208, pp. 53-4.

49 Roger of Hoveden, *Chronica*, vol. 3, p. 165.

50 Venice, Archivio di Stato, MSS *Senato misti*, Reg. 40, fols 40r (12.08.1386), 80v (04.07.1387); Reg. 43, fol. 4r (15.05.1394).

51 Caffaro de Caschifelone, *De liberatione civitatum Orientis liber*, p. 118.

52 Saewulf, *Peregrinatio*, in R. B. C. Huygens, ed., *Peregrinationes tres: Saewulf, John of Würzburg, Theodericus* [CC *ContMed*, vol. 139], (Turnhout, 1994), 58-77, p. 59; Al-Idrīsī, *Géographie*, IV.3 [(Jaubert), vol. 2, p. 121; (Bresc and Nef), p. 345]; Fulcher of Chartres, *Historia Hierosolymitana*, III.xiv.1 (p. 656); Nikētas Chōniatēs, *Nicetae Choniatae historia*, ed. J.-L. van Dieten (Berlin, 1975), p. 541.

The evidence for the Venetians besieging the town in 1122-3 is in the *Historia ducum Veneticorum*, ed. H. Simonsfeld (MGH *SS*, vol. 14), 72-89, p. 73 ("..., castrum illud fortissimum fortiter expugnates obsidere ceperunt") and Andrea Dandolo, *Chronica*, p. 233. It is not supported by Fulcher of Chartres.

Venetian Senate issue instructions to build a cistern of 8,000 *botti* inside the town. Until then, water could be obtained only outside the city.[53]

The Venetian port of Corfu north of the medieval town was built only in the fifteenth century and before that ships anchoring in Corfu Channel were exposed in the winter to both north-easterlies and south-easterlies which from December to March average five knots in the morning, reaching six to seven knots by noon. During the summer the Channel is exposed to north-westerlies that reach six knots from June to September.[54] Frederick II stayed at Corfu from early morning till evening in 1228 and may have had calm weather. Alternatively, he may have moored south of Cape Sidhero in Garitsas Bay which was shallow, however, and could be used only by galleys and small vessels. Before the 1380s there was some kind of a marine construction there, a "*porporella*", that could harbour 15-20 galleys and other craft of 2-300 *milliaria*.[55] Both the date of its construction and what exactly it was is unknown. The Venetians found it in ruins in the 1380s.[56] Quite probably the Venetian fleet of 1122-23 and that of the Fourth Crusade moored here since their armies must have camped near water sources somewhere near Cape Sidhero. In 1122 the horses must have been disembarked from the *naves* by barges or boats, as at Damietta in 1249. In 1204 the Venetian *taride* would have been backed onto the beaches to disembark the horses and then beached, as the Genoese beached their galleys at Jaffa in 1101.[57]

South of Corfu the route passed between Paxos and Andipaxos to Levkas, thence to Ithaka and Keffalinia, the main island of the Byzantine theme, whose capital had been Kastron or Hagios Georgios in the south of the island. The main port had been in the Gulf of Argostoli in the south-west beneath Kastron. However, in 1228 Frederick II put in to *Portus Guiscardus*, so named from Robert Guiscard, who came to a small bay south of Cape Atheras and died there in 1085. It had a

---

53 Venice, Archivio di Stato, MSS *Senato mar*, Reg. 12, fol. 140r (22.03.1488); Reg. 14, fol. 86r (05.12.1495); Pietro Casola, *Viaggio a Gerusalemme di Pietro Casola*, ed. A. Paoletti (Alessandria, 2001), p. 184.

54 *Mediterranean Pilot. Vol. III*, §1.213 [Kerkira Climatic Table] (p. 41), §5.82 (p. 152).

55 The Venetian *milliarium* or *migliaio* was a measure of weight equivalent to 1,000 Venetian pounds or approximately 477 kilogrammes. See L. Martini, *Manuale di metrologia ossia misure, pesi e monete* (Turin, 1883), p. 818.

56 The Venetian Senate first discussed the construction of a port at Corfu in April 1387 but did not make a decision until September 1401. See Venice, Archivio di Stato, MSS *Senato misti secreta*, Reg. 40, fol. 78r (22.04.1387); C. N. Sathas, *Documents inédits relatifs à l'histoire de la Grèce au moyen-âge*, 9 vols (Paris, 1880-90), no. 265 (vol. 2, pp. 50-51).

For the Bay of Garitsas see Venice, Archivio di Stato, MSS *Senato misti secreta*, Reg. 40, fol. 87r (22.04.1387), fol. 74v (23.04.1387), fol. 79r (09.05.1387); Sathas, *Documents*, nos 482 (vol. 2, p. 230), 588 (vol. 3, pp. 45-6), 593 (vol. 3, p. 48). See also *Mediterranean Pilot. Vol. III*, §5.60 (p. 149).

57 *Rothelin continuation*, §59 (p. 590); Joinville, *Vie de saint Louis*, §154 (pp. 76-7); Fulcher of Chartres, *Historia Hierosolymitana*, II.viii.1 (pp. 393-5).

secure, small natural harbour whose entrance was only around 100 metres wide and which ran back only around 200 metres from the entrance to the shore. Apparently there was no water and Frederick may have used it only to recover after a difficult voyage from Corfu through the shoals of *Le Formigue* off Cape Asprokavos at the southern tip of Corfu and those between Ithaka and Keffalinia, or to wait for a favourable wind to carry him to Kastron.[58]

From Keffalinia shipping proceeded via Zakynthos, which was fertile and plentifully supplied with water, and then turned south-east past the bay of Avarinos towards Methone and Korone on the peninsula of Cape Akritas. Among these places only Methone was a port of call for Crusader fleets: the Venetians in 1123, Bohemond II in 1226, and Frederick II in 1228.[59]

Since Methone's exposed anchorage was poor, we may assume that it was weather conditions that compelled the Venetians in 1123, Bohemond II in 1126, and probably Roger of Lauria in 1292, to anchor there: a calm such as encountered by Belisarios in 533 or contrary winds such as encountered by Frederick II in 1228.[60] Moreover, Methone's water sources were a long way outside the town and its inhabitants relied on rain.[61] The town could not satisfy the demands of such fleets and Belisarios had to water at Zakynthos while Lauria watered at Avarinos or Navarino, by then known as Ionchium/Zonchio, which was on the route between Methone and Corfu, had an anchorage offering shelter from all winds and adequate depth for all ships, and plentiful water. Control of its water was one of the crucial factors behind the battle of Sapienza in 1403 between the Genoese under Marshal Boucicault and the Venetians.[62] Methone provided shelter only against north-westerlies in the lee of the town's walls and prior to the construction of an artificial

---

58 Anna Komnēnē, *Alexiade*, VI.vi.1-2 (vol. 2, pp. 55-6); Al-Idrīsī, *Géographie*, IV.3 [(Jaubert), vol. 2, p. 121; (Bresc and Nef), p. 345]. See also Malamut, *Iles de l'Empire Byzantine*, pp. 186-8; H. Ahrweiler, *Byzance et la mer: la marine de guerre, la politique, et les institutions maritimes de Byzance aux VIIe-Xve siècles* (Paris, 1966), pp. 95, 160. For the dangers between Corfu and Keffalinia, see J. H. Pryor, "The voyages of Saewulf", in Huygens, ed., *Peregrinationes tres*, 34-57, p. 39; idem, "Winds, waves, and rocks: the routes and the perils along them", in K. Friedland, ed., *Maritime aspects of migration* (Cologne, 1989), 71-85, p. 79.

59 Fulcher of Chartres, *Historia Hierosolymitana*, III.lvii.3 (p. 807); *Breve chronicon de rebus Siculis*, p. 899; Al-Idrīsī, *Géographie*, IV.4 [(Jaubert), vol. 2, pp. 121, 124; (Bresc and Nef), pp. 345, 348].

60 Prokopios, *History of the wars*, III.xiii.10-11 (vol. 2, pp. 120-21); *Breve chronicon de rebus Siculis*, p. 899; G. Airaldi, "Roger of Lauria's expedition to the Peloponnese", *MHR*, 10 (1995), 14-23, p. 17.

61 C. Schefer, ed., *Le voyage de la sainte cyté de Hierusalem avec la description des lieux, portz, villes, citez et aultures passaiges fait l'an mil quatre cens quatre vingtz* (Paris, 1882), p. 47.

62 See R. Gertwagen, "The Venetian colonies in the Ionian and Aegean Seas in Venetian defensive policy in the fifteenth century", *JMS*, 12 (2002), 351-84, pp. 354-5.

port by the Venetians in 1358, all shipping had had to anchor in the open bay exposed to prevailing winds. By "*bom porto*", *Lo compasso de navegare* referred to the adequate water depth at Methone rather than to shelter.[63] Shelter could be found, however, east of Sapienza island at the south-west entrance to the bay and it may have been there that Frederick II moored because of contrary winds. Galleys may have become awash if attempting to reach the lee of the town from there.[64]

Methone was an almost obligatory port of call for ships entering the Aegean or the Ionian and for that reason attracted corsairs who became so dangerous that in 1125 the Venetian fleet returning from *Outremer* demolished the town.[65] The Norman fleet of 1147 which raided the Ionian and Aegean and occupied Corfu may have occupied Methone for the same reasons also and, according to the Greek version of the *Chronicle of the Morea*, the Venetians again demolished its walls in 1149 after they had assisted the Byzantines in expelling the Normans from Corfu.[66]

Because of its importance, Venice acquired Methone in the Treaty of Partition of 1204. In the same year a Genoese corsair named William Porco captured at Methone a ship carrying relics from Constantinople to the Papacy. Three years later, Leo Vetrone, a Genoese corsair who had settled in 1199 in northern Corfu, moved to Methone. Chasing him from Corfu to Methone and then to Korone, Ranieri Dandolo and Rogerio Permarino occupied these places, thus realizing the Treaty of Partition.[67]

---

[63] At Methone in summer, from May to August, the wind averages 78.5% from the west to north-west at 12 knots. However, in winter, from November to February, it averages 51.75% from the north around to east at 12 knots. See *Mediterranean Pilot. Vol. III*, §1.210 [Methóni Climatic Table] (p. 38).

[64] Motzo, *Compasso da navigare*, p. 34. On Methone see R. Gertwagen, "Venetian Modon and its port, 1358-1500", in A. Cowan, ed., *Mediterranean urban culture, 1400-1700* (Exeter, 2000), 125-48, 248-54.

[65] *Historia ducum Veneticorum*, pp. 73-4; Dandolo, *Chronica*, pp. 234-5. See also A. Bon, *Le Péloponnèse byzantin jusqu'en 1204* (Paris, 1951), p. 170.

[66] Otto of Freising, *Ottonis et Rahewini Gesta Friderici I. imperatoris*, 3rd ed., ed. G. Waitz, in MGH *ScriptRerGerm.Cont.* (Hanover, 1912), I.xxxiiii (p. 53); *Annales Cavenses*, ed. G. H. Pertz, in MGH *SS*, vol. 3, 185-97, C.E. 1147 (p. 192); J. Schmitt, ed., *The chronicle of the Morea/Τὸ Ξρονικὸν τοῦ Μορέως* (1904; rpt, New York, 1979), ll. 1690 ff. (p. 115).

[67] *Historia ducum Veneticorum*, pp. 73-4; *Annales Ianuenses* [FstI], vol. 2, p. 93 [MGH *SS*, p. 122]; G. L. Fr. Tafel and G. M. Thomas, eds, *Urkunden zur älteren Handels- und Staatsgeschichte der Republik Venedig*, 3 vols (1856-57; rpt, Amsterdam, 1964), no. 121 (vol. 1, pp. 469, 490); Dandolo, *Chronica*, pp. 282-3. See also Bon, *Péloponnèse byzantin*, p. 170 & n. 2; J. Fotheringham, "Genoa and the Fourth Crusade", *EHR*, 25 (1910), 26-57, p. 31 & n. 21; R. Gertwagen, "Does naval activity – military and commercial – need artificial ports? The case of Venetian harbours and ports in the Ionian and Aegean till 1500", forthcoming in *Graeco-Arabica*.

Leaving the Ionian, fleets headed for Rhodes. No accounts detail the exact route eastwards; however, Philip Augustus's return voyage in 1192 and Nicholaus de Marthono's voyage on a *navis* from Rhodes to Venice in January 1395 are instructive since made against adverse winds and currents.[68] As reconstructed in reverse, from Methone to Monemvasia ships crossed the Gulf of Messinia past Korone to Cape Tainaron and then the Gulf of Lakonia through Elafonisi Channel to round Cape Malea. Only Korone, Mainē, and Monemvasia were mentioned by Roger of Hoveden along this difficult route. Korone was a fortified town on a lofty hill, described by Al-Idrīsī as a "coastal town" or a "great fortress on the sea, next to a prosperous suburb". Until the Venetians constructed a mole in the early fourteenth century, its bay, which had inadequate depth in any case, was exposed to both northerlies and southerlies and *Lo compasso de navegare* referred to the castle only.[69] Al-Idrīsī described Mainē on the peninsula of Cape Tainaron as prosperous with a thriving market and an important craft industry and Malea, a village also somewhere on Cape Tainaron, as a coastal fortress with a thriving market.[70] The route, which was dangerous in inclement weather or calms, most probably forced ships to anchor in the bays of either Mainē or Quallie, called a *portus* by *Lo compasso de navegare*,[71] and of Cape Tainaron, and local markets and industries developed there. Philip Augustus did not put in anywhere until Monemvasia, after rounding Cape Malea into the Aegean. Monemvasia's bay did not provide shelter against prevailing winds and *Lo compasso de navegare* ignored it and referred to the place merely as a "*castello*". However, because of its location at the crossroads of sea routes leading to Constantinople and the Cyclades, it was central to the Empire and Monemvasiotes served in Byzantine fleets and also operated as corsairs.[72] Most probably because of the poor anchorage as well as the

---

68 Roger of Hoveden, *Chronica*, vol. 3, pp. 160-61, 165-6; Nicolas de Martoni, *Liber peregrinationis ad Loca Sancta*, ed. L. Legrand, "Relation du pèlerinage a Jérusalem de Nicolas de Martoni notaire italien (1394-1395)", *ROL*, 3 (1895), 566-669, pp. 643-8.

69 Roger of Hoveden, *Chronica*, vol. 3, p. 160; *Chronicle of the Morea* (Schmitt), p. 115; Motzo, *Compasso da navigare*, p. 35; Al-Idrīsī, *Géographie*, IV.4 [(Jaubert), vol. 2, p. 124; (Bresc and Nef), p. 348]. The two translations were made from different editions of different manuscripts. See also *Mediterranean Pilot. Vol. III*, §3.47 (pp. 62-3).

On the construction of the mole, see Venice, Archivio di Stato, MS. *Maggior Consiglio, Clericus civicus*, fol. 22r.

70 Al-Idrīsī, *Géographie*, IV.4 [(Jaubert), vol. 2, pp. 124-5; (Bresc and Nef), p. 348]. The mention of a market and artisans at Mainē and a market at Malea is found only in the critical edition by G. Tucci et al., eds, Al-Idrīsī, *Opus geographicum sive «Liber ad eorum delectationem qui terras peragrare studeant»*, 9 fasc. (Leiden, 1970-84) as trans. Bresc and Nef. Jaubert used different manuscripts.

71 Motzo, *Compasso da navigare*, p. 36.

72 Motzo, *Compasso da navigare*, p. 38. See also Great Britain, Admiralty, Hydrographic Office, *Mediterranean Pilot. Volume IV: the islands of the Grecian archipelago with the adjacent coasts* ..., 9th ed. (London, 1968), pp. 89-90; H. Kalligas, *Byzantine Monemvasia:*

local piratical activity, al-Idrīsī mentioned only the extremely high bluff on which the Byzantine town was built, from which, he claimed, one could see Crete.[73]

From Monemvasia the route to Rhodes ran through the Cyclades via the islets of Karavi, Falkonera, Sifnos, Paros, and Naxos thence, most probably, via Amorgos, Astipalaia, and Kos. Bays of these islands provided moorings and shelter and al-Idrīsī wrote that they were prosperous and well populated. According to Roger of Hoveden, however, they were haunts for pirates and Nicholas de Marthono was in fact seized by pirates near Fermia in 1395.[74] Naxos lay at the heart of the Aegean sea lanes and Venice claimed the island in the Treaty of Partition of 1204. Together with Paros, Melos, Sifnos and Siros, it was occupied in 1206 by Marco Sanudo, who made it the capital of his Duchy of the Aegean Sea, Αἰγαῖον Πέλαγος, Aigaion Pelagos, the Archipelago.[75]

The port of Rhodes was a major port of call both *en route* to *Outremer* and on the way home. Roger of Hoveden explained why Philip Augustus put in there on his way home: "And it should be known, that when the island of Rhodes is reached, a third part of the sea lane between Acre and Brindisi is passed."[76] On the eve of the First Crusade Rhodes, together with Kos and Cyprus, was an important naval base under the command of an *archōn abydikos* with several warships. Alexios Komnēnos had begun to reconstruct the Byzantine navy in response to the depredations of the Turkish corsair Tzachas, who had pillaged the Aegean and the Dodecanese, including Rhodes itself, between 1089 and 1093.[77] Byzantine galleys must have used the enclosed and shallow northern harbour of Mandraki, which had an arsenal. The island became an important provisioning base for the sieges of Antioch in 1097-8 and Tripoli in 1099.[78] The Venetian fleet wintered on Rhodes in 1098-9, albeit that Cyprus was closer for provisioning the Crusaders at Antioch;[79]

---

the sources (Monemvasia, 1990), pp. 98-9, 102-7; idem, "Monemvasia, seventh-fifteenth centuries", in Laiou, ed., *Economic history of Byzantium*, vol. 2, 879-97, pp. 883, 894-5.

[73] Al-Idrīsī, *Géographie*, IV.5 [(Jaubert), vol. 2, p. 125; (Bresc and Nef), p. 349].

[74] Roger of Hoveden, *Chronica*, vol. 3, p. 159; Nicolas de Martoni, *Liber peregrinationis*, p. 646.

[75] Al-Idrīsī, *Géographie*, IV.4 [(Jaubert), vol. 2, pp. 126-8; (Bresc and Nef), pp. 349-51]. See also Gertwagen, "Concept of ports", pp. 187-9.

[76] Roger of Hoveden, *Chronica*, vol. 3, p. 160: "Et est sciendum, quod cum perventum fuerit ad insulam de Rodes, praeteritur tertia pars viae maris inter Accon et Brundusium."

[77] Anna Komnēnē, *Alexiade*, VII.viii (vol. 2, pp. 110-16). See also Ahrweiler, *Byzance et la mer*, pp. 160, 165, 170, 184.

[78] William of Tyre, *Willelmi Tyrensis archiepiscopi chronicon*, ed. R. B. C. Huygens, 2 vols, in CC *ContMed*, vols 63 & 63A [paginated continuously], VI.9.23-6, VII.21.41-3 (vol. 63, pp. 318, 371); Ralph of Caen, *Gesta Tancredi in expeditione Hierosolymitana, auctore Radulfo Cadomensi, ejus familiari*, in RHC *HOcc*, vol. 3, 587-716, §54 (p. 647).

[79] Anonymous monk of the Lido, *Historia*, §§4-8 (pp. 256-9). For Cyprus see J. France, "The First Crusade as a naval enterprise", *MM*, 83 (1997), 389-97, pp. 390-91; idem, "The Western Mediterranean powers and the First Crusade", *JMS*, 10 (2000), 265-74, pp. 269-70.

however, where it could have moored is unclear. Any galleys could have used Mandraki harbour, although further research is needed to establish whether the depth would have been sufficient for galley horse transports. However, it would certainly have been insufficient for *naves* unless, perhaps, they were totally emptied of all cargo and equipment. Perhaps they used the outer Emborikos harbour, even though it was exposed to winds from all directions until the construction of the modern moles and the *Sirocco* still raises heavy surf at its entrance. In such conditions, pilot books recommend that ships seek shelter at Trianda bay in the north-west of Rhodes.[80]

Rhodes would have provided water and food, as it did later for Richard Cœur de Lion, whose *naves* must have anchored in Emborikos harbour. However, al-Idrīsī mentioned only the location of Rhodes, ignoring its fertility and commercial advantages and Ambroise described the town as tumbling down from old age and the *Itinerarium peregrinorum* as it being depopulated and in ruins. *Lo compasso de navegare* did not describe Rhodes as a "*porto*", no doubt because of the shallow depth of Mandraki harbour and the exposed nature of Emborikos,[81] and although Frederick II stayed in the port for almost 48 hours, he did not disembark.

From Rhodes ships made their way east along the Lycian coast towards Cape Chelidonia. Frederick II's fleet reached Patara with its ruins of Roman and Byzantine installations, quays, storehouses, and a market, along a beach between Cape Ince and the river Eşen, classical Xanthus, around 6.5 kilometres to the north-west. The beach is fringed by shoal water and galleys could beach overnight. Next morning he reached Phoinikous at the mouths of the rivers Basgöz and Tara, which medieval ships could enter to take shelter and to water behind reefs creating a natural haven. Due to its location and water, Phoinikous became a haunt for Pisan corsairs, two of whose ships Philip Augustus seized, and known as *Portus Pisanorum.* Richard Cœur de Lion had already provisioned in Rhodes; however, Frederick II had not done so and therefore landed at Phoinikous.[82]

---

80 See *Mediterranean Pilot. Vol. IV*, pp. 335-8.

81 Al-Idrīsī, *Géographie*, IV.4 [(Jaubert), vol. 2, p. 128; (Bresc and Nef), p. 351]; *Breve chronicon de rebus Siculis*, p. 899; Ambroise, *Estoire de la Guerre Sainte*, ll. 1299-1300; *Itinerarium peregrinorum* (Stubbs), II.xxvii (pp. 179-80); Motzo, *Compasso da navigare*, p. 51; K. Kretschmer, *Die italienischen portolane des Mittelalters: ein Beitrag zur Geschichte der Kartographie und Nautik* (1909; rpt, Hildesheim, 1962), Portolan Parma-Magliabecchi, §167 (p. 329), Portolan des Gratiosus Benincasa, §58 (p. 376). See also Malamut, *Iles de l'Empire Byzantine*, pp. 242, 281-2.

82 Fulcher of Chartres, *Historia Hierosolymitana*, III.lvii.3-lix.1 (pp. 807-10); Kretschmer, *Italienischen Portolane*, "Portolan des Marino Sanudo", §24 (p. 245); Ambroise, *Estoire de la Guerre Sainte*, ll. 1315-32; *Itinerarium peregrinorum* (Stubbs), II.xxviii (pp. 180-81); Roger of Hoveden, *Chronica*, vol. 3, pp. 157-8; Saewulf, *Peregrinatio*, ll. 56-69, 588-95.

Cape Chelidonia was the departure point for two alternative routes to *Outremer*. The first lay across the Gulf of Antalya to Cilicia and Syria. The Gulf was notorious, however, for strong squalls descending suddenly from the Tauros Mountains and no one would ever take this route unless bound for Cilician Armenia or Antioch, although Bohemond II did take it to Antioch in 1126. Fulcher of Chartres wrote of the Gulf, "Sicut in terra, sic et in mari pericula", in discussing the navigational and pilotage skills required. Although there were rivers, bays, and havens, the route provided no shelter from the strong southern, western and northern winds. Squalls off the land often reach force 5-6 on the Beaufort scale and winds of force 3-5 from seaward send swells into harbours, making anchorages unsafe. Ludolf of Suchem's story of a squall wrecking a galley sailing by night along the coast makes the point.[83] No wonder that the legendary Saint Nicholas, supposedly born in Patara and Bishop of Myra, became the patron saint of seamen and saviour of ships in distress.

For return voyages, however, this route offered the assistance of a current and onshore/offshore diurnal winds against contrary winds prevailing further out to sea. Philip Augustus took it in 1191, perhaps also to avoid English-controlled Cyprus.[84] In 1103 the ship carrying Saewulf on his return voyage reached Cape Andreas in Cyprus from Palestine and then put back to St Symeon, its track suggesting that it was pushed by an E.S.E. wind to Cape Andreas, which had only a bay with surrounding reefs. *Lo compasso de navegare* referred to the cape only as a landmark. However, a ship could find shelter there against east-north-easterlies.[85]

South from Antioch to Palestine the coastal route may not have been much used by *naves* before the conquest of Tripoli in 1109 and Sidon and Beirut in 1110, and even after that they would have had to detour out to sea around Tyre, which remained a Fāṭimid naval base until conquered in 1124. *Naves* would have had

---

See also Great Britain, Admiralty, Hydrographic Office, *Mediterranean Pilot. Volume V: the coasts of Libya, Egypt, Syria, Lebanon and Israel; ...*, 6th ed. (London, 1976), p. 88; Pryor, "Voyages of Saewulf", pp. 57-8.

83 Fulcher of Chartres, *Historia Hierosolymitana*; III.lvii.3 (pp. 807-8); Ludolph of Suchem, *Description of the Holy Land and of the way thither. Written in the year A.D. 1350*, trans. A. Stewart in PPTSL, vol. 12.C (London, 1895), §§6-7 (p. 13); *Itinerarium peregrinorum* (Stubbs), II.xxviii (pp. 180-81).

See also *Mediterranean Pilot. Vol. V*, Climatic Table - Antalya (p. 27) and pp. 84-100, esp. p. 91; R. Heikell, *Turkish waters and Cyprus Pilot: a yachtsman's guide to the Mediterranean and Black Sea coasts of Turkey with the island of Cyprus*, 6th ed. (London, 2001), pp. 276-92; R. Gertwagen, "Maritime activity concerning the ports and harbours of Cyprus from the late 12th to the 16th centuries (1191-1571)", in N. Coureas and J. Riley-Smith, eds,*'Η Κύπρος καὶ οἱ Σταυροφορίες / Cyprus and the Crusades* (Nicosia, 1995), 511-38, p. 512.

84 Roger of Hoveden, *Chronica*, vol. 3, pp. 126, 155-61.

85 Saewulf, *Peregrinatio*, p. 76; Motzo, *Compasso da navigare*, p. 127. See also *Mediterranean Pilot. Vol. V*, p. 110.

difficulty defending themselves against galleys stationed there. Similar circumstances of Muslim control of the Syro-Palestinian coast, except for the fortress of Margat and for Tripoli and Tyre, prevailed between 1187-1191 and Richard Cœur de Lion's *naves* sailed direct from Cyprus to Acre before following winds whereas his galleys crossed to Margat and then coasted down to Acre.[86]

Except for Aradus, Sidon, and Tyre, whose ancient Phoenician moles can still be seen, ports along the littoral were merely natural roadsteads.[87] Further research is needed to confirm the extent to which Aradus and Sidon still functioned; however, the picture for Tyre is clearer. Its northern port had been created by a sea wall and closed by an iron chain running from a tower at the eastern edge of the northern mole to another on land and was well described by Ibn Jubayr. An arsenal had existed there from Umayyad times. In the ninth century, the Egyptian *amīr* Aḥmad ibn Ṭūlūn had ordered that Acre be developed on the same layout as Tyre, which had both substantial marine installations protecting it against prevailing winds and also adequate water depth, both lacking at Acre. Ships therefore had to anchor in the natural open bay of Acre. When there were strong winds at Acre, ships sought refuge at Tyre. In calm weather ships could also anchor in the lee of reefs extending north of the port. Had Tyre not lain too far north, it rather than Acre may have become the main port and capital of the Kingdom of Jerusalem, as it was during the short period of Ayyūbid dominion from 1187 to 1191.[88] *En route* from Famagusta to Acre, Richard Cœur de Lion tried to put into Tyre but was refused admission. He had to camp outside the walls and anchor his galleys in the lee of the northern reefs, unless perhaps he beached them as the Venetians had during the siege of Tyre in 1124.[89]

Philip Augustus stayed at Tripoli for several days on his way home. Roger of Hoveden referred only to the town but the port was composed of natural reefs and had adequate water depth according to *Lo compasso de navegare* and was described as a *bonus portus* by Marino Sanudo Torsello.[90] Latakia had similar topographic features and in 1101 accommodated Genoese and Italian galleys for a whole winter.[91]

---

86 *Itinerarium peregrinorum* (Stubbs), II.42 (pp. 204-5).

87 *Mediterranean Pilot. Vol. V*, pp. 119 (Rouâd), 124 (Sidon), 126 (Tyre).

88 See R. Gertwagen, "The Crusader port of Acre: layout and problems of maintenance", in M. Balard, ed., *Autour de la Première Croisade: actes du Colloque de la Society for the Study of the Crusades and the Latin East (Clermont-Ferrand, 22-25 juin, 1995)* (Paris, 1996), 553-82.

89 Roger of Hoveden, *Chronica*, vol. 3, p. 112; *Historia ducum Veneticorum*, p. 74.

90 Roger of Hoveden, *Chronica*, vol. 3, p. 156; Motzo, *Compasso da navigare*, p. 61; Kretschmer, *Italienischen Portolane*, "Portolan des Marino Sanudo", §3 (p. 238). See also *Mediterranean Pilot. Vol. V*, pp. 120-21.

91 Fulcher of Chartres, *Historia Hierosolymitana*, II.viii (pp. 393-4). See also *Mediterranean Pilot. Vol. V*, pp. 115-16.

From Cape Chelidonia the natural, and substantially shorter, route to *Outremer* ran south to Cyprus at Cape Arnauti and Paphos and then along the south coast to Limassol, from there perhaps around Cape Greco to Famagusta, and from either Limassol or Famagusta across to the coast around Tortosa or Tripoli. This was the route taken by Saewulf, Richard Cœur de Lion, and Frederick II. By the thirteenth century, coast watchers were stationed south of Beirut to report on shipping coming down the coast.[92] However, Paphos's artificial port had decayed during the Byzantine period, the anchorage silted and reduced in size, and its installations suffered great damage in an earthquake in 1159. Only the castle, the *phrourion* or "castellum quod dicitur Baffes" handed over to Richard when he conquered the island, survived. Visiting Paphos in 1211, Wilbrand of Oldenburg described it as a small town inside the Roman remains but completely ignored the port. Only later was a new castle, Saranda Kolonnes, built on a mound behind the anchorage to protect the town. An earthquake in 1222 destroyed Paphos completely, giving the port its final blow as it dried up together with the town's springs. Subsequently the shoreline advanced seawards and Paphos became a coastal town. Waves deposited sand in the shallow water, creating sand bars and endangering ships. *Lo compasso de navegare* warned of these sand bars and recommended anchoring 450 metres east of Moulia Rocks, four kilometres south-east of the ruined Roman eastern mole, where ships would have been sheltered from westerlies although exposed to southerlies. Paphos no longer had a protected anchorage. In 1249 a strong westerly broke the anchor cables of the *nef* of the Latin Empress of Constantinople and drove her ship from Paphos to Acre. The earthquake of 1222 also affected the layout of the former port. Saranda Kolonnes castle now lay inland, lost its strategic importance, and therefore was not reconstructed. It was replaced by the first phase of a harbour castle at the landward end of the ancient Roman western mole. Another fortification was constructed at the seaward end.[93]

---

92 *Breve chronicon de rebus Siculis*, p. 900; *Itinerarium peregrinorum* (Stubbs), II.28, 31 (pp. 180-82, 187-8); Saewulf, *Peregrinatio*, ll. 56-91 (pp. 60-61); Ernoul, *Chronique*, c. XXVIII (pp. 315-16).

93 Joinville, *Vie de saint Louis*, §137 (pp. 68-9); Roger of Hoveden, *Chronica*, vol. 3, p. 111; Wilbrand of Oldenburg, *Wilbrandi de Oldenborg peregrinatio*, c. xxviii, in J. C. M. Laurent, ed., *Peregrinationes medii aevi quatuor: Burchardus de Monte Sion, Ricoldus de Monte Crucis, Odoricus de Foro Julii, Wilbrandus de Oldenborg*, 2nd ed. (Leipzig, 1873), p. 180, as cited in C. D. Cobham, trans., *Excerpta Cypria: materials for a history of Cyprus* (Cambridge, 1908), p. 14; Oliver of Paderborn, *Historia Damiatina*, ed. H. Hoogeweg, *Die Schriften des Kölner Domscholasters, späteren Bischofs von Paderborn und Kardinal-Bischofs von S. Sabina, Oliverus* (Tübingen, 1894), 159-282, §86 (p. 279); Motzo, *Compasso da navigare*, p. 127: "A la dicta Baffa da levante à II isolecte, che so lontano a terro entorno meczo millaro. En la dicta isolecta à bono parago per vento da ver ponente. La quale isola devete onorare da ver levante entorno III prodesi, e de ver la città entorno meczo millaro. En lo capo de la dicta città de Baffa da ver lo ponente, è una secca en mare, entorno

After Richard's conquest of Cyprus and the foundation of Limassol, it became the main port of call and centre for international commerce in Cyprus until destroyed by earthquake in the early fourteenth century. The port lay north-east of Cape Gata, which sheltered it from westerlies, and *Lo compasso de navegare* described it as having a good roadstead and sufficient depth for all ships. However it was only a natural anchorage exposed to winter winds and to southerlies in all seasons. Furthermore, ships had to negotiate shoals along the coast and two of Richard's ships driven to Limassol were wrecked on them. On his way home in 1254 Louis IX also ran aground on them, probably because fog obscured visibility and the sailors thought they were further offshore. In 1271 a Mamluk fleet assaulting Cyprus also ran aground on them. However, having negotiated them successfully, ships could then dock at a wooden quay along the shore which would have been able to disembark horses from *naves* such as those of Richard's and Louis's fleets. Unloaded *naves* may have been beached for repairs and caulking, perhaps on wooded scaffolds as is still done in the Aegean. Alternatively, they could have been raised by barge pontoons such as Breydenbach showed in his woodcut of caulking a *navis* in Rhodes.[94]

The third port of call in Cyprus was Famagusta, whose harbour was sheltered by a line of islets and reefs forming a natural breakwater parallel to the shore. These created two connected ports, access to the inner southern one being only from the outer northern one. The inner port had a wooden quay in front of the city gate and the castle. In the far south there was an arsenal, accessed through a gate in the sea walls, in which small vessels could be constructed, most probably including those ordered by Louis IX in 1249 before departure for Damietta. However, because of the shallow depth of the inner harbour, they could be loaded only in the outer one. The same would have been true of the galleys that carried Richard from Famagusta to Acre in June 1191. The rocky entrance to the outer port was also shallow and only galleys could enter it. In 1232 *salandria* horse transports and small boats, *navilia*, loaded with the horses of John of Ibelin attempting to regain

---

meczo millaro per garbino ver lo ponente." See also Gertwagen, "Maritime activity", esp. pp. 516-20.

[94] *Itinerarium peregrinorum* (Stubbs), II.28, 30, 31 (pp. 182, 184, 188); Roger of Hoveden, *Chronica*, vol. 3, pp. 105-6; Motzo, *Compasso da navigare*, p. 127: "Lo dicto Limesso è bono sorgidore e fondo plano. E ben pod'onqua remanere nave, zoè a ssavere denanti la villa, entorno meczo millaro en mare"; Kretschmer, *Italienischen Portolane*, "Portolan Parma-Magliabecchi", §170 (p. 331); Joinville, *Vie de saint Louis*, §§618-22 (pp. 306-9); Pietro Casola, *Viaggio*, p. 162 for the quay; Bernard of Breydenbach, *Peregrinatio in Terram Sanctam* (Speyer, 1502), p. 15; Ibn al-Furāt, Nāṣir al-Dīn Muḥammad ibn 'Abd al-Raḥīm ibn 'Alī al-Miṣrī al-Ḥanafi, *Tarīkh al-duwal wa 'l mulūk*, trans. U. and M. C. Lyons, *Ayyubids, Mamlukes and Crusaders: selections from the Tārīkh al-Duwal wa'l Mulūk*, 2 vols (Cambridge, 1971), A. H. 669 (vol. 2, p. 152). See also Gertwagen, "Maritime activity", esp. pp. 520-22.

Cyprus from Frederick II's forces, scraped their bottoms here: "... from where one could cross to the land when the waters are lowest on the land side and the sea is calm. No wonder, would anyone have believed that the host could have landed there." The horses having been landed, the rest of the troops were sent at midnight into the port aboard barges or boats or *naviglietti* which could cross the shallows. *Naves*, including horse transports, had to anchor in the outer bay exposed to winds from all directions. Probably for these reasons, Famagusta was not mentioned in *Lo compasso de navegare*. In the fifteenth century, the *Parma-Magliabecchi portolan* advised approaching the northernmost islet of the outer port by sounding and to be on guard while anchoring. The *Portolan of Gratiosus Benincasa* of 1435-45 advised anchoring in a small inlet shaped like a bottle north-west of the outer port which had depths of 14 Venetian feet (4.87 metres) inside and 16 feet (5.56 metres) outside. No measures were taken during the Middle Age to dredge the port, notwithstanding Famagusta's status as the main Cypriot emporium under Genoese rule in the fourteenth and fifteenth centuries. The Genoese did dredge the harbour of Chios and they made efforts from the 1380s to maintain Famagusta as the principal port of Cyprus; however, they did not dredge its port.[95]

After Richard's conquest of Cyprus, Crusading fleets began to sail direct to Cyprus via the Sicilian Channel. Louis IX's forces wintered there and left from there for Egypt in 1249. Seamen could now offer a shorter voyage without interim ports of call and the warnings that their predecessors had made to Roger of Hoveden were no longer of such importance. Even so, Ludolph of Suchem still had reservations as late as 1350 while nevertheless claiming that *naves* could sail direct to *Outremer* via the Sicilian Channel. Noting Cyprus as the first port of call after setting forth, he nevertheless wrote that *naves* might have to take refuge due to "contrary winds, want of food, or some such matter of prime necessity". He named a series of islands from Corsica to Rhodes but only as being islands of which passengers might get a distant sighting.[96] Any islands along the North African

---

95 Kretschmer, *Italienischen Portolane*, "Portolan Parma-Magliabecchi", §170 (p. 331), "Portolan des Gratiosus Benincasa", §124 (p. 399); *Itinerarium peregrinorum* (Stubbs), II.38, 41, 42 (pp. 189, 204-5). See Francesco Amadi, *Chronique*, ed. R. de Mas Latrie, in *Chroniques d'Amadi et de Strambaldi*, 2 vols (Paris, 1891-93), vol. 1, pp. 164-6, esp. p. 166: "... vide un scoglio, ch'è fora del porto de Famagosta, qual ha poccissima acqua in la banda della terra, tal che si po da quello andar in terra benissimo a cavallo, quando maxime le acque sonno sminiute; nel qual scoglio descese 'hoste con gran discommodo, per le roche et grebani che ivi sono, dove no se haria mai creduto che un hoste si havesse messo in quel loco." See also J. Richard, *Chypre sous les Lusignans. Documents chypriotes des archives du Vatican (XIVe et XVe siècles)* (Paris, 1962), pp. 38-9; Gertwagen, "Maritime activity", esp. pp. 522-6.

96 Ludolph of Suchem, *Description of the Holy Land*, §XIII (pp. 20-21). Ludolph himself may have sailed on a merchant galley through the Straits of Messina since he describes the coastal galley route "round the whole of the northern part of the world" and the crossing of the Ionian from Sicily to the Peloponnesos. See §§XIII, XVII (pp. 20, 28).

coast in Muslim possession were dangerous for the Franks, a fact well stressed to Louis IX by his seamen. Of these only Lampedusa was available, since it was depopulated in the mid thirteenth century. Louis IX found a deserted hermitage there and also water and wild rabbits. The king, however, was ready to attack Muslim Pantelleria, where three of his galleys sailed to fetch fruit for his children.[97] A victory would have undoubtedly softened the pain of defeat in Egypt. Taking into account the delay, distress, and losses of the outward voyage on the one hand and, on the other hand, the fact that the King followed the same track for the return voyage in 1254 in spite of his previous experience, it suggests that by this time Crusader leaders aimed to make the most rapid and shortest voyages possible.

97 Joinville, *Vie de saint Louis*, §§638-9 (pp. 316-19), 640-44 (pp. 318-21).

Chapter 7

Provisioning Peter the Hermit: from Cologne to Constantinople, 1096[1]

Charles R. Glasheen

In the lead-up to his famous speech urging help for the Holy Land made at Clermont on 27 November 1095, Pope Urban II may have been encouraged by an itinerant preacher known as Peter the Hermit. Beginning in Berry after Clermont,[2] Peter preached the Crusade and his efforts attracted many followers including Walter Sans-Avoir, a knight of the family of Poissy.[3] Peter came together with his followers at Cologne on 12 April 1096 and shortly after left for Constantinople, the staging place for the war against the infidel. Walter and a small group had departed a few days earlier. The distance was around 2,400 kilometres and, according to Nesbitt, Peter's forces covered it in 104 days.[4] This paper considers how Peter was able to feed his followers on that journey.

Albert of Aachen, who was a contemporary of the First Crusade, described Peter's journey. Although "fired with longing" to make the journey, he himself could not do so because of "various hindrances"; however, he listened to the reports of those who had done so and thus even with his "poor mental powers" could relate what had transpired.[5]

Four issues require consideration. Who, and how many, went? At what time of the year did the journey occur and what route was taken? Responses to these considerations suggest how food was provided.

Urban urged initially that people of all ranks should go. His letter to the Flemish specifically said that he had enjoined "subjects" as well as "lords" to go.[6] Fulcher of Chartres wrote that he envisaged those who were "sad" and "poor", as

---

1 Consult Map 10.

2 Albert of Aachen, *Historia Hierosolymitana*, in RHC *HOcc*, vol. 4, 265-713, I.ii (p. 272). I am grateful to Susan Edgington for allowing me access to her forthcoming translation.

3 Albert of Aachen, *Historia Hierosolymitana*, I.vi (p. 274).

4 J. W. Nesbitt, "The rate of march of Crusading armies in Europe", *Traditio*, 19 (1963), 167-81, p. 181. Nesbitt actually says 103 days but fails to count inclusively.

5 Albert of Aachen, *Historia Hierosolymitana*, I.i (p. 271).

6 H. Hagenmeyer, ed., *Epistulae et chartae ad historiam primi belli sacri spectantes quae supersunt aevo aequales ac genuinae: Die Kreuzzugsbriefe aus den Jahren 1088-1100* (1901; rpt, Hildesheim, 1973), no. 2 (p. 136): "... principes et subditos ...".

well as those who were "joyous and wealthy", participating.[7] Guibert of Nogent noted that "the common people, poor in resources but copious in number, attached themselves to a certain Peter the Hermit."[8] Albert of Aachen observed that a wide variety joined in, mentioning clergy, nobility, and princes, as well as common people: those who were sinful as well as pious, those who were adulterers, murderers, thieves, perjurers, robbers, "that is to say every sort of people of Christian faith, indeed even the female sex."[9] While Albert did not mention specifically those with a military background, he recorded that in the battle at Zemun, Godfrey Burel, a "commander and standard-bearer of two hundred infantry", led the attack.[10] He was assisted in this by Rainald of Broyes, a "distinguished cavalryman".[11] Albert also told of trumpeters and archers, including non-combatants who were slow and at the "tail end" of the army, as well as women, girls, and young boys. There were also two thousand "youngsters of excessive irresponsibility, a wild and undisciplined set of people."[12] There are many views about the make-up of Peter's forces but Albert certainly characterized them as being variegated.

It is difficult to know how many people followed Peter. Albert of Aachen wrote that the army was as innumerable as the sands of the sea,[13] and that when Walter wrote to Alexios Komnēnos asking for aid he said that he and Peter had "thousands of men".[14] The two thousand youngsters of excessive irresponsibility comprised only a small part of the army and when the army was routed no more than a few of its "forty thousand" survived.[15] Anna Komnēnē reported that Peter arrived in Constantinople with eighty thousand infantry and one hundred thousand horsemen.[16] Guibert wrote that Peter's forces were most numerous in number.[17]

---

7 Fulcher of Chartres, *Historia Hierosolymitana (1095-1127)*, ed. H. Hagenmeyer (Heidelberg, 1913), I.iii.7 (p. 137): "... quinimmo hic tristes, hic pauperes, illic autem laeti et locupletes, ...". There is no evidence that Fulcher of Chartres was actually present at Clermont and whether this is what Urban said or whether such a mix of people actually did go is unknown.

8 Guibert of Nogent, *Guibert de Nogent Dei gesta per Francos, et cinq autres textes*, ed. R. B. C. Huygens, in CC *ContMed*, vol. 127A, 76-352, II.viii (p. 121): "... tenue illud quidem substantia sed numero frequentissimum vulgus Petro cuidam Heremitae cohesit eique interim, ..."

9 Albert of Aachen, *Historia Hierosolymitana*, I.ii (p. 272).

10 Ibid., I.vii (p. 277).

11 Loc. cit.

12 Ibid., I.xi (p. 280).

13 Ibid., I.vii (p. 276).

14 Ibid., I.vi. (p. 275).

15 Ibid., I.xii (p. 281).

16 Anna Komnēnē, *Alexiade*, ed. and Fr. trans. B. Leib, 3 vols (Paris, 1937-45), X.v.10 (vol. 2, p. 210). The account of the passage of the Crusaders as far as Nicaea has been identified by Howard-Johnston as having been drafted by Nikēphoros Bryennios. See J.

Modern authors have considered that the medieval use of large numbers is inaccurate, although some have defended Albert's accuracy, and those same authors have themselves made many estimates.[18] It is sufficient for our purpose to assume that Peter's forces were comprised of many thousands, perhaps 15-20,000.[19]

Peter and his army left Cologne on 20 April 1096 and arrived in Constantinople on 1 August. The seasons of travel were thus spring and summer. Of the 104 days' journey, 17 were given over to fighting, resting or preparing food, leaving 87 for travel.[20] The average daily progress on the march was around 28 kilometres.

From Cologne, Peter's forces moved up the Rhine and Neckar rivers and down the Donau to Regensburg. Following the Danube through Linz and Vienna, they entered Hungary at Sopron, crossed the Sava river into the Empire at Belgrade, and moved on to Niš, Sofia, Plovdiv, Edirne, and finally Constantinople.

Food could have been taken with them from Cologne or even before. They could also have purchased it *en route*. They could have pillaged or stolen provisions. And finally, Alexios Komnēnos could have provided for them. In each of the major segments of the trip two or more of these methods were used but one dominated. During the march through Germany food taken was probably the major source. By the time that they reached Hungary this would have been consumed, and most would have had to have been bought. During the stage from Zemun though Belgrade to Sofia, they pillaged. From Sofia to Constantinople Alexios provided food by gift or sale.

Let us first consider individuals or groups taking food with them for the entire trip. It would not have been unusual to require that individuals bring at least some of their own food on a campaign nor that the forces would take collective supplies. However, this method would have had at least three serious limitations. Firstly, it is likely that many participants had neither the necessary access to supplies of

---

Howard-Johnston, "Anna Komnene and the Alexiad", in M. Mullett and D. Smythe, eds, *Alexios I Komnenos. I: Papers* (Belfast, 1996), 260-302, p. 294.

17 Guibert of Nogent, *Dei gesta per Francos*, II.viii (p. 121).

18 Mulinder, A., "Albert of Aachen and the Crusade of 1101", in A. V. Murray, ed., *From Clermont to Jerusalem: the Crusades and Crusader societies 1095-1500. Selected proceedings of the International Medieval Congress, University of Leeds 10-13 July, 1995* (Turnhout, 1998), 69-77.

19 S. Runciman, *A history of the Crusades*, 3 vols (1951-54; rpt, Harmondsworth, 1965), vol. 1, pp. 114, 123, 340. France points out that the Norman army that conquered England in 1066 numbered around 14,000, that Robert Guiscard's army that attacked the Byzantine empire in 1081 had about 15,000 men, and he estimates the People's Crusades at over 20,000. See J. France, *Victory in the East: a military history of the First Crusade* (Cambridge, 1994), pp. 58, 75, 136.

20 Nesbitt, "Rate of march", p. 172.

food, nor the money with which to buy it, to provide their own food for an extended period, nor the knowledge of how to transport it for long distances without spoiling.

Secondly, Peter departed in mid April, which was at the end of the annual food cycle, the time when the local food surplus available for sale or contribution would have been at its lowest.[21] To make matters even worse, recent crops had been poor and this was a time of famine.[22] Cologne is believed to have been a city of approximately twenty thousand in 1096,[23] and we must ask whether its region would have been capable of providing food for an almost equal number of people for three months. In any case, even if food was available, there would have been little time to acquire it since Peter arrived at Cologne only eight days before he left.[24]

Thirdly, the greatest limitation on carrying food is that because of its bulk and difficulty of transportation, armies can not carry sufficient for more than a finite number of days, perhaps as few as ten for large armies and certainly less than 104. Assuming that the basic food for a march of this type was grain that could be made into bread, biscuit, or porridge,[25] a marching man would need at least one and perhaps two kilogrammes of milled grain each day. It is obvious that an individual could not carry one kilogramme of grain per day to cover even the minimal level

---

21 In another context Albert wrote that grain and corn was harvested in July. See Albert of Aachen, *Historia Hierosolymitana*, I.xii (p. 281). Even winter corn was not yet available in April and May in northern Europe. See S. A. Matz, *The chemistry and technology of cereals as food and feed* (Westport, 1959), pp. 5, 17, 19. It is possible that grain stored from earlier crops was still available since Guibert wrote that at the time of Peter's journey the Hungarians had wheat stored from previous years. See Guibert of Nogent, *Dei gesta per Francos*, II.viii (pp. 121-2). Presumably those who went with Peter had some food stored to last until the next harvest and they may well have taken grain they would normally have stored for seed for the next crop.

22 Guibert of Nogent, *Dei gesta per Francos*, II.vi (p. 118).

23 J. C. Russell, *Late Ancient and Medieval Population* [TAPS, 43, no. 3] (Philadelphia, 1958), p. 111.

24 Peter's journey did not begin in Cologne. It is merely that Albert of Aachen picked up the story there. See Albert of Aachen, *Historia Hierosolymitana*, I.vii (p. 276). Participants could have acquired provisions at previous stops. Indeed, they could have made plans prior to arriving in Cologne to have food available on the planned route. E. O. Blake and C. Morris, "A hermit goes to war: Peter and the origins of the First Crusade", in W. J. Shiels, ed., *Monks, hermits and the ascetic tradition* [SCH, vol. 22] (Oxford, 1985), 79–107, pp. 84-5.

25 See Albert of Aachen, *Historia Hierosolymitana*, I.xii (p. 281), where we are told that Peter's forces carried "grain, barley, and meat". The barley can only have been for horses, if we accept the testimony. Cf. *Gesta Francorum et aliorum Hierosolimitanorum*, ed. and trans. R. Hill, *The deeds of the Franks and the other pilgrims to Jerusalem* (Oxford, 1962), I.iv (p. 8) where the provisions found by Bohemond's forces in "Bulgaria" (actually Albania) included grain (*frumentum*), wine, and other food.

for 104 days. A pack animal could carry the load, but it would itself need to be fed. While it might graze grass along the way, for which it would have to compete with many other pack animals, it would need substantial amounts of grain. When the animal's grain was added to the load, the traveller and animal would consume what was carried in a finite period.[26]

In addition to problems of transportation, there was the issue of spoilage. Grain, fresh meat, and vegetables would decay rapidly, thus limiting the period for which provisions could be taken.[27] It was possible to drove live animals on a campaign and to slaughter them as needed; however, driven animals tend to travel more slowly than marching men and Peter's quite rapid rate of march does not suggest that there were substantial numbers of driven animals.[28] Albert of Aachen mentioned herds and flocks on several occasions,[29] but it appears that these were the spoils of pillage *en route*.

Even though Peter could not have carried from Cologne very much of the food needed for the whole trip, it is probable that as much as possible was taken to feed the forces during the German segment of the journey until supplies ran out. That food could have been carried by individuals, pack animals, or four-wheeled wagons. Two-wheeled carts and draft animals would have also been important to carry forward food obtained from sources *en route*. The forces had some such capacity because the food obtained in the battle at Zemun was loaded on carts and wagons; although, they may have been ones acquired there. Nevertheless, Albert

---

26 Discussion of the limitations of the ability of armies to carry food and fodder for men and animals is extensive and complex. I have relied on B. S. Bachrach, "Logistics in pre-Crusade Europe", in J. A. Lynn, ed., *Feeding Mars: logistics in Western warfare from the Middle Ages to the present* (Boulder, 1993), 57-78 [rpt in B. S. Bachrach, *Warfare and military organization in pre-Crusade Europe* (Aldershot, 2002), No. V]; D. W. Engels, *Alexander the Great and the logistics of the Macedonian army* (Berkeley, 1978); J. France, *Victory in the East: a military history of the First Crusade* (Cambridge, 1994); idem, *Western warfare in the age of the Crusades, 1000-1300* (Ithaca, 1999); J. F. Haldon, "The organisation and support of an expeditionary force: manpower and logistics in the Middle Byzantine period", in N. Oikonomides, ed., *Τό εμπόλεμο Βυζάντιο (9ος - 12ος αι.)/Byzantium at war (9th-12th c.)* (Athens, 1997), 111-51; idem, *Warfare, state and society in the Byzantine world 565-1204* (London, 1999).

27 Medieval sources often referred to dried or salted meat and perhaps this could have been taken, although it does not appear that meat was a major source of food. See B. S. Bachrach, "Animals and warfare in early medieval Europe", in *L'uomo di fronte al mondo animale nell'alto medioevo* [SSCISAM, XXXI] (Spoleto, 1985), 707-64, p. 714.

28 Leyser has commented on the slowness of droving herds on military campaigns. See K. Leyser, "Money and supplies on the First Crusade", in idem, *Communications and power in medieval Europe: the Gregorian revolution and beyond*, ed. T. Reuter (London, 1994), 77-95, p. 85.

29 Albert of Aachen, *Historia Hierosolymitana*, I.vii-ix (pp. 277-80).

recorded that there were sufficient wagons that over 2,000 were lost in a later battle.[30]

Purchase *en route* was the main method by which Peter provided for his forces in Hungary and it probably had been significant in Germany and and was then again later in the Empire also.[31] This raises several issues. Local rulers or governors had to give consent and this might be denied if there was fear, for example, that the forces were not under control. Walter and Peter were able to obtain permission from King Coloman of Hungary to purchase food in that country and Albert noted that in Peter's case he "travelled through the kingdom of Hungary peacefully, giving and obtaining everything necessary for their use in quantity, justice, and fair measure."[32] After an initial misunderstanding, Walter asked for and received permission from Alexios to purchase in the Empire, as also did Peter.[33]

Travellers also had to have sufficient acceptable coins or other valuables to buy the food. Albert of Aachen narrated that Peter had a treasure chest "full of countless gold and silver" even after he had passed through Germany and Hungary.[34] Perhaps he raised this money while preaching the Crusade in the months after Clermont, although he is reported to have been quite generous giving money to the poor and to the prostitutes.[35] The army was the beneficiary of gifts of money during the march, for example from the people of Plovdiv and later from

---

30 Ibid., I.xii (p. 281).

31 There is no indication that Peter asked for permission to purchase in Germany although it is likely that he did make purchases there. Louis VII obtained permission to purchase in Germany some fifty years later in 1146. Odo of Deuil, *De profectione Ludovici VII in orientem*, ed. and trans. V. G. Berry (New York, 1948), bk I (p. 10).

32 Albert of Aachen, *Historia Hierosolymitana*, I.vi, I.vii (pp. 274, 276). Cf. II.iii (p. 300), II.vi (p. 303). Louis VII was promised "suitable exchange" by Manuel Komnēnos in 1146. See Odo of Deuil, *De profectione Ludovici VII*, bk II (p. 28). Either gouging or inflated exchange rates could undermine participants' rights to food markets. Bohemond's forces learned about supply and demand when a loaf of bread sold for twenty or thirty pence. See *Gesta Francorum*, II.vii (p. 14).

Guibert of Nogent told a different story as to Peter's peaceful nature, writing that his army took food from the Hungarians by theft and murder, burning granaries and committing other atrocities. See Guibert of Nogent, *Dei Gesta per Francos*, II.viii (p. 122). However, Guibert was vehemently hostile to Peter because he regarded the participation of such forces as his in such an expedition as contrary to the divine ordering of society in Three Orders. It was not the function of such people to participate in such an expedition and their ultimate denouement testified to that. Morris describes Guibert as the "champion of the anti-Peter cause". C. Morris, "Peter the Hermit and the chroniclers", in J. Phillips, ed., *The First Crusade: origins and impact* (Manchester, 1997), 21-34, p. 30.

33 Albert of Aachen, *Historia Hierosolymitana*, I.vi, I.xiii (pp. 275, 282).

34 Ibid., I.xii (p. 281).

35 Guibert of Nogent, *Dei gesta per Francos*, II.viii (p. 121).

Alexios Komnēnos.[36] It has also been argued that many members of the army were people of property and thus may have brought money on the trip to purchase their own victuals.[37]

Finally, for Peter's forces to have purchased food there must have been sellers. Guibert of Nogent wrote that Hungary had produced abundant wheat in the years prior to Peter's passage.[38] When Godfrey of Bouillon passed through Hungary a few months later his followers were able to purchase bread, wine, grain, and barley, as well as farm animals and fowl. Although Albert was not specific it appears that Godfrey traded with a widespread group of sellers.[39] While some purchases may have been from local markets or individual peasants, much must have been from someone with a great deal of grain at hand. One kilogramme of grain per day for 15,000 participants would have been 15 tonnes. Sufficient for the forces to reach the next source, say ten days away, would have been 150 tonnes, without providing anything for animals. In response to this need for a seller, Albert wrote that during the passage through the Empire Alexios Komnēnos ordered cities to sell to Peter.[40] After entering the Empire, Alexios, or those he could control, were the sellers.[41]

Armies *en route* normally attempted to carry sufficient food or money to avoid pillage in friendly territories and all of Peter's journey to Constantinople was through such territories, thus making a considered plan of pillage untenable. In any case, he would certainly have wanted to avoid treating fellow Christians in such a

36 Albert of Aachen, *Historia Hierosolymitana*, I.xiv, xv (pp. 282, 283).

37 Guibert wrote that some people sold "everything for a low number of pennies". Guibert of Nogent, *Dei gesta per Francos*, II.vi (p. 118). See also F. Duncalf, "The Peasants' Crusade", *AHR*, 26 (1920-21), 440-53, p. 440.

38 Guibert of Nogent, *Dei gesta per Francos*, II.viii (pp. 121-2).

39 Albert of Aachen, *Historia Hierosolymitana*, II.vi (p. 303).

40 Ibid., I.xiii (p. 282).

41 On the southern route, after Bohemond's forces had crossed the Vardar at Thessalonikē, Alexios provided them with an imperial household official of some kind, who ordered the "... inhabitants of the land to provide a market for us.". See *Gesta Francorum*, II.v (p. 10).

Laiou describes the situation as: "Alexios, like his successors, established special markets for the provisioning of the Crusaders, and that he, as emperor, had to guarantee that he would do so.". She also says that the "... Komnenoi had, in 1081, great storage spaces for wheat and other victuals.". A. E. Laiou, "Exchange and trade, seventh–twelfth centuries", in idem, ed., *The economic history of Byzantium from the seventh through the fifteenth century*, 3 vols paginated continuously (Washington, 2002), vol. 2, 681-770, p. 741. Magdalino suggests that Alexios had something of a monopoly in the trading of grain. See P. Magdalino, "The grain supply of Constantinople, ninth–twelfth centuries", in C. Mango and G. Dagron, eds, *Constantinople and its hinterland: papers from the Twenty-seventh Spring Symposium of Byzantine studies, Oxford, April, 1993* (Aldershot, 1995), 35-47, p. 44.

manner.[42] Additionally, from a logistical point of view, pillage was not a particularly efficient way to obtain to feed an army. Food to be stolen had to be located, seized from often militant possessors, and transported back. Pillage might occur when purchase was not possible because of lack of money or sellers or because the ruler of an area denied the right to buy, but gratuitous pillaging in friendly territories would have been inefficient and therefore uncommon.

In assessing the availability of food when Peter entered the Empire, the initial question is whether Alexios knew that the Crusaders were coming and if so whether he had made provision for them. The early thirteenth-century French paraphrase of William of Tyre known as the *Eracles* reported that Bohemond sent a message to Alexios announcing his coming and the late thirteenth-century *Estoire de Jérusalem et d'Antioche* recorded that he sent a delegation to him to ask for safe conduct.[43] Anna Komnēnē, or Nikēphoros Bryennios, wrote that Alexios was heavily involved in another project when he heard a rumour of the approach of the Frankish armies.[44] He was an experienced military campaigner who would have known of the food needs of such large forces and he issued orders "to receive courteously those who had crossed and to provide large stockpiles along their route brought from all regions." [45] Alexios certainly had motives other than simple good will for assisting in this way. He may not have wanted to lose face with those coming and he certainly would have wished to avoid destruction caused by pillage.[46] On the southern route, along the course of the *Via Egnatia*, he anticipated

---

42 When approaching Zemun Peter refused to believe that he would be attacked since the other parties were "fellow Christians". Albert of Aachen, *Historia Hierosolymitana*, I.vii (p. 276). Alexios Komnēnos agreed to rights of purchase for Peter "... because you are Christians and your fellow Christians should not impede your journey beyond.". Albert of Aachen, *Historia Hierosolymitana*, I.xiii (p. 282). When he began his journey through the Balkans, Bohemond warned against plundering lands that belonged to Christians. *Gesta Francorum*, I.iv (p. 8).

43 Anonymous, *L'estoire de Eracles empereur et la conqueste de la terre d'Outremer*, in RHC *HOcc*, vols 1.1, 1.2, 2, II.xi (vol. 1.1, p. 87); Anonymous, *Li estoire de Jérusalem et d'Antioche*, in RHC *HOcc*, vol. 5, 621-48, pp. 627-8. As it stands, the record of the delegation is not really credible, but it may well have reflected a popular memory of such delegations.

44 Anna Komnēnē, *Alexiade*, X.v.4 (vol. 2, pp. 206-7). Thomas has claimed that it was "... quite inconceivable that the Pope would not have informed Alexios of the arrival of thousands of armed pilgrims in Byzantine territory.". See R. D. Thomas, "Anna Comnena's Account of the First Crusade", *BMGS*, 15 (1991), 269-312, p. 274. France says that Anna was "mendacious" in denying Alexios' knowledge, although perhaps he did not know of the early arrival of Peter. See France, *Victory in the East*, p. 97.

45 Anna Komnēnē, *Alexiade*, X.v.9 (vol. 2, p. 209). See also X.x.3 (vol. 2, p. 228).

46 Ibid., X.v.9, X.x.3 (vol. 2, pp. 209, 228). That Alexios was fully aware of the dangers of pillage was demonstrated when he finally offered aid to Peter and restricted his forces' stay in any town to three days. See Albert of Aachen, *Historia Hierosolymitana*, I.xiii (p. 282).

pillage and directed his commanders to watch for raids or plunder and to check them by "appropriate skirmishes".[47]

Notwithstanding Alexios's intentions, provisions were not always provided to the armies arriving in the south and pillage did ensue. According to the *Gesta Francorum*, when Bohemond arrived at Kastoria the inhabitants refused to sell to him because they were afraid of his forces. As a result his army "seized oxen, horses, asses and everything which we found."[48] According to Raymond of Aguilers, Raymond of St Gilles expected that once his army reached the Empire Alexios would provide assistance but in fact the army was attacked when it arrived at Dyrrachion and its property stolen. While Raymond initially passed up a chance for vengeance, his army seized the city of Roussa after some other dispute with its inhabitants.[49]

Peter's forces may have encountered even more difficulties than others did since it is probable that Alexios would not have anticipated the time of his arrival. He reached the Empire even before the date of 15 August set for departure from the West which Urban had laid down. Even if that date was unknown in the East, Peter and his forces had departed much earlier in the year than would have been expected of a normal military expedition and had come by the northern route rather than the more common southern one. Whatever Alexios may have known about the schedule of the other arriving armies, anticipating the arrival of Peter would have been most difficult.

Walter Sans-Avoir had arrived in Zemun, across the Sava river from Belgrade and the last city in Hungary, several days ahead of Peter. According to Albert of Aachen, while his main force crossed over to Belgrade, 16 men stayed behind to buy arms. Attacked by the Hungarians, their arms, clothes, and gold and silver were taken from them and Walter was denied permission to buy food at Belgrade because the authorities feared his forces spreading out and committing crimes. His followers then took herds of cattle and sheep and a battle ensued in which 60 of them were burned alive. Moving on to Niš, Walter again made his case and here they were given arms and money and permission to buy. As he marched to Constantinople he sent ahead requesting time to rest and permission to buy food there.[50]

Peter reached Zemun a few days after Walter had. Again according to Albert, Peter then learned that the authorities in Zemun and Belgrade had conspired to attack him, although he refused to believe it because they were Christians.

---

47 Anna Komnēnē, *Alexiade*, X.v.9 (vol. 2, p. 209).

48 *Gesta Francorum*, I.iii (p. 8). In fact, more probably they simply had no spare food to sell because it was the depths of winter.

49 Raymond of Aguilers, *Historia Francorum qui ceperunt Iherusalem*, in RHC *HOcc*, vol. 3, 231-309, §§I, II (pp. 236, 237); idem, *Liber Raimundi de Aguilers*, ed. J. H. and L. L. Hill (Paris, 1969), §§1, 2 (pp. 38, 39).

50 Albert of Aachen, *Historia Hierosolymitana*, I.vi (p. 275).

However, when his forces saw the arms and spoils of Walter's men hanging on Zemun's walls, restraint disappeared and they attacked the town. Four thousand Hungarians were supposedly killed, against 100 of Peter's men. They stayed for five days in the fortress of Zemun because of the abundance of provisions which they found there in grain, flocks of sheep, cattle, plenty of drink, and an infinite number of horses.[51]

The governor of Belgrade, an imperial Duke by the name of Nikētas, seeing the devastation in Zemun retired from Belgrade to Niš. He sent the inhabitants into the mountains with their herds until reinforcements could arrive from Constantinople. Six days later Peter received word that the Hungarians sought revenge and crossed the Sava, taking with him the provisions gathered at Zemun. After having pillaged Belgrade, he moved on towards Niš, carrying the pillaged provisions on carts into the forests.[52]

At Niš Peter asked Nikētas for permission to buy food. This was granted subject to hostages being required. Unfortunately, after they had begun to move out from Niš, some unruly members of Peter's army began a fracas and in the end his forces were routed and widely dispersed.[53] To make matters even worse, the provisions obtained at Zemun and Belgrade were lost, together with Peter's treasure chest.[54] Their desperate plight was relieved only when they reached Bela Palanka, which had been abandoned, where they harvested the ripe crops to sustain them a little longer.[55]

During the 104 days of the journey, in addition to what the forces had brought with them or could buy, pillage provided an important source of food for over a month from the arrival at Zemun on 5 June to that at Sofia on 7 July.[56] During this time there was no right of purchase and much had been lost when the army was routed after leaving Niš. Even if the provisions had been retained, Peter had lost his 2,000 carts and in any event it would have been unlikely that the roads could have borne traffic that heavy.[57] It is difficult to know whether Peter's skirmishes began for other reasons and provisions then followed naturally or whether they occurred because the food had run out. France argues that it was the "... inability of

---

[51] Ibid., I.vii (p. 277).

[52] Ibid., I.viii (pp. 277-8).

[53] Ibid., I.ix-xii (pp. 278-81).

[54] Ibid., I.xii (p. 281).

[55] Loc. cit.

[56] Nesbitt, "Rate of march", p. 181. According to Nesbitt, Peter arrived at Zemun on 5 June and approached Sofia on 7 July.

[57] K. Belke, "Roads and travel in Macedonia and Thrace in the middle and late Byzantine period", in R. Macrides, ed., *Travel in the Byzantine world: papers from the Thirty-fourth Spring Symposium of Byzantine studies, Birmingham, April 2000* (Aldershot, 2002), 73-90, esp. pp. 79, 80-82, 90.

the Byzantines to procure supplies which caused the real trouble."[58] His followers were suffering from lack of food and Peter must have been gravely worried about how the journey could be completed.

As he approached Sofia he received a message from Alexios Komnēnos reproaching him for the discord he had caused in Zemun and thereafter. However, Albert of Aachen reported that the message also said that since Peter and his companions were Christians the Byzantines would not stand in their way. To facilitate their passage, the emperor ordered all cities to henceforth sell provisions to the army. Peter wept with joy.[59] This period in which purchase was permitted by Alexios was to continue for 25 days until the pilgrims reached Constantinople on 1 August.[60]

While the language attributed to the emperor was in terms of permitting sales to Peter's forces, it may have been that some of the food acquisitions were more in the nature of gifts than the arm's-length puchases by Peter in Hungary. By this time Peter had lost his treasure chest and we know that at least some of his forces were without money.[61] The people of Plovdiv are known to have made gifts to Peter out of pity,[62] and when he arrived in Constantinople Alexios made a gift of two hundred gold bezants, some of it in small coins which could have been distributed for the purchase of food.[63] Had Peter's forces been the beneficiary of gifts it would have been consistent with the way the emperor treated Godfrey of Bouillon, who followed Peter down the northern route, arriving at Constantinople on 23 December. An imperial envoy met him on the frontier and granted permission to trade and purchase supplies, subject to good behaviour. Albert reported that at Niš Godfrey was given a great abundance of grain, barley, wine and olive oil, and that these gifts continued.[64]

In the end there is no evidence that Peter planned in any significant way the monumental task of feeding 15,000 people in foreign lands for 104 days. Indeed advance logistics would have been very difficult because of the distances involved and lack of knowledge of who would join the forces and when. But sufficient food was critical. Without it the forces would have withered. Might, money, and sympathy were Peter's allies. Perhaps good leadership along the way, either by Peter himself or by others, turned the tide.

---

58 France, *Victory in the East*, p. 91.

59 Albert of Aachen, *Historia Hierosolymitana*, I.xiii (p. 282).

60 Nesbitt, "Rate of march", p. 181.

61 Albert of Aachen, *Historia Hierosolymitana*, I.ix (p. 278).

62 Ibid., I.xiv (p. 282).

63 Ibid., I.xv (p. 283). Even though 200 Byzantine gold *hyperpyra*, bezants as they were known in the West, were a great deal of money, they would still not have fed forces numbering 15,000 people for very long.

64 Ibid., II.vii (p. 304).

Chapter 8

Roads and communications in the Byzantine Empire: wagons, horses, and supplies[1]

John Haldon

The need for medieval states or polities to collect and distribute resources to maintain armies affected all aspects of political organization and processes, determining fiscal systems and methods of extracting surplus. It impacted directly upon political structures as well as upon the structures and memberships of political elites and, according to region and period, regional cultures and social life. But logistical infrastructures which evolved in different parts of the medieval world from the later Roman empire until the twelfth century are subjects which remain almost entirely neglected: road-systems, communications in general, and the supportive capacity of lands in which armies were based or through which they passed; issues of relative availability of resources, including relationships between sizes of populations, levels of agrarian and other production, and the size of armies and their supporting structures within and between medieval societies. Virtually no research has been carried out to place the social logistics of the medieval Byzantine and Islamic worlds within the broader comparative context which they clearly need if the directions and tendencies of their different development is to be understood properly. Although several of these aspects have occasionally been subject to scholarly attention, they have also been the subject of unproven assumptions rarely tested against evidence.

Studies devoted to such questions have often been flawed. A number in which conclusions reached are at the least questionable have appeared on various aspects of one or other of the themes above. This applies particularly to issues of the size of early medieval armies and their logistical support, where the latter is taken into account at all, and the organizational inheritance of the late Roman world. Constant argument over the same sparse references in a few medieval texts advances us no farther in efforts to understand such matters, and results produced by even the most competent and wide-ranging scholars generally remain almost entirely subjective, unfalsifiable, and non-testable. Determining the parameters within which the agricultural output of a region at a given time can be quantified, together with the producing and surplus population it could support, are essential

---

1 Consult Maps 1, 3, 8, 11.

elements in attempts to resolve this and related issues. To begin to address such issues we need to place them within a framework set by broader issues of land-use, climatic and seasonal changes, productivity of land and its supportive potential and capacity, and demography.

Armies are, by definition, amongst the simplest of human social groups. As agents, they are deterministic and demonstrate highly directed behaviour. Any model of their impact essentially revolves around motion and provisioning since they move within parameters dictated by military goals, usually by moving from one offensive action to another in conditions of relative safety. Except under duress their progress is a balanced procedure offsetting required speed and acknowledging the necessity of maintaining troops in battle-ready physical condition, fed as adequately as possible. Armies therefore are goal seeking and seek to be self-sustaining in an efficient manner. In this, they sidestep many problems associated with the mathematical modelling of complex human societies and approximate more closely the behaviour of predators or, perhaps, meta-entities such as ant colonies. Understanding the logistics of medieval warfare therefore requires a methodology entirely different from that traditionally employed. Until this is available, however, discussion of Byzantine logistics will of necessity have to rely on limited textual and archaeological evidence.

The subject of Byzantine logistics has attracted relatively little attention, which is somewhat surprising since there is some quite useful material to draw upon from the late Roman era through to the Crusades. In addition, most Byzantine historians have shown little interest in the subject of military infrastructures in general if we leave aside topics such as the so-called theme system, which was neither thematic nor a system. Yet acquaintance with middle Byzantine and contemporary Islamic sources suggests that warfare in the Middle East from the sixth-seventh centuries to the eleventh at least was very different in scale from the earlier Roman period and the later Ottoman period, and it is worth asking why this should be the case. Is it merely a reflection of lack of sources? Is it a reflection of the tendentiousness of some sources or ignorance of military matters on the part of the writers of many narrative sources? Or is it a combination of these and other factors?

If we survey the history of military campaigning across Asia Minor from the Roman conquest of the various Hellenistic kingdoms and principalities up to the wars between the Ottomans and Ṣafavids in the sixteenth century, one feature in particular strikes us: the size of the armies employed by the different states across this period varied very considerably. Armies averaging 20-30,000 plus supporting elements were not unusual at the height of the Republic and during the Principate and early Dominate but declined fairly quickly from the sixth century to reach a low in the seventh to later tenth centuries, when Byzantine and Arab armies numbered from 4-12,000, and exceptionally as many as 25-30,000 for large-scale and carefully-planned expeditionary forces.[2] Thereafter they increased very gradually through to the twelfth century until, with a temporary collapse in the fourteenth century resulting from pandemics, numbers similar to those found in the legionary forces of the first and second centuries B.C.E. and C.E. were found once again in European and Ottoman armies of the pre-Reformation and Reformation periods, the era of the so-called "gunpowder empires", when regular campaigning armies of over 30,000 became not unusual.[3] We should differentiate between the theoretical standing manpower available, those nominally registered and fit for active service in the state or society in question, and the actual size of forces that could be put into the field at any given moment and, more importantly, maintained for the duration of a campaign. However, our sources do not always enable us to do this with any degree of confidence. I am concerned here with the range applicable to the latter.

Evidence for fluctuation in the size of armies is equivocal, of course, and we do not have clear and reliable figures. Nevertheless, examining the logistical arrangements of the middle Byzantine Empire and comparing them to those of both the Caliphate and the Empire's western neighbours, there is a relatively clear downward trend until the twelfth century which appears to accord with the demography and economy of the western edge of the Eurasian world: Europe and the Near and Middle East. We should not use Byzantium as a generalizable example, however, since recovery from the demographic slump after the late Roman period appears to have been a little slower in the Empire than in either the medieval West or the Muslim world, although again the issue is debated.

This trend in the size of field armies seems to reflect fairly closely the curve in demographic change across the same period: a population of approximately 67-70 million for Roman Europe around 200 C.E., falling to around 27-30 million in the early eighth century, and rising again to some 73 million by 1300, with a particularly noticeable rise around 1200. All the evidence suggests a similar curve

---

2 See H. Kennedy, *The armies of the Caliphs: military and society in the early Islamic state* (London, 2001), esp. pp. xiii, 97-9, 126-8; J. Haldon, *Warfare, state and society in the Byzantine world 565-1204* (London, 1999), pp. 99-106.

There are some known exceptions, such as the so-called first Cretan expedition to northern Syria in 910-911 in which over 40,000 personnel were involved including all oarsmen and soldiers. The logistics of this remain problematic, however, because it was seaborne but relied heavily on foraging after landing. Moreover, not enough is yet understood about the water-supply of Byzantine vessels or the transportation of horses. See J. F. Haldon, "Theory and practice in tenth-century military administration: chapters II, 44 and 45 of the *Book of Ceremonies*", *Travaux et Mémoires*, 13 (2000), 201-352.

3 See R. Murphey, *Ottoman warfare, 1500-1700* (London 1999), pp. 5-6, 9-10, 35-49 and for late Roman and early medieval warfare and military organization the review article by J.-M. Carrié and S. Janniard, "L'armée romaine tardive dans quelques travaux récents, 1: l'institution militaire et les modes de combat", *L'antiquité tardive*, 8 (2000), 321-41.

in the Muslim world. The catastrophic slump of the mid fourteenth century, which saw the population of Europe drop to somewhere in the region of 45 million, was made good within a century.[4] While these figures are necessarily crude approximations because of the nature of the sources and problems of interpretation, and while one can point to a number of exceptions, a differential rate of change from east to west, and important regional and local variations, they are generally agreed at least in their broad outlines. As population increased hand in hand with the extension or intensification of agrarian production, more resources available for military consumption became available in regions affected by such changes, in theory at least. Since the capacity of a given area to support an additional transient population such as soldiers and their support train for any length of time depended on its demography and rate of agrarian production, all logistical systems were designed to extract both maximum resources from such areas as well as to facilitate movement of supplies to them. However, as human populations declined and land exploited for arable production declined proportionately, so the number of livestock tended to expand as more grazing became available. This may have impacted upon both availability of pack-animals as well as a tendency to employ animals rather than wheeled vehicles.

The size of armies was thus directly proportional to the sum of many factors which could vary dramatically from campaign to campaign: manpower resources; agrarian production, dependent in turn upon season, region, climate and types of crops grown; types of livestock employed for transportation; population density of areas in which campaigning took place; roads and other aspects of communications infrastructure; and so on. There were so many variables and our data for the ancient and medieval periods is so fragmentary that providing satisfactory and generally-agreed answers is always contentious, as is attested to by the great variety in numbers proposed by historians for the size of ancient and medieval armies. For some periods differences can be minimized when we have good information about the numbers of corps and their size. For example, when a source narrated that three legions with attached auxiliaries were operating in south-eastern Britain in 43 and 44 C.E., we have a reasonable chance of estimating total numbers

---

4 For general trends throughout this period, see F. Braudel, *The Mediterranean and the Mediterranean world in the age of Philip II*, trans. S. Reynolds, 2 vols (London, 1972), vol. 1, p. 394 ff.; J. C. Russell, *Late ancient and medieval population* [TAPS, 43, no. 3] (Philadelphia, 1958); B. H. Slicher van Bath, *The agrarian history of western Europe, A.D. 500-1850*, trans. O. Ordish (London 1963); and with a summary of more recent evidence, M. McCormick, *Origins of the European economy: communications and commerce, A.D. 300–900* (Cambridge, 2001), pp. 31-6. For the impact of these changes on Byzantium see A. Harvey, *Economic expansion in the Byzantine empire, 900-1200* (Cambridge 1989), pp. 244-63, and for the situation in Ottoman Asia Minor from the fourteenth century, H. Inalcık, ed. with D. Quataert, *An economic and social history of the Ottoman empire, 1300-1914* (Cambridge 1994), pp. 25-32.

involved and total amounts of provisions required, even if some uncertainty concerning support units, numbers of servants and slaves, and so on, remains.[5]

Information from all periods is not necessarily comparable, however. In the fourth to seventh centuries a species of hard wheat which brought certain advantages appears to have been introduced and spread throughout the eastern provinces of the Empire. Hard wheats have a protein content of some 11%-15%, compared with the 8%-10% of soft wheats, and produce better flour for bread.[6] This means that the volume and weight of the bread ration for Roman armies can be compared to that of later Byzantine or Western medieval armies, or indeed to modern handbook statistics on animal feed, only with difficulty.

By the same token, there are cultural assumptions about factors such as the weight-bearing ability of mules, which affected the ways in which such animals were actually employed. Roth offers a set of persuasive statistics about capabilities of animals based on nineteenth-century United States army manuals; however, it is not clear whether the mules on which he based his calculations were the specially-bred large "Missouri" mules, or the traditional smaller European/Asian mules. There was a real difference in their load-bearing abilities.[7] We must treat physiological evidence about loads which animals could carry derived from more modern sources, such as nineteenth-century military handbooks, with caution. Here I want to sketch the physical and logistical framework for understanding the Byzantine middle period, drawing in particular on Byzantine sources, and to offer some thoughts on the possibilities open to polities and military commanders for achieving military ends.

One of the recognized achievements of Roman armies from the first century B.C.E. to the late second century C.E. was the construction of a network of major arterial roads suitable for rapid movement of men and materials from inner provinces to frontiers and connecting these provinces laterally to one another and to major political centres. In large part, this road system was what made Roman armies so efficient in their response to external threats and their use of resources. During the later third century this system was further expanded in certain frontier districts, linking military bases and forts to sources of supply and facilitating the movement of both small and large bodies of troops to meet external threats. The

---

5 See, for example, the discussion in J. P. Roth, *The logistics of the Roman army at war (264 B.C.-A.D. 235)* (Leiden, 1999), pp. 16-67.

6 See A. M. Watson, *Agricultural innovation in the early Islamic world: the diffusion of crops and farming techniques, 700-1100* (Cambridge 1983), p. 20. It is not clear that Roth has taken this into account when calculating average feeds and loads for animals in his otherwise excellent book on Roman army logistics.

7 Roth, *Logistics of the Roman army*, pp. 62 ff., 206-7.

network associated with the *strata Diocletiana* in Arabia, for example, has received a great deal of attention from historians.[8]

From the later fourth to the seventh century, for reasons which remain unclear, a gradual decline in the standard of many if not most major public roads set in. In terms of maintenance and upkeep this seems to have reflected in part a shift in priorities in the allocation of resources and an unwillingness of provincial cities to devote the necessary time and money to such matters except on a highly localized basis. In part, it must be associated with transformations taking place in urbanism. Longer-term changes, exacerbated by constant devastation and raiding in the Balkans from the later sixth century and in Anatolia from the middle of the seventh, led to the near total collapse of the late Roman urban network. Since local municipalities had borne the chief responsibility for maintaining roads in their administrative territories, the implications of this for road maintenance were considerable.[9] In the *Codex Theodosianus*, laws of the later fourth and early fifth centuries lamented the poor state of many roads.[10] The fifth-century historian Malchos of Philadelphia noted that western sections of the *Via Egnatia*, the major route from Constantinople to the Adriatic, were in such a state of disrepair that travellers could barely pass along it and in the sixth century Prokopios described one section as almost impassable in wet weather.[11] In the last years of the sixth century general Komentiolos supposedly had to rely on an aged local man to find the *Via militaris* leading south from the Danube plain through Trajan's Gate. But the route was known and regularly used throughout the Byzantine period and the *De administrando imperio* noted that it took an individual, namely a single man such as an imperial courier travelling at speed and with ability to change horses, eight days to travel along it from Thessalonikē to Belgrade including pauses.[12]

---

8 For the *strata Diocletiana* and its strategic significance see J. Eadie, "The transformation of the eastern frontier, 260-305", in R. W. Mathisen and H. S. Sivan, eds, *Shifting frontiers in late antiquity* (Aldershot, 1996), 72-82.

9 See J. F. Haldon, *Byzantium in the seventh century: the transformation of a culture*, 2nd ed. rev. (Cambridge 1997), pp. 92-124.

10 For the general state of the roads by the late fourth and fifth centuries, see CTh, XV.3.4 (C.E. 399), which refers to "the immense ruin of the highways" throughout the prefecture of the East.

11 Malchos of Philadelphia, *Byzantiaka*, ed. and trans. R. C. Blockley, *The fragmentary classicising historians of the Later Roman Empire: Eunapius, Olympiodorus, Priscus and Malchus. Vol. 2: text, translation and historiographical notes* (Liverpool, 1983), vol. 2, 402-70, §20 (p. 440); Prokopios of Caesarea, *Buildings*, in H. B. Dewing, ed. and trans., *Procopius*, 7 vols (London, 1914-40), vol. 7, IV.viii.5 (p. 284).

12 Constantine VII Porphyrogennētos, *De administrando imperio*, vol. 1, ed. G. Moravcsik, trans. R. J. H. Jenkins (Budapest, 1949); R. H. Jenkins, *Volume II: commentary* (London, 1962), §42.15-18 (vol. 1, p. 182). For Komentiolos see Theophylaktos Simokattēs, *Theophylacti Simocattae historiae*, ed. P. Wirth (Stuttgart, 1972), VIII.4.3-8 (pp. 289-90).

Even if the story of Komentiolos has been exaggerated, it is indicative of what contemporaries thought of the road system in the region.

Only around a few major cities, chiefly Constantinople, is there evidence for road maintenance or repairs undertaken by the state, and much of this is associated with Justinian I's building schemes in the mid sixth century. Prokopios reported that efforts were made to resurface short stretches of road along the *Via Egnatia* between Constantinople and Rhēgion, along the main highway from Bithynia to Phrygia, and on the road from Antioch northwards across the mountains into Cilicia, which he noted was a dangerous route at the best of times. He also credited Justinian with repairing or building several bridges, across the Sangarios and Drakōn rivers in Bithynia, and the Siberis in Galatia.[13] An inscription from Sofia dated to 580 recorded repairs to an aqueduct made under a *kandidatos* Julianus and a number of such local inscriptions for the fifth and sixth centuries from across the Empire show that work of this sort was carried out fairly regularly. This was particularly the case for Constantinople, of course, where emperors frequently committed substantial resources to upkeep of defensive structures, cisterns, aqueducts and so forth.[14] After the age of Justinian I, however, there is no evidence for any central direction of road-building or maintenance in the provinces except on a purely ad hoc basis, even in laudatory accounts of building programmes of emperors such as Basil I in the second half of the ninth century.

Roads were certainly maintained, however, and some bridges survived well into the middle Byzantine period or were newly constructed. Several historians reported the existence of a bridge at Zompos or Zompē over the Sangarios in the eleventh and early twelfth centuries.[15] Late Roman and Byzantine sources make it

---

The differential between the speed of an individual and a body of soldiers is clear from the remark of Theophylaktos Simokattēs that a somewhat shorter journey from Drizipera in Thrace to Dorostolon on the Danube could take a field army 20 days. Theophylaktos Simokattēs, *Historiae*, VI.6.5 (pp. 230-31).

13 Prokopios, *Buildings*, IV.viii. 4-9 (pp. 284-6); V.ii.12-13 (p. 324); iii.4-6, 8-10, 12-15 (pp. 324-8); iv.1-4 (p. 330); v.1-7 (pp. 334-6). There are a number of problems with Prokopios's account, and his reports should not always be taken at face value. See M. Whitby, "Justinian's bridge over the Sangarius and the date of Procopius' *De Aedificiis*", *Journal of Hellenic studies*, 105 (1985), 129-48; G. Greatrex, "The dates of Procopius' works", *BMGS*, 18 (1994), 101-114.

14 V. Beševliev, *Spätgriechische und spätlateinische Inschriften aus Bulgarien* [BBA, 30] (Berlin, 1964), p. 2, no. 3. For Constantinople see C. Mango, "The water supply of Constantinople", in idem and G. Dagron, eds, *Constantinople and its hinterland: papers from the Twenty-seventh Spring Symposium of Byzantine studies, Oxford, April, 1993* (Aldershot, 1995), 9-18.

15 Michael Attaleiates, *Michaelis Attaliotae Historia*, eds W. Brunet de Presle and I. Bekker [CSHB, vol. 34] (Bonn, 1853), p. 145, l. 20; Nikēphoros Bryennios, *Nicephori Bryenni Historiarum libri quattuor*, ed. and Fr. trans. P. Gautier [CFHB, vol. 9] (Brussels,

clear that road and bridge maintenance was achieved through compulsory duties imposed upon communities by local military or provincial authorities and sanctioned by the central government. Obligations such as *viarum et pontium sollicitudo*, which was known certainly from the ninth century and probably earlier by a variety of terms such as *odostrōsia* or *gephyrōsis*, occur in various sources from the late Roman period up to the twelfth century and beyond.[16] But the results were patchy in the extreme. And while military treatises regularly included sections on bridging rivers by constructing pontoons, a tactic referred to also in the histories and chronicles, such constructions were, of course, very temporary.[17]

Not all roads were of the same standard, nor were they constructed for the same purposes. Byzantine sources often differentiated between wide roads and narrow roads or paths, between paved and unpaved roads, and between roads suitable for wagons or wheeled vehicles and others. The anonymous tenth-century treatise on skirmishing known as *De velitatione* distinguished clearly between "public roads", maintained at least irregularly by the local administration through compulsory services imposed on local communities, and paths and tracks of a humbler nature.[18] Prokopios noted with pride that the repaved sections of road between the capital and Rhēgion were wide enough to permit two wagons to pass each other.[19] Roads that were strategically important to the state may in general have been maintained more regularly. Yet the majority of these routes, even major arterial roads, were often mere tracks and, even where formerly well-paved, had generally decayed substantially by the eighth and ninth centuries. In 877 Basil I led his army through mountain passes on foot, so narrow and inaccessible was the track he followed. The coastal road from Lopadion to Edremit taken by most of the French army during the second Crusade was so decayed and overgrown that many

---

1975), II.14 (p. 169, l. 8); Anna Komnēnē, *Alexiade*, ed. and Fr. trans. B. Leib, 3 vols (Paris, 1937-45), XV.4 (vol. 3, p. 201).

16 See the imperial prescriptions incorporated into the *Codex Theodosianus*, repeated or replaced in sixth-century legislation of Justianian, and repeated again in the early tenth-century *Basilika*. CTh, XI.16.15 (p. 601); 16.18 (pp. 602-3); XV.3.6 (p. 818); CJ, I.2.5 (p. 12); JN, 131.5 (p. 656); *Basilicorum libri LX*, ser. A, eds H. J. Scheltema et al., 8 vols (Groningen, 1953-88), V.1.4, V.3.6 (vol. 1, pp. 125, 142). For local military authorities responsible see Leo VI, *Leonis imperatoris tactica*, ed. J. Meursius, in PG, vol. 107, coll. 669-1094, XX.71 (coll. 1031-2).

17 The emperor Herakleios is reported to have employed a large pontoon bridge to cross the Bosporos on horseback in 638. See Nikēphoros I, Patriarch, *Ἱστορία σύντομος*, ed. and trans. C. Mango, *Short history* [CFHB, vol. 13] (Washington, 1990), §25 (pp. 73-5). Anna Komnēnē described a substantial pontoon construction in an account of one of her father's campaigns. See *Alexiade*, VIII.4.4 (vol. 2, p. 137).

18 See the commentary of Dagron and Mihăescu to *De velitatione*, eds and Fr. trans. G. Dagron and H. Mihăescu, *Le traité sur la guérilla (De velitatione) de l'empereur Nicéphore Phocas (963-969)* (Paris 1986), p. 219.

19 Prokopios, *Buildings*, IV.viii.4-9 (p. 284).

troops wandered off it and became lost. By contrast, a smaller contingent followed the shorter, broader, and more accessible, but less well-supplied route over the Mysian hill-country to Edremit, illustrating the variations between routes.[20]

Seasonal weather variations clearly affected unpaved or deteriorated roads and tracks more dramatically than they did properly paved roads and therefore the flexibility and mobility of large Byzantine forces must necessarily have been affected adversely by comparison to earlier Roman armies, which had had a much better strategic road network at their disposal. During the summer wheeled vehicles could have moved relatively easily, even if not very rapidly, along broader tracks, even in fairly hilly regions, as the evidence for Manuel I's expedition which ended at Myriokephalon in 1176 demonstrates. However, during the winter and in rain the reverse would have been the case and this would have made winter campaigns less viable on the one hand and but potentially more effective on the other since they were so unusual. Michael Psellos remarked specifically on Basil II's refusal to be bound by the seasons, to which his military success was partly attributed, and the military treatises generally included some words of advice on campaigning seasons and what precautions to take at particular times of the year.[21] A need for reliable scouts with local knowledge of roads, passes, and river crossings was constantly emphasized in both military treatises and historians' accounts, and this highlights the uncertainty and difficulty attendant upon any military undertaking in a context in which both major routes and local tracks were generally of such poor quality.[22]

---

20 *Theophanes continuatus*, in I. Bekker, ed., *Theophanes continuatus, Ioannes Cameniata, Symeon magister, Georgius monachus*, 3 vols [CSHB] (Bonn, 1838), vol. 1, *De Basilio Macedone*.48 (p. 280, ll. 13-14). See J. G. C. Anderson, "The campaign of Basil I against the Paulicians in 872 A.D.", *Classical review*, 10 (1896), 136-40. For the Second Crusade see Odo of Deuil, *De profectione Ludovici VII in orientem*, ed. and trans. V. G. Berry (New York, 1948), bk 6 (pp. 102-6).

21 For Basil II see Michael Psellos, *Chronographie ou histoire d'un siècle de Byzance (976-1077)*, ed. and Fr. trans. E. Renauld, 2 vols (Paris, 1926-28), I.xxxii (vol. 1, p. 20). For advice on campaigning seasons, Nikēphoros Ouranos, *Taktika*, trans. E. McGeer, *Sowing the dragon's teeth: Byzantine warfare in the tenth century* (Washington, 1995), 89-163, §63.1 (p. 142) and see McGeer's comments, pp. 255-6. Arab geographers in the ninth and tenth centuries recommended staying only 20 days in Roman territory during winter raids because of the lack of forage. See Qudāma ibn Ja'far al-Katīb al-Baghdādī, Abū 'l-Faraj, *Kitāb al-Kharaj*, ed. and Fr. trans. in de Goeje, *BGA*, vol. 6, 184-266 (trans. 144-208), p. 259 (trans. pp. 199-200).

22 Treatises on imperial expeditions note the importance of sending both an advance division to prepare the way and of employing suitably qualified scouts. See Constantine VII Porphyrogennētos, *Constantine Porphyrogenitus: three treatises on imperial military expeditions*, ed. and trans. J. F. Haldon [CFHB, vol. 28] (Vienna, 1990), Text B, ll. 116-21 (p. 90); Text C, ll. 564-5 (p. 130). Commentary with other sources, p. 171.

One result of these changes was an increased reliance upon pack animals rather than on wheeled vehicles drawn by draught-animals for the movement of goods and people.[23] In this respect, the archaeological evidence is clear. In hilly country inclines were coped with by shallow steps which were impassable for wheeled transport and the surface area of paved roads was reduced.[24] Except where the terrain was unsuitable, long-distance Roman roads in eastern provinces were generally wider than 6.50 metres, having a central spine and a rather uneven surface of comparatively small stones. Many of them show clear evidence of wheeled vehicles.[25] According to one study, however, Byzantine roads were generally much narrower, effectively reduced to non-vehicular "roadways", and were stepped in mountainous stretches, a development which dated from the sixth century.[26] There are three clear examples: the road from Antioch to Aleppo and Chalkis in northern Syria, the road from Tarsos to the Cilician Gates along a surviving section near Sağlıklı, and the *Via Sebaste* through the Döşeme pass from Pamphylia to Lycia. The last provides a clear illustration, with up to four layers of pavement from consecutive repairs or partial rebuildings, the earliest observable stratum dating to the *Via Sebaste* built under Augustus in 6 B.C.E. This road was more than 6 metres wide and has a wheel-rutted but irregular surface. It seems to have been completely rebuilt in the early Byzantine period and used, with further repairs which cannot be accurately dated, until the late Ottoman period. But the late Roman road was much narrower, having a maximum width of around 3.50 metres, and was stepped on steeper inclines. The edges were made of comparatively large blocks and the carriageway itself of smaller blocks, laid more carefully than the Roman surface, although not as regular as other examples of road surfaces which can be dated to this period.[27] While the surfaces of the other two roads dated to the early Byzantine period were slightly different, they too were stepped on most ascending sections. The implication is clear. Even on substantial, important long-distance routes, wheeled traffic had effectively ceased during the sixth century and was replaced by pack-animals. Importantly, similar observations have been made of the *Via Egnatia*, which was also rebuilt on a much narrower

---

23 R. W. Bulliet, *The camel and the wheel* (Cambridge Mass., 1975).

24 For the physical evidence see P. Gazzola, *Ponti Romani. Vol. 2: contributo ad un indice sistematico con studio critico bibliografico* (Florence, 1963); C. O'Connor, *Roman bridges* (Cambridge, 1993).

25 See D. H. French, "The Roman road-system of Asia Minor", in ANRW, 7.2 (1980), 698-729, pp. 703 ff., 713; idem, *Roman roads and milestones of Asia Minor. Fasc. 1: the Pilgrim's Road* [BAR *IS*, 105] (London, 1981), pp. 19–22.

26 See Prokopios, *Buildings*, IV.viii.4-9 (pp. 284-6), V.ii.6-8, 12-14 (p. 322), V.v.1-3 (pp. 334-6).

27 D. H. French, "1989 Roma Yolları Mıltaşları ve Yazıtları Araştırması", *Araştırma Sonuçları Toplantısı*, 8 (1990), 229-40, pp. 233-5; idem, "A road problem: Roman or Byzantine?", *Istanbuler Mitteilungen*, 43 (1993), 445-54, p. 448.

plan with a smoother surface and was stepped on inclines.[28] Crusader armies could still traverse the *Via militaris* with their carts, but the sources describe stretches of this road as rocky or mountainous or swampy, where wheeled transport became a clear hindrance.[29]

From the mid seventh century a clearly-defined system of predominantly military routes evolved in Asia Minor, along which imperial and provincial marching camps, *aplēkta*, were established. A similar pattern emerged in the Balkans also, although without the marching camps. In both cases, while based on the pre-existing Roman network, the new emphasis reflected specifically Byzantine strategic responses to invasion in both regions. Although there is little direct evidence, the situation of many settlements and the continued, even if much reduced, occupation of most late Roman urban sites suggests that late Roman roads continued in use in spite of gradual dilapidation until they became so pot-holed and irregular that even pack-animals and soldiers could not pass, as was the case with the stretch of the *Via Egnatia* in the sixth century noted above. Since maintenance was localized and infrequent, many must have become little more than paths or tracks unsuitable for any wheeled vehicles by the sixth and seventh centuries, although the upkeep of major roads does appear to be attested archaeologically.[30]

In order to reach a meaningful starting point for the analysis of logistical systems, we need to agree on basic statistical comparators. We may need to discuss and compare such figures but so that we can establish some parameters for logistical possibilities, I wish to propose a brief set of statistics based on physiological remains and documentary sources of what is known about medieval animals and their needs and about human resources and speed of movement and related issues.

Let us begin by setting out some generally-agreed statistics for men and animals. First, speed of movement for large forces varied according to a range of variables from 11-12 kilometres per day to 29 or 32. The emperor Nikēphoros II Phōkas considered a march of 16 Byzantine *milia*, approximately 24 kilometres, to

---

28 See French, "Road problem", p. 449.

29 See K. Belke, "Roads and travel in Macedonia and Thrace in the middle and late Byzantine period", in R. Macrides, ed., *Travel in the Byzantine world: papers from the Thirty-fourth Spring Symposium of Byzantine studies, Birmingham, April 2000* (Aldershot, 2002), 73-90, pp. 76–9.

30 For a good overview of the Roman road system in Asia Minor see W. M. Calder and G. E. Bean, eds, *A classical map of Asia Minor* (London, 1958). For discussion of Byzantine material, see K. Belke, "Von der Pflasterstrasse zum Maultierpfad? Zum kleinasiatischen Wegenetz in mittelbyzantinischer Zeit", in N. Oikonomides, ed., *Byzantine Asia Minor (6th-12th cents)/Η Βυζαντινή Μικρά Ασία· 6.-12. αι*, ed. S. Lampakis (Athens, 1998), 267-84.

be both long and tiring for men and horses in the mountainous and broken terrain of the Tauros and Anti-Tauros regions.[31]

Unaccompanied cavalry could achieve distances of up to 65-80 kilometres in a day, provided that the horses were rested regularly and well nourished and watered. Nomadic societies could move even more rapidly than this. Mules travelled at around 7.5 kilometres per hour and could move for 10-12 hours continuously, therefore covering 75-90 kilometres per day. They have been known to cover up to 130-165 kilometres in a day.

Small units could move much faster than large divisions. Distances of up to 48-50 kilometres per day for infantry have been recorded in various pre-modern historical contexts. However, average marching speeds for infantry were three miles or 4.8 kilometres per hour on even terrain and around two miles or 3.2 kilometres per hour on uneven, broken, or hilly ground.

Distances at which supply dumps could be established or stops made to feed and water men and animals were directly related to distances covered in a day's march and how much provisions and water could be carried before re-supply was necessary.[32]

The numbers of men, animals, and carts in a column, and the space they occupied, played fundamental roles. Large forces moved more slowly than small ones because men and animals did not start off all at the same moment, but one after the other. The longer the column, the longer it took for the rearmost files to set off. Correspondingly, the rearmost groups would arrive at the next camp later than the foremost groups and the delay between the arrival of first and last was proportional to the length and breadth of the column. For example, an army of 5,000 infantry ordered five abreast and with each row occupying two metres would be strung out over two kilometres. Marching at the standard infantry rate of about 4.8 kilometres per hour over good ground, and assuming a three-second delay between each row setting off, and in reality it would probably have been greater, there would be a gap of some 50 minutes between foremost and rearmost ranks setting out and as the column reached marching speed and closed up there would be 25 minutes between the first and last ranks passing any one spot. This is, in fact, an exceedingly optimistic set of assumptions.

Where terrain was broken and mountainous, marching rates would have been slower. Where tracks were very narrow, columns would have been longer. A column of 1,000 cavalry in double file would stretch over two kilometres if each rank took up approximately four metres. An army of 10,000 infantry and 5,000

[31] See the discussion in McGeer, *Sowing the Dragon's teeth*, pp. 340-41. See also *De velitatione*, ed. Dennis, §XIII, ll. 8-11 (p. 188).

[32] It is worth noting for comparative purposes that Agricola's marching camps during his campaign in eastern Scotland in 83-84 C.E. were established between 17 and 21 kilometres apart. See D. J. Breeze, "The logistics of Agricola's final campaign", *Talanta*, 18-19 (1986-7), 7-28.

cavalry would string out over 14 kilometres even with the infantry five abreast and the cavalry two abreast. The rearmost ranks would be some 175 or so minutes behind the first at the very least. On a road surface three metres wide, the infantry column would be reduced to three abreast at the most, so this force would stretch over 6.6 kilometres for the infantry and ten kilometres for the cavalry.

The larger an army and the narrower its front, the longer and more exposed the column would have been, the more difficult the maintenance of regular marching time and discipline, and the greater the delay between the first and last ranks arriving at a given destination.

Various additional and ever-present factors such as fording rivers, passing through defiles, and stopping to rest periodically, as well as changes in marching order imposed by all such movements, would further slow down rates of march and exacerbate delays between starting off and arrival. Bottlenecks, bridges, narrow passes, and so forth, would have been especially problematic in this respect.

For example, when Basil I led his army on foot, probably in ranks of two or three abreast at the most, through the defiles of the Anti-Tauros on his way from Koukousos to the siege of Maraş in 877,[33] and assuming that the army was 8,000 or so horse and foot, although it was probably larger than this, the column would have strung out for eight kilometres at least. Moving at the slower rate of 3.2 kilometres per hour and with at least a three-second delay between each rank starting off, the rearmost ranks could not have set out from the camp before more than three hours after the first ranks started marching. This is to take no account of baggage animals, of course. Such figures are entirely hypothetical but are based on established ratios of movement to space and provide some idea of the practical problems which faced commanders on the march.[34]

An average day's march for infantry or combined forces was probably rarely more than 19.5-22.5 kilometres, the average for most infantry forces throughout recorded history.[35] The average could be increased when no accompanying

---

33 See *Theophanes continuatus, De Basilio Macedone*.48 (p. 280). The road is discussed in F. Hild, *Das byzantinische Strassensystem in Kappadokien* [Veröffentlichungen der Kommission für die *TIB* 2. Denkschriften der Österreichischen Akad. der Wiss., phil.-hist. Kl., 131] (Vienna, 1977), pp. 134-5 and map 14.

34 For similar exercises, see F. Maurice, "The size of the army of Xerxes in the invasion of Greece, 480 B.C.", *Journal of Hellenic studies*, 50 (1930), 210-35; D. W. Engels, *Alexander the Great and the logistics of the Macedonian army* (Berkeley, 1978), pp. 131-3, 154-5. I have on the whole erred on the side of caution in estimating both the space occupied by a rank of men or horses and the time-lapse between rows setting out.

35 On such issues in general see Engels, *Alexander the Great*, pp. 154-6 for detailed figures and averages derived from the marches of Alexander's army across Greece, Anatolia, Iran, and into northern India. See also M. Van Creveld, *Supplying war: logistics*

baggage train was present. Thus Roman legionaries of the first century C.E., carrying most of their immediate requirements in equipment and provisions, were supposed to maintain a rate of 20 Roman *milia*, 30 kilometres, in five hours on paved roads or good tracks and in good weather. A faster pace, intended to cover 24 *milia*, 35.5 kilometres, in five hours, was also practised. These rates were repeated by Vegetius in the fifth century.[36] They could be increased even further by forced marches, although there was an inverse relationship between the length and speed of such marches and the loss of men and animals from exhaustion.

Sources record a variety of rates of march determined by conditions of roads and types of troops involved. Prokopios reported a march of some 10-11 kilometres per day over a seven-day period during the Vandal war, a figure which reflected the presence of infantry, a baggage train, and a supply train.[37] Although the sources disagree on details, one example of a forced march by Byzantine forces is provided by Basil II's expedition of 995 from Constantinople to relieve Aleppo. The emperor supposedly set out with an estimated 40,000 men and the journey, which would normally have taken some 60 days, was completed in a quarter of that time, although only 17,000 men and their mounts or pack-animals reached Aleppo.[38]

Horses and mules required regular rest and breaks for grazing, at least one day in six or the equivalent, if they were not to develop sores and damage to their feet and backs making them temporarily useless, and permanently so if not rested and cared for. They were also relatively inefficient animals in economic terms, requiring considerably more in weight of provisions proportional to carrying capacity than men.[39] Roman military mounts in rest conditions may have required between 6.8 and 9 kilogrammes of fodder per day composed of 2.2-2.7

---

from Wallerstein to Patton (Cambridge, 1977), pp. 28-9, particularly for the campaigns of Frederick II of Prussia in 1757-8.

36 Flavius Vegetius Renatus, *Epitoma rei militaris*, ed. C. Lang (Stuttgart, 1872), I.9, 19 (pp. 13, 21). See also G. Watson, *The Roman soldier* (London, 1969), pp. 54-5; H. Elton, *Warfare in Roman Europe, AD 350-425* (Oxford 1996), pp. 244-5 for further discussion and evidence.

37 Prokopios of Caesarea, *History of the wars*, in H. B. Dewing, ed. and trans. *Procopius*, 7 vols (London, 1914-40), vols 1-6, IV.xiii.32-33 (vol. 2, p. 322).

38 Al-Anṭākī, Yaḥyā ibn Sa'īd ibn Yaḥyā, *Kitāb al-ta'rīkh*, part. Fr. trans. I . Kratchkovsky and A. Vasiliev, *Histoire de Yaḥyā ibn Sa'īd d'Antioche continuateur de Sa'īd-ibn-Biṭrīq*, 2 vols, in PO, 18.5, 23.3, vol. 23.3 (pp. 442-3). See J. H. Forsyth, *The Byzantine-Arab chronicle (938-1034) of Yaḥyā b. Sa'īd al-Antākī*, 2 vols (Ph.D. thesis, University of Michigan, 1977), vol. 2, p. 492 ff. Although the Arabic sources differ widely on the size of Basil's original force, the rate of attrition attributed to the column is at least indicative.

39 Virtually every authority offers slightly different figures for the absolute amounts of feed required by animals, as well as the proportional relationship between the different elements.

kilogrammes of barley and 4.5-6.8 kilogrammes of hay or grazing,[40] or for normal rations per day based on modern figures 5.5-6.5 kilogrammes of hard fodder and 6.5-7.5 kilogrammes of green or dry fodder.[41] Based on ancient statistics and evidence, the ancient and medieval fodder ration would have been smaller for the smaller animals of those times, perhaps 2.5 kilogrammes of hard fodder and 7 kilogrammes of green or dry fodder.

In addition, horses required between 22.75 and 36.4 litres of water per day according to temperature, nature of work, and other factors.[42] In wartime, rations tended to be smaller and animals could manage on as little as 1.5 kilogrammes of hard fodder per day. Horses could derive much of their nutrition from grazing, later medieval horses 50%, ancient and early medieval horses 66%; however, a portion of grain was essential for maintaining them in health. Twice as much forage or grazing was needed as dry fodder, 11 kilogrammes per day, and the area required for grazing depended on quality of pasturage, seasonal variations, and so forth. Given that the basic requirements per horse were some 4-5 hours' grazing every 24 hours, 20 horses would graze out an acre of medium-quality pasture per day, although less on campaign.[43] The availability of grazing obviously depended upon regional and seasonal variations. Where fodder had to be transported in addition to grain, mobility would be drastically limited and transport costs increased.

Mules needed approximately 75% of the ration of horses, generally hay and barley. Comparative evidence suggests a daily ration of 2.3-4 kilogrammes of hard fodder and 6 kilogrammes of dry fodder or 11 kilogrammes of green fodder. However, they could work well on less: 2.3 kilogrammes of barley and 4.5 kilogrammes of straw per day in Wellington's mule train, for example. Modern manuals recommend a daily ration for donkeys of 1.5 kilogrammes of hard fodder and 5 kilogrammes of green fodder plus 20 litres of water, although they can survive on very mean rations as well as grazing thistles, poor grass, and other herbage. Oxen converted protein more efficiently than either horses or mules, but had a greater requirement per day: 6.8 kilogrammes of hay, 11 kilogrammes of mash, and 15-30 litres of water.

---

40 See A. Hyland, *Equus: the horse in the Roman world* (New Haven, 1990), p. 90, followed by Dixon, *Roman cavalry*, pp. 208-17, esp. 210-11. Hyland underestimates the amount of barley required. See Haldon, "Expeditionary force", pp. 126-7 and n. 57.

41 Roth, *Logistics of the Roman army*, p. 62.

42 Hyland, *Equus*, p. 96; Engels, *Alexander the Great*, p. 127. See also C. Gladitz, *Horse breeding in the medieval world* (Dublin 1997), pp. 127-8 and further literature. Roth, *Logistics of the Roman army*, p. 67 agrees in general with these figures.

43 Roth, *Logistics of the Roman army*, pp. 64-5. The modern equivalent of 12 horses per acre reflects different priorities for animals bred under modern conditions. See I. P. Roberts, *The horse* (New York, 1905), p. 360 ff.

Table 8.1: Feed requirements of animals using Roth's lower estimates

	hard fodder	dry or green fodder	or	pasturage	water
Donkeys	1.5 kgs	5.0 kgs	or	10.0 kgs	20 litres
Mules	2.0 kgs	6.0 kgs	or	12.0 kgs	20 litres
"Pack-animals"	2.0 kgs	5.5 kgs	or	11.0 kgs	20 litres
Oxen	7.0 kgs	11.0 kgs	or	22.0 kgs	30 litres
Horses	2.5 kgs	7.0 kgs	or	14.0 kgs	30 litres

The carrying capacities of pack-animals may be tabulated as in Table 8.2.

Table 8.2: Carrying capacities of animals[44]

	man plus accoutrements, arms, clothing	military saddle	pack saddle	load	total
Donkeys[45]			16-19 kgs	56.8 kgs	76 kgs
Mules			16-19 kgs	85 kgs	104.5 kgs
Pack-horses			16-19 kgs	85 kgs	104.5 kgs
Horses[46]	70 kgs	12 kgs		30 kgs	112 kgs

Roth suggests more generous average loads on the basis of modern authorities, Diocletian's *Edict of Prices*, and Egyptian papyri: for donkeys 100 kilogrammes and for mules 135 kilogrammes.[47]

As far as two-wheeled carts and four-wheeled wagons are concerned, if we accept Bachrach's calculations, two-wheeled carts could carry 500 kilogrammes and four-wheeled wagons 650 kilogrammes. Mule carts could cover 30 kilometres per day and ox carts 15 or 19-24 kilometres per day.[48]

Using Roth's averages of 135 kilogrammes maximum weight including pack-harness for horses and mules and my own of 113.6 kilogrammes, loads for

---

44 All figures rounded up and the higher or a mean taken where a range of options is available. All figures must be taken as approximations to within one or two kilogrammes.

45 Ancient data suggest a load equal to two-thirds that of a mule.

46 Extra weight without damaging a mount, around four kilogrammes, sufficient for three-four days' supplies for a horseman, or about one day's worth for both himself and his mount. The *Stratēgikon* attributed to Maurice and the *De velitatione* recommended that cavalry soldiers carry three-four days' supply with them in their saddle bags. See Maurice, *Das Strategikon des Maurikios*, eds and Germ. trans. G. T. Dennis and E. Gamillscheg [CFHB, vol. 17] (Vienna, 1981), I.2.40-45 (p. 80); *De velitatione*, ed. Dennis, §16, ll. 8-13.

47 Roth, *Logistics of the Roman army*, pp. 205-6, 206-7.

48 B. S. Bachrach, "Animals and warfare in early medieval Europe", in *L'uomo di fronte al mondo animale nell'alto medioevo* [SSCISAM, XXXI] (Spoleto, 1985), 707-64. Cf. Roth, *Logistics of the Roman army*, pp. 208-12.

campaign purposes may be calculated within the range based on a series of equivalences. The standard weight of an armed cavalryman with harness was 112 kilogrammes plus a ration of 1.3 kilogrammes per day. The standard load of an unridden remount was 68 kilogrammes plus a ration of 2.2 kilogrammes hard fodder as well as 6.8 kilogrammes of fodder or 14 kilogrammes of grazing. The standard load of a mule or horse pack-animal was either 96 kilogrammes plus a ration of 2.2 kilogrammes hard fodder as well as 6.8 kilogrammes of fodder or 14 kilogrammes of grazing, or 114 kilogrammes plus a ration of 2.0 kilogrammes hard fodder as well as 5-6 kilogrammes of fodder or 11-12 kilogrammes grazing.

Furthermore, since the *Taktika* of Leo VI noted that infantry should be accompanied by one pack-animal for each 16 soldiers, and *in extremis* one for every 32, to carry provisions for up to ten days should they be sent on ahead of the main column, we will assume that the proportion of remounts to ridden horses was approximately 1:4 and that the ratio of pack-mules carrying tents, kitchen utensils, hand-mills, etc. to soldiers was approximately 1:32.[49] Since the maximum load such animals could carry was about 96 or possibly 100 kilogrammes excluding pack-saddles, this suggests a rather small basic ration allowance for the men. 96 kilogrammes for 16 men equalled 6 kilogrammes per man over 8 days: a mere 0.75 kilogrammes per day or just over half the 1.3 kilogrammes ration generally assumed. This suggests that men must also have been expected to forage for themselves.[50] Roman guidelines and descriptions allocated one pack-animal to each *contubernium* or tent-group of eight men.[51]

The *Stratēgikon* attributed to Maurice recommended that cavalry carry three to four days' supply in their saddle bags. The same treatise asserted that troops encamped between six and ten days' march from the enemy should take 20-30 *litrai*, approximately 6.5-9.5 kilogrammes, of biscuit each when they marched to battle. The *Codex Theodosianus* included regulations stipulating the ration to be carried for 17-20 days, and Prokopios corroborated this for the sixth century.[52]

---

49 Leo VI, *Tactica* (PG), VI.29 (coll. 729-30).

50 Following Engels, *Alexander the Great*, pp. 17-18, based on Roman arrangements: one tent for eight men, each tent weighing about 18.2 kilogrammes, thus permitting a pack-animal to carry up to five tents. This would be an extremely heavy load, however, and the proportion of animals to men may well be too low. The standard Roman arrangement of one mule for each *contubernium* was repeated by the *Stratēgikon* of Maurice, and Leo VI, *Taktika*. See Maurice, *Strategikon*, XII.B.6 (pp. 422-3); Leo VI, *Leonis imperatoris tactica*, ed. R. Vari, 2 vols (Budapest, 1917-22) [incomplete], VI.28 (vol. 1, p. 123) [= Leo VI, *Tactica* (PG), VI.29 (coll. 729-30)].

51 See J. Peddie, *Invasion: the Roman invasion of Britain in the year AD 43 and the events leading to their occupation of the West Country* (New York, 1987), p. 27 ff.; Roth, *Logistics of the Roman army*, p. 21 ff.

52 For Roman rates, see Vegetius, *Epitoma*, I.19 (p. 21); Ammianus Marcellinus, *Historia*, in J. C. Rolfe, trans., *Ammianus Marcellinus*, 3 vols (London, 1935-40), XVI.2.8, XVII.9.2 (vol. 1, pp. 206, 354). See also Dixon, *Roman cavalry*, pp. 91-3; Elton, *Warfare in*

In the fifth and sixth centuries rations were eaten on a three-day rotation: bread for one day in three, *bucellatum* (biscuit) for two days in three, salt pork for one day in three, mutton for two days in three, wine and sour wine on alternate days. Other foods such as fish, cheese, and oil, could also be eaten depending on context and availability. The weight of such rations varied, but the figure of one Roman pound (327 grammes) of meat and/or two-to-three Roman pounds (654-981 grammes) of bread per day per man given in one document for stationary troops seems to have been standard into the seventh century in Egypt. The limited evidence suggests that this was more-or-less constant in the preceding and following periods. The *De velitatione* noted that basic rations for soldiers in the tenth century consisted of bread and either cheese or dried and salted meat.[53]

Even though the amount of meat proportional to the rest of the diet might frequently be reduced to a minimum or to nothing under some campaign conditions, this diet would still provide a reasonable amount of nutrition since ancient strains of wheat and barley had considerably higher protein content than modern strains. The bread ration of soldiers in ancient and medieval times provided adequate nutrition for the duration of a campaign season even without meat.[54]

On this basis, the maximum 60 Roman pound, 19.65 kilogramme, load per man specified by Vegetius,[55] would last about 20 days. Under normal marching conditions much of an individual's supplies would have been transported by pack-animal or wagon, however.[56] An army of 15,000 would thus require a minimum of

---

Roman Europe, pp. 115-16; Watson, *Roman soldier*, pp. 62-6. For the sixth century, Prokopios, *History of the wars*, III.xiii.15 (vol. 2, pp. 120-22); Maurice, *Strategikon*, I.2, ll. 42-4 (p. 80), V.4. ll. 5-7 (p. 212), and cf. I.9, ll. 15-17 (p. 102).

53 I will take the figure of 1.3 kilogrammes as the average. See *De velitatione*, ed. Dennis, §8, ll. 9-14 (p. 164).

54 See K. D. White, *Roman farming* (Ithaca, 1970) and for Byzantium from the seventh century, J. L. Teall, "The grain supply of the Byzantine Empire, 330-1025", *DOP*, 13 (1959), 89-139, pp. 117-32. On nutritional values see P. J. Reynolds, *Iron-Age farm: the Butser experiment* (London, 1979). On the size and value of Roman military rations see L. Foxhall and H. A. Forbes, "Σιτομετρεία : the role of grain as a staple food in classical antiquity", *Chiron*, 12 (1982), 41-90; C. E. P. Adams, "Supplying the Roman army: O. Petr. 245", *Zeitschrift für Papyrologie und Epigraphik*, 109 (1995), 119-24. For comparative statistics see Van Creveld, *Supplying war*, pp. 24, 34; V. Aksan, "Feeding the Ottoman troops on the Danube, 1768-1774", *War and Society*, 13 (1995), 1-14; Engels, *Alexander the Great*, pp. 123-6.

Note that the figures given by Engels, *Alexander the Great*, at pp. 126-7 overestimate the weight of grain required since he overlooks the considerable differences between nutritional values of ancient grains which would have been used and modern grains, upon which his calculations were based.

55 Vegetius, *Epitoma*, I.19 (p. 21).

56 See Nikēphoros Phōkas, *Praecepta militaria* [*Στρατηγικὴ Ἔκθεσις καὶ Σύνταξις*

some 288,400 kilogrammes of provisions for between two or, in exceptional cases, three weeks, excluding water, wine, and extras such as lard and/or oil, cheese or fish, and fodder for animals. Evidence suggests that while the state provided grain and on occasions dried meat for soldiers' diets, their households generally catered for items such as oil, cheese, and so forth, at least for provincial troops in the *themata*.[57] Assuming a rate of march for infantry and cavalry together of 19-22.5 kilometres per day, an optimistic figure when compared to most known pre-industrial military marches,[58] such a force could travel some 400-475 kilometres in three weeks. This provides a guide to distances at which supply dumps would have had to be established. In friendly territory, and whenever provisions could be garnered readily, it would have replenished supplies much more frequently in order to avoid having to take along large numbers of pack-animals until that became absolutely essential, a question which also raises a number of problems.

An anonymous tenth-century treatise on campaign organization confirmed this estimate of maximum provisions that could be carried for a three-week cavalry march, noting that: "... it is not feasible, in turn, for an army to transport more than a twenty-four days' supply of barley from its own country for its horses".[59]

In fact, for the 8,200 cavalry envisaged by the treatise, assuming that each animal carried between one and two days' supply of barley, and assuming at least 1,000 remounts each carrying the 68 kilogrammes of barley feed stipulated in a tenth-century treatise on imperial military expeditions accompanied the corps, providing feed for four to five days,[60] some 6,460 extra pack-animals would be required to enable this force to stay in the field for a further 18 or so days without being re-supplied with hard fodder.[61] If there were 2,000 remounts similarly loaded, over 1,000 fewer pack-animals would be required, thus reducing proportionally the necessary amount of pack-animal feed. The context of this passage suggests that in practice much of this feed would be carried by a separate supply train, including carts as well as pack-animals, and that the cavalry itself

---

Νικηφόρου δεσπότου], ed. and trans. McGeer, *Sowing the Dragon's teeth*, 12-59, II.1 (p. 22).

57 Ibn Khurradādhbīh, Abū 'l-Qāsim 'Ubayd Allāh ibn 'Abdallāh, *Kitāb al-Masālik wa 'l-mamālik*, ed. and Fr. trans. in De Goeje, *BGA*, vol. 6, 3-183 (trans. 1-144), pp. 111, 112 (trans. pp. 83, 85).

58 Engels, *Alexander the Great*, p. 112 ff.

59 Anonymous, *'Ανωνύμου Βιβλίον τακτικόν*, in G. T. Dennis, ed. and trans., *Three Byzantine military treatises: text, trans. and notes* [CFHB, vol. 25] (Washington, 1985), 241-335, §21, ll. 22-3 (pp. 302-4)): "... καὶ τῷ στρατεύματι ἀμήχανον αὖ πάλιν πλεῖον τῶν κδ΄ ἡμερῶν κριθὴν τοῖς ἰδίοις ἀλόγοις ἀπὸ τῆς ἑαυτῶν χώρας βαστάσαι, ...".

60 Constantine VII, *Three treatises*, Text C, ll. 411-14 (pp. 120-21). See also J. Haldon, "The organisation and support of an expeditionary force: manpower and logistics in the Middle Byzantine period", in N. Oikonomides, ed., *Τό εμπόλεμο Βυζάντιο (9ος - 12ος αι.) / Byzantium at war (9th-12th c.)* (Athens, 1997), 111-51, pp. 125-6.

61 The calculation is based on figures discussed below.

would be less heavily burdened and accompanied by far fewer pack-animals as it moved along.

A simple equation for establishing the relationship between the size of an army, the number of pack-animals and quantity of provisions it requires, and the number of days it can be kept in the field, was established by Engels in his analysis of the strategy of Alexander the Great, in which he took into account the need to transport water also since much of Alexander's campaigning took place in relatively arid regions.[62] I have simplified this by excluding the water ration:

$$N = \frac{(a + b + c + d)\,y}{(x - z)\,y}$$

where:

N = the number of pack-animals required
a = the sum of the soldiers' provisions in kilogrammes
b = the sum of the horses' rations in kilogrammes
c = the sum of the rations of the pack-mules in kilogrammes
d = the sum of the rations of the remounts that also carry provisions in kilogrammes
x = the average load carried in kilogrammes
z = the standard ration of the animals carrying the provisions in kilogrammes
y = the duration of the expedition in days.[63]

In what follows I set out a number of hypothetical examples intended to show possible ranges of requirements and loads that could be carried over longer distances.[64] It should be emphasized, however, that these are hypothetical and in most senses unrealistic since they represent only one type of combination under a constant set of conditions, neither of which will usually have prevailed in the past.

Example I: a small cavalry force of 1,000 men accompanied by 250 remounts

Such a force would have required 1,000 x 1.3 kilogrammes + 1,250 x 2.2 kilogrammes per day: = 4,050 kilogrammes. This would then be multiplied by the number of days and divided by weight in kilogrammes carried by each pack-animal, minus its own daily ration. Calculations run to the 24th day since that was the length beyond which tenth-century evidence suggested that cavalry could no

62 For detailed discussion see J. Haldon, *Warfare, state and society in the Byzantine world 565-1204* (London, 1999), ch. 5 and the Appendices.

63 Cf. Engels, *Alexander the Great*, p. 22 and n. 35.

64 See Haldon, *Warfare, state and society*, Appendix 3, the results of which have been modified in the light of these considerations.

longer supply themselves. How many pack-animals would have been available to armies is impossible to assess but it is unlikely that many thousands would have been acquired easily unless campaigns were planned well in advance.

Assuming that remounts carried no provisions, the result would be as in Table 8.3.

Table 8.3: Provisioning a cavalry force of 1,000 men and 250 remounts

Days	Total of provisions / Load minus ration per animal	Total no. of mules required[65]	No. of mules per day extra to needs after rations consumed
1	$N = \frac{4{,}050 \times 1 = 4{,}050}{96 - (2.2 \times 1)}$	= 43 (40)	1 (1)
2	$N = \frac{4{,}050 \times 2 = 8{,}100}{96 - (2.2 \times 2)}$	= 88 (83)	2 (2)
5	$N = \frac{4{,}050 \times 5 = 20{,}250}{96 - (2.2 \times 5)}$	= 238 (220)	6 (5)
10	$N = \frac{4{,}050 \times 10 = 40{,}500}{96 - (2.2 \times 10)}$	= 547 (500)	12 (10)
15	$N = \frac{4{,}050 \times 15 = 60{,}750}{96 - (2.2 \times 15)}$	= 964 (864)	22 (18)
20	$N = \frac{4{,}050 \times 20 = 81{,}000}{96 - (2.2 \times 20)}$	= 1,558 (1,446)	35 (30)
24	$N = \frac{4{,}050 \times 24 = 97{,}200}{96 - (2.2 \times 24)}$	= 2,250 (1,900)	51 (39)

Two hundred and fifty remounts, each carrying 68 kilogrammes, could carry 17,000 kilogrammes, maintaining men and horses for up to four days. Soldiers themselves could also carry four days' supplies for themselves or one day's for themselves and their mounts, thus extending the initial period to five days.[66] With additional provisions carried by 1,523 mules, they could provision themselves for

---

[65] The figures in brackets in Tables 8.3, 8.4, 8.5, and 8.7 give the result using Roth's lower estimate for feed (2.0 kilogrammes), but retaining also the more conservative estimate for loads of 96 kilogrammes rather than 114 kilogrammes.

[66] See *De velitatione*, ed. Dennis, §16, ll. 11-13 (p. 200). Cf. Maurice, *Strategikon*, I.2, ll. 40-45 (p. 80).

about 24 days. However, whether such forces would have had access to such large numbers of pack-animals is debatable, since even the imperial ranches, *mētata*, were unable to provide all the animals needed for the imperial baggage train of over 1,000 horses and mules in the ninth and tenth centuries.[67]

Using Roth's equivalents, requiring 1,000 x 1.3 kilogrammes + 1,250 x 2.0 kilogrammes per day, i.e. 3,800 kilogrammes per day, but with the heavier load of 114 kilogrammes per animal, gives results as in Table 8.4.

Table 8.4: Provisioning a cavalry force of 1,000 men and 250 remounts, using Roth's figures

Days	Total of provisions / Load minus ration per animal	Total no. of mules required	No. of mules per day extra to needs after rations consumed
1	$N = \frac{3,800 \text{ x } 1 = 3,800}{114 - (2.0 \text{ x } 1)}$	= (40)	(1)
2	$N = \frac{3,800 \text{ x } 2 = 7,600}{114 - (2.0 \text{ x } 2)}$	= (69)	(2)
5	$N = \frac{3,800 \text{ x } 5 = 19,000}{114 - (2.0 \text{ x } 5)}$	= (182)	(3)
10	$N = \frac{3,800 \text{ x } 10 = 38,000}{114 - (2.0 \text{ x } 10)}$	= (404)	(7)
15	$N = \frac{3,800 \text{ x } 15 = 57,050}{114 - (2.0 \text{ x } 15)}$	= (680)	(12)
20	$N = \frac{3,800 \text{ x } 20 = 76,000}{114 - (2.0 \text{ x } 20)}$	= (1,027)	(18)
24	$N = \frac{3,800 \text{ x } 24 = 91,200}{114 - (2.0 \text{ x } 24)}$	= (1,382)	(25)

Using spare horses, and with men carrying provisions, the force could stay in the field for 24 days with 1,132 mules. If we assume that a number of mules was set aside for carrying pack-animal rations, these could have been left behind as rations were consumed, thus reducing the overall numbers required as noted above.

67 Constantine VII, *Three treatises*, Treatise C, ll. 59-66, 84-5.

Example II: an army of 4,000 cavalry with 1,000 remounts

Applying these calculations to larger armies, we obtain the following results for 4,000 cavalry with 1,000 remounts: 4,000 men x 1.3 kilogrammes per day + 5,000 horses x 2.2 kilogrammes per day = 5,200 + 11,000 = 16,200 kilogrammes.

Table 8.5: Provisioning an army of 4,000 cavalry with 1,000 remounts

Days	Total of provisions / Load minus ration per animal	Total no. of mules required	No. of mules per day extra to needs after rations consumed
1	$N = \frac{16{,}200 \times 1 = 16{,}200}{96 - (2.2 \times 1)}$	= 173 (162)	4 (4)
2	$N = \frac{16{,}200 \times 2 = 32{,}400}{96 - (2.2 \times 2)}$	= 354 (330)	8 (7)
5	$N = \frac{16{,}200 \times 5 = 81{,}000}{96 - (2.2 \times 5)}$	= 953 (893)	22 (19)
10	$N = \frac{16{,}200 \times 10 = 162{,}000}{96 - (2.2 \times 10)}$	= 2,190 (2,000)	50 (42)
15	$N = \frac{16{,}200 \times 15 = 243{,}000}{96 - (2.2 \times 15)}$	= 3,858 (3,455)	88 (72)
20	$N = \frac{16{,}200 \times 20 = 324{,}000}{96 - (2.2 \times 20)}$	= 6,231 (5,428)	142 (113)
24	$N = \frac{16{,}200 \times 24 = 388{,}800}{96 - (2.2 \times 24)}$	= 9,000 (7,600)	206 (158)

If provisions were carried by the 1,000 remounts, each horse could carry some 68 kilogrammes, a total of 68,000 kilogrammes. Men and horses could have been maintained for up to four days from these. At the same time, the soldiers themselves could carry a day's supplies for themselves and their mounts, extending this initial period to some five days. We should then reduce the number of mules by a proportion each day as rations for pack-animals were consumed and animals could be dispensed with. With provisions carried by 6,091 mules in addition, they could provide themselves with food to last the full 24 days noted in the sources.

Using Roth's lower estimate for feed and higher estimate for loads, the same force would require pack-animals as shown in Table 8.6.

Table 8.6: Provisioning an army of 4,000 cavalry with 1,000 remounts, using Roth's figures			
Days	Total of provisions / Load minus ration per animal	Total no. of mules required	No. of mules per day extra to needs after rations consumed
1	N = (16,200 x 1 = 16,200) / (114 - (2.0 x 1))	= (145)	(3)
2	N = (16,200 x 2 = 32,400) / (114 - (2.0 x 2))	= (295)	(5)
5	N = (16,200 x 5 = 81,000) / (114 - (2.0 x 5))	= (779)	(14)
10	N = (16,200 x 10 = 162,000) / (114 - (2.0 x 10))	= (1, 7232)	(30)
15	N = (16,200 x 15 = 243,000) / (114 - (2.0 x 15))	= (2,893)	(51)
20	N = (16,200 x 20 = 324,000) / (114 - (2.0 x 20))	= (4,378)	(77)
24	N = (16,200 x 24 = 388,800) / (114 - (2.0 x 20))	= (5,891)	(104)

Again, if provisions were carried on the remounts, the number of pack-animals required at the outset could be reduced by some 600.

Example III: an army of 10,000, made up of 6,000 infantry and 4,000 cavalry, accompanied by 1,000 remounts

Applying the same calculations and conditions to larger armies, we obtain the following results for an army of 10,000, made up of 6,000 infantry and 4,000 cavalry, accompanied by 1,000 remounts: 10,000 men x 1.3 kilogrammes per day + 5,000 horses x 2.2 kilogrammes per day = 13,000 + 11,000 = 24,000 kilogrammes (Haldon). 10,000 men x 1.3 kilogrammes per day + 5,000 horses x 2.0 kilogrammes per day = 13,000 + 10,000 = 23,000 kilogrammes (Roth).

Table 8.7: Provisioning an army of 10,000, made up of 6,000 infantry and 4,000 cavalry, accompanied by 1,000 remounts

Days	Total of provisions / Load minus ration per animal	Total no. of mules required	No. of mules per day extra to needs after rations consumed
1	N = 24,000 x 1 = 24,000 / 96 - (2.2 x 1)	= 256 (255)	6 (5)
2	N = 24,000 x 2 = 48,000 / 96 - (2.2 x 2)	= 524 (521)	12 (11)
5	N = 24,000 x 5 = 120,000 / 96 - (2.2 x 5)	= 1,412 (1,395)	32 (32)
10	N = 24,000 x 10 = 240,000 / 96 - (2.2 x 10)	= 3,244 (3,158)	74 (72)
15	N = 24,000 x 15 = 360,000 / 96 - (2.2 x 15)	= 5,715 (5,455	130 (125)
20	N = 24,000 x 20 = 480,000 / 96 - (2.2 x 20)	= 9,231 (8,572)	210 (197)
24	N = 24,000 x 24 = 576,000 / 96 - (2.2 x 24)	= 13,334 (12,000)	300 (275)

The figures assume that infantry carried arms and clothing only, which is unlikely to have been often so. With infantry and remounts carrying provisions, a march of four days could have been covered by the horses alone. Infantry could carry up to 20 days' supplies if necessary. Subtracting the infantry portion (6,000 x 1.3 kilogrammes per day = 7,800 kilogrammes), we reach the same figures as for 4,000 cavalry and 1,000 remounts. Adding mules carrying siege equipment, tents, and other paraphernalia, as well as camp followers, servants, and armourers, an expedition of 20 days would have been logistically feasible. How many attendants for pack-animals would have been needed is unclear. For imperial expeditions in the later ninth and tenth centuries, one attendant per ten mules or 20 pack-horses was provided by the *kōmes* of the stable from Malagina and mules of the imperial baggage train were each accompanied by a muleteer drawn from the special corps of the *Optimatoi*, although these were exceptional circumstances.[68] Armies substantially larger than this would have lost flexibility and speed rapidly.

Again, the higher estimates for loads of 114 kilogrammes per animal give a more advantageous logistical option.

Table 8.8: Provisioning an army of 10,000, made up of 6,000 infantry and 4,000 cavalry, accompanied by 1,000 remounts, using Roth's figures

Days	Total of provisions / Load minus ration per animal	Total no. of mules required	No. of mules per day extra to needs after rations consumed
1	$N = \frac{24{,}000 \times 1 = 24{,}000}{114 - (2.0 \times 1)}$	= (214)	(4)
2	$N = \frac{24{,}000 \times 2 = 48{,}000}{114 - (2.0 \times 2)}$	= (436)	(8)
5	$N = \frac{24{,}000 \times 5 = 120{,}000}{114 - (2.0 \times 5)}$	= (1,154)	(20)
10	$N = \frac{24{,}000 \times 10 = 240{,}000}{114 - (2.0 \times 10)}$	= (2,553)	(45)
15	$N = \frac{24{,}000 \times 15 = 360{,}000}{114 - (2.0 \times 15)}$	= (4,286)	(75)
20	$N = \frac{24{,}000 \times 20 = 480{,}000}{114 - (2.0 \times 20)}$	= (6,486)	(114)
24	$N = \frac{24{,}000 \times 24 = 576{,}000}{114 - (2.0 \times 24)}$	= (8,727)	(153)

Contexts in which armies will have needed to provide all hard fodder for the horses and all rations for the men from supplies accompanying the expedition will usually have involved leaving imperial territory. But even inside the Empire there existed a relationship between troops and pack-animals which determined the possibilities for long-distance movement. In the next example, I have calculated the figures for an army in which all food, forage, and hard feed for the animals was available along the route, conditions which might be met on occasion within the empire, or in enemy territory when the Byzantine commander had the advantage of surprise

68 Constantine VII, *Three treatises*, Treatise C, ll. 332-346 (p. 116).

and the appropriate season. The variants in Example IV show that armies requiring no feed for animals could be supported for longer with fewer pack-animals needed.

Example IV: armies requiring no feed for animals.

(a) First, an army of 1,000 men. 1,000 men x 1.3 kilogrammes per day = 1,300 kilogrammes per day. 250 remounts each loaded with 68 kilogrammes = 17,000 kilogrammes and would provide enough rations for about ten days (17,000 ÷ 1.3 = 13,076 ÷ 1,000 = 13 kilogrammes per man, i.e. 1.3 per man per day). Thereafter extra mules with rations would be required.
(b) Second, a force of 4,000 cavalry with 1,000 remounts carrying rations for the men only would be able to transport some 96,000 kilogrammes of provisions, enough for the men for about 14 days.
(c) Third, a similar cavalry force, accompanied by 6,000 infantry, and with the infantry carrying their own provisions, would have needed fewer extra pack-animals to stay in the field for up to three weeks or more.

Examples I and IV (a) would probably have been most typical of the defensive warfare in the frontier regions of Anatolia, where the Byzantine forces could rely on local support and provisioning. By contrast, there must have been occasions when imperial armies had to take not only their basic rations in grain feed but also green fodder or hay, and possibly water as well. But under these conditions the army would have been able to provide for only a couple of days at the outside, as example for an army of 4,000 cavalry and 1,000 remounts, accompanied by 80 mules with tents and other equipment.

Calculations can be made on the same basis as above for any permutation or combination and the results would enable us to provide some statistical parameters within which to determine how plausible is information derived from medieval sources. The relationship between the size of armies and available pack-animals would clearly have been a crucial consideration and the availability of pack-animals would have placed very clear limits on the size of armies operating in enemy territory where supplies were inadequate. These points are fundamental to any consideration of the issue of numbers and the logistical feasibility of the various forces mentioned by the sources in different situations.

As noted above, all these examples are somewhat forced and represent very abstract situations. We should probably take both the lower rations and also the lower average loads for animals as the basis for any calculations. Medieval figures generally suggest lower tolerances and load capacities than modern figures, although nineteenth-century British army regulations would support the lower weights carried. On the other hand, an army accompanied by pack-animals would certainly shed numbers of these as supplies were consumed. Although this has been written into the calculations, I suspect that animals would have been shed more quickly and at a higher rate than represented here.

In conclusion, it is important to emphasize that it was normal for armies to forage for the needs of both men and animals. This would have given much greater leeway in terms of duration but also assumes careful scouting and reasonable local knowledge of areas to be crossed, as numerous historical examples illustrate.[69] The discussion above illustrates the value of assessing more carefully the requirements in provisions, fodder, and livestock for medieval armies, of assessing more carefully the carrying capacities of men and animals under different circumstances, and of relating these more closely to the physical and geographical context in which forces operated. As noted at the outset, however, such calculations nevertheless remain subject to a wide range of problematic variables. Until we are able to assess more accurately the demographic context, the rate of agrarian production and types of crops, as well as the direct impact of transient military populations upon particular regions, we will continue to be bogged down in circular arguments based on subjectivities. More closely mapping the movement of armies to the terrain through which they pass, the possible tracks, roads and paths they followed, and associating this data with climate and seasonal factors will also be necessary.[70] As a consequence, this paper is less a statement of what we can do with respect to pre-modern logistics than of a list of desiderata and a sketch for future work.

---

[69] See, for example, the advice in Constantine VII, *Three treatises*, Treatise B, ll. 5-17 (p. 82).

[70] Using traditional sources and categories of evidence, some very provisional attempts were made in chs. 5 and 6 of *Warfare, state and society*. Other essays appear in a number of articles by Bernard Bachrach and John France. See the literature cited in Bachrach, "Animals and warfare"; J. France, *Victory in the East: a military history of the First Crusade* (Cambridge, 1994); idem, *Western warfare in the age of the Crusades, 1000-1300* (Ithaca, 1999). The literature on the Roman army before the second century C.E., on the one hand, and on early modern European warfare, on the other, also includes similar attempts. See the literature in Roth, *Logistics of the Roman army* and Elton, *Warfare in Roman Europe*.

Chapter 9

Reflections on maps, Crusading, and logistics

Benjamin Z. Kedar

Historical maps are a major tool for the study of the logistics of the past. On their basis we make calculations and offer reconstructions. Yet such maps are modern maps in reality. We superimpose on them data derived from our sources. Based on state-of-the-art geography, they contain a wealth of accurate information about distances, elevations, terrain, river courses, coastal configurations, and much more, that the people we are studying simply did not possess. Consequently, reliance on them may mislead us into ascribing to these people logistical and other calculations they would not have been able to generate themselves.

Now I do not advocate doing away with historical maps, merely that we should sharpen our awareness of their inherent anachronism. We should also try to ascertain as much as possible about the ways in which the people we are studying perceived their surroundings and their world. Specifically, we should routinely have recourse to truly historical contemporary maps. We should then be able to understand better the thinking of those we are studying, including their thinking about what we call logistics.

All students of Crusading are familiar with the scene in which in his account of the People's Crusade of 1096 Guibert of Nogent described condescendingly poor people yoking their oxen, shod like horses, to two-wheeled carts, loading them with their few belongings and small children, and these little ones eagerly asking whether this was the Jerusalem toward which they were travelling whenever they saw a castle or a town.[1] Much less well known is that the anonymous author of an Old French verse account based on Baudri of Bourgueil, dated to no earlier than the reign of Philip Augustus, suggested remarkably better geographical knowledge, at least among the leaders. Describing Godfrey of Bouillon's tent, he asserted that it contained "a world map to show the kingdoms".[2]

---

[1] Guibert of Nogent, *Guibert de Nogent Dei gesta per Francos, et cinq autres textes*, ed. R. B. C. Huygens, in CC *ContMed*, vol. 127A, 76-352, II.vi (p. 120).

[2] P. Mayer, "La chanson de Doon de Nanteuil. Fragments inédits", *Romania*, 13 (1884), 1-26, p. 24, n. to l. 205: "*La mapamunde i fu as regnes demostrer*". See also P. Mayer, "Un récit en vers français de la première croisade fondé sur Baudri de Bourgeuil", *Romania*, 5 (1876), 1-63. On the appearance of the term *mappa mundi* in Carolingian times see P. Gautier Dalché, *La "Descriptio mappe mundi" de Hugues de Saint-Victor: texte inédit avec introduction et commentaire* (Paris, 1988), pp. 92-4.

We are intuitively inclined to regard Guibert's children, naively innocent of all geographical knowledge, as more credible than the poet's *mapamunde* in Godfrey's tent. Indeed, the latter may have been a literary topos. From the mid twelfth century onward world maps in leaders' tents became a recurring literary motif. In the *Roman de Thèbes*, dated to the 1150s, we read that on a pane of King Adrastus of Argos's tent there was a round *mappamonde* showing the earth's five zones as well as its towns, kingdoms, and seas, the four rivers of Paradise, mount Etna, birds and beasts, black men of Ethiopia, and the Ocean flowing through the warmest zone.[3] The somewhat later *Romans d'Alixandre* related that on a pane of Alexander the Great's tent there was a *mapemonde* showing the earth's three parts of Asia, Europe, and Africa, mountains, rivers, and towns, surrounded by the sea, which the world conqueror often contemplated:[4] a prefiguration of the famous scene in Charlie Chaplin's *The Great Dictator*. In Walter of Châtillon's *Alexandreis*, completed around 1181, a depiction of the world appeared on the tomb of Alexander's defeated foe Darius. Here also was seen the tripartite world, with rivers, peoples, towns, forests, and mountains, with legends characterizing each region: "The Nile makes Egypt fruitful", "India abounds in ivory", "Love of possession consumes the Ligurians", "The Teuton retains his customary fury".[5]

While world maps recurred in twelfth-century fiction, maps are conspicuously absent from twelfth-century Crusade chronicles; although, in describing Richard Coeur de Lion's quandary in January 1192, Ibn al-Athīr (d. 1233 C.E.) claimed that the king requested that the Syrian Franks "Draw for me the town of Jerusalem". Once they had drawn it, he realized that but for a small stretch in the north the town was surrounded by a valley.[6] Yet we cannot be sure that Ibn al-

---

The various forms of *mappamunde, mappamonde, mappe mundi, mappa mundi, mundi mappa*, etc. have been used as they have been found in the various sources. They have not been regularized.

3 *Le Roman de Thèbes*, ed. G. Raynaud de Lage, 2 vols (Paris, 1968-69), vv. 4223-62, (vol. 1, pp. 132-3) and p. xxvi. L. G. Donovan, *Recherches sur le Roman de Thèbes* (Paris, 1975), pp. 217-27 has shown that the description of the *mappamonde* reflected familiarity with contemporary world maps. Another *mapemonde* appears in or on a tent in the *Chanson de Doon de Nanteuil*. See Mayer, "Doon de Nanteuil", p. 24, l. 205.

4 *Li Romans d'Alixandre par Lambert li Tors et Alexandre de Bernay*, ed. H. Michelant (BLVS, vol. 13) (Stuttgart, 1846), p. 55.

5 M. L. Colker, ed., *Galterus de Castellione, Alexandreis* (Padua, 1978), 7.393-420, (pp. 190-92). For the date see D. Townsend, trans., *The Alexandreis of Walter of Châtillon. A twelfth-century epic* (Philadelphia, 1996), p. xv. With one exception, these texts have been noted in P. Abrahams, ed., *Les oeuvres poétiques de Baudri de Bourgueil (1046-1130)* (Paris, 1926), pp. xliii, 233.

6 Ibn al-Athīr, 'Izz al-Dīn Abu 'l-Ḥasan 'Alī, *Al-Kāmil fī 'l-ta'rīkh*, ed. C. J. Tornberg, 13 vols (1851-76; rpt, Beirut, 1965-66), A.H. 587 (vol. 12, p. 75); Part. Fr. trans., *Extrait de la chronique intitulée Kamel-Altevarykh par ibn-Alatyr*, in RHC *HOr*, vol. 1, 189-744 & vol. 2, 1-180, here vol. 2, part 1, p. 55, where the French trans. is inexact.

Athīr actually knew what went on in Richard's council and in any case his description suggests that the Franks drew merely a rough sketch of Jerusalem's walls and immediate surroundings. The first attested Crusader recourse to a true map is found in William of Nangis's account of Louis IX's expedition to Tunis in 1270. He wrote that when the fleet was hit by a great storm *en route* from Aigues-Mortes to Cagliari with no land in sight, the king asked the ship's masters for their opinion on where the ship was. Saying that they believed themselves close to land and puzzled as to why it had not yet been sighted, a *mappa mundi* was produced and they showed the king the location of Cagliari and how close was the shore. Maritime historians consider this to be the first written mention of a portolan chart; however, this resort to a map was not decisive. The Crusaders became suspicious and the king's son, Philip, expressed apprehension that the ship's masters might be leading the fleet into the unknown, although it did ultimately reach Cagliari.[7]

Probably because of the late date in the annals of Crusading of this first mention of a true map, the question of possible recourse to maps in earlier Crusades has not been addressed. Crusade historians assume that the First Crusaders relied on Byzantine and Eastern Christian guides as well as on Westerners who had acquired familiarity with the East either as pilgrims or as mercenaries in Byzantine service and that later Crusaders depended on Franks who had grown up in *Outremer*.[8] Such dependence differed qualitatively from resorting to maps, of course. Western forces relying on oral guidance could lose their way in far more familiar surroundings than those of Asia Minor or Syria.[9]

Yet can we be certain that our preference for Guibert's naïve children over the *mapamunde* in Godfrey's tent is justified and that no early Crusaders were aware of, or relied on, maps? One fallacy that historians should avoid is to equate reality with documentation. The first written mention of recourse to a map should not be equated with the first actual use of one. Reality is only partially documented and the earliest documented appearance of a phenomenon may postdate considerably

---

7 William of Nangis, *Vita Sancti Ludovici regis Franciæ / Vie de Saint Louis*, in RHGF, vol. 20, 309-465, Anno 1270 (pp. 444-5).

8 Recourse to maps during the Crusades is not raised in the one book seemingly pointing to it: J. K. Wright, *The geographical lore of the time of the Crusades: a study in the history of medieval science and tradition in Western Europe* (New York, 1925).

9 See, for example, J. Viard and E. Déprez, eds, *Chronique de Jean le Bel*, 2 vols (Paris, 1904-5), cc. 11-12 (vol. 1, pp. 53-63). See also M. Van Creveld, *Technology and war: from 2000 B.C. to the present* (New York and London, 1989), p. 45. For Byzantine guides accused of having deliberately misled the Crusade of 1101, see Matthew of Edessa, *Chronicle*, trans. A. E. Dostourian, *Armenia and the Crusades tenth to twelfth centuries: the chronicle of Matthew of Edessa* (Lanham, 1993), III.5 (p. 186).

its first appearance in reality.[10] The chance mention of a *mappa mundi* en route to Cagliari in 1270 should not rule out the possibility that earlier Crusaders had also had recourse to maps. James of Vitry wrote in his *Historia Orientalis* that he based his account of Amazons, Gymnosophists, Brachmans (*sic*), and other barbarous and monstrous peoples on various written sources and on a *mappa mundi.*[11]

In due course I shall conjecture about the possible acquaintance of some First Crusaders with contemporary cartography but first let me turn to a more simple problem allowing for more definite answers. What kind of graphic representations of the world existed at the time of the First Crusade? If we posit late eleventh-century Muslims or Latin Christians attempting to visualize their world or parts of it in a two-dimensional way with the help of maps, what resources were there in their cultures which they could use?

In the Muslim world literary descriptions of early maps predate the earliest surviving maps.[12] Al-Mas'ūdī (d. ca 956 C.E.) described a large map constructed by a group of scholars at the prompting of the 'Abbāsid caliph al-Ma'mūn (813-33 C.E.). The map represented "the world with its spheres, stars, land, and seas, the inhabited and uninhabited regions, settlements of peoples, cities, etc.".[13] Yet only from much later have maps going back to tenth-century prototypes survived. They form a collection that Miller labelled the *Islam-Atlas* and which more recently has been considered as the set of the Balkhī school of geographers. A Persian scholar who lived most of his life in Baghdad, Abū Zayd Aḥmad ibn Sahl, al-Balkhī, (d. 934 C.E.) represented the earth in a set of maps which he described. While his maps did not survive independently, they influenced directly or indirectly those of

---

[10] For discussion of the relationship between documentation and reality in a similar context see B. Z. Kedar, "The subjected Muslims of the Frankish Levant", in J. M. Powell, ed., *Muslims under Latin rule, 1100-1300* (Princeton, 1990), 135-74 [rpt in Kedar, *The Franks in the Levant, 11th to 14th centuries* (Aldershot, 1993), No. XVIII], pp. 137-43.

[11] James of Vitry, *Historia Orientalis*, in *Iacobi de Vitriaco ... libri duo. Quorum prior Orientalis sive Hierosolymitanae, alter, Occidentalis Historiae nomine inscribitur*, ed. F. Moschius (Douai, 1597), 1-258, c. 92 (pp. 198, 215).

[12] What follows is based mainly on G. R. Tibbetts, "The beginnings of a cartographic tradition", "The Balkhī school of geographers", and "Later cartographic developments", in J. B. Harley and D. Woodward, eds, *The history of cartography. Vol. Two, Book One: cartography in the traditional Islamic and South Asian societies* (Chicago and London, 1992), 90-107, 108-36, 137-55 respectively.

[13] See Tibbetts, "Beginnings", p. 95. Sezgin claims that a part of the Ma'mūnic map has been found recently and hypothesizes that in the fourteenth century it influenced the Byzantine reconstruction of Ptolemy's map. See F. Sezgin, *The contribution of the Arabic-Islamic geographers to the formation of the world map* (Frankfurt, 1987). A rectangular world map forming part of a late eleventh-century Arabic treatise may be related to the lost map prepared for al-Ma'mūn. See J. Johns and E. Savage-Smith, "The book of curiosities: a newly discovered series of Islamic maps", *Imago Mundi* (forthcoming). Thanks to Dr Savage-Smith for making this article available to me.

al-Iṣṭakhrī (mid tenth century), Ibn Ḥawqal (d. ca 977 C.E.), and al-Muqaddasī (d. ca 990 C.E.). All three presented the same set, usually consisting of 21 units: a world map, maps of the Mediterranean and Caspian seas and the Indian Ocean, and maps of 17 regions of the Muslim world. No less than 13 of these regional maps represent parts of the former Sāsānid Empire and may go back to Sāsānid sources.

Tibbetts's select list of manuscripts containing sets of Balkhī school maps contains 49 items;[14] however, only one dates from the eleventh century: a manuscript of Ibn Ḥawqal copied in 1086, now in the Ṭopḳapı Palace Museum in Istanbul. Five manuscripts date from the twelfth century, four from the thirteenth, seven from the fourteenth, eight from the fifteenth, and eleven from the sixteenth to nineteenth. They are, however, relevant to the discussion because by the end of the tenth century at least two different versions by al-Iṣṭakhrī, two by Ibn Ḥawqal, and one by al-Muqaddasī must have existed. The post-1100 sets depend on at least five different tenth-century prototypes but there is no way of knowing how many tenth- and eleventh-century copies have not survived.

A recently discovered Arabic treatise offers a different set: two world maps (one rectangular, the other circular), maps of Sicily, al-Mahdīyya, Tinnīs, and Cyprus, and maps depicting the courses of the Nile, Euphrates, Tigris, Indus, and Āmū Daryā. The surviving copy dates from the thirteenth century but the original appears to have been composed at the end of the eleventh century in Fāṭimid Egypt.[15] It is reasonable to assume that a Muslim commander could have acquired some such set or a specific regional map. In one case a map woven in gold on silk became part of a ruler's treasure: a map prepared in 964 C.E. for the caliph al-Mu'izz. Showing in different colours the climates, mountains, rivers, cities, seas and routes, and valued at 22,000 *dinars*, it was looted from the Fāṭimid palace in 1067-8.[16]

---

14 Tibbetts, "Balkhī School", pp. 130-35.

15 See Johns and Savage-Smith, "Book of curiosities".

16 D. A. King, *World-Maps for finding the direction and distance to Mecca: innovation and tradition in Islamic Science* (London and Leiden, 1999), p. 35 and n. 65. In his history of the Fāṭimids al-Maqrīzī wrote: "There were found [among the looted treasures] thousands of silk hangings of different colours and sizes woven with gold. Among them depictions of countries and their kings and the famous things in them. The name of the king was written on each one, and the length of his reign, and explanations of his affairs". See al-Maqrīzī, Takī al-Dīn Abū 'l-'Abbās Aḥmad ibn 'Alī, *Itti'āẓ al-ḥunafā'*, ed. J. al-Dīn al-Shayyal and M. H. M. Aḥmad, 3 vols (Cairo, 1967-73), vol. 2, p. 285 [trans. Y. Lev]. See also O. Grabar, "Imperial and urban art in Islam: the subject matter of Fatimid art", in *Colloque international sur l'histoire du Caire, 27 mars-5 avril 1969* (Cairo, 1972), 1-57 [rpt in his *Studies in medieval Islamic art* (London, 1976), No. VII], p. 39.

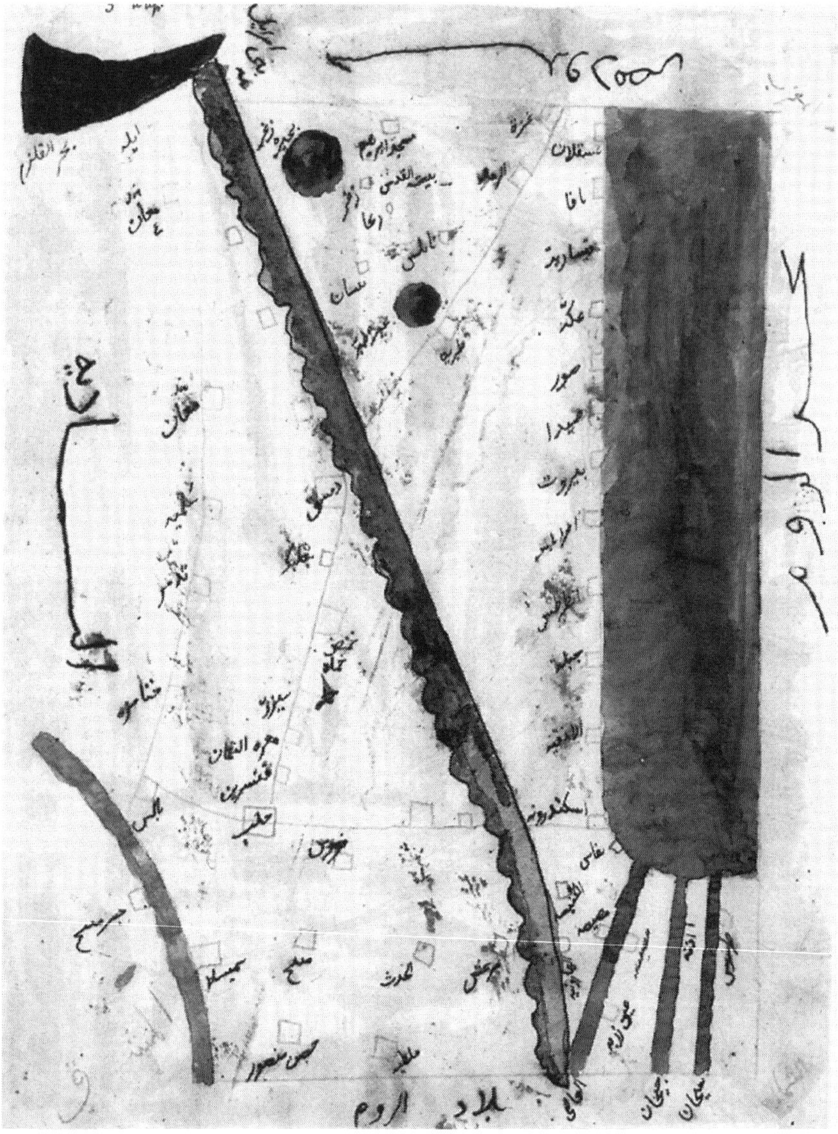

Fig. 9.1: Syria according to al-Iṣṭakhrī (1193 C.E.)
Berlin, Staatsbibliothek zu Berlin, preussischer Kulturbesitz, Orientabteilung, MS. Sprenger 1, fol. 28r
Reproduced with permission

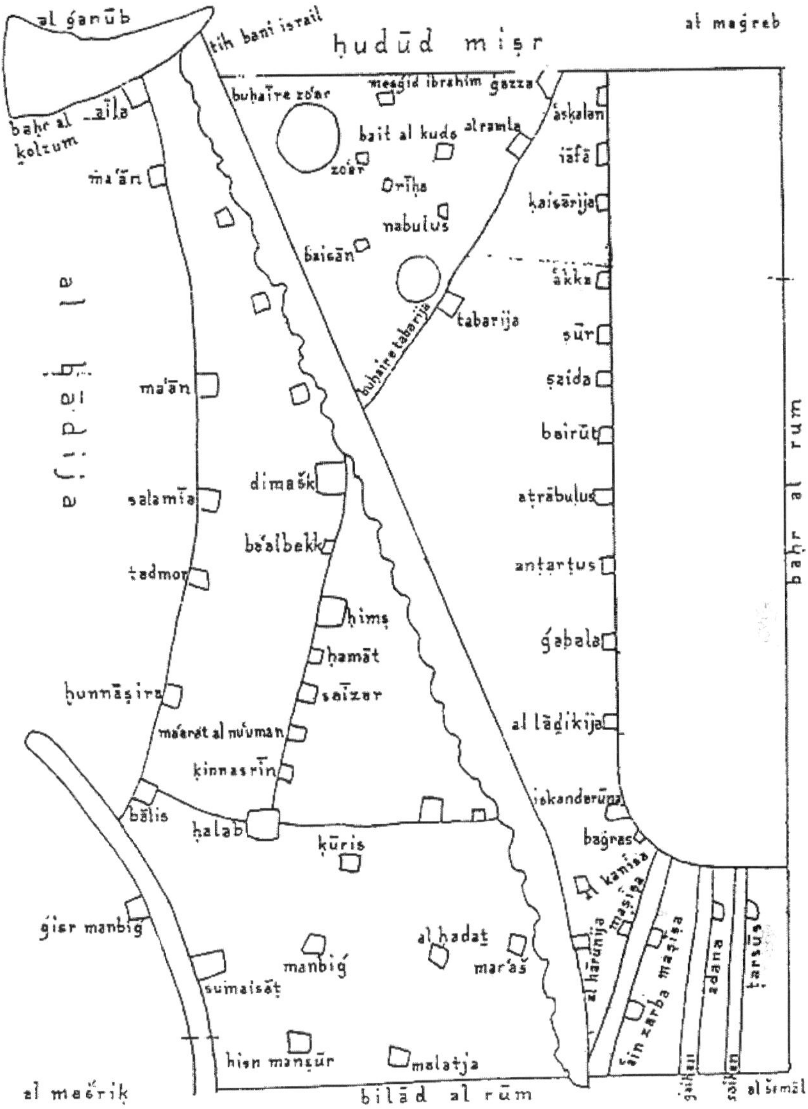

Fig. 9.2: Sketch of al-Iṣṭakhrī's map of Syria (1840 C.E.)
From K. Miller, *Mappae arabicae: Arabische Welt- und Länderkarten des 9.-13. Jahrhunderts*, 6 vols (Munich, 1926-31) [rpt; Frankfurt am Main, 1994], Bd 3, Beiheft 3, no. V, Tafel 11.2

Looking more closely at such maps, the likes of which may have been in the hands of a Kerbogha, a Zangī, or a Ṣalāḥ al-Dīn, almost all the regional ones were rectangles oriented toward the south in either the upper left or upper right corner.

When south was in the upper left corner, west was in the upper right, north in the lower right, and east in the lower left, with the south-north axis a diagonal running from upper left to lower right. Rivers were represented by wide parallel lines, mountain ridges by small, interlinked, round-topped heaps or interlinked triangles, towns by small squares or circles or rectangles topped by triangles, and routes by simple lines.

Al-Iṣṭakhrī's map of Syria is from a copy made in 1840 of an original drafted in 1193. In Miller's edition, the map, which is printed in negative, is accompanied by a sketch giving Latin transliterations of the Arabic legends. South is marked in the upper left. The upper end of the Red Sea (*baḥr al-qulzum*) is next to it, while a continuous mountain ridge runs from there diagonally to the lower margin. The bend of the Euphrates east of Aleppo stands out in the lower left corner, while the Mediterranean (*baḥr al-Rūm*) occupies most of the right margin. In the lower right corner, three rivers flow south into it. Along its eastern coast several towns are marked, the southernmost of which is *'Asqalān* (Ascalon). Inland in Palestine, *Ṭabariya* (Tiberias), *Baisān* (Beit Shean), *Nābulus* (Nablus), *Rīḥa* (Jericho), *Bayt al-Quds* (Jerusalem), *al-Ramla* (Ramlah), *Masjid Ibrāhīm* (Hebron) and *Ghazza* (Gaza) are marked.

The *Buḥayrat Ṭabariya* (Sea of Tiberias/Galilee) and the *Buḥayrat Zu'ar* (Dead Sea) are represented by circles but no Jordan connects them. A route runs from *Ḥalab* (Aleppo) to *Dimashq* (Damascus) and from there, via Tiberias and Ramlah, to Gaza. Another links *Bālis* on the Euphrates with *Ayla* on the Red Sea. Despite inaccuracies, the map provided a wealth of information about Syria-Palestine and the relative location of many of its towns.

Compared to regional maps, most early Muslim world maps and maps of the Mediterranean were far less exact. For example, the Leiden manuscript of al-Iṣṭakhrī dated to 1173 C.E. [Figure 9.3] depicted the Mediterranean schematically as an almost circular sea. Three equal-sized, perfectly-round islands equidistant from each other, captioned as Sicily, Crete, and Cyprus, appear in its centre. A triangular mountain island is situated at the place where the Mediterranean opens up to the Ocean. A large circle bordering the middle of the Mediterranean's left half marks the Nile, whilst a rectangle bordering the middle of the right half stands for the "Gulf of Constantinople". The coastal towns of Syria-Palestine as well as those situated west from Alexandria follow a sensible order, but the area in which one would look for Europe is mostly taken up by Spain, with the Franks limited to a small circle along the Mediterranean and Rome located far inland, much to the west of the Franks. This map was inferior to the regional maps exemplified by that of Syria.

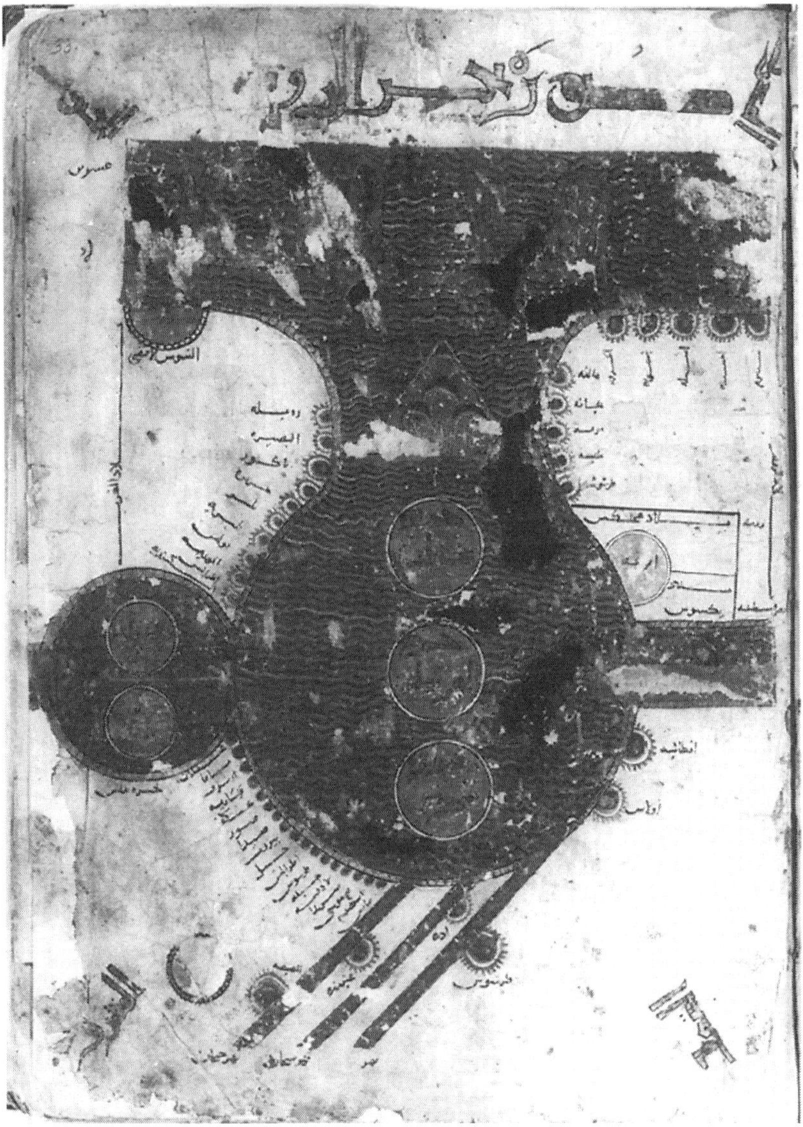

Fig. 9.3: The Mediterranean according to al-Iṣṭakhrī
Leiden, University Library, Legatum Warnerianum, MS. Or. 3101, fol. 33
Reproduced with permission

Some later maps of the Mediterranean were far less schematic. Ibn Ḥawqal, whose first patron was a Syrian ruler and who travelled in Muslim Africa and Sicily, pro-

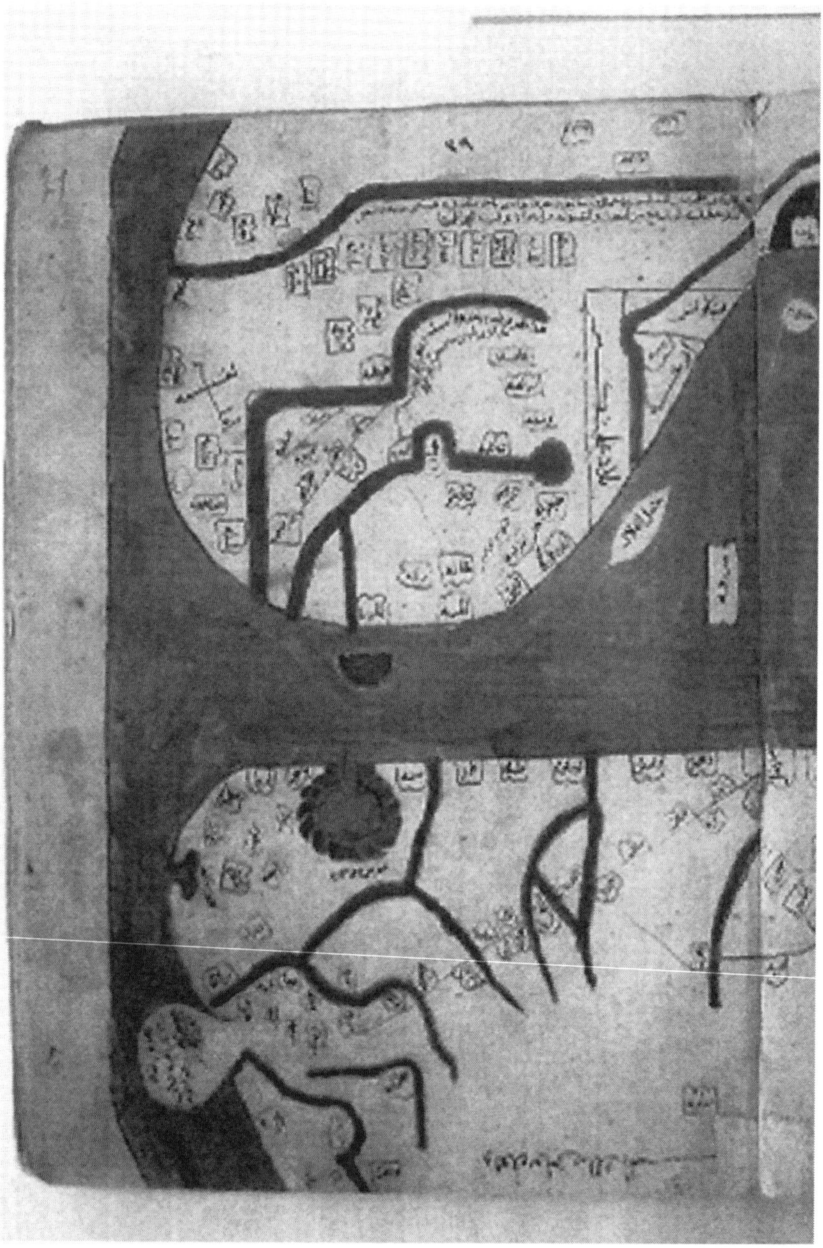

Fig. 9.4: The Western Mediterranean
Istanbul, Ṭopkapı Sarayı Müzesi,
Reproduced

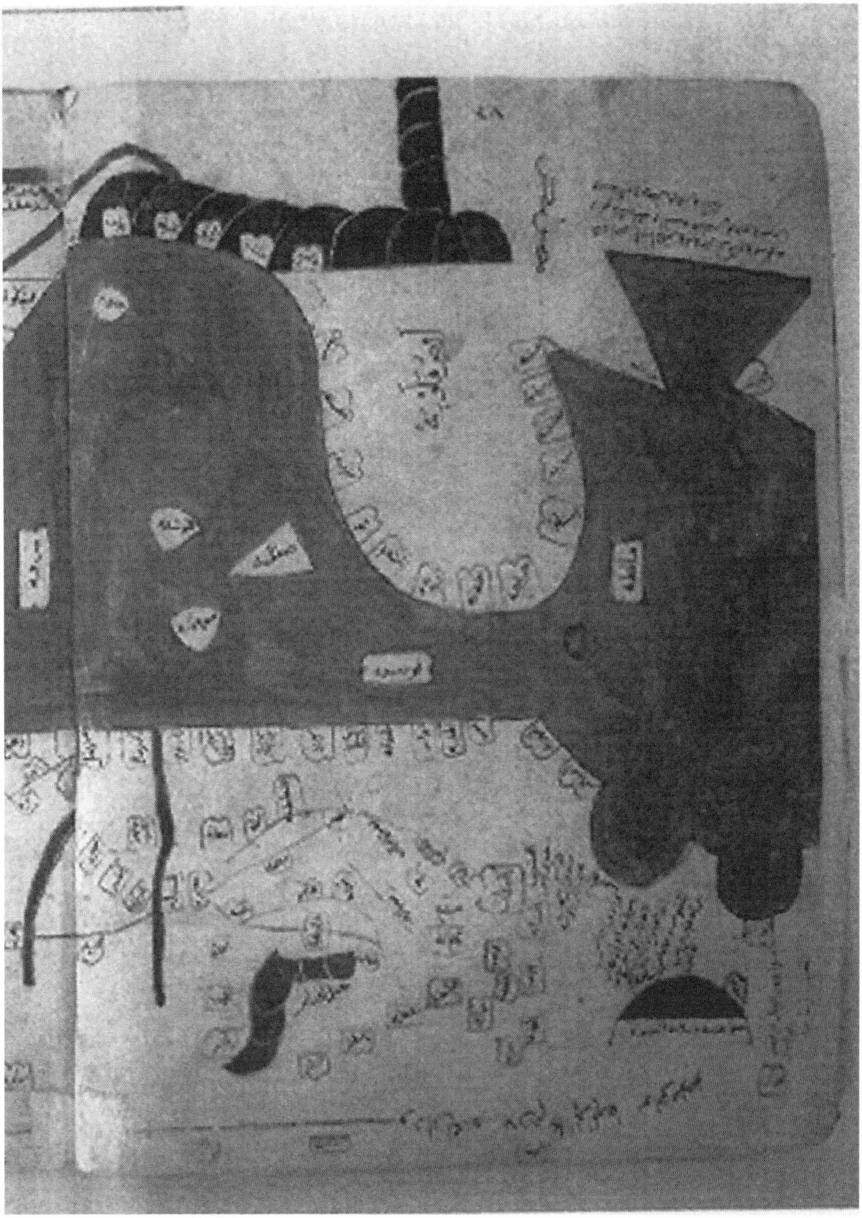

according to Ibn Ḥawqal (1086 C.E.)
MS. A. 3346, fol. 21r
with permission

duced maps of the Mediterranean that presented its islands more realistically and its European coast in greater detail. His map of the Maghrib, in reality of the entire Mediterranean, in a Ṭopkapı manuscript copied in 1086, is a case in point. The part of the map depicting the Western Mediterranean shows a triangular Sicily and the islands of Malta, Pantellaria, Sardinia, Corsica, Majorca, and Minorca. Surprisingly, Genoa is presented as yet another island. A trapezoid Adriatic Sea, with the Gulf of Venice at its top, figures near the map's upper right corner. Sicily, Pantellaria, and Malta flank a disproportionately large Calabria, almost half the size of Spain. Only a river separates Spain from France and Rome.[17] Thus the map was superior to that by al-Iṣṭakhrī, but its accuracy was still limited.

A circular world map forming part of a recently discovered treatise apparently composed in Fāṭimid Egypt showed a much better grasp of European geography. Spain, Italy, and Greece were now distinct from one another, and an attempt was made to chart the northern parts of the continent also. Indeed, this map was almost identical to the well-known one of al-Idrīsī. If indeed drawn before 1100, some eleventh-century Muslims must have had a better knowledge of Europe's geography than hitherto assumed. A rectangular world map appearing in the same treatise stands out for the graphic scale that figures prominently at its top.[18] This scale, which had no precedent in any extant earlier map, affords further proof of the sophistication attained by some cartographers in the Muslim world.

The state of cartography in the Latin West at the time of the First Crusade was strikingly different to that of the Muslim world. There was no *Christentum-Atlas* to match the *Islam-Atlas*, no Western school to match the Balkhī school of geographers. There were no regional maps, to say nothing of standardized sets of them. The West's lag behind the Muslim world is unmistakable in this sphere.

As in the Muslim world, in the Latin West references exist to early world maps no longer extant, such as those of Pope Zachary (741-52) in the Lateran and of Charlemagne at Aachen.[19] World maps did exist in the West in considerable numbers. About 180 dated to between the eighth and eleventh centuries have

---

17 For a slightly schematized rendering of all of this map, including the Eastern Mediterranean, accompanied by a sketch in which Arabic names are replaced by numbers referring to those names as they appear in the French translation, see J. H. Kramers and G. Wiet, *Ibn Hauqal, Configuration de la terre (Kitab surat al-ard),* 2 vols paginated continuously (Paris, 1964), Map 4 and pp. 59-62. For Syria see Map 7 and pp. 163-4.

18 Johns and Savage-Smith, "*Book of curiosities*". For a reproduction of al-Idrīsī's world map in the manuscript Oxford, Bodleian Library, MS. Pococke 375, fols 3v-4r, see Harley and Woodward, *History of cartography. 2.1*, plate 11.

19 See K. Miller, *Mappaemundi: die ältesten Weltkarten*, 6 fasc. (Stuttgart, 1895-98), fasc. 3, p. 151; A. Vidier, "La mappemonde de Théodulphe et la mappemonde de Ripoll (IXe-XIe siècles)", *Bulletin de géographie historique et descriptive*, 26 (1911), 285-313, pp. 289-92.

survived,[20] and these can have been only a fraction of those which actually existed around 1096. The vast majority represented the world in a strictly schematic manner, constituting diagrams rather than maps. One basic schema was the O-T (*Orbis Terrarum*) or T-O map, in which a perfect circle O encloses a T. Asia occupies the upper half of the circle and is divided from Europe and Africa by the horizontal bar of the T, representing the rivers *Tanais* (the Don) and Nile. Europe in the circle's lower left quarter is divided from Africa in the lower right by the T's vertical bar, representing the Mediterranean. The schema may have borne a Christian message, the T as cross and the O as orb evoking the world's salvation through Christ's crucifixion.[21] Nevertheless, about a dozen such representations of the world attempted to depict geographical features such as seas, rivers, mountains, and towns, in this respect resembling modern maps. At the same time they differed fundamentally from them and from contemporary Muslim world maps inasmuch as one of their aims was to provide information about major events in sacred history. They mapped time as well as space, or time superimposed on space, thereby communicating a blend of cosmological, historical, and theological notions.[22]

The surviving world map of the period in question with the most details was prepared by a certain Stephanus Garsia in the mid-eleventh century at the monastery of St-Sever in Gascony as part of a copy of the commentary on the Apocalypse compiled by Beatus of Liébana in 776. In other manuscripts of his commentary considered to be closer to the now-lost archetype, the map illustrated

---

20 Destombes's catalogue records 177 maps: 11 of the eighth century, 44 of the ninth, 50 of the tenth, and 72 of the eleventh. See M. Destombes, *Mappemondes A.D. 1200-1500* [*Imago Mundi*, Supplement no. IV] (Amsterdam, 1964), pp. 21-3. Destombes's dating of some maps has been challenged. For example, that of Rome, Biblioteca Apostolica Vaticana, MS. Reg. Lat. 571, fol. 71v, dated by Destombes to the twelfth century, is considered a product of the second half of the eleventh by M. A. Castiñeiras González, "Topographie sacrée, liturgie pascale et reliques dans les grands centres de pèlerinage: Saint-Jacques de Compostelle, Saint-Isidore-de-Léon et Saint-Étienne-de-Ribas-de-Sil", *Cahiers de Saint-Michel de Cuxa*, 34 (2003), 26-49, pp. 26-7.

21 See E. Edson, *Mapping time and space: how medieval mapmakers viewed their world* [The British Library studies in map making, 1] (London, 1999 [1997]), pp. 4-5, 14-15, 44-7 *et passim*. Note the similarity between the T's horizontal bar and the exact juxtaposition of the Nile estuary with the Straits to the Black Sea (the Dardanelles to the Bosporos), on some Bal<u>kh</u>ī school maps. See for example Fig. 9.4 above.

22 See M. C. Andrews, "The study and classification of medieval mappae mundi", *Archaeologia or miscellaneous tracts relating to antiquity*, 75 (1926), 61-76, pp. 65-6; P. D. A. Harvey, "Medieval maps: an introduction", in J. B. Harley and D. Woodward, eds, *The history of cartography. Vol. One: cartography in prehistoric, ancient and medieval Europe and the Mediterranean* (Chicago and London, 1987), 283-5, pp. 283-4; D. Woodward, "Medieval *mappaemundi*", in Harley and Woodward, *History of cartography. 1*, 286-370, pp. 286, 326, 342; M. Kupfer, "Medieval world maps: embedded images, interpretive frames", *Word & image*, 10 (1994), 262-88, pp. 262-4; Edson, *Mapping time and space*, *passim*.

assignment of mission areas to the Apostles by placing their busts at sites where they were said to have been buried.[23] The St Sever map shows an oval earth surrounded by an ocean full of fish, boats, and islands. Paradise appears at the top in the farthest east. Depicted in profile, naked and with long flowing hair, Eve is about to pick an apple from the Tree of Knowledge. The serpent curled around the tree's trunk gazes at her expectantly and Adam faces frontally with hands crossed over his genitals.[24] Rivers flow from Paradise in various directions. As on O-T maps, Asia occupies much of the upper half while Europe is in the lower left quarter and Libya, of which Africa is a part, in the lower right. The Mediterranean, however, is shown far more realistically than on O-T maps and the *Tanais*/Don and Nile are not aligned with each other.[25] Seven oval islands occupy much of the Mediterranean. Sardinia, Sicily, Crete, and Cyprus are more or less the same size and a small Corsica is closer to Sicily than Sardinia. In the south, beyond a red-coloured Red Sea with its Arabian and Persian gulfs and a large oval Sri Lanka marked *Tapaprone*, a fourth part of the world looms. The legend describes this part as unknown and claims that the Antipodes are said to dwell in it, repeating verbatim a statement of Isidore of Seville. This is true also of long legends abounding in Asia and Libya. Dependence on authoritative texts is evident. At the same time, the map-maker relied on personal knowledge when dealing in very great detail with the vicinity of his own monastery. The *Ecclesia Sancti Severi* is almost as large as the building symbolizing Rome and the same size as that representing Constantinople. The small river Adour, on whose southern bank St Sever stood, is as prominent as the Loire, with the legend: "Flumen Adurris qui alio nomine Alirris[?] dicitur". The town of Lescar is to the right of St Sever, and a castle, whose name is difficult to decipher, possibly Oloron, is to the left. *Bigorra* (Tarbes), *Aquis* (Dax), *Laburdum* (Bayonne), *Sancta Maria Mirsanensis* (Mont-de-Marsan) and *Basata* (Bazas), all small places less than 80 kilometres from St Sever, also appear. However, Paris, the Seine, and Marseilles are all absent.[26]

---

23 The map in the Beatus manuscript Burgo de Osma, Archivo de la Catedral, MS. Cod. 1, fols 34v-35r is regarded as the closest to the archetype surviving. For Beatus maps and reproductions, see J. Williams, *The illustrated Beatus: a corpus of the illustrations of the commentary on the Apocalypse*, 5 vols (London, 1994-2002), pp. 50-53 and Ill. 21; Edson, *Mapping time and space*, pp. 149-59 and Plate XI, opp. p. 117. For an opinion challenging the view that the Apostles' heads appeared in the archetype and regarding the manuscript New York, Pierpont Morgan Library, MS. 644 as closer to it, see J. Williams, "Isidore, Orosius, and the Beatus map", *Imago Mundi*, 49 (1997), 7-32.

24 Adam and Eve were also shown frontally with hands crossed over their genitalia in the genealogical tables of the St Sever Beatus manuscript. Adam is bearded and Eve's hair cascades down in strips. See Williams, *Illustrated Beatus*, vol. 3, Ill. 370.

25 On other Beatus maps the O-T scheme was much more pronounced. See Williams, "Isidore, Orosius, and Beatus", Figs 1, 2, 7 (pp. 8, 9, 19).

26 For transcription and discussion of the 270 place names and the various legends, see Miller, *Mappaemundi*, fasc. 1, pp. 41-70.

Wasconia (Gascony), the region of St Sever, occupies a huge portion of the map, far exceeding the space allotted to Iberia. An intention to highlight St Sever and its surroundings and to draw attention to their location in the world is clearly transparent.

Stephanus Garsia also depicted the Holy Land in considerable detail and allotted a disproportionate part of his map to it. The rivers Jor and Dan begin at *Mons Libanus*, running parallel to the Mediterranean, and combine into a Jordan flowing south into a rhomboid *Mare Tiberiadis* and continuing southwest to an oval Dead Sea extending, surprisingly, from east to west. Legends give lengths and widths of these seas. South of the Dead Sea is a mountain chain with a very high peak at its centre, bearing the legend: "Mons Sinai ubi filii Israel legem acceperunt". Like all others, the chain is represented by interlinked triangles, a sign used also on other Western maps as well as on some Arabic ones. The parts of the country are named *Iudea, Palestina, Galilea Superior*, and *Galilea Inferior*, the last two located east of the Jordan. Jerusalem, closer to the Sea of Galilee than the Dead Sea, is depicted as a large building with five doors, while Rome and Constantinople have only three each and Carthage four. As on all maps antedating the First Crusade,[27] Jerusalem is not at the map's centre. The second large town of the region is Tyre, which stands on a sizable peninsula jutting out into the Mediterranean. Other towns are Sidon, "Samaria", Caesarea, and Jericho, Sidon being located between Tyre and "Samaria". Caesarea is inland, separated from the Mediterranean by Mount Carmel and situated between Jerusalem and the Nile. Jericho, depicted as the smallest town, is close to the Sea of Galilee and Damascus, the same size as Jericho, is on its far side. The Orontes wends its way to the Mediterranean from a mountain chain northeast of Damascus. This detailed but rather chaotic representation of the Holy Land appears to have stemmed from a wish to somehow project Biblical and classical place names on the map rather than from knowledge based on experience.[28]

On the whole, Catholic Europe was depicted less accurately than the Holy Land, Egypt, or North Africa. Its Mediterranean shore is a single straight line that runs

---

[27] An observation made by Miller, *Mappaemundi*, fasc. 1, p. 30 and Wright, *Geographical lore of the time of the Crusades*, p. 259.

[28] The exaggerated size of the Holy Land on an eighth-century world map in a manuscript of the *Etymologiae* of Isidore of Seville, Rome, Biblioteca Apostolica Vaticana, MS. Vat. Lat. 6018, fols 64v-65r, led Smith to conclude that its two halves were regional maps on two scales. See C. D. Smith, "Geography or Christianity? Maps of the Holy Land before AD 1000", *Journal of theological studies*, 42 (1991), 143-52, pp. 150-52. For a reproduction see Destombes, *Mappemondes*, Ill. XIX [planche U]. However, the depiction of Africa as a single block extending in both halves on the same scale, as well as a comparison with the St Sever Beatus map, suggests otherwise.

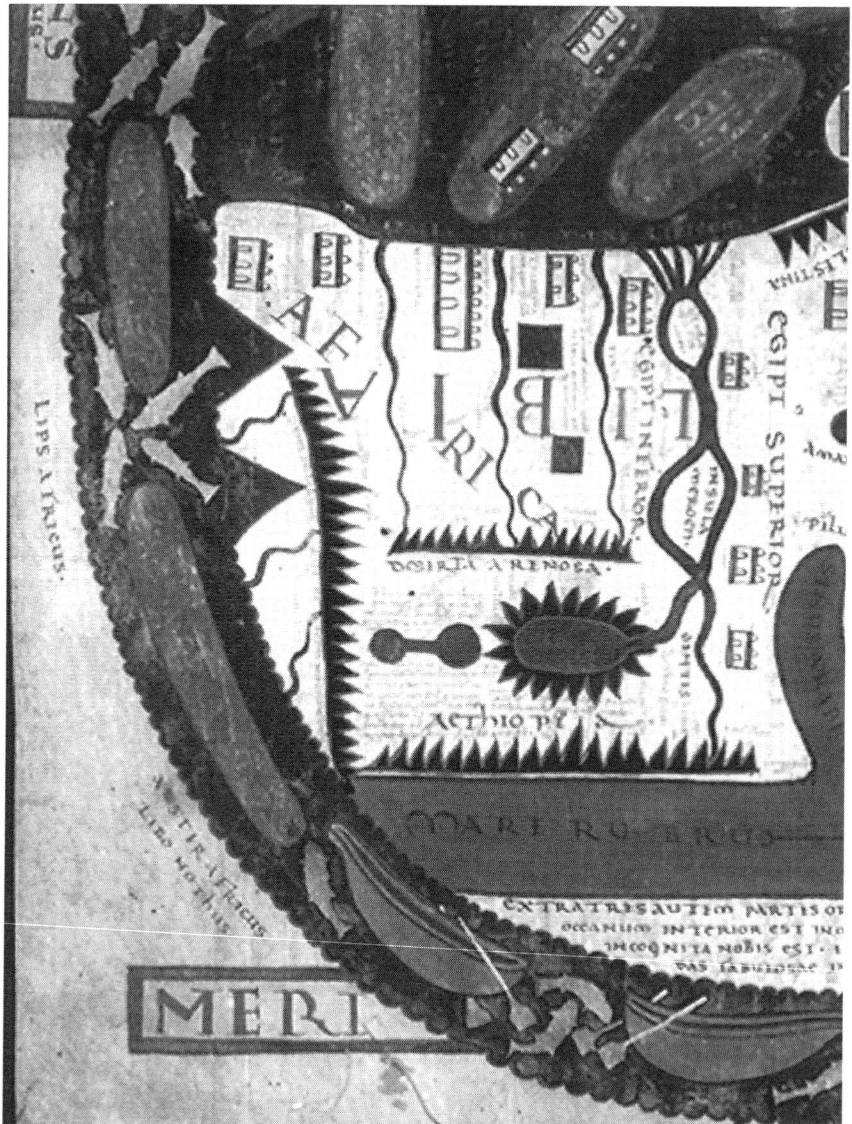

Fig. 9.5: Southern half of the St Sever
Paris, Bibliothèque Nationale,
Cliché Bibliothèque Nationale de France.

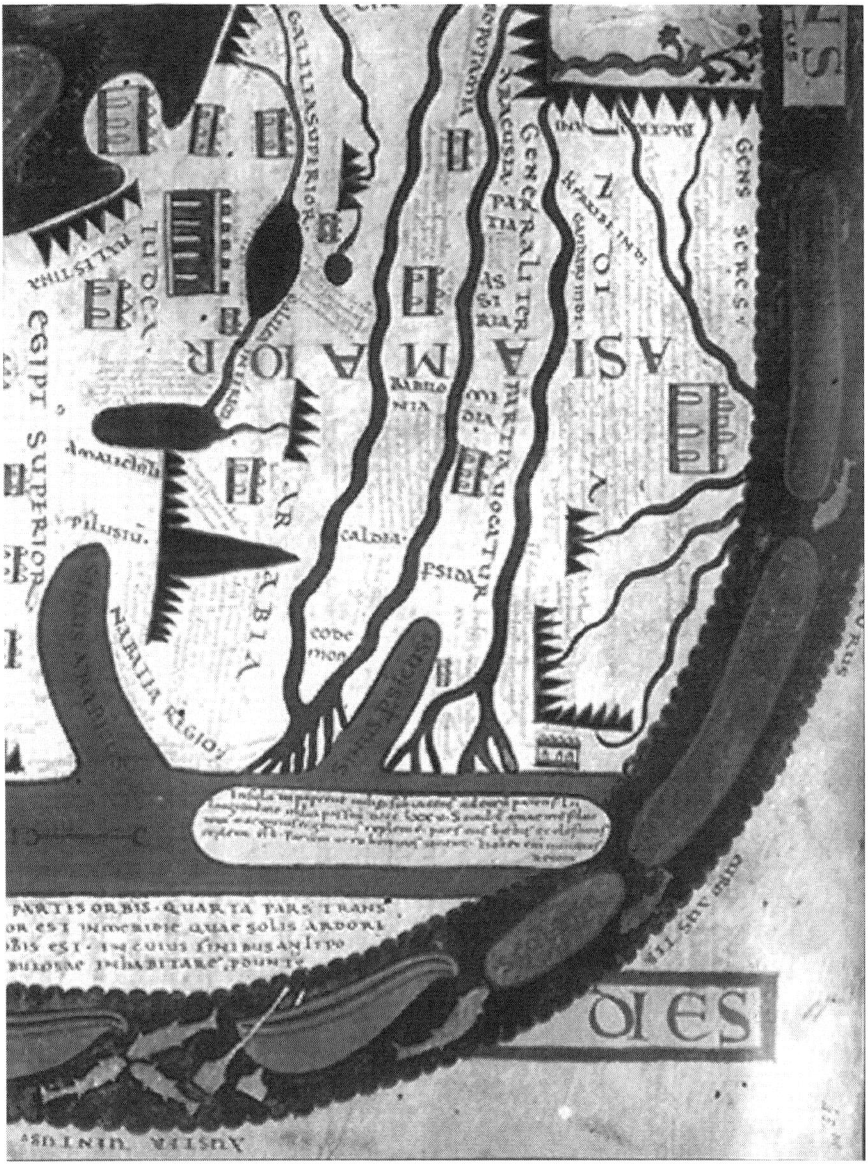

Beatus map (mid eleventh century).
MS. Lat. 8878, fol. 45 ter.
Reproduced with permission.

continuously from Lusitania (Portugal) to Benevento, amounting to a counter-part to Libya's straight northern coast.[29] The two straight coastlines and the schematic Mediterranean between them may go back to the O-T scheme's vertical bar but the result is that Spain, France, and Italy constitute parts of a single huge block. Obviously intent on pinpointing his monastery and its region on his world map, together with major events in sacred history, Stephanus Garsia was unable to arrive at a graphic representation of the lands of Latin Europe that would approximate the exactitude of the Holy Land or Egypt. The Pyrenees, the Rhône, and the Alps, all converging closely at Stephanus's straight Mediterranean coast, lie between Spain and Italy, calling to mind the river separating Spain from France and Rome on Ibn Ḥawqal's map copied in 1086.[30] This, however, showed, a Mediterranean coast that curved northeast from Cartagena to Carrara and thence southeast to Reggio. His Calabria was disproportionately large but jutted southward into the Mediterranean and there was a small, triangular Sicily to its west. On Stephanus Garsia's map Calabria was located northeast of Rome bordering the central part of the *Sinus Adriaticus*, while a large oval Sicily lay between Tuscany and Carthage. Ibn Ḥawqal's north Mediterranean coast was considerably more exact than Garsia's and the depiction of Latin Europe and its Mediterranean shores on the rectangular and circular world maps that formed part of the treatise presumably composed in Egypt in the late eleventh century was even more exact than his. At the time of the First Crusade some Muslims had a more precise idea of Latin Europe's geography than some of the finest map-makers in Europe itself.[31]

And yet the St Sever Beatus map and others resembling it could offer Latin Europeans some general idea about the relation of the continents to one another, about the Mediterranean and its islands, about countries, rivers, mountains, and towns. Educated men intent on making their way from France to the Holy Land could obtain from such maps a rough notion of what lay ahead. For while they differed from modern, or medieval Muslim, maps because of their theological

---

29 For colour reproductions of the St Sever Beatus map showing its northern section also, see Miller, *Mappaemundi*, fasc.1, back inset; Williams, *Illustrated Beatus*, vol. 1, pl. 8 opp. p. 49. On the eighth-century Isidore map the northern Mediterranean coast was depicted more realistically.

30 I am unable to deal here with the issue of whether European and Muslim world maps shared elements derived from common prototypes in antiquity. As for Ptolemy, see the cogent argument against the assumption that his *Geography* included maps in J. L. Berggren and A. Jones, *Ptolemy's Geography: an annotated translation of the theoretical chapters* (Princeton, 2000), pp. 46-50.

31 Note that Stephanus Garsia did not place in his Asia and Africa any town founded after the rise of Islam. For him there was no Baghdad, no Cairo, no Ramlah. This contrasted with the appearance of the land of the Franks and of Genoa on Ibn Ḥawqal's map. However, in the geographical treatise of Hugh of St Victor, *Bogdada*, *Maroch* and *Toflit* were mentioned. See Gautier Dalché, *Descriptio mappe mundi*, p. 78.

outlook, they attempted to visualize geographical reality also. As Kupfer has observed rightly: "... to define medieval *mappaemundi* solely as pictorial allegories of Christian cosmology, as theologized simulacra, would be inadequate." Surely they related to the physical world also. Hugh of St Victor wrote of the *mappa mundi* as a tool that allowed people to behold the ocean, islands, and distant regions without seeing them in the flesh.[32]

Monks and nuns presumably used world maps when setting out on spiritual pilgrimages.[33] Now, at least six abbots took part in the First Crusade and one of them, a certain Baldwin, travelled with Godfrey of Bouillon.[34] A number of monks and nuns also joined the expedition. Is it too far-fetched to assume that some of them, wont to have recourse to *mappae mundi* for imaginary pilgrimages, glanced at one before departure in 1096 or even took along a copy? Perhaps the versifier's notion of a *mapamunde* on Godfrey's tent was not utterly divorced from reality.

Moreover, we may conjecture that at least one leader of the First Crusade had seen a *mappa mundi* before departing for the East.

Baudri, abbot of St-Pierre de Bourgueil and later archbishop of Dol (1046-1130), is well known for his chronicle of the First Crusade. But Baudri was also an original and prolific poet. His longest and best-known poem, amounting to a panegyric of her, was dedicated to Countess Adela of Blois and contained a minute description of her bedchamber, whose walls were said to have been covered by tapestries showing Biblical and classical scenes as well as the Battle of Hastings. The bed was adorned with statues representing Philosophy, the Seven Liberal Arts, and Medicine. The sky with its constellations was depicted on the ceiling and on the floor was a *mundi mappa.*[35]

---

[32] Kupfer, "Medieval world maps", p. 263. See also Woodward, "Medieval mappaemundi", pp. 288-90; Edson, *Mapping time and space*, p. 163; Gautier Dalché, *"Descriptio mappe mundi"*, p. 133 and pp. 95-101 for evidence that maps were used for teaching.

The Crusades provide two illustrations of the amalgam of pragmatism and spirituality in medieval maps. If the *mappa mundi* which the Genoese seamen showed to Louis IX in 1270 was indeed in fact a portolan chart, then William of Nangis's description of it as a *mappa mundi* would signify that he considered such an accurate drawing prepared for practical purposes as fundamentally alike to traditional, theology-laden world maps. Similarly the Venetian Marino Sanudo Torsello who appended a map of the Holy Land to his plan for a new Crusade, bequeathed such maps to two Venetian churches in his testament of 1343. See R. Pallucchini, *La pittura veneziana del Trecento* (Venice and Rome, 1964), p. 8.

[33] Edson, *Mapping time and space*, pp. 14 and 169 n. 32.

[34] See J. Riley-Smith, *The First Crusaders, 1095-1131* (Cambridge, 1997), p. 200. On abbot Baldwin see Guibert of Nogent, *Dei Gesta per Francos*, 4.17, 7.32 (pp. 197, 330); William of Tyre, *Willelmi Tyrensis archiepiscopi chronicon*, ed. R. B. C. Huygens, 2 vols, in CC *ContMed*, vol. 63 & 63A paginated continuously, 10.15 (vol. 63, p. 472).

[35] See Baudri of Bourgueil, *Carmina*, ed. K. Hilbert, *Baldricus Burgulianus, Carmina*,

Baudri devoted 230 of the poem's 1,367 lines to this map.[36] Scholars have compared his description of it with earlier texts, mainly by Isidore of Seville, and with later mosaics;[37] however, it is also worth comparison to roughly contemporary world maps. Such a comparison reveals some striking similarities to the St Sever map. The round earth, enclosed by the ocean, was shaped like an egg, wrote Baudri.[38] The St Sever map showed an oval earth surrounded by the ocean. Asia, wrote Baudri, occupied about one-half of the orb, while Europe and Libya, a part of which was called Africa, occupied the other. Of these, Asia would be described first because it was honoured and ennobled by the Lord's Paradise.[39] With Paradise in its upper part, Asia was also given pride of place on the St Sever map and there again Africa formed part of Libya, spelled *Libia* both in the poem and on the map. Baudri wrote that towns and islands were marked on the map in Adela's bedchamber, although he did not give any of their names.[40] It would appear that on Adela's map the Holy Land figured about as prominently as on the St Sever map for of the ten mountains of Asia that Baudri enumerated, "Montes Riphei, Caucasus et reliqui, / Montes Israel, Sina, Carmelus et Ermon, / Taurus, mons Libani, Gelboe, mons Sinai", no less than seven were situated in the Holy Land.[41] Elsewhere he dealt at length with the Dead Sea.[42]

In Europe, Baudri concentrated on the region of his monastery, Bourgueil, to about the same extent that Stephanus Garsia highlighted that of St Sever. The Loire, which flowed through Adela's town of Blois as well as being near to Bourgueil, he presented as Europe's fourth-largest river and the most charming, its

---

[Editiones Heidelbergenses, 19] (Heidelberg, 1979), No. 134 (pp. 149-85). This and all earlier editions are based on the manuscript Rome, Biblioteca Apostolica Vaticana, MS. Reg. Lat. 1351. More recently a fourteenth-century manuscript, Paris, Bibliothèque Nationale, MS. Lat. 4126, has come to light. See P. Gautier Dalché and J.-Y. Tilliette, "Un nouveau document sur la tradition du poème de Baudri de Bourgueil à la comtesse Adèle", *BEC*, 144 (1986), 241-57. On Baudri as a homosexual poet see J. Boswell, *Christianity, social tolerance, and homosexuality: gay people in Western Europe from the beginning of the Christian era to the fourteenth century* (Chicago and London, 1980), pp. 244-7, 251 n. 31, 261-2 n. 66.

36 Baudri of Bourgueil, *Carmina*, No. 134, ll. 719-948 (pp. 168-74). For a translation of these lines into French prose see X. Barral i Altet, "Poésie et iconographie: un pavement du XIIe siècle décrit par Baudri de Bourgueil", *DOP*, 41 (1987), 41-54, pp. 42-4.

37 For the most meticulous discussion of the map's description see C. Ratkowitsch, *Descriptio Picturae. Die literarische Funktion der Beschreibung von Kunstwerken in der lateinischen Großdichtung des 12. Jahrhunderts* [Wiener Studien, 15] (Vienna, 1991), pp. 86-101. See also Barral i Altet, "Poésie et iconographie", pp. 42, 46-54.

38 Baudri of Bourgueil, *Carmina*, No. 134, ll. 741, 743 (p. 168).

39 Ibid., ll. 786-92 (pp. 169-70).

40 Ibid., ll. 901-2 (pp. 172-3).

41 Ibid., ll. 802-4 (p. 170).

42 Ibid., ll. 762-78 (p. 169).

water the healthiest, rendering whiter the bodies of young girls who bathed in it. "The fish of no other river equal those of the divine Loire and their very taste demonstrates it". He then mentioned two of the Loire's tributaries, the Vienne, which joins it 12 kilometres southwest of Bourgueil, and the Mayenne, which does so about 60 kilometres northwest of it. "There is also", he went on to say, "a river of no little name, whose waters unite with those of the Loire: *Cambio Burgullii*".[43] This was evidently the rivulet Changeon, which flows south and then changes direction near Bourgueil and continues west parallel to the Loire.[44] It does not matter that the Changeon is a little-known stream found only on detailed present-day maps. Just as Stephanus Garsia chose to highlight the small river Adour on whose banks stood the monastery of St Sever, so did Baudri pinpoint on Adela's world map the rivulet that flowed near his own monastery of Bourgueil. The rationale was the same in both cases.

Baudri related that when he was about to proceed mentally from Europe to Libya on Adela's map, he was seized by fear because the depiction of the Mediterranean was so lifelike that it seemed to be in motion.[45] Nevertheless he crossed the sea and was able to observe the mountains, rivers, and animals of Libya and Africa, as well as "the Ethiopians, the remotest people of Libya".[46] On the St Sever map also, Ethiopia occupied the continent's southernmost part. The details of Libya which Baudri mentioned having seen on Adela's map were far less numerous than those of Europe and here, again, the map resembled that of St Sever.

Alongside these resemblances there were some striking differences. Baudri wrote that on Adela's map the seas and rivers were marked in green.[47] On the St Sever map they were blue and on Muslim maps such as that of Ibn Ḥawqal they were denoted by different colours. To encounter seas and rivers marked in green on an extant map, one has to wait for the twelfth-century map of Lambert of St Omer.[48] Adela's map also resembled far more closely the O-T scheme than the St

---

43 Ibid., ll. 877-92 (p. 172).

44 Barral i Altet does not offer a modern equivalent of *Cambio Burgullii* and adds a question mark after it. Ratkowitsch speaks of the *Cambio* as the last and most prominent of the tributaries, which suggests Bourgueil as the ideal place. She draws attention also to verse 313 where William the Conqueror declares that he spared Bourgueil. See Barral i Altet, "Poésie et iconographie", p. 44; Ratkowitsch, *Descriptio Picturae*, p. 95.

45 Baudri of Bourgueil, *Carmina*, No. 134, ll. 905-6 (p. 173).

46 Ibid., l. 935 (p. 173).

47 Ibid., ll. 733-4, 865-6 (pp. 168, 172).

48 On the colours of seas and rivers on extant maps see A.-D. Van den Brincken, "Die Ausbildung konventioneller Zeichen und Farbgebungen in der Universalkartographie des Mittelalters", *Archiv für Diplomatik, Schriftgeschichte, Siegel- und Wappenkunde,* 16 (1970), 325-49, pp. 336-8.

Sever map.[49]

Several scholars have assumed that the map Baudri described was a mosaic.[50] But Baudri's statement that the map was covered by a glassy surface in order to prevent dust from damaging the *pictura*, which rested on a layer of marble,[51] suggests that he had in mind a painting. A mosaic would not have needed protection by a glassy surface.[52] Baudri's observation that the surface consisted of a material more transparent (*lucida*) than glass recalls the claim of Theophilus, the contemporary writer on the techniques of painting, glassmaking, and goldsmithing, that if one prepared a sticky varnish according to his instructions and applied a coating to a *pictura*, it became "transparent [*lucida*], decorous, and absolutely durable".[53]

A weightier issue on which scholars are divided is whether Baudri's description of Adela's bedchamber and of its map should be regarded as factual or not. Some have argued with varying degrees of conviction that the bedchamber was imaginary, pointing out that Baudri himself stated that he was describing "... plus, quod decuit, quam quod erat."[54] Others have claimed that while the bedchamber was fictitious, the description of some of its components, including the map, was based on Baudri's acquaintance with objects seen elsewhere.[55] Still

---

49 Baudri of Bourgueil, *Carmina*, No. 134, ll. 793-6 (p. 170).

50 J. Von Schlosser, "Ein veronesisches Bilderbuch und die höfische Kunst des XIV. Jahrhunderts", *Jahrbuch der kunsthistorischen Sammlungen des Allerhöchsten Kaiserhauses,* 16 (1895), 144-230, p. 159 [Italian trans., *L'arte di corte nel secolo decimoquarto* (Cremona, 1965), p. 25]; M. Manitius, *Geschichte der lateinischen Literatur des Mittelalters*, 3 vols. (1911-31; rpt, Munich, 1965-74), vol. 3, p. 894, n. 1; E. De Bruyne, *Etudes d'esthétique médiévale. Vol. II: l'époque romane* (Bruges, 1946), p. 73; N. Bartolomucci, "Tecnica e poesia in Balderico de Bourgueil: c. CXCVI", *Annali della Facoltà di Lettere e Filosofia della Università di Bari*, 17 (1974), 137-57, pp. 152-3; Ratkowitsch, *Descriptio Picturae*, p. 86.

51 "Ne uero puluis picturam lederet ullus / Tota fuit uitrea tecta superficie. / ... / Sustentabatur marmore supposito". Baudri of Bourgueil, *Carmina*, No. 134, ll. 727-8, 732 (p. 168).

52 Woodward, "Medieval *mappaemundi*", p. 339 n. 253, reporting O. A. W. Dilke. As early as 1926, Abrahams raised the question of whether Baudri described a mosaic or a painting protected by a transparent substance. See Abrahams, *Baudri de Bourgueil*, p. 249, n. 79.

53 Baudri of Bourgueil, *Carmina*, No. 134, ll. 729-30 (p. 168); Theophilus, *De diversis artibus / The various arts*, ed. and trans. C. R. Dodwell (London, 1961), c. 21 (p. 19). Thanks to Tim Dawson for drawing my attention to this passage.

54 Baudri of Bourgueil, *Carmina*, No. 134, l. 1354 (p. 184). See also De Bruyne, *Etudes d'esthétique médiévale. 2*, p. 72; J.-Y. Tilliette, "La chambre de la comtesse Adèle: savoir scientifique et technique littéraire dans le c. CXCVI de Baudri de Bourgueil", *Romania*, 102 (1981), 145-71, pp. 147-9, 153, 170; Ratkowitsch, *Descriptio Picturae*, pp. 24-6; Kupfer, "Medieval world maps", pp. 276-7.

55 O. K. Werckmeister, "The political ideology of the Bayeux Tapestry", *Studi Medievali*,

others consider the description of the bedchamber as realistic, or at least partially so.[56] As for the map, Baudri's mention of such details as the protective glassy surface has been adduced as an argument for some degree of factuality.[57]

Assuming that a world map whose basic contours resembled those of the *mundi mappa* described in the poem did exist in the bedchamber of Countess Adela, who was *sagacissima feminarum* according to Guibert of Nogent and who knew Latin,[58] and that Baudri amplified his recollection of it by studying a map such as the St Sever one and drawing on his knowledge of Isidore of Seville; if so, when would Baudri have seen the map on the chamber's floor? Scholars agree that the poem was written sometime between 1096 and 1102 because Baudri praised Adela for her chastity, for her love of her husband, and for preserving her marriage covenant inviolate despite many would-be tempters. In other words, her husband was alive but absent and Adela was chastely awaiting his return.[59] Her husband was, of course, Stephen of Blois (ca 1045-1102), who had married the much younger Adela (ca 1067-1137) around 1083 and became count of Chartres, Blois, and Meaux in 1089. In September 1096 he went east as one of the leaders of the First Crusade but he left the Crusader camp at Antioch in June 1098 and returned home. Perhaps prodded by Adela, he left again for Jerusalem in 1101 and lost his life in the Second Battle of Ramlah in May 1102.

We may further assume that the world map had adorned Adela's chamber for

---

ser. 3, 17 (1976), 535-96, pp. 555-6, 560-62; X. Barral i Altet, "Architecture, sculpture et mosaïque", in *Le monde roman 1060-1220: Les royaumes d'Occident*, ed. F. Avril, X. Barral i Altet and D. Gaborit-Chopin (Paris, 1983), 8-158, p. 146; Idem, "Poésie et iconographie", pp. 41, 53-4.

56 Von Schlosser, "Veronesisches Bilderbuch", p. 159; Manitius, *Geschichte der lateinischen Literatur*, vol. 3, pp. 892-95; N. Bartolomucci, "Il registro epico di Baudri de Bourgueil. *Adelae comitissae*, vv. 207-582", *Annali della Facoltà di Lettere e Filosofia della Università di Bari*, 39 (1996), 73-87, p. 77. See also Ratkowitsch, *Descriptio Picturae*, p. 25, nn. 23-24.

57 Woodward, "Medieval *mappaemundi*", p. 339.

58 Guibert of Nogent, *Dei Gesta per Francos*, 2.15 (p. 132). Orderic Vitalis called Adela "mulier sagax et animosa". See Orderic Vitalis, *The Ecclesiastical History of Orderic Vitalis*, ed. and trans. M. Chibnall, 6 vols. (Oxford, 1968-80), 10.20 (vol. 5, p. 324). On Adela's education and rulership see K. A. LoPrete, "Adela of Blois as mother and countess", in *Medieval mothering*, ed. J. C. Parsons and B. Wheeler (New York and London, 1996), 313-33, pp. 314-16; eadem, "The Anglo-Norman card of Adela of Blois", *Albion* 22 (1990), 569-89. See also eadem, *Adela of Blois, countess and lord* (Dublin, 2003) [non vidi!].

59 Baudri of Bourgueil, *Carmina*, No. 134, ll. 61-74 (pp. 150-51). For a translation of the passage, see J. H. Pryor, "Stephen of Blois: sensitive new-age Crusader or victim of history?", *Arts: the journal of the Sydney University Arts Association*, 20 (1998), 26-74, p. 29. For dating the poem on its basis see Abrahams, *Baudri de Bourgueil*, p. 232; Ratkowitsch, *Descriptio Picturae*, pp. 19-20. Abrahams's argument that the poem was written after Stephen's second departure to the East but before his death is less convincing.

some time before Baudri described it and that Stephen had seen it before leaving in 1096. Two of the letters he wrote to Adela while en route have survived and have been characterized rightly as "the most human of the documents regarding the Crusade".[60] In the first, written at Nicaea in June 1097, Stephen stated that those who claimed that the Bosporos was stormy and dangerous were wrong. He had crossed it swiftly and found it calm, safer than the Marne or the Seine. In the second, written on 29 March 1098 in camp at Antioch, he observed that those who claimed that the heat of the sun throughout Syria was hardly bearable were wrong, for Syria's cold winter resembled that of the West.[61] Evidently Stephen, whom Baudri described as "homo facundus et singularis scientiae",[62] had some notions about the East before he left on Crusade and subsequently compared them to his actual experience. To take a good look at a *mappa mundi* would have been in character for such a man.

In what sounds like an extremely unrealistic estimate, Stephen expressed confidence in his first letter that the Crusaders would reach Jerusalem from Nicaea in five weeks, unless held up at Antioch.[63] Even if they had taken the Pilgrims' Route across Asia Minor, which was substantially shorter than the route they actually did follow,[64] they would have had to cover about 1,400 kilometres. To accomplish this in five weeks they would have had to have made around 40 kilometres a day without any rest. So rapid an advance would clearly have been impossible. When they left Tripoli on 16 May 1099 their number was considerably smaller than it had been two years earlier at Nicaea. Yet even these smaller forces took 23 days, including 8 days of rest, to cover the 360 kilometres from Tripoli to Jerusalem, despite meeting virtually no resistance.[65] At the same rate of march the

---

60 S. Runciman, *A history of the Crusades*, 3 vols (1951-54; rpt, Harmondsworth, 1965), vol. 1, p. 333.

61 H. Hagenmeyer, ed., *Epistulae et chartae ad historiam primi belli sacri spectantes quae supersunt aevo aequales ac genuinae: Die Kreuzzugsbriefe aus den Jahren 1088-1100* (1901; rpt, Hildesheim, 1973), IV.5 (p. 139): "quidam illud Constantinopolitanum maris brachium saeuum ac periculosum dicebant, quod falsum est: nam in eo plus quam in Marna uel Sequana minime dubitandum est."; X.8 (p. 150): "quod quidam dicunt, uix posse pati aliquem in tota Syria solis ardorem falsum est: nam hiems apud eos Occidentali nostrae similis est.". The similar wording of the two observations is noteworthy. Hagenmeyer and Brundage believe that the first observation refered to the Sea of Marmara but the context leaves no doubt that the Bosporos was meant. See Hagenmeyer, *Epistulae*, p. 53; J. A. Brundage, "An errant Crusader: Stephen of Blois", *Traditio*, 16 (1960), 380-95, p. 382, n. 10.

62 Baudri of Bourgueil, *Baldrici, episcopi Dolensis, Historia Jerosolimitana*, in RHC *HOcc*, vol. 4, 1-111, III.xiii (p. 71).

63 Hagenmeyer, *Epistulae*, IV.14 (p. 140).

64 For these routes see J. France, *Victory in the East: a military history of the First Crusade* (Cambridge, 1994), Fig. 2 (p. 94).

65 France, *Victory in the East*, p. 327 and Fig. 16 (p. 320).

Crusaders would have needed at least 90 days to cover the 1,400 kilometres from Nicaea via the Pilgrims' Route to Jerusalem, not the 35 days believed by Stephen. Still, Stephen's estimate sounds extremely unrealistic only when confronted by present-day calculations of distances involved, calculations based on exact, scale-equipped maps. But if Stephen obtained his notion of the East from a map similar to that of St Sever, which lacked a scale and showed a disproportionately large Holy Land and a relatively small Asia Minor, his estimate becomes reasonable.

In the second letter to Adela written during the siege of Antioch, Stephen announced that the leaders of the Crusade, with the common counsel of the entire army, had made him "dominus, provisor atque gubernator" of all affairs.[66] "Provisor" has been interpreted to mean that Stephen was entrusted with administration and provisioning.[67] "Gubernator" may be understood as meaning "helmsman", someone capable of steering a way, while "provisor" may refer to an ability to see forwards. Supposing this to have been the case, and assuming that Baudri's poem was not totally divorced from reality, the map in Adela's chamber may have had some bearing on the way Stephen, in charge of logistics or of steering the Crusade's course at a crucial juncture, perceived his world.

---

66 Hagenmeyer, *Epistulae*, X.3 (p. 149).

67 Runciman, *History,* vol. 1, p. 232, n. 2; Brundage, "An errant Crusader", pp. 386-9; P. Rousset, "Étienne de Blois, croisé fuyard et martyr", *Genava,* n.s., 11 (1963), 183-95, pp. 186 n. 21, 192. Pryor, however, believed it more probable that *provisor* meant something like "overseer". See Pryor, "Stephen of Blois", p. 37 [but Pryor is impressed by this new interpretation! Ed.].

Chapter 10

Infantry in Muslim armies during the Crusades[1]

Yaacov Lev

Infantry had traditionally been an integral component of medieval Muslim armies but it seems that in the twelfth century a shift occurred in military organization which relegated infantry to an inferior position within overall military structures. The primary aim of this paper is to establish how widespread this shift really was, what led to it, and its consequences, especially in the context of campaigns against Franks and Crusaders. It then discusses the logistical problems involved in moving large infantry forces across the desert from Egypt to Palestine and Syria.

Contrary to popular belief, in pre-Islamic Arabia armies consisted primarily of infantry rather than cavalry, reflecting the local society and environment and the fact that Arabian society at that time was composed of sedentary populations as well as of nomadic tribes and semi-nomadic groups. In the South Arabian state of Sabā' in the second and third centuries C.E., for example, cavalry forces were small and camels were used for mobility rather than for fighting. Establishment of large cavalry forces was limited by a harsh environment in which most basic resources, such as water and agricultural and even grazing land, were scarce. As a result the number of camels, let alone horses, was low and this limited seriously any possibility of creating large cavalry forces.[2] During the Muslim conquest of the Middle East in the seventh century it was infantry that played the central role that led to victories at the battsles of the Yarmūk in 636 and at al-Qādisiyya in 637.[3]

---

1 Consult Maps 2, 7, 12.

2 For armies in pre-Muslim Arabia and nascent Islam, see A. F. L. Beeston, *Qahtan: studies in old South Arabian epigraphy. Fasc. 3: Warfare in ancient South Arabia (2nd – 3rd centuries A.D.)* (London, 1976), pp. 1-21; R. Hill, "The role of the camel and the horse in the early Arab conquests", in V. J. Parry and M. E. Yapp, eds, *War, technology and society in the Middle East* (London, 1977), 32-43. Other aspects of the military history of early Islam are discussed by F. M. Donner, "Centralized authority and military autonomy in the early Islamic conquests", in A. Cameron et al., eds, *The Byzantine and early Islamic Near East*, 3 vols (Princeton, 1992-5), vol. 3: *States, resources and armies*, 337-60; H. Kennedy, "The financing of the military in the early Islamic state", ibid., 361-78; and E. Landau-Tasseron, "Features of the pre-conquest Muslim armies in the time of Muḥammad", ibid., 299-336.

3 J. W. Jandora, "The battle of the Yarmūk: a reconstruction", *JAH*, 19 (1985), 8-21; idem, "Developments in Islamic warfare: the early conquests", *SI*, 64 (1986), 101-13. For

The most famous infantry forces in the medieval Muslim world were the *abnā'*, descendants of the 'Abbāsid revolutionary force recruited in Khurāsān in the mid-eighth century, who settled in Baghdad after its foundation and became identified with it. They were a versatile force capable of fighting as cavalry but who won a name as infantry. Infantry regiments historically formed part of Muslim armies along the Byzantine-'Abbāsid frontiers and both infantry and cavalry took part in the seasonal raids launched across the frontiers from the main towns of the border regions. In this context it is necessary to distinguish between cores of professional infantry led by officers and larger but more amorphous bodies of volunteers.[4] Another famous infantry force of the High Middle Ages was the Daylamites, an Iranian tribal group from the Elburz mountain region south of the Caspian Sea who entered Baghdad in 946. As infantry they were armed with typical infantry weapons such as swords, battle axes, and shields; however, their most famous weapon was the *župin*, a two-pronged spear also used as a javelin.[5]

In Egypt and Tunisia infantry became identified with black African slaves, especially during the ninth and tenth centuries. Ibrāhīm ibn al-Aghlab, the founder of the Aghlabid dynasty in Tunisia, bought and released from bondage 5,000 African slaves and settled them in al-Abbāsiyya, the capital which he built close to al-Qayrawān. The number of black slave infantry expanded greatly towards the end of the 870s and the term *'abīd* (a plural form of *'abd*, for a slave) began to be used to identify them.[6] The ability of the Aghlabid rulers to create an army based

---

wider military aspects of the Muslim conquest of the Middle East see F. M. Donner, *The early Islamic conquests* (Princeton, 1981); W. E. Kaegi, *Byzantium and the early Islamic conquests* (Cambridge, 1992).

4 For Ibn al-'Adīm's description of how troops, cavalry, infantry, and volunteers were recruited in Tarsos, see Ibn al-'Adīm, Kamāl al-Dīn Abu 'l-Qāsim 'Umar ibn Aḥmad, *Bughyat al-ṭalab fī ta'rīkh Ḥalab*, ed. S. Zakkār, 11 vols (Damascus, 1988-9), vol. 1, pp. 188-9. For a translation of relevant fragments, see C. E. Bosworth, "Abu 'Amr 'Uthmān al-Ṭarsūsī's *Siyar al Thughūr* and the last years of Arab rule in Tarsus (fourth/tenth century)", *Graeco-Arabica*, 5 (1993), 183-95.

5 Daylamites had also served as infantry in Sāsānid armies. Together with cavalry regiments they deserted and were incorporated into early Muslim armies. See D. Ayalon, "Preliminary remarks on the Mamlūk military institution in Islam", in Parry and Yapp, *War, technology and society*, 44-58; P. Crone, "The 'Abbāsid abnā' and Sāsānid cavalrymen", *JRAS*, 8 (1998), 1-19; I. Hasson, "Les mawālī dans l'armée musulmane sous les premièrs Umayyades", *JSAI*, 14 (1991), 176-213. For Daylamite infantry in the tenth-eleventh-century 'Abbāsid Caliphate see C. E. Bosworth, "Military organization under the Būyids of Persia and Iraq", *Oriens*, 18-19 (1965-6), 143-67. For other infantry troops in Islamic medieval armies see idem, "Ghaznavid military organization", *Der Islam*, 36 (1960), 37-77, pp. 59-60.

6 Al-Balādhurī, Abu 'l-'Abbās Aḥmad ibn Yaḥyā ibn Jābir, *Futūḥ al-Buldān*, ed. M. J. de Goeje, *Liber expugnationis regionum, auctore Imámo Ahmed ibn Jahja ibn Djabir al-*

on *'abīd* reflected trade routes and the local economy. Two trade routes linked Tunisia with Africa south of the Sahara and supplied the country with gold ore and black slaves. The long-prolonged conquest of Sicily was another source of slaves for labour on the large and medium-sized agricultural estates of Tunisia, on which olives were cultivated. The Aghlabid *dīnār* was a fine gold coin maintained at a normative weight of 4.25 grammes and an excellent standard of fineness.[7] The Aghlabid case is a significant example of the basic preconditions needed for successful establishment of slave armies in the medieval Muslim world: a prosperous economy and steady supply of slaves.

In ninth-century Egypt corps of black slave troops were formed by Aḥmad ibn Ṭūlūn and, when he died in 884, his army consisted of 24,000 military slaves and 7,000 *mawālī*, non-Arab converts to Islam. Ibn Ṭūlūn's army was a multi-ethnic force composed of both white slaves (*rūm* of Byzantine Greek or European origin) and blacks. Al-Qalqashandī wrote that it included 12,000 blacks, which is a reasonable estimate by comparison to other much higher and inflated figures provided by al-Maqrīzī.[8] In 905 the 'Abbāsids invaded Egypt, put an end to Ṭūlūnid rule, and destroyed the *'abīd*, who were the only element of the army to resist. To what extent racial prejudice against blacks played a role in the

---

Beládsorí (1866; rpt. Leiden, 1968), p. 234. See also M. Talbi, *L'émirat Aghlabide* (Paris, 1966), pp. 136, 284-5.

7 For Saharan trade routes in the eight to tenth centuries, see M. Brett, *The rise of the Fatimids: the world of the Mediterranean and the Middle East in the fourth century of the Hijra, tenth century CE* (Leiden, 2001), pp. 165-6; E. Savage, *A gateway to hell, a gateway to paradise: the North African response to the Arab conquest* (Princeton, 1997), ch. 4: "Ibāḍī slavers: an eighth-century transformation". For Aghlabid *dīnārs* and agriculture see A. S. Ehrenkreutz, "Byzantine tetartera and Islamic dīnārs", *JESHO*, 7 (1964), 183-90, p. 186; idem, "Studies in the monetary history of the Near East in the Middle Ages II. The standard of fineness of western and eastern dīnārs before the Crusades", *JESHO*, 6 (1963), 243-77, pp. 251, 256; M. Talbi, "Law and economy in Ifriqiya (Tunisia) in the third Islamic century: agriculture and the role of slaves in the country's economy", in A. L. Udovitch, ed., *The Islamic Middle East, 700-1900: studies in economic and social history* (Princeton, 1981), 209-49, pp. 214, 215, 237-8.

8 See al-Balawī, Abū Muḥammad 'Abd Allāh, *Sīrat Aḥmad ibn Ṭūlūn*, ed. M. K. 'Alī (Damascus, 1939), p. 349; al-Maghribī, Abu 'l-Ḥasan 'Alī ibn Mūsā ibn Sa'īd, *Kitāb al-Mughrib fī ḥulā 'l-Maghrib*, ed. and trans. K. Vollers as *Fragmente aus dem Muġrib des Ibn Sa'īd*, in *Semitistische Studien*, Heft 1 (Berlin, 1894), p. 76. For late medieval sources see al-Qalqashandī, Shihāb al-Dīn Abu 'l-'Abbās Aḥmad ibn 'Alī, *Ṣubḥ al-a'shā fī ṣina'at al-inshā'*, 14 vols (Cairo, n.d.), vol. 3, p. 467; al-Maqrīzī, Taqī al-Dīn Abū 'l-'Abbās Aḥmad ibn 'Alī, *Al-Mawā'iẓ wa 'l-i'tibār fī dhikr al-khiṭat wa 'l-āthār*, 4 vols (Būlaq, 1906), vol. 1, p. 152. Al-Maqrīzī wrote that Ibn Ṭūlūn's army consisted of 24,000 Turkish military slaves, 40,000 blacks, and 7,000 free-born mercenaries, apparently meaning *mawālī*. See also Y. Lev, "Regime, army and society in medieval Islam, 9th-12th centuries", in idem, ed., *War and society in the Eastern Mediterranean, 7th-15th centuries* (Leiden, 1997), 115-52, pp. 128-35.

'Abbāsids' severe treatment of them is difficult to ascertain.[9] Apart from the above, black slave infantry also played a role in Baghdad but the last reference to them dates from 929.[10]

The Fāṭimids employed *'abīd* from the inception of their rule in Tunisia in 909 and together with the Berbers of Kutāma they fought the rebel Abū Yazīd between 943 and 947. With their transfer to Egypt in 972, *'abīd* served them in both Egypt and Syria, for example against a rebel known as 'Allāqa in Tyre in 998.[11] There were quarters of *'abīd* in and around Cairo and under al-Ḥākim their corps were expanded vastly.[12] The creation of military quarters inhabited by certain ethnic socio-military groups or retainers of high ranking courtiers and *wazīrs* was typical of Fāṭimid Cairo and the *'abīd* quarters reflected this pattern.[13] Certain groups, especially the Berbers of Kutāma and Turks, also settled in adjacent Fusṭāṭ and married into the local population. Some Kutāma, who were the mainstay of the Fāṭimid army in Tunisia and in Egypt during the early decades of Fāṭimid rule, eventually lost their military character and became fully integrated into the local civilian society. But this option was not open to Fāṭimid *'abīd* because of racial prejudice and they remained an isolated foreign socio-military group. Their estrangement is best exemplified by the events of Dhū 'l-Qa'da 410 A.H. (February-March 1020 C.E.) when al-Ḥākim unleashed his *'abīd* against the population of Fusṭāṭ, which resisted them with the help of Turks and Berbers. Driven by hunger as a result of a famine which affected everyone gravely in 1024-5 and the failure of authories to pay them, the *'abīd* rioted and attacked Fusṭāṭ again in 1025. The population was authorized to defend itself and the regime sent other units, including naphtha-hurlers, to protect the town.[14]

---

9 Christian sources provide information about the establishment of a corps of black slaves in Egypt between 815 and 826 by the son of the governor Sarrī ibn al-Ḥakam. See P. Crone, *Slaves on horses: the evolution of the Islamic polity* (Cambridge, 1980), p. 255, n. 578.

10 See J. L. Bacharach, "African military slaves in the medieval Middle East: the cases of Iraq (869-955) and Egypt (868-1171)", *IJMES*, 13 (1981), 471-95.

11 Al-Anṭākī, Yaḥyā ibn Sa'īd ibn Yaḥyā, *Kitāb al-ta'rīkh*, ed. L. Cheikho et al., in CSCO *ScriptAr*, 3rd ser., vol. 7 (Paris, 1909), p. 108; Ibn al-Qalānisī, Abū Ya'lā Ḥamza ibn Asad, *Dhayl ta'rīkh Dimashq*, ed. H. F. Amedroz (Beirut, 1908), p. 50; Idrīs 'Imād al-Dīn, *'Uyūn al-akhbār*, ed. M. Ghālib, 6 vols (Beirut, n.d.), vol. 5, pp. 114, 118, 128, 184, 194, 195, 202, 214, 233, 258, 276, 311.

12 Al-Anṭākī, *Kitāb al-ta'rīkh*, p. 108; al-Maghribī, Abu 'l-Ḥasan 'Alī ibn Mūsā ibn Sa'īd, *Al-Nujūm al-zāhira fī ḥula haḍrat al-Qāhira*, ed. Ḥ. Naṣṣār (Cairo, 1969), p. 67; al-Musabbiḥī, al-Amīr al-Mukhtār 'Izz al-Mulk, *Al-Juz' al-arba'ūn min Akhbār Miṣr*, eds A. F. Sayyid and T. Bianquis (Cairo, 1978), p. 87.

13 Al-Musabbiḥī, *Akhbār Miṣr*, p. 80.

14 The events of A.H. 410 were the outcome of al-Ḥākim's controversial religious policies. See al-Anṭākī, *Kitāb al-ta'rīkh*, pp. 25-6; al-Musabbiḥī, *Akhbār Miṣr*, pp. 87-8, 89;

A glimpse into the strength and composition of Fāṭimid infantry was provided by the Persian traveller and Fāṭimid sympathiser Nāṣir-i Khusraw in his description of a military review held in Cairo in 1048. His numbers seem to be highly inflated but other details are more reliable. According to him, the Fāṭimid infantry was composed of 20,000 Maṣāmida or Maṣmūda, Berbers from the High Atlas mountains, 30,000 purchased black slaves, *'abīd al-shirā'*, and 30,000 East African, *Zanj*, troops armed only with swords. He made the important observation that "each ethnic group uses its own type of weaponry". Although bows were not mentioned among their weapons, there is no reason to believe that there were no archers among the black infantry. African archers became famous in the Muslim world and Arabic sources attributed the failure to conquer Nubia and the Sudan in the seventh century to superior African archery.[15] There were some references to Fāṭimid infantry archers of unknown ethnicity as well as to a few hundred crossbowmen who were detailed for fighting aboard warships.[16]

Competition between different socio-military groups within the multi-ethnic Fāṭimid armies and the relative position of each to the regime lay behind the outbreak of the civil war of 1066-73, in which *'abīd* played a crucial role. Isolated clashes between them and Turks had already taken place in 1062 but had been contained. The years 1063-6, however, were marked by a rapid turnover in the posts of *wazīr* and chief *qāḍī* which created internal instability. From 1067 onwards Turks became the strongest and most highly paid socio-military group and during the civil war they defeated those *'abīd* stationed in Cairo and Alexandria, although *'abīd* in Upper Egypt held their ground. The civil war exacted a heavy toll in human life, brought about the encroachment of nomads into the agricultural lands of Egypt, caused destruction and depopulation of the capital, and led to the regime's loss of the wealth and treasures that had accumulated in the palaces of Cairo. The process of recovery which began in 1073 was rapid and successful, however, and *'abīd* continued to play a role in Fāṭimid armies between 1073 and

---

Ibn Ẓāfir, Jamal al-Dīn Abu 'l-Ḥasan 'Alī, *Akhbār al-Duwal al-munqati'a*, ed. A. Ferré (Cairo, 1972), pp. 54, 55-6.

15 See Nāṣir-i Khusraw, Abū Mu'īn, *Safar-Nāmah*, trans. W. M. Thackston, *Nāser-e-Khosraw's Book of Travels (Safarnama)* (New York, 1985), p. 49. See also G. M. A. al-Tahir, "The Nubian archers in pre-Islamic and Islamic periods", *Graeco-Arabica*, 5 (1993), 139-52; D. Ayalon, "The Nubian dam", *JSAI*, 12 (1989), 372-90 [rpt in his *Islam and the Abode of War: military slaves and Islamic adversaries* (Aldershot, 1994), No. XII].

16 Ibn al-Ṭuwayr, Abū Muḥammad 'Abd al-Salām, *Nuzhat al-muqlatayn fī akhbār al-dawlatayn*, ed. A. F. Sayyid (Beirut, 1992), pp. 165-6, 187. For the different types of bows used by Fāṭimid troops see A. F. Sayyid, *Al-Dawla al-fāṭimiyya fī Miṣr. Tafsīr jadīd*, 2nd ed. (Cairo, 2000), pp. 702-3.

1171. Their number was estimated at 36,000-40,000 troops in the second half of the twelfth century, about 10,000 less than before the civil war.[17]

Fāṭimid *'abīd* fought the Franks at the Battle of Ascalon on 12 August 1099; however, there are no Arabic accounts contemporary with the events and later ones are not helpful for understanding the battle. The battle ended in a debacle, which explains why later historians minimized references to this dismal performance of a Muslim army. Ibn al-Qalānisī, the historian of Syria in the tenth to twelfth centuries, wrote that the Fāṭimid army camped outside Ascalon awaiting the arrival of Bedouin and a fleet. In the battle the Muslims suffered 10,000 casualties, 2,700 of which were from Ascalon, including merchants, court witnesses, urban militia, and the local garrison. The impression is that they were surprised and that civilians visiting the camp perished in the ensuing fighting. Other Arabic sources stated explicitly that the army was caught off guard and added that the Franks set on fire orchards to which the Muslims had fled.[18]

The anonymous author of the *Gesta Francorum* wrote that the Fāṭimid army "stood ready for battle" and attributed the Frankish victory to a decisive attack against what appears to have been the Fāṭimid centre, one of the flanks, and the camp. A direct attack on the camp would explain the losses sustained by members of the civilian elite of Ascalon who had apparently been present there. The Fāṭimid cavalry fled and was destroyed while others, most probably the infantry, were killed seeking refuge in trees or after throwing themselves to the ground.[19] Fulcher of Chartres, contemporary with the events but not a participant, wrote that the Frankish rear was endangered by the Fāṭimids and had to be rescued. The impression conveyed by his report is of a battle in which the Franks had the upper hand and destroyed the Fāṭimid infantry: the militia, the volunteers from Ascalon, and the *'abīd*. According to Albert of Aachen, the Fāṭimid black infantry fought gallantly but was defeated.[20]

---

[17] Al-Maqrīzī, Taqī al-Dīn Abū 'l-'Abbās Aḥmad ibn 'Alī, *Itti'āẓ al-ḥunafā'*, ed. J. al-Dīn al-Shayyal and M. H. M. Aḥmad, 3 vols (Cairo, 1967-73), vol. 2, p. 275; vol. 3, pp. 265-6; idem, *Khiṭaṭ*, vol. 1, p. 152.

[18] Ibn al-Qalānisī, *Dhayl ta'rīkh Dimashq*, p. 137. See also W. J. Hamblin, *The Fatimid army and the early Crusades: 1099-1124* (Ph.D., University of Michigan, 1985), pp. 236-8, 241-4. For Latin sources see J. France, *Victory in the East: a military history of the First Crusade* (Cambridge, 1994), pp. 360-65.

[19] Anonymous, *Gesta Francorum et aliorum Hierosolimitanorum*, ed. and trans. R. Hill, *The deeds of the Franks and the other pilgrims to Jerusalem* (Oxford, 1962), X.xxxix (pp. 95-6); Peter Tudebode, *Historia de Hierosolymitano itinere*, in *RHC HOcc*, vol. 3, 1-117, XVI.1-8 (pp. 111-17).

[20] Albert of Aachen, *Historia Hierosolymitana*, in RHC *HOcc*, vol. 4, 265-713, IX.xlix (pp. 622-3); Fulcher of Chartres, *Historia Hierosolymitana (1095-1127)*, ed. H. Hagenmeyer (Heidelberg, 1913), I.xxi.6-8 (pp. 314-16). See also Hamblin, *Fatimid army*, pp. 32, 150.

In other battles the scenario of Ascalon was repeated. Fulcher of Chartres wrote that in the third battle of Ramlah of 27 August 1105 the Franks defeated the Fāṭimid cavalry and destroyed the black infantry which "could not flee" and consequently was "slaughtered in the fields".[21] No less disastrous for the Fāṭimids was the battle of Ibelin of 29 May 1123 where, according to Fulcher, the Fāṭimid cavalry fled the battlefield like men "bewitched, going into a panic instead of using good sense" and where the infantry was massacred. The Fāṭimid army had been 16,000 strong but suffered 6,000 casualties. Ibn Muyassar wrote that the Fāṭimids were defeated without a battle, which can be interpreted as meaning that the cavalry fled.[22]

Examination of their battles with the Franks shows clearly that the Fāṭimids failed because in crucial battles their armies crumbled due to lack of cooperation between cavalry and infantry. Following the defeat at Ascalon, the Fāṭimid *amīr al-juyūsh*, al-Afḍal Shāhānshāh, initiated reforms to improve the cavalry.[23] The defeat of 1123 reflected the failure of these reforms but says nothing about the quality of the Fāṭimid *'abīd* and there is no reason to assume that they were either inferior or unreliable.[24] Destruction of infantry when separated from cavalry occurred many times during the Crusades but the separation was usually due to dynamics of fighting rather than to desertion of cavalry. For instance, in 1105 defeat of Riḍwān of Aleppo's cavalry by Tancred of Antioch resulted in the destruction of his infantry and his army suffering 3,000 casualties. Such misfortunes happened to Franks also. In 1164 Nūr al-Dīn's siege of the fortress of Harim led to the mobilization of the Franks and Byzantines. In the ensuing battle, although one wing of Nūr al-Dīn's army was defeated and pursued by the Franks, troops stationed by him to lie in ambush attacked the Frankish infantry. The pursuing Frankish cavalry were aware of the danger to which they had exposed the

---

21 Fulcher of Chartres, *Historia Hierosolymitana*, II.xxxii.11 (p. 500). See also M. Brett, "The Battles of Ramla (1099-1105)", in U. Vermulen and D. de Smet, eds, *Egypt and Syria in the Fatimid, Ayyubid, and Mamluk eras. I: Proceedings of the 1st, 2nd, and 3rd international colloquium organized at the Katholieke Universiteit Leuven in May 1992, 1993, and 1994* (Louvain, 1995), 17-39, including English translations of relevant Arabic accounts.

22 Fulcher of Chartres, *Historia Hierosolymitana*, III.xviii.4 (pp. 665-7); Ibn Muyassar, Taj al-Dīn Muḥammad ibn Yūsuf, *Akhbār Miṣr*, ed. A. F. Sayyid as *Choix de passages de la chronique d'Egypte d'ibn Muyassar ... selection faite par Taqiyy ad-Dīn Aḥmad ibn 'Alī al-Maqrīzī en l'anneé 814 H.* (Cairo, 1981), p. 95.

23 Ibn al-Ṭuwayr, *Nuzhat al-muqlatayn*, pp. 3-4.

24 For Fāṭimid preparations for the campaign see Ibn al-Ma'mūn al-Baṭā'iḥī, Abū 'Alī Mūsā, *Akhbār Miṣr*, ed. A. F. Sayyid (Cairo, 1983), pp. 61-2. For the desertion of the cavalry and the destruction of the infantry see al-Maqrīzī, *Itti'āẓ al-ḥunafā'*, vol. 3, p. 100.

infantry but returned too late to save them and the battle ended in a Frankish defeat with many leaders captured.[25]

Military failure of Fāṭimid armies reflected a social problem: the position of blacks in Muslim armies in particular and society as a whole. When examined within a socio-military context, the Battle of Ḥaydarān in Tunisia in 1052 provides insight into events of the wars between the Fāṭimids and the Franks. In that battle Sharaf al-Dawla al-Mu'izz, the Zīrid sultan of Tunisia, fought the invading Bedouin of the Banū Hilāl from Egypt; however, his Berber army crumbled and abandoned him and his *'abīd* to their fate.[26] Multi-ethnic armies such as his were very difficult to handle on the battlefield since such heterogeneous forces were ridden with ethnic animosities intensified by the different status of free-born troops vis-à-vis military slaves.

Such internal breakdown of multi-ethnic armies occurred not only when they were composed of blacks and other ethnic elements. In August 978 in Palestine, the Fāṭimid ruler al-'Azīz faced a coalition of Turks and Daylamites who were refugees from ethnic feuds within the military in Baghdad. After a brief battle, the Daylamites sought quarter, causing the collapse of the whole force.[27] In the case of forces composed of blacks and other ethnic groups, the problem became more intense because of racial prejudice against the blacks, who were regarded as inferior. For many centuries after its inception, Muslim society knew blacks only within the context of slavery, whether military, domestic, or agricultural. Only during the late Middle Ages did Muslims encounter free-born African Muslims, especially as pilgrims and traders, although even then such encounters were limited for the most part to North Africa and Egypt.[28]

Examining the careers of blacks in Muslim society, two facts stand out clearly. There is not a single reference to a free-born black person who made an administrative or military career and those who were mentioned were always eunuchs who rose from servitude to eminence. These realities applied to the Fāṭimid period in Egypt also, where some black eunuchs rose to prominence and even commanded military expeditions but their high position had no repercussions on broader military realities. The twelfth-century Fāṭimid army was composed of a bewildering assortment of corps created by various rulers and their *wazīrs* and consisting of different ethnic groups, some of whom were free-born and others

---

25 Ibn al-'Adīm, Kamāl al-Dīn Abu 'l-Qāsim 'Umar ibn Aḥmad, *Zubdat al-ḥalab fī ta'rīkh Ḥalab*, ed. S. Dahan, 3 vols (Damascus, 1951-68), vol. 2, pp. 50-51, 320.

26 M. Brett, "The military interest of the battle of Ḥaydarān", in Parry and Yapp, *War, technology and society*, 78-88.

27 Al-Maqrīzī, *Itti'āẓ al-ḥunafā*, vol. 1, pp. 243-4; Ibn al-Qalānisī, *Dhayl ta'rīkh Dimashq*, p. 19. In this case, al-Maqrīzī should be preferred to Ibn al-Qalānisī, who gave the improbable figure of 20,000 as the losses suffered by al-'Azīz's opponents.

28 See B. Lewis, *Race and color in Islam* (New York, 1970), pp. 38-74; idem, *Race and slavery in the Middle East: an historical enquiry* (New York, 1990), pp. 92-102.

military slaves. On the one hand, the cavalry was composed of some free-born Turks, Kurds, and Turkmen (Oghuz), as well as limited numbers of Turkish military slaves. On the other hand, the infantry was composed of despised blacks who were frequently abandoned on battlefields.

The question that needs to be asked is for what purpose the Fāṭimids maintained such large *'abīd* forces. Between 1073 and 1171, and perhaps even earlier, the corps of *'abīd* consisted of both slave and indigenous troops. In order to maintain the black slaves' social isolation African female slavery was also necessary and black men and women were continually shipped to Egypt to fill the ranks of the *'abīd*. Some of these were indeed slaves but others were their offspring born in Egypt, on whose soil an indigenous black slave army had evolved, a process indicated by the names of the *'abīd* corps. For example in the war that Ṣalāḥ al-Dīn waged against the blacks in Cairo in 1169 three corps, the *Rayḥāniyya*, *Juyūshiyya*, and *Farḥiyya*, were mentioned. The name Rayḥāniyya was derived from a black eunuch Rayḥān who had been a high-ranking courtier in the mid-eleventh century. The Farḥiyya corps was active during al-Ḥākim's reign and the Juyūshiyya corps derived its name from Badr al-Jamālī's title *amīr al-juyūsh*. The creation of personal corps designated by the name or title of their masters was common in the Muslim world but their continued existence long after their creators' deaths indicates their evolution into indigenous regiments which retained only the original name. Both importation of slaves and indigenous breeding of their offspring explains the high numbers of *'abīd*. To reduce costs, the Fāṭimids settled some *'abīd* on lands declared military zones closed to the local population so that they could support themselves by agriculture.[29] The large number of black *'abīd* settled in and around Cairo, however, was assigned to protect the regime and serve as a Caliphal guard. Their social estrangement was instrumental in achieving the aim and consequently their fate became interlocked with that of the regime that fostered them. Repeated failures of their multi-ethnic armies against the Franks in Palestine were of secondary importance from the point of view of the Fāṭimid regime. More significant was the eventual twofold failure of the blacks to fulfil their role during the civil war and again against Ṣalāḥ al-Dīn. In 1169 Ṣalāḥ al-Dīn annihilated the two military buttresses of the Fāṭimid regime in Cairo, the blacks and the Armenians, making extensive use of naphtha, setting on fire his opponents' positions, and razing the blacks' quarters outside the city walls and taking the land for other purposes. Many were killed and the rest were dispersed.[30]

The First Crusade encountered two different types of Muslim forces. In Asia Minor and north Syria they fought the Saljūqid *amīrs* whose armies were composed of Turkmen, nomadic Turkish tribes fighting as mounted archers, and in

---

29 Al-Maqrīzī, *Khiṭaṭ*, vol. 3, p. 29.

30 Ibid., vol. 3, pp. 2, 3-4.

Palestine the Fāṭimid army, which was a more complex force composed of both infantry and cavalry.[31] However, in contrast to our ample knowledge of the Fāṭimid army and its composition, our knowledge of Muslim armies of Asia Minor and Syria in the eleventh and twelfth centuries is rather limited. They were recruited from two very different social milieux: the nomadic world and urban populations. Following the Saljūqid invasion of the Middle East, Turkmen settled in the region and served as a reservoir of military manpower. After the Franks established themselves in Syria and Palestine warfare became endemic in the region and those Turkmen who fought the Franks in Syria came from the al-Jazīra region between the Euphrates and Tigris and sometimes from as far as Diyār Bakr, north of Mosul.[32] They were also prominent in northern Syria around Aleppo and were involved in campaigns against Franks there. One of their most successful raids took place in 1136 against the region of Latakia. As a local historian noted, 100 villages were attacked and 7,000 people and many thousands of animals captured and taken back to Aleppo. In this case those captured were Christians but Muslims also suffered from the incursions of the Turkmen. In the thirteenth century they joined the notorious Khʷārazmians in pillaging northern Syria.[33] As nomads like the Bedouin, however, the Turkmen were exposed to counter raids by Franks and suffered losses in people and stock.[34] In 1149 the Franks attacked them and the army of Aleppo had to come to the rescue. They fought as mounted archers and when in the service of Muslim rulers were rewarded with money and grain and were allowed to keep plunder they had taken.[35]

The Turkmen supplanted the Bedouin, whose service became less in demand, and Muslim armies of Syria rarely included Bedouin, although a notable exception to this was the battle of 1104 in the region of Ḥarrān east of Aleppo when the army of Mu'īn al-Dīn Sökmen I, *amīr* of Mardin, was composed of the three nomadic elements from the region: Turkmen, Bedouin, and Kurds.[36] Bedouin did fight for

---

[31] A glimpse into the composition of the Fāṭimid army and its activity against Franks is provided by a fragment of a report of a Fāṭimid commander from the beginning of the twelfth century. He described a siege operation, apparently successful, and his own contribution to the campaign in maintaining troops' morale and closely supervising cavalry and infantry. See G. Khan, ed. and trans., *Arabic legal and administrative documents in the Cambridge Genizah collections* (Cambridge, 1993), doc. 111 (pp. 428-30).

[32] Ibn al-Athīr, 'Izz al-Dīn Abu 'l-Ḥasan 'Alī, , *Al-Kāmil fī 'l-ta'rīkh*, ed. C. J. Tornberg, 13 vols (1851-76; rpt, Beirut, 1965-66), vol. 10, p. 639; vol. 11, pp. 7-8.

[33] For Turkmen raids on Muslim populations in 1120-21 and 1236-7 see Ibn al-'Adīm, *Zubdat al-ḥalab*, vol. 2, pp. 195-6; vol. 3, pp. 229-30. For their joining the Khʷārazmians in 1242-4 see vol. 3, pp. 264, 267.

[34] Ibn al-'Adīm, *Zubdat al-ḥalab*, vol. 2, pp. 268, 276, 276, 302, 311-2.

[35] Ibn al-Qalānisī, *Dhayl ta'rīkh Dimashq*, pp. 213, 225, 226.

[36] Ibn al-Athīr, *Al-Kāmil*, vol. 10, p. 374. For the wider consequences of this battle see T. K. el-Azhari, *The Saljuqs of Syria during the Crusades 463-549 AH / 1070-1154 AD* (Berlin, 1997), pp. 122-4.

Ṣalāḥ al-Dīn during the protracted struggle for Acre and the Third Crusade but, as far as can be ascertained, they acted as auxiliaries, in one case being detailed off to lay in ambush and in another being sent to attack and lure the Franks into an ambush set up by Ṣalāḥ al-Dīn's personal bodyguard, the *ḥalqa*. Bedouin robbers referred to by the offensive expression *shlūḥ* ("base") regularly raided the Frankish camp kidnapping and looting.[37] Fāṭimid forces in Palestine did, however, employ Bedouin regularly because of the geographic proximity of the war zone to the different nomadic groups in the region.[38] Egypt was outside the range of the Turkmen as also, apparently, was southern Palestine.[39] Another question that needs to be addressed is whether the military skills of Bedouin were inferior to those of Turkmen. Unlike Turkmen, Bedouin were not mounted archers but fought with the lance and proved themselves quite capable of defending themselves against detachments of Franks.[40]

Either as free-born warriors or as military slaves, Turks also served rulers of Syrian towns in the eleventh and twelfth centuries, most probably as mounted archers, but they also wielded other weapons and were capable of fighting with swords and possibly lances as well. In these armies some professional infantry, *rajjāla*, did exist but it must be emphasized that the meaning of *rajjāla* as professional infantry is not expressly stated; even though this may be deduced from what are quite suggestive sources. For example, in 1105 Riḍwān of Aleppo mobilized cavalry and 7,000 *rajjāla*, including 3,000 volunteers, to relieve a fortress besieged by the Franks.[41] Infantry were also drafted for inter-Muslim wars. For example, in the campaign of 1094 between Kerbogha, *amīr* of Mosul, and Aksungur al-Ḥājib both sides employed foot soldiers. Aksungur's army was described as consisting of Bedouin horsemen of the Kilāb tribe, the urban militia or *aḥdāth* of Aleppo, Daylamites and Khurāsānians, who fought as infantry. Kerbogha's army, on the other hand, was described as being composed of Bedouin, *rajjāla*, and a small number of Turks.[42]

On many occasions Muslim forces facing Franks in twelfth-century Syria included diversified infantry elements such as *mujāhidūn*, volunteers for the Holy

---

[37] Ibn Shaddād, Bahā' al-Dīn Abū 'l-Maḥāsin, *Al-Nawādir al-sulṭāniyya wa 'l-maḥāsin al-Yūsufiyya* (published under the title *Sīrat Ṣalāḥ al-Dīn*) (Cairo, n. d.), pp. 84, 91, 185, 193.

[38] Ibn al-Ma'mūn, *Akhbār Miṣr*, pp. 13, 20-1; Ibn Muyassar, *Akhbār Miṣr*, p. 75; Ibn al-Athīr, *Al-Kāmil*, vol 10, pp. 394-5.

[39] Five thousand Kurdish troops are referred to in Egypt in the eleventh century but it seems that these were mercenaries who were not part of the Kurdish nomadic population or its family and clan structure.

[40] Ibn al-Athīr, *Al-Kāmil*, vol. 10, p. 556.

[41] Ibid., vol. 10, pp. 394-5. For other references to *rajjāla* see vol. 10, pp. 399-400, 465, 467.

[42] Ibn al-'Adīm, *Zubdat al-ḥalab*, vol. 2, pp. 111-12.

War, *ghuzāh*, fighters of the Holy War, rural people, *aḥdāth* and other urban men with no military background. For instance, Muslim losses in the struggle for Antioch in 1098 were described as including professional troops or military slaves, *ghilmān*, *mujāhidūn*, and urban dwellers called *sūqa*, that is, men earning livelihoods in the market and ranging from small shopkeepers to manual labourers.[43] Occasionally *sūqa* and *'āmma*, meaning simple people, suffered heavy casualties while professionals escaped unscathed.[44] Active involvement of the Muslim rural population seems to have been marginal but during the siege of Damascus by the Second Crusade archers from the al-Buḳay'a valley reinforced the town. Whether recruited from the local rural population or whether a more permanent military force remains unresolved.[45]

Examining Usāma ibn Munqidh's *Kitāb al-'Itibār*, two facts emerge clearly. The first is that *rajjāla* bowmen were part of Shayzar's army and were quite capable of withstanding Franks, although they fared badly against Byzantines. However, Usāma's experience with other Muslim infantry forces was not particularly encouraging. Those of Ascalon who engaged Franks against his best advice were defeated easily. Another interesting military aspect is the devastating effect of mangonels used in siege warfare. Byzantine mangonels threw rocks of about 12-15 kilogrammes far beyond the range of bows.[46] However, details of Byzantine and Muslim siege troops are lacking. No information is available concerning specialized units such as the teams that operated the mangonels, *ḥajjārūn*, sappers, *naqqābūn*, and naphtha-throwers, *naffāṭūn*. Some of these were of Khurāsānian origin but how they and other siege corps were recruited and trained is unknown.[47]

The second fact is that Usāma and the military of Shayzar fought as cavalry employing the lance in exactly the same way their Frankish opponents did. In a short passage Usāma explained how a horseman should employ his lance when facing his enemy. The Arabic translates as follows:

> I say [that] it is necessary for one who enters into combat against a *fāris* (horseman) on horseback with a lance (*ṭa'n*) that he [should hold] his lance firmly with his hand (*yad*) and forearm (*dhirā'*) close to his side,

---

43 Ibid., vol. 2, p. 137.

44 For example, during the Muslim siege of 'Azāz in 1124-1125. See Ibn al-'Adīm, *Zubdat al-ḥalab*, vol. 2, p. 231.

45 Ibn al-Qalānisī, *Dhayl ta'rīkh Dimashq*, p. 299 and cf. pp. 201, 213, 225-6, 239, 265, 299, 340.

46 Usāma ibn Munqidh, ibn Murshid ibn 'Alī ibn Muqallad ibn Naṣr, *Kitāb al-I'tibār*, ed. P. Hitti (Princeton, 1930) [trans. P. Hitti, *An Arab-Syrian gentleman and warrior in the period of the Crusades* (1929; rpt, London, 1987)], text pp. 15-16, 67, 92, 113 [trans. pp. 40-41, 96, 122, 143].

47 Ibn al-Qalānisī, *Dhayl ta'rīkh Dimashq*, pp. 236-7.

> and should let his horse do what is done in combat against a knight on horseback with a lance. For if he should move his hand while holding the lance or stretch it out with the lance then his thrust will have no effect whatsoever and will result in no harm.[48]

Whether this passage actually described the couched technique in which the lance was held under the armpit is doubtful. According to Usāma, the lance should be held by the hand and under the forearm but here the word *dhirā'* should perhaps be understood as meaning the arm and not the forearm only. Usāma's description of the use of the lance is but one of many ways in which it could be used and it is important to bear in mind that only "... towards the end of the Middle Ages [did] the European couched technique became so refined that a knightly cavalryman could hardly hope to use this weapon in any other way."[49] Whatever the case, that Usāma and the military of Shayzar fought in the same way that Franks did means that they were engaged in small-scale warfare, fighting as horsemen with lances and in cooperation with infantry. Neither mounted archery nor mounted shock combat were mentioned by Usāma and many of his descriptions were of duels between two horsemen or engagements between small cavalry detachments.

In the twelfth century Egyptian and Syrian armies reflected two different types of societies. The Fāṭimids ruled over the richest country in the Middle East, one which produced large annual surpluses of grain in all but a few exceptional years and which was also a bridge for trade between India and the Mediterranean. Fāṭimid rulers had vast financial resources, access to merchants and commerce, and could create large armies in which the largest component, the *'abīd*, served as Caliphal guards. Syrian rulers' economic resources, no matter whether their territories were small or large, were only a fraction of those of the Fāṭimids. Shayzar was perhaps an extreme case of a small *amīrate* whose rulers could afford neither military slaves nor hired Turkmen. Other Syrian rulers did hire Turkmen and Kurds and their armies tended to be composed of cavalry fighting as mounted archers augmented by small, and frequently insignificant, infantry supplements.

Following the overthrow of the Fāṭimids in 1171, Ṣalāḥ al-Dīn established an Egyptian army similar to Syrian forces, with a strong preference for cavalry over infantry. His military build-up in Egypt represented a decisive shift away from Fāṭimid military organization in line with broader Muslim military tradition. From a military review conducted in 1171 and from information concerning his army in 1181, it is clear that the army became an all-cavalry force made up of two types: *ṭawāshī*, who were highly paid and well equipped, and lightly armed *qarāg͟hulams*. In the review of 1171, 14,000 troops paraded but the composition of the army is

---

48 Usāma ibn Munqid͟h, *Kitāb al-I'tibār*, p. 42 (trans. Lev).

49 D. Nicolle, "The impact of the European couched lance on Muslim military tradition", *The journal of the arms and armour society*, 10 (1980), 6-40, p. 6.

not specified while, in 1181, there were 6,976 *qarāghulams* and 1,553 *ṭawāshī*. Bedouin levies are mentioned on both occasions but Ṣalāḥ al-Dīn tried to limit their numbers.[50]

However, when accounts of Ṣalāḥ al-Dīn's campaigns against the Franks are examined, it is clear that infantry troops belonging to two different groups, *mujāhidūn*, and *rajjāla*, took part in at least some engagements. In 1179 forces mobilized to besiege the fortress of Chastellet near Jacob's Ford included cavalry, *rajjāla*, and *mujāhidūn*.[51] In 1189, while engaging the Third Crusade on a bridge between Tyre and Sidon, Ṣalāḥ al-Dīn's forces launched an unsuccessful infantry attack, involving *rajjāla*, *mujāhidūn*, and *sūqa*.[52] On 26 December 1189 ships from Egypt entered Acre bringing 10,000 sailors and "Muslim *rajjāla*" to the besieged city.[53]

Ṣalāḥ al-Dīn's example was followed by later Ayyūbids. Infantry were not an integral part of their military organization in either Egypt or Syria. The reason for this break with military tradition seems to have been related to the spread of the *iqṭā'* as the exclusive financial tool for paying the standing armies, *'askar*, of Ayyūbid rulers. The *iqṭā'* system became "the most crucial factor for maintaining the Ayyūbid state order".[54] The right to grant or cancel *iqṭā's* was the most important prerogative of the sultan and the cooperation of family members and *amīrs* was secured through sultanic control of the system. The size of iqṭā' grants was related to the number of cavalry an *amīr* was obliged to provide. Revenues from *iqṭā's* maintained the cavalry that became the mainstay of Ayyūbid armies but the spread of *iqṭā's* diminished tax revenues available to the sultans. In the Fāṭimid period the state rather than *iqṭā'* holders had collected most rural taxes and these had been used to maintain the court and armies. Armies had been paid in money and the state maintained stables for military needs and workshops for production of weapons. On the eve of a campaign against the Byzantines in Syria in 995, there had been 12,000 horses, 30,000 camels, and 1,800 two-humped Bactrian camels in the state stables.[55] Following the demise of the Fāṭimids, court personnel was dispersed and palaces used for other purposes. Ṣalāḥ al-Dīn was impoverished by comparison to Fāṭimid rulers who had large liquid resources at

---

50 Al-Maqrīzī, *Khiṭaṭ*, vol. 1, 139. Cf. H. A. R. Gibb, "The armies of Saladin", *Cahiers d'histoire égyptéenne*, ser. 3, 4 (1951), 304-20 [rpt in his *Studies on the civilization of Islam* (London, 1962), 74-90], rpt pp. 76-7; Y. Lev, *Saladin in Egypt* (Leiden, 1999), pp. 143-7.

51 Abū Shāma, Shihāb al-Dīn Abu 'l-Qāsim 'Abd al-Raḥmān, *Kitāb al-rawḍatayn fī Akhbār al-Dawlatayn*, ed. I. al-Zaybaq, 2 vols (Beirut, 1997), vol. 3, pp. 36-7.

52 Ibid., vol. 4, p. 72; Ibn Shaddād, *Al-Nawādir al-sulṭāniyya*, pp. 81-2.

53 Abū Shāma, *Kitāb al-rawḍatayn*, vol. 4, p. 103.

54 T. Sato, *State and rural society in medieval Islam: sultans, muqtās and fallahun* (Leiden, 1997), pp. 46, 50.

55 Al-Maqrīzī, *Itti'āẓ al-ḥunafā'*, vol. 1, pp. 287-8.

their disposal to the very end of their rule.[56] Ṣalāḥ al-Dīn held personal *iqṭā'*s but had very little money and the spread of the *iqṭā'* diminished the financial resources available to his administrators. These realities are well illustrated by his budget of 1181 in which the troops maintained through the *iqṭā'* system included a cadre of 111 *amīrs* and the *ṭawāshī* and *qarāghulams*. 3,670,500 *dīnārs* were allocated for them, but the state had further military expenditures as well since Bedouin levies were maintained by a special type of *iqṭā'* and former Fāṭimid troops were still on the military payroll. The only civilian expenditures mentioned were the salaries of jurists, *qāḍīs*, and mystics, and the expenditures on these budgetary items amounted to only a million *dīnārs*.[57]

Ṣalāḥ al-Dīn spent vast sums on his armies but they were still much smaller than those of the Fāṭimids and infantry were eliminated. In reports of the strength of Ayyūbid armies references were almost always to cavalry only. For example, the army of al-Mu'aẓẓam of Damascus was said to consist of 3,000 high-quality cavalry. Although his own army was 12,000, al-Kāmil of Egypt was reported to be afraid to engage al-Mu'aẓẓam for fear that it might desert him. Another example was sultan al-Ashraf of Damascus who became involved in the wars of the Kh[w]ārazmians in 1230 with a force of 5,000 high-quality cavalry in terms of accoutrements and horses from al-Jazīra, Ḥomṣ, and Ḥamā. A third was that of sultan al-Ṣāliḥ Najm al-Dīn who left Damascus in 1239 with 5,000 cavalry to conquer Egypt.

The disappearance of infantry from Ayyūbid payrolls did not mean that they were not available and the participation of *mujāhidūn*, urban elements, and *rajjāla* in Ayyūbid campaigns against the Franks is well attested. At al-Manṣūra in 1249, Bedouin, volunteers, *ghuzāh*, and urban riff-raff (*ḥarāfish*) fought alongside regular regiments.[58] However, such reports of *rajjāla* referred to very small contingents. When Damascus was besieged by al-Kāmil in 1237, the *amīr* of Ḥomṣ, al-Mujāhid Asad al-Dīn, sent a detachment of a mere 50 *rajjāla* to reinforce the town and their fate was a sorry one, being captured and executed. Execution of prisoners was not standard conduct but such atrocities did take place and in this case al-Kāmil's behaviour can be explained by deep animosity toward al-Mujāhid.

---

[56] In 1169-70, while defending Damietta against the Franks and Byzantines, Ṣalāḥ al-Dīn received one million *dīnārs* from the Fāṭimid *imām*. See Tāj al-Dīn Shāhanshāh ibn Ayyūb, *Kitāb al-ta'rīkh*, in Ibn Shaddād, *Al-Nawādir al-sulṭāniyya*, 253-310, p. 262, quoting Ṣalāḥ al-Dīn himself.

[57] Al-Maqrīzī, *Khiṭaṭ*, vol. 1, p. 139. Cf. H. Halm, *Ägypten nach den mamlukischen Lehensregistern*, 2 vols (Wiesbaden, 1979-82), vol. 1, pp. 14-16.

[58] A.-M. Eddé, "Saint Louis et la Septième Croisade vus par les auteurs arabes", *Cahiers de recherches médiévales (XIIIe-XVe s.)*, 1 (1996), 65-92, pp. 82-4.

On another occasion, al-'Adil brought a force of 500 cavalry and 100 *rajjāla* from Aleppo to Acre in 1203.[59]

When required for campaigns, infantry and siege troops were enlisted and paid in the same way as Turkmen. However, rulers deciding on who to hire were inclined to favour Turkmen over *rajjāla*. As standing armies maintained through *iqṭā'* systems, Turkmen fought as mounted archers and augmented rulers' own forces. On a more fundamental level, disappearance of infantry was the result of socio-military realities and the armies of Saljūqid, Zangīd, and Ayyūbid rulers reflected the way Turkish and Kurdish societies fought. The spread of *iqṭā's* was instrumental in sustaining this socio-military tradition during the transition from nomadism to statehood. Removal of infantry from the military payroll was the price paid for continuing the nomadic military tradition of mounted archery under the auspices of the state. Information available about Saljūqid armies in Asia Minor in the thirteenth century suggests that in those, as in those of the Ayyūbids, emphasis was placed on cavalry and infantry were usually drafted only to answer needs of specific campaigns.[60]

Broadening the investigation, we must question whether the poor Fāṭimid record against the Franks was due to difficulties of equipping and dispatching to Palestine armies including large numbers of infantry. Rather surprisingly, Arabic sources, including travel and geographic literature, contain very few references to logistical problems involved in travel, pilgrimage, and warfare. Reading them, one is left with the impression that moving people and goods across land or sea was neither worthy of special attention nor needed comment. People had the organizational skills and means to provide for armies crossing deserts, or caravans going to Arabia or for commerce. William of Tyre, most probably relying on the *Gesta Francorum*, did mention water shortages and the difficulties they caused during the siege of Jerusalem in 1099; however, when reporting Amalric's expeditions to Egypt, he barely alluded to logistical concerns. He did refer to the army's need to carry food when crossing the desert between Gaza and Bilbays, but water seems to have posed no problem. In October moderate autumn temperatures made faster advance possible and the journey took only ten days, while in August progress was slower and the army had to move from stopping place to stopping

59 Ibn Wāṣil, Abu 'Abd-Allāh Jamāl al-Dīn Muḥammad, *Mufarrij al-kurūb fī akhbār Banī Ayyūb*, eds M. Shayyal, H. Rabie, and S. Ashour, 5 vols (Cairo, 1954-77), vol. 4, pp. 209, 298-9; vol. 5, pp. 151, 211; al-Maqrīzī, Taqī al-Dīn Abū 'l-'Abbās Aḥmad ibn 'Alī, *Al-Sulūk li-ma'rifat al-mulūk*, eds M. M. Ziyādah and S. 'A. 'Ashūr, 10 vols (Cairo, 1934-73), vol. 1, pt. 1, p. 160.

60 A. Bombaci, "The army of the Saljuqs of Rum", *Annali del Instituto orientale di Napoli*, 38 (1978), 343-69, p. 353. Ayalon emphasizes the continued role of Turkish military slaves. See D. Ayalon, "The mamlūks of the Seljuks: Islam's military might at the crossroads", *JRAS*, 6 (1996), 305-33.

place where there was abundant water and by moderate stages in order not to exhaust the infantry.[61]

The route from Egypt to Palestine proceeding from Fusṭāṭ to Bilbays, Fāqūs, and either Jurjīr or al-Qanṭara, to al-Faramā', al-Baqqāra, al-Warrāda, and then to al-'Arīsh and southern Palestine was well known. In 1168 the Franks reached Bilbays from Ascalon in about two weeks from 20 October to 5 November. In the tenth century Bilbays had been an important provisioning centre for pilgrimages to the holy cities of Arabia and many mills providing flour for biscuit, *ka'k*, operated there. The infrastructure serving pilgrimages was also available for armies. At the other end of the route, al-Faramā' was a different place altogether and in his geographic dictionary of the Muslim world Yāqūt provided a detailed account of it copied from the tenth-century Fāṭimid court geographer, al-Muhallabī. According to this, al-Faramā' was not a town but rather a fortress on the sea surrounded by salt marshes notorious for their stench, whose only potable water was rainwater and that imported from the Nile. The surrounding region was, however, kinder and did provide fresh water which made the region famous for date palms. The Coptic and Bedouin population's main occupation was trading in fodder and equipping caravans going to Egypt. In both 1115 and 1150 al-Faramā' was destroyed by the Franks and on their way to Egypt in 1169 they found the place deserted. The dikes that had prevented flooding from the sea had collapsed and the littoral was covered by a shallow sea, rich in fish. The destruction of al-Faramā' did not mean, however, that the Franks had no access to water in the surrounding area.[62] In any case they invaded Egypt with the same ease with which the Fāṭimids moved to Palestine prior to the destruction of Bilbays and al-Faramā'. The sources do not suggest that the desert between Egypt and southern Palestine posed any serious problems for medieval armies.

William of Tyre's remark that armies moved from one watering point to another indicates that water was available along the desert route, but fodder and food were another matter and had to be carried. After entering Palestine, Fāṭimid armies lived off the land rather than being supplied from the rear. This does not necessarily mean that they obtained supplies by force for Fāṭimid commanders were supplied with money to buy supplies en route. The presence of herds of

---

[61] *Gesta Francorum*, X.xxxvii-xxxviii (pp. 88-90); William of Tyre, *Willelmi Tyrensis archiepiscopi chronicon*, ed. R. B. C. Huygens, 2 vols, in CC *ContMed*, vols 63 & 63A paginated continuously, 8.4, 8.7, 19.14, 20.5-6, 20.14 (pp. 388-90, 393-4, 883, 917-18, 927-8).

[62] Abū Shāma, *Kitāb al-rawḍatayn*, vol. 2, p. 47; Yaqūt al-Rūmī, Shihāb al-Dīn Abu 'Abd Allāh, *Mu'jam al-buldān*, ed. F. 'Abd al-'Aziz, 5 vols (Beirut, 1990), vol. 4, p. 290; Ibn al-Ma'mūn, *Akhbār Miṣr*, p. 13; Ibn Muyassar, *Akhbār Miṣr*, p. 144. For maps and geography see G. Cornu, *Atlas du monde Arabo-Islamique à l'époque classique, IXe-Xe siècles* (Leiden, 1985), map 11; J.-M. Mouton, *Le Sinaï médiéval: un espace stratégique de l'Islam* (Paris, 2000), pp. 148-53.

animals around the Fāṭimid camp outside Ascalon in 1099 was probably not accidental. Most probably, these herds belonged to local Bedouin who had come to provide the army with meat.

The Fāṭimids' ability to supply their armies in Palestine by sea is also questionable. While fleets sailed to Palestine in conjunction with land campaigns there is no evidence that they carried supplies for armies. They operated on their own and launched naval raids unconnected to land operations. They were, nevertheless, very important in providing grain to coastal towns besieged by the Franks. In the summer of 1103 a Fāṭimid fleet brought grain to a number of coastal towns in Palestine and another did so in 1108 to Tripoli even though it did not prevent the fall of the town. This pattern of events whereby Fāṭimid ships could supply a town in peacetime but could not prevent its fall when besieged was repeated in the cases of Tyre and Ascalon.[63]

Ascalon was important and the Fāṭimids stored great quantities of grain there and maintained a garrison of regular troops and Bedouin of the Kinānīyya tribe. The Franks invested great efforts in attempting to cut it off from Egypt and conquer it. Against this, the record of the Fāṭimids and of Ṣalāḥ al-Dīn in southern Palestine does not point to Ascalon having any great military significance. In 1099 the Fāṭimids held Ascalon but their army camped outside the town suffered a crushing defeat. In 1123 a Fāṭimid army of cavalry and infantry regiments reinforced by Bedouin moved into Palestine splendidly equipped with tents and arms and accompanied by physicians, muezzins and *Qur'ān* reciters. The grain stores of Damietta and Ascalon were to supply the army while the fleet was to be equipped and supplied in Alexandria. But in spite of elaborate preparations and attention to logistics the army was defeated by the Franks on 29 May 1123.[64] Fāṭimid involvement, or lack of involvement, in campaigns against the Franks was not apparently related to possession or loss of Ascalon. Neither can it be explained by logistics since water was available along the route from Egypt to Palestine and in southern Palestine. However, Ascalon was lost in 1153 and when Ṣalāḥ al-Dīn was defeated at Mont Gisard on 25 November 1177, the fact that his army had entered hostile territory held by the Franks as soon as it had crossed the desert must have created difficulties for him.

Although the ancient *via maris*, the coastal route connecting Egypt to Palestine, was relatively easy to use, there was another more southerly route connecting Suez and Eilat. In 1164, when Nūr al-Dīn sent his first expedition to Egypt, his forces entered deep into the desert around the great fortress of Krak de Montréal and eventually reached Suez via Eilat on the Gulf of 'Aqaba. How they crossed the Sinai peninsula between Eilat and Suez is not specified but a clue is supplied in another account that recorded that in 1169, during the third of Nūr al-

---

[63] Ibn al-Qalānisī, *Dhayl ta'rīkh Dimashq*, pp. 142-3, 164, 207, 321-2.

[64] Ibn al-Ma'mūn, *Akhbār Miṣr*, pp. 61-2.

Dīn's expeditions to Egypt, his army passed through Ṣadr on its way to Egypt. Qalat Ṣadr was an important point on the route from Suez that went through 'Uyūn Mūsā, the Springs of Moses, to Qalat Ṣadr and then across the desert to Eilat via Qalat Nakhl. In Qalat Ṣadr, which is 435 metres above sea level, there was a perennial spring and Ṣalāḥ al-Dīn carried out much construction, known from the many inscriptions that have survived there. He built a fort, Qa'lat al-Jundī, a mosque, and a dam at the confluence of two *wādīs*, thus creating a water reservoir.[65] Until the fall of the Kingdom of Jerusalem this route was important for moving armies between Egypt and southern Palestine but the Suez-Eilat crossing must have been arduous and a logistical nightmare. In 1168-9, Nūr al-Dīn's cavalry expeditionary force of between 6,000 and 8,500 men left Syria, probably Damascus, around 18 December 1168 and reached Cairo by this route on 8 January 1169. They must have been assisted by Bedouin along the way or at least have procured supplies from them. Each horseman was given twenty *dīnārs* for expenses en route, suggesting that the supply train was minimal and that they bought supplies en route.[66]

People had the organizational skills and means to equip armies for crossing deserts and some glimpses of how this was done can be gleaned from Ibn Baṭṭūṭa, probably the only author to pay attention to such problems. Although he described in his travels a pilgrim caravan to Arabia in the fourteenth century, the experience gained in organizing and leading pilgrims was applicable to military purposes also and certainly the water vessels would have been identical. Animals were watered by pouring water into a large container made of buffalo hides so that a number of camels could drink at the same time. Water was stored in what Ibn Baṭṭūṭa called "large waterbags and ordinary water skins" carried on camels. During one 'Abbāsid campaign in the ninth century, *ka'k* and water skins were supplied to frontline troops by a special support corps and a personal water skin became standard equipment for Fāṭimid infantry. Food cooked in large brass pots was served to the pilgrims of Ibn Baṭṭūṭa's caravan, indicating that fuel for cooking was also carried. In other cases, certainly when troops moved from Egypt to Palestine, biscuit was most probably the preferred staple food, and this explains the importance of Bilbays providing it for both pilgrims and armies. Medieval people were, however, capable of readjustment and Egyptian pilgrimage caravans or armies dispatched to Palestine continued to make their way along this route even after the destruction of Bilbays by the Franks.[67]

---

65 Al-Maqrīzī, *Itti'āẓ al-ḥunafā'*, vol. 3, pp. 266, 299. See J.-M. Mouton et al., "La route de Saladin (tariq Sadr wa Ayla) au Sinaï", *Annales Islamologiques*, 30 (1996), 41-71.

66 Abū Shāma, *Kitāb al-rawḍatayn*, vol. 2, pp. 18, 50, 51, 55.

67 Ibn Baṭṭūṭa, Shams al-Dīn Abū 'Abd Allāh Muḥammad, *Riḥla*, ed. and Fr. trans. C. Defrémery and B. R. Sanguinetti, 4 vols, *Voyages d'Ibn Batoutah* (Paris, 1853-9) [trans. H. A. R. Gibb, *The travels of Ibn Baṭṭūṭa A.D. 1325-1354*, 3 vols, in HSW, 2nd ser., vols 110, 117, 141 (London, 1956-71)], vol. 1, pp. 258, 405, 406. The trans. is Gibb's, vol. 110, p.

I would argue that there was greater availability of water for animals and people in the Middle Ages than in modern times. Medieval populations were smaller, overall water consumption was lower, and the environment was healthier. Due to lower human consumption, rivers and streams carried greater quantities of water and subterranean water was neither exhausted nor polluted. This made it possible for medieval armies to find the water they needed more easily than the present-day landscape might suggest and this was especially true for armies moving along well-established routes such as the desert crossings between Egypt and Palestine or operating in familiar territories. However, on campaigns in unfamiliar territories, or because of the evolving dynamics of particular campaigns, circumstances might deteriorate and in such cases armies might indeed find themselves in dire shortage of water and supplies.

Are there wider implications? Harari has recently posed the question of how one can explain "... the continuing military success of the Franks throughout most of the twelfth century, and their survival in the Levant for close to two hundred years." His question is based on the assumption that, except in hand-to-hand combat, Muslim "... light cavalrymen had a clear strategic, operative and tactical superiority ..."[68] His explanation of the seeming paradox of why Muslim military superiority did not materialize rests on the argument that the Franks adopted massively the use of light cavalry in the form of Turcopoles as part of their military structure and so created an effective counterbalance to Muslim mounted archery. They became capable of fighting Muslim armies in their own way and thus neutralized the latter's main military advantage.

Evidence presented here, however, suggests that Frankish and Muslim armies of the Ayyūbid-Mamlūk period were constructed differently and fought differently. The main military asset of the Franks was the close cooperation that existed between infantry and cavalry and the cavalry charge delivered by knights.[69] Muslims fought as mounted archers shooting massive volleys of arrows and tried to surround enemies and lure them by feigned flight into planned ambushes. These two different ways of fighting demanded cohesive armies on the part of Franks but

---

161. A large waterbag made of several skins sewn together and carried by a camel was a *rāwiya* and an ordinary waterskin a goat-skin, *qirba*.

68 Y. Harari, "The military role of the Frankish Turcopoles: a reassessment", *MHR*, 12 (1997), 75-116, p. 75.

69 For the role of infantry and cavalry in European medieval warfare see M. Bennett, "The myth of the military supremacy of knightly cavalry", in M. Strickland, ed., *Armies, chivalry and warfare in medieval Britain and France: proceedings of the 1995 Harlaxton Symposium* (Stamford, 1998), 304-16; idem, "La *Règle du Temple* as a military manual, *or* how to deliver a cavalry charge", in C. Harper-Bill et al., eds, *Studies in medieval history presented to R. Allen Brown* (Woodbridge, 1989), 7-19; C. J. Marshall, "The use of the charge in battles in the Latin East, 1192-1291", *HR*, 63 (1990), 221-6; J. France, *Western warfare in the age of the Crusades, 1000-1300* (Ithaca, 1999), chs 5-6.

large numbers on the part of Muslims. Each method had its limitations and problems but, more importantly, both were barely able to defeat, let alone crush, enemies.[70] Muslims could counteract Frankish charges by disengaging and regrouping while the Franks could withstand Muslim attacks and keep mounted archers at bay with their infantry and simply refuse to be dragged into battle. Their different styles of warfare were well understood and explicitly noted by contemporary Muslim historians. Nūr al-Dīn, for example, was quoted as saying that the arrows of the Turks were the only effective weapon against the knights and their way of fighting. Ibn Shaddād, who accompanied Ṣalāḥ al-Dīn during the siege of Acre and the Third Crusade, commented that the infantry protected the cavalry like a "wall" and on several occasions the cavalry refrained from attacking. The ability of Byzantine and Frankish armies to move from place to place while facing and repelling attacks is well known and clearly borne out by Ibn Shaddād's description of the battle of Arsuf.[71] Victory on the battlefield depended on external factors not directly related to the military capabilities of the armies and of these external factors, the quality of human leadership and the ability to avoid mistakes were the most crucial. High-quality leadership could forge a cohesive force from disparate elements while poor leadership would only magnify internal divisions within forces. Quality leadership also meant paying sufficient attention to logistical problems.

Having said that medieval people had the organizational skills and means to equip their armies does not mean that the issue was insignificant. While both logistical preparations and an understanding of what the environment had to offer a passing army were crucial for the success of a campaign, mastering these aspects was not self-evident. To better understand the logistical dilemmas of Muslim and Frankish armies we should strive to reconstruct the medieval Middle Eastern environment, which is not an unfeasible project. The vast medieval and pre-modern travel literature has not been examined from this point of view and remains an untapped source for studying the history of ecology. Travellers did not often

---

[70] For two different assessments of Muslim mounted archers and their impact on Byzantine and Crusading armies, see W. E. Kaegi, "The contribution of archery to the Turkish conquest of Anatolia", *Speculum*, 39 (1964), 96-108; C. R. Bowlus, "Tactical and strategic weaknesses of horse archers on the eve of the First Crusade", in M. Balard, ed., *Autour de la Première Croisade: actes du Colloque de la Society for the Study of the Crusades and the Latin East (Clermont-Ferrand, 22-25 juin, 1995)* (Paris, 1996), 159-66. For a broader treatment of the subject, see P. B. Golden, "War and warfare in the Pre-Cinggisid western steppes of Eurasia", in N. Di Cosmo, ed., *Warfare in inner Asian history (500-1800)* [Handbuch der Orientalistik: Abt. 8, Zentralasien, vol. 6] (Leiden, 2002), 105-72.

[71] Ibn Shaddād, *Al-Nawādir al-sulṭāniyya*, pp. 49-50 (referring to events of 1183-4), p. 134 (referring to the struggle for Acre), pp. 170, 175 (referring to the battle of Arsuf and preceding events). For the fighting march, see Bennett, "Crusaders' fighting march".

mention logistical problems involved in their journeys, but they did describe lands they travelled through.

Logistics in the medieval Middle East should be discussed against a background of medieval ecology, agriculture, and seasonal patterns of nomadism. Another aspect that should be considered is the agricultural calendar and the availability of crops in the fields. The author of the *Gesta Francorum* remarked that the decision to end the futile siege of 'Arqa was influenced by the onset of the harvest season.[72] The Franks preferred to make their way to Jerusalem during harvest time and this, of course, meant they intended to live off the land.

Mutual adaptations, whereby Muslims incorporated some heavy cavalry into their forces and Franks employed some light mounted archers, did not alter the basic patterns of military organization and style of warfare on either side. I agree with France's observation that "... the actual make-up of Middle Eastern armies was, however, radically different from that in the West."[73] A discussion of Turcopoles or infantry alone is inadequate to deal with the question of the success and long presence of the Franks in the Middle East as this approach is too narrow to provide sufficient insight into what was a complex problem. The battle of the Horns of Ḥaṭṭīn exemplifies the limits of a discussion that focuses on a single aspect of military history. Whatever the shortcomings of Muslim armies with no infantry component may have been, Ṣalāḥ al-Dīn's all-cavalry force did inflict a crushing defeat on the Franks at Ḥaṭṭīn and one must ask where the Turcopoles were in this decisive battle? Focusing on Turcopoles or infantry may provide material for broader discussion of the overall military structures of Muslim and Frankish polities and differences in their military structures were obviously notable. There was no Muslim equivalent of the Military Orders and the ability of the Kingdom of Jerusalem to recruit a large and quality force from urban populations had no parallel in the Muslim states of the period.[74] The Franks had Turcopoles while Muslims, with the exception of Fāṭimid Egypt, lacked professional infantry in adequate numbers. Nevertheless, a truly meaningful discussion must go beyond the limits of military history and deal with broader issues: the political make-up of the Muslim world of the eleventh-thirteenth centuries, Muslim perceptions of and attitudes toward the Franks, and questions of demography and the composition of Muslim and Frankish societies.

Muslims made war in ways that reflected the economic resources available to their rulers and the socio-military characteristics of ethnic groups involved. The Fāṭimids made war as befitted a Caliphal power which could maintain armies of

---

72 *Gesta Francorum*, X.xxxvi (p. 85).

73 J. France, "Crusading warfare and its adaptation to eastern conditions in the twelfth century", *MHR*, 15 (2000), 49-66, p. 54.

74 For example, in the Kingdom of Jerusalem, 2,000 of the 5,025 sergeants came from the cities. See France, *Western warfare*, p. 67.

tens of thousands whose main purpose was to keep the regime in power. To achieve this aim black slave infantry was instrumental but the armies as a whole were riven by ethnic animosities and lacked cohesion. Muslim armies in Syria and Egypt from 1171 onwards reflected the centrality of ethnic considerations in the political and military thinking of rulers. In order to maintain Turks and Kurds as the most privileged ethnic elements in the state, the *iqṭāʿ* system was employed and expanded and rulers expended their limited financial resources on hiring nomadic Turkmen and Kurds to supplement their armies. Armies of this type, although small, were cohesive when fighting the Franks and were, on the whole, successful.

Chapter 11

Food and the Fourth Crusade: a new approach to the "Diversion Question"[1]

Thomas F. Madden

Historians like a good puzzle, so it comes as no surprise that over the centuries many have tackled the so-called "Diversion Question" of the Fourth Crusade. The question is simple enough. How did a Crusade of this magnitude veer wildly off course, ultimately conquering Constantinople, the greatest Christian city in the world? The answer, of course, is anything but simple. The intricacies of motives, errors, and miscommunications require one to understand not only the Crusade, but also the histories of the Papacy, the German Empire, Venice, and the Byzantine Empire. The sources for the Fourth Crusade are many, often contradictory, and of differing quality.

The historiography of the Crusade has generally fallen into two camps: "treason theorists", who argue that one or more players conspired to divert the enterprise to Constantinople, and "accident theorists", who maintain that the tragic end of the Crusade was the result of unforeseen events and human error. In recent decades accident theories have dominated academic scholarship, although treason theories are still common in popular literature. Both, however, agree that money was one of the principal causes of the diversion. Treason theorists usually point to the Venetians' avaricious appetite for profit, which led them to embrace any plan that would put Byzantine wealth into their own pockets. Their commercial guile also allowed them to entrap the Frankish Crusaders in a contract that the latter never really understood.[2] Thus they acquired not only the Franks' silver but their obedience as well. Accident theorists generally reject this portrayal of the Venetians, arguing instead that it was the Franks' grossly exaggerated estimates of numbers that led to their chronic poverty and indebtedness, which in turn forced their leaders to accept the mission to Constantinople rather than to sail to Egypt as planned. Both sides of this question would agree that money truly was the root of all evil in the Fourth Crusade.

Despite an abundance of scholarly investigation, no one has considered the question of provisioning and what role it may or may not have played in the course and outcome of the Fourth Crusade. It is generally assumed that the Franks, while

---

1 Consult Map 13.

2 I have used "Frankish Crusaders" or "Franks" throughout as the most convenient shorthand to distinguish them from the Venetian Crusaders.

penniless, were well-fed, and there is a seemingly good reason for this. They lacked money precisely because they had given all they had to the Venetians trying to fulfil their contract. That contract obligated the Venetians to provide a fleet and provisions for one year, beginning on 29 June 1202. Scholars may attribute to medieval Venetians a host of sins, but failing to meet contractual obligations is not among them. All contemporary sources noted how scrupulously the Venetians kept their part of the bargain. Acquiring food during the Crusade, therefore, must not have been a problem. Although outside observers, such as Pope Innocent III and Emperor Alexios III Angelos, offered solutions to this problem, this is taken as evidence of how little they understood the internal dynamics of the Crusade.[3]

Yet Innocent and Alexios were not alone in noting the Franks' need for provisions. Indeed, virtually every contemporary source, including all of those written by participants, mentioned scarcity of food as a frequent problem. Yet none of them, not even those hostile to Venice, blamed her for the situation. There are, in other words, some glaring contradictions. Venice was contractually responsible for provisioning the Crusade although it was apparently not sufficiently supplied. Yet Venice was uniformly praised for fulfilling her contractual responsibilities. Something is amiss. Part of the puzzle does not fit correctly or is missing. Resolution of this contradiction may elucidate the provisioning of the Fourth Crusade and the role food played in its tragic outcome.

Any discussion of provisions must begin with the Treaty of Venice, entered into by Doge Enrico Dandolo on behalf of Venice with the representatives of the Frankish barons in April 1201. The treaty reads:

> And so the aforesaid envoys requested that we provide for you vessels to transport 4,500 well-armed knights and as many horses, and 9,000 squires (concerning which squires, if they are lacking in numbers, the below-said money ought not nevertheless to be reduced for us), and 20,000 infantry well armed, with provisions for up to one year, which we promised to give to them.[4]

The first point to note is that, contrary to modern accounts of the Crusade (including my own), the provisions were to be supplied not for an entire year, but

---

[3] For example, D. E. Queller and T. F. Madden, *The Fourth Crusade: the conquest of Constantinople*, 2nd ed. (Philadelphia, 1997), p. 91.

[4] G. L. Fr. Tafel and G. M. Thomas, eds, *Urkunden zur älteren Handels- und Staatsgeschichte der Republik Venedig*, 3 vols (1856-57; rpt, Amsterdam, 1964), no. 92 (vol. 1, p. 365): "Petierunt itaque supradicti viri, ut vobis navigium daremus ad traffretandos quatuor millia quingentos millites bene armatos, et totidem equos, et novem milia scutiferos (de quibus scutiferis, si defecerint, argentum inferius dictum nobis tamen minui non debet), et viginti milia pedites bene armatos, cum victualibus ad annum unum, quod eis dare promissimus."

for up to one year (*ad unum annum*). This is a crucial point, and one that makes the following clause, which has been virtually ignored, much more understandable.

> Provisions for each and every man will be thus: for each man six *sextaria* of bread, flour, grain, and legumes and a half amphora of wine. For each horse three *modia* [of grain] according to the measure of Venice, and of water there will be sufficient amounts.[5]

The Venetians were to provide a set amount of provisions *during the course of one year*. They were not responsible for feeding the army for an entire year.

When Philip II Augustus bought passage for his Crusade army from the Genoese in 1190 it included provisions sufficient for eight months and wine for four months.[6] How did this compare with the contract between the Venetians and Franks in 1201? The *sextarium* of ancient Rome, which measured about a half a litre, had only its name in common with the one used in medieval Italy.[7] The Venetian *sextarium* or *staio* was the equivalent of 83.3 litres.[8] Six *staia*, therefore, would equal 500 litres. Like all medieval measurements, these were in practice only approximates.[9] This is all the more true given that the items being measured were of different sizes and weights and there is no indication of their relative ratios in the whole. The presence of both bread and flour on the list of provisions is problematic, since flour baked into bread would take up more space and weigh more due to water content. Six *staia* of grain would weigh about 400

---

5 Loc. cit.: "Victualia unius cuiusque hominis talia erunt: pro uno quoque homine sextaria sex inter panem et farinam et blavam et legumina et dimidiam anforam vini. Pro uno quoque equo modia tria ad mensuram Venetie, et de aqua, quantum sufficiens erit." In another copy of the treaty, the phrase "modia blaue tria" has been transposed from the horses' feed to replace "et blavam" in the Franks' provisions. It also supplies the missing "blave" between "modia" and "tria" in the horses' feed. Ibid., no. 93 (vol. 1, p. 371).

6 C. Imperiale di Sant' Angelo, ed., *Codice diplomatico della repubblica di Genova*, 3 vols [FStI, nos. 77, 79, 89] (Rome, 1936-42), vol. 2, no. 191 (p. 365): "... per octo menses sufficienti ..."

7 Using the ancient measurements, one might conclude that the food amounts in the treaty were daily rations for the full year. This may be one reason why this clause has escaped much scrutiny.

8 J. E. Dotson, trans., *Merchant culture in fourteenth century Venice: the Zibaldone da Canal* (Binghamton, 1994), p. 204; H. Doursther, *Dictionnaire universel des poids et mesures anciens et modernes contenant des tables des monnaies de tous les pays* (1840; rpt, Amsterdam, 1976), p. 505; L. Martini, *Manuale di metrologia ossia misure, pesi e monete* (Turin, 1883), p. 818; R. E. Zupko, *Italian weights and measures from the Middle Ages to the nineteenth century* (Philadelphia, 1981), p. 276.

9 As Dotson has noted sagely: "One must always keep in mind that the attitude toward measurement in the medieval world was practical and commercial, not precise and scientific.". Dotson, *Merchant culture*, p. 207.

kilogrammes.[10] Bread and legumes would weigh more. Milled flour would weigh less. It seems reasonable, then, that a man's provisions weighed approximately 400 kilogrammes, which would sustain him for roughly 300 days or about ten months if he consumed a standard 1.3 kilogrammes per day at low activity.[11] At higher activity levels the food would, of course, be consumed more quickly.[12] This suggests that the amount of food purchased by the Fourth Crusade was quite similar to that purchased by Philip II.

The wine poses a problem. Since it could be stored without spoiling, wine was provided on voyages commonly to reduce the need for fresh water.[13] Nevertheless, the horses would have needed daily watering so the Venetians would need to take on supplies at least every week or so.[14] According to the treaty, each man was to receive half an amphora of wine. However, the size of the measure known as "amphora" varied greatly across place and time in the Middle Ages.[15] There is no evidence to suggest the size of the Venetian amphora in the thirteenth century. In the fourteenth century it measured approximately 680 litres.[16] At that size each Crusader would have had 340 litres of wine. If the wine was meant to last as long as the food, this would provide a little over one litre of wine per day for ten months, an amount well in keeping with wine allowances for other medieval voyages. If it was meant to last for only half as long as the food, as was the case for Philip II's voyage, each Crusader would have received well over two litres per day. That also was not an unusual amount.

---

10 One hundred litres of wheat weighs approximately 80 kilogrammes.

11 On food requirements, see D. W. Engels, *Alexander the Great and the logistics of the Macedonian army* (Berkeley 1978), pp. 123-9; J. F. Haldon, "The organisation and support of an expeditionary force: manpower and logistics in the Middle Byzantine period", in N. Oikonomides, ed., *Τό εμπόλεμο Βυζάντιο (9ος - 12ος αι.) / Byzantium at war (9th-12th c.)* (Athens, 1997), 111-51, pp. 124-5.

12 Cf. B. S. Bachrach "Logistics in pre-Crusade Europe," in J. A. Lynn, ed., *Feeding Mars: logistics in Western warfare from the Middle Ages to the present* (Boulder, 1993), 57-78 [rpt in his *Warfare and military organization in pre-Crusade Europe* (Aldershot, 2002), No. V], p. 71.

13 U. Tucci, "L'alimentazione a bordo delle navi veneziane", *Studi veneziani*, n. s., 13 (1987), 103-45, pp. 110-11.

14 J. H. Pryor, *Geography, technology, and war: studies in the maritime history of the Mediterranean, 649-1571* (Cambridge, 1988), pp. 85-6. Horses at low activity levels would require approximately 36 litres per day. See Engels, *Alexander the Great*, pp. 128-9; A. Hyland, *Equus: the horse in the Roman world* (New Haven, 1990), p. 96.

15 Zupko, *Weights and Measures*, p. 5. It should be understood that the word "amphora" had become merely the name of a measure and no longer referred to a pottery container.

16 J.-C. Hocquet, "Tonnages ancien et moderne: botte de Venise et tonneau anglais", *Revue historique*, 281 (1989), 349-60, pp. 354-6. By the sixteenth century the Venetian amphora measured 518 litres. See Dotson, *Merchant culture*, p. 204; Doursther, *Dictionnaire universel*, p. 14; Martini, *Manuale di metrologia*, p. 818; Zupko, *Weights and measures*, p. 6.

Each horse was to receive three *modia* of grains. The Venetian *moggio* was equivalent to 4 *staia* or 333 litres.[17] Thus, each horse was to have 1 cubic metre of grain, which would weigh approximately 800 kilogrammes. A horse at low activity levels requires on average about 9 kilogrammes of food per day, with about a quarter of that in grain and the rest in hay or grazing.[18] The treaty mentioned only the grain. Assuming a departure date of 29 June, hay could have been loaded aboard before departure and acquired during watering stops before crossing over to Egypt. When disembarked, the horses could graze. The grain allotment would be sufficient for about 355 days, essentially the entire year if the horses remained at low activity levels, which was obviously not what the Franks had in mind.

The preparation of the fleet and assembling of sailors, marines, and provisions was a formidable task that forced Venice to suspend overseas commerce for more than a year. All merchant vessels were required to return home so that they could be pressed into service as troop and materiel transports.[19] Fifty galleys and 150 transport galleys had to be purchased or constructed.[20] The food also needed to be purchased. Set in the lagoon, Venice produced virtually none of her own food but never went hungry. By the time of the Fourth Crusade she had established herself as the hub for wheat and other produce marketing for all of north-eastern Italy. All vessels carrying wheat in the Adriatic were required to unload at Venice and the Venetians had the patrols to make certain that it happened. Venetian merchants regularly obtained food from markets in central Italy as well as other locations across the Mediterranean. On the northern Italian mainland local merchants would ship food along the rivers to Venice in order to secure the lucrative price guarantees available there. Inevitably this pitted the desire of local authorities to secure cheap agricultural produce from local farmers against merchants' desire to make a profit by shipping it to Venice. In years of bad harvests this could result in a tense tug of war.[21]

---

17 Dotson, *Merchant culture*, p. 204; Doursther, *Dictionnaire universel*, p. 285; Martini, *Manuale di metrologia*, p. 818; Zupko, *Weights and measures*, p. 168.

18 Haldon, "Expeditionary force", p. 126, n. 57.

19 Robert of Clari, *La conquête de Constantinople par Robert de Clari*, Fr. trans. A. Micha, texte de P. Lauer [1924] (Paris, 1991), §§7, 11 (pp. 129-32). On the return of one such ship, the *Paradiso*, see R. Morozzo della Rocca and A. Lombardo, eds, *Documenti del commercio veneziano nei secoli XI-XIII*, 2 vols (Turin, 1940), no. 461 (vol. 1, pp. 451-2); A. Lombardo and R. Morozzo della Rocca, eds, *Nuovi documenti del commercio veneto nei secoli XI-XIII* (Venice, 1953), nos 53-4 (pp. 59-60).

20 J. H. Pryor, "The Venetian fleet for the Fourth Crusade and the diversion of the Crusade to Constantinople", in M. Bull and N. Housley, eds, *The experience of Crusading*, 2 vols (Cambridge, 2003), *Vol. 1: Western approaches*, 103-123, pp. 115-18.

21 F. C. Lane, *Venice: a maritime republic* (Baltimore, 1973), pp. 58-60.

By the terms of the Treaty of Venice, the Venetians were ultimately responsible for securing 21,250 cubic metres of provisions,[22] weighing approximately 17,000 metric tonnes.[23] In addition, they would need to acquire provisions for their own sailors and marines. Pryor has estimated that the fleet contracted for would have required something in the neighbourhood of 31,000 men.[24] Assuming that the Venetians and other Italians ate as much as their northern colleagues, that would mean finding an additional 15,500 cubic metres of provisions weighing about 12,400 metric tonnes. The average displacement tonnage of a very large round ship or *navis* in the thirteenth century was about 800 metric tonnes.[25] The food for the Franks and the grain for their mounts, therefore, would have required approximately 21 such vessels, while the food for the sailors and marines would fill another 15. Of course very few of the fleet's vessels would have been that large, so many more ships would have been needed. It is not clear how much of the provisions would have been acquired before the planned departure date and how much (if any) would have been purchased along the way,[26] but, given that they were planning to sail directly to Egypt to begin military operations, it seems likely that all of the provisions were obtained in advance. Venetian agents would have needed to have been sent into the Po valley to make certain that the necessary produce would make its way to Rialto. Depending on the harvest, that could put a strain on the region, requiring higher guaranteed prices in Venice.[27]

---

22 33,500 men at 500 litres per man plus 4,500 horses at 1 cubic metre per horse.

23 33,500 men at 400 kilogrammes per man plus 4,500 horses at 800 kilogrammes per horse.

24 Pryor, "Venetian fleet", pp. 118-19. The Venetian government instituted a lottery in the city to draft one-half of all able-bodied men. See Robert of Clari, *Conquête de Constantinople*, §11 (pp. 130-31). The population of Venice in 1200 was approximately 100,000. If we assume that the standard one-quarter of those were able-bodied men, then the lottery would gather only about 12,000 men. The fleet that actually sailed would not have required 31,000 men, but it would certainly have needed more than 12,000, so the Venetians must have hired or drafted men from other Italian or Dalmatian towns. Andrea Dandolo implied this in his chronicle. See Andrea Dandolo, *Chronica per extensum descripta, aa. 46-1280 d. C.*, ed. E. Pastorello [RISS *NS*, tome 12, part 1] (Bologna, 1938), X.3 (p. 276).

25 Pryor, *Geography, technology, and war*, p. 86.

26 U. Tucci, "L'impresa marittima: uomini e mezzi", in G. Cracco and G. Ortalli, eds, *Storia di Venezia* (Rome, 1995), vol. 2, 627-59, pp. 631-2. The provision figures in this study, however, are defective.

27 It is often said that the Venetians' efforts were hampered by poor harvests in Italy and across the Mediterranean. See, for example, F. Hurter, *Storia di Papa Innocenzio III*, trans. T. G. Gliemone, 4 vols (Milan, 1857-8), vol. 2, p. 149; Queller and Madden, *Fourth Crusade*, p. 17. However, the only source for these poor harvests is Robert of Auxerre, who mentioned a great scarcity in the context of poor Crusaders seeking food while on the Lido. See Robert of Auxerre, *Chronicon*, ed. O. Holder-Egger, in MGH *SS*, vol. 26, 219-76, Annus Domini 1202 (p. 261). None of the participants mentioned this problem.

We have a hint of these dynamics in the treaty itself. When describing the scheduled rendezvous at Venice in 1202, it stated: "Nor should it be passed over, that you ought not to purchase foodstuffs at Cremona and beyond towards Venice, nor at Bologna, Imola, and Faenza and beyond towards Venice, except with our consent."[28] These cities marked a line along the plains below the Apennines. In essence, this clause forbad Franks from buying food in the Po valley from Cremona eastward. After coming through the Alpine passes, they could, of course, purchase provisions west of Cremona at places such as Milan, Pavia, or Piacenza, but the Venetians apparently planned to acquire some or all of the army's provisions in the off-limits area and they did not want the Franks driving up prices. Given the extraordinary Venetian purchases, local authorities in the Po valley would also not have been pleased to have thousands of travelling Crusaders relying on their markets for sustenance. Indeed, when the Crusaders did begin to traverse the region in the following year, local authorities decreed that no one should sell provisions to them nor should hospitality be offered for more than a single night. This left the impression among some Crusaders that they were being hurried along from town to town, which in fact they were.[29]

Despite the prohibitions, the various bands of Franks appear to have had no problems finding provisions elsewhere for their journey to Venice. Abbot Martin of Pairis remembered the support and friendship that he and the others received along the way, as well as the food available in markets at fair prices.[30] When the Franks began to arrive in Venice they were warmly welcomed and ferried out to the Lido, a long and largely uninhabited island not far from the city. There they camped and waited for the 33,500 men to arrive. It was to be a long wait. After an initially good showing, arrivals began to taper off during the month of July. When it became increasingly clear that the Crusade was in trouble, those bound for Venice waited to see how it would all come out. Even Louis of Blois, whose seal was affixed to the Treaty of Venice, delayed in Verona before a delegation of leaders was able to shame him into joining his comrades. He was the exception. Most of the latecomers, unwilling to encumber themselves with an impossible contract, headed to other ports for transportation to the east.

Some Franks may have arrived at Venice with limited provisions but after they were consumed, the army became completely reliant on Venice for food and water.

---

28 Tafel and Thomas, *Urkunden*, nos 92, 93 (vol. 1, pp. 367, 371): "Nec est pretermittendum, quod victualia comparare non debetis a Cremona et infra versus Veneciam, et a Bononia, Immola, Faventia et infra versus Venetiam, nisi verbo nostro."

29 *Devastatio Constantinopolitana*, ed. and trans. A. J. Andrea, "The *Devastatio Constantinopolitana*, a special perspective on the Fourth Crusade: an analysis, new edition, and translation," *Historical Reflections*, 19 (1993), 107-49, fol. 253v (p. 132).

30 Gunther of Pairis, *Hystoria Constantinopolitana: Untersuchung und kritische Ausgabe*, ed. P. Orth (Hildesheim and Zurich, 1994), §5 (p. 118).

Transportation on and off the Lido was available, but there was probably not much food to be had on the mainland.[31]

It is not precisely clear at what point the Franks ceased purchasing food in Venetian markets and began to receive it as their due under the terms of the treaty. The first of them began to arrive in the lagoon in early June and were transferred to the Lido, where, according to Geoffrey of Villehardouin, "... the Venetians held a market, rich and abundant, of all things necessary for horses and men."[32] This market (*mercié*) may have been a distribution centre for provisions or a place to purchase them. Indeed, it was probably both at different times. According to the anonymous author of the *Devastatio Constantinopolitana*, who appears to have been one of the early arrivals, the Venetians charged 50 *solidi* for one *staio* (about 83 litres) of grain.[33] Yet he was the only source to record paying for food while at Venice. Robert of Clari, a poor knight from Picardy who also arrived early, said nothing about purchasing provisions, although he did describe the Venetians bringing food and water for the Franks.[34] Before the terms of the treaty commenced on 29 June 1202, the Franks were probably responsible for purchasing their own supplies. Given that the promised payment was not forthcoming on 29 June, it may be that the Venetians continued to charge for food. It seems more likely, however, that they began supplying the troops on that date, assuming, together with everyone else, that the arrival of more Frankish Crusaders would allow the bills to be paid.

A month or more passed before it became clear that no more Franks were coming to Venice. When Doge Dandolo demanded payment the individual passage costs were collected throughout the host. Of course, with only one-third the projected army the money collected was only one-third the required amount. Dandolo demanded full payment in accordance with the treaty. Most of the Franks balked, refusing to pay passage for non-existent soldiers, and the stalemate was a dangerous one. Given the enormous expenses they had incurred, the Venetians could not renounce payment. Venetian law and custom made contractual obligations sacrosanct. The Franks must fulfil those obligations or face the penalty.

---

31 *Devastatio Constantinopolitana*, fol. 253v (p. 132); L. De Mas Latrie, ed., *Chronique d'Ernoul et de Bernard le Trésorier* (1871; rpt, Famagusta, 1974), ch. 32 (p. 349); Gunther of Pairis, *Hystoria Constantinopolitana*, §6 (p. 122).

32 Geoffrey of Villehardouin, *La conquête de Constantinople*, ed. and Fr. trans. E. Faral, 2 vols (Paris, 1938-9), §56 (vol. 1, p. 58): "Et li Venisien lor firent mercié si plenteüros con il convint de totes les choses qu'il convient a chevaus et a cors d'omes."

33 *Devastatio Constantinopolitana*, fol. 253v (p. 132). The author described lodging in the city before being moved out to the Lido. This could only have occurred in the early days, before sizeable numbers had arrived.

34 Robert of Clari, *Conquête de Constantinople*, §11 (p. 132).

To make that case more forcefully, the Doge threatened to cut off food and water shipments to the Lido.[35] This is usually construed as a threat to starve the army into submission. Yet it could have been that only if the Franks were unable to leave the island. In truth, although several participants grumbled about being kept there "like prisoners" they also recorded large numbers of Franks who simply tired of waiting and returned home.[36] Getting a boat to the mainland was obviously not a problem. Dandolo's threat makes more sense from a Venetian commercial perspective. The Franks' failure to meet their obligations meant that the Venetians no longer had a responsibility to meet theirs. The army's provisions were one of those obligations. Cutting off food and water shipments would not starve the Crusade, but it would require the Franks to decide quickly between paying what they owed and leaving Venice altogether.

As it happened, Dandolo did not need to carry out his threat.[37] The Crusade leaders asked everyone in the host to give all they had so that the enterprise could get underway. Some did, others did not. When all was collected they still lacked 34,000 of the 85,000 marks owed. More time passed as the Doge and his councils assessed the situation and sought a solution.[38] Finally, in late August, the Crusade leaders agreed to help the Venetians conquer Zara in return for the deferment of the outstanding debt. So September was spent outfitting the vessels once again and loading weapons, siege machinery, and provisions. The fleet consisted of approximately 50 large transports, 100 horse transport galleys, and 60 war galleys as well as numerous auxiliary craft.[39] Although not all of the ships and sailors assembled were actually needed, most were. Many of the sailing ships had been assigned to certain "high men", who would not willingly share with their peers. Although a particular lord might not have had as many men as he had hoped, he would nonetheless want his own ship. It appears, then, that about half of the transports and one-third of the horse transports were not needed and were left behind. The total number of Venetians and Italians on the Crusade was probably around 21,000, almost double the number of Frankish Crusaders.[40]

---

35 Robert of Clari, *Conquête de Constantinople*, §11 (p. 132).

36 Anonymous of Soissons, *De terra Iherosolimitana*, ed. and trans. A. J. Andrea and P. I. Rachlin, "Holy war, holy relics, holy theft: the Anonymous of Soissons's *De terra Iherosolimitana*: an analysis, edition, and translation", *Historical reflections*, 18 (1992), 147-75, fol. 212r (p. 159); Gunther of Pairis, *Hystoria Constantinopolitana*, §6 (p. 122).

37 Robert of Clari, *Conquête de Constantinople*, §12 (p. 133).

38 For a more detailed analysis of the dynamics at work in the stalemate, see T. F. Madden, *Enrico Dandolo and the rise of Venice* (Baltimore, 2003), pp. 130-41.

39 Queller and Madden, *Fourth Crusade*, pp. 68-9.

40 According to Pryor, the large transport vessels perhaps required approximately 4,500 sailors. If only one-half were needed, that number would decrease to around 2,000. The horse transports had a crew of about 130 each, thus requiring 13,000 men. The war galleys,

When the fleet finally sailed out of the Venetian lagoon it was already October.[41] For more than three months the army of approximately 11,000 warriors and many hundreds of horses had been eating and drinking. Since the Venetians were responsible for providing between nine and ten months' provisions for the army, the Franks were thus far consuming food at a faster rate than they were consuming time on their lease. The Anonymous of Soissons noted that the Franks left Venice with their "supplies almost consumed".[42] Had they departed on schedule on 29 June they could reasonably have arrived in the Nile delta in eight to ten weeks, leaving them some six to eight months of food to sustain them until they could secure provisions locally. But as matters stood, they would be unable to sail for Egypt until the spring, some five months hence. The fleet, therefore, took its time moving down the Adriatic coast, exacting promises of allegiance to Venice along the way. It did not arrive at Zara until the following month and the city surrendered on 24 November.

From a logistical standpoint, the conquest of Zara solved the problem of where to spend the winter. But the problem of provisions only worsened. Zara and its environs belonged to Venice and was not a Crusade conquest to be shared equally with the Franks. They had agreed to help the Venetians recover their property, but they had no claim to half the spoils. That meant that they remained reliant on them for provisions. By the end of December 1202 they would have been consuming those provisions for six months. When the fleet left Zara on 20 April 1203 the Franks had consumed a full ten months' provisions, all or perhaps more than they had purchased.

The food problem is well attested to in the sources. The Anonymous of Halberstadt remarked that the Franks at Zara were "destitute of goods".[43] Robert of Clari explained more fully:

> In the meantime, while the Crusaders and the Venetians spent the winter there, they considered how much they had already spent; and they said to one another that they could not go to Babylon or to Alexandria or to Syria, for they had neither provisions nor money for going there, for they had already used up nearly everything, either during the sojourn that they had made or in the great price they had paid for the vessels. And they said

---

which apparently now numbered 60, needed at least 6,000 men and probably more. See Pryor, "Venetian fleet", pp. 116-18.

41 Geoffrey of Villehardouin, *Conquête de Constantinople*, §§75-6 (vol. 1, pp. 76-8); Robert of Clari, *Conquête de Constantinople*, §13 (pp. 133-4).

42 Anonymous of Soissons, *De terra Iherosolimitana*, p. 159.

43 Anonymous, *Gesta episcoporum Halberstadensium*, ed. L. Weiland, in MGH *SS*, vol. 23, 73-123, p. 118.

> that they could not go, and if they did go they would accomplish nothing, since they had neither food nor money sufficient to sustain them.[44]

It was at this point, when the army was penniless, low on provisions, and under the ban of excommunication, that the envoys of the Byzantine pretender, Alexios Angelos, sailed into Zara's harbour.[45] It is well known that the young man offered the Crusaders a large amount of money, military support in the Holy Land, and reunification of the churches in exchange for their championing his imperial claims in Constantinople. What has been overlooked, however, is the important role that food played both in the offer and its acceptance.

Geoffrey of Villehardouin, who was present when the envoys addressed the Crusade leaders, heard them relate that Alexios "... knows that you have spent all of your money, and that you are poor: he will therefore give you 200,000 silver marks, and food for everyone in the army, both small and great."[46] In a letter written a few months later, Hugh of St Pol explained that Alexios "... enticed us with the promise that he would liberally provide food for the entire army for a year."[47] Robert of Clari, who was not privy to the discussions of the leaders, nevertheless confirmed that the promise of provisions was seen by the rank and file as a central argument for a diversion to Constantinople. He recorded (or rather imagined) that Doge Dandolo told the leaders:

> Lords, in Greece there is a very rich land, bountifully supplied with all good things; if we had a just cause for going there and taking food in the

---

44 Robert of Clari, *Conquête de Constantinople*, §16 (p. 136): "Entrementiers que li croisié et li Venicien sejornerent illueques l'iver, si se pourpenserent qu'il avoient molt despendu; et parlerent ensanle et disent qu'il ne pooient mie aler en Babyloine, ne en Alixandre, ne en Surie, car il n'avoient mie viandes ne avoir dont il y peussent aler, car il avoient ja pres de tous despendu, que ens u sejorner qu'il avoient fait, que ens u grant loier que'il avoient donné au navie. Et disent qu'il n'i pooient me aler, et s'il y aloient, n'i feroient il nient, qu'il n'avoient ne viande ne avoir dont il se peussent soustenir."

45 The envoys arrived in mid-December 1202. Geoffrey of Villehardouin, *Conquête de Constantinople*, §91 (vol. 1, p. 90).

46 Geoffrey of Villehardouin, *Conquête de Constantinople*, §93 (vol. 1, p. 92): "... il set que vos avez mis le vostre et que vos iestes povre: si vos donra .II. .C. .M. mars d'argent, et vïande a toz cels de l'ost, a petiz et a granz."

47 Hugh of St Pol, *Epistola* [to Henry of Louvain] in *Annales Colonienses maximi*, ed. G. Waitz, in MGH *SS*, vol. 17, 723-847, pp. 812-24, here p. 812: "... promisit nobis, quod toti exercitui nostro per annum victualia largiretur integrum." Cf. Hugh of St Pol, *Epistola* [to R. de Balues], ed. R. Pokorny, in "Zwei unedierte Briefe aus der Frühzeit des lateinischen Kaiserrecihs von Konstantinopel", *Byzantion*, 55 (1985), 180-209, pp. 203-9, here p. 205: "... ecan(t)avit nobis, quod toti exercitui per annum victualia largiretur ..." [trans. A. J. Andrea, *Contemporary sources for the Fourth Crusade* (Leiden, 2000), p. 189].

> land and other things, so that we may be well restored to health, that would seem to me a good plan. Then we could very well go over the sea.[48]

On cue, Boniface of Montferrat then stood up and told the story of Alexios Angelos, concluding: "Whoever could get this young man ... would be well able to go to the land of Constantinople and take food and other things; for this young man is the rightful heir."[49] Money and military support were important to the Crusaders, but food was an immediate need. By recognizing Alexios as the true Byzantine emperor and supporting his bid for the throne, the Crusade would be justified in extracting provisions from Byzantine domains. This immediate solution to a pressing problem played an important role in the final decision to sail to Constantinople.

The problem of provisions also argued against the planned voyage to Egypt or Palestine. In his letter to Henry of Louvain, Hugh of St Pol reported that heading straight to the Holy Land was "... fruitless and injurious for everyone, insofar as they were destitute and low on provisions."[50] In a letter of August 1203, the Crusade leaders reported to the Pope that they were making the detour because "... lacking all foodstuffs and supplies, we appeared to be bearing a burden to the Holy Land ... rather than bringing some sort of aid, nor did we believe that, given such extreme poverty, we could effectively land in the territory of the Saracens."[51] In his less accurate, but more colourful way, Robert of Clari recalled the supporters of the diversion arguing: "What are we going to do at Babylon or Alexandria when we have neither food nor money enough to go? Better for us, before we go there, to obtain food and money by means of a just cause than to go there and die of hunger."[52]

---

[48] Robert of Clari, *Conquête de Constantinople*, §17 (p. 137): "Seigneur, en Grece a molt rike tere et molt plentive de tous biens; se nous poiemes avoir raisnavle acoison d'aler y et de prendre viandes en le tere et autres coses, tant que nous fuissiemes bien restoré, che me sanleroit boins consaus, et si porriemes bien outre mer aler."

[49] Robert of Clari, *Conquête de Constantinople*, §17 (p. 137): "Qui chu vaslet porroit avoir ... il porroit bien aler en le tere de Constantinoble et prendre viandes et autres coses, car li vaslés en est drois oirs."

[50] Hugh of St Pol, *Epistola* [To Henry of Louvain], p. 812: "... inutilis et dampnosa, cum ipsi essent inopes et victualibus inmuniti." Cf. Hugh of St Pol, *Epistola* [to R. de Balues], p. 204: "... cum ipsi essent inopes et vic(t)ualibus imminuti ... erat omnibus (inut)ilis et (da)mnosa." [trans. Andrea, *Contemporary sources*, p. 189].

[51] *Literae Crucesignatorum*, in Innocent III, *Innocentii III Romani pontificis Regestrorum sive epistolarum libri*, ed. E. Baluze, et al., in PL, vols 214-216, and 217, coll. 9-308, VI.211 (vol. 215, col. 238): "... cum victualibus omnibus et rebus egentes. Terrae sanctae videremur gravamen potius illaturi, ... quam juvamen aliquod allaturi, nec terrae Sarracenorum in tanta egestate nos crederemus applicare potentes, ..." [trans. Andrea, *Contemporary sources*, pp. 79-85, here p. 81].

[52] Robert of Clari, *Conquête de Constantinople*, §33 (p. 152): "Que ferons nous en

The Crusade's lack of provisions was also known to the Pope. In a letter of June 1203, Innocent explicitly forbad the leaders to take up Alexios Angelos's cause, ordering them to sail immediately to the Holy Land. Realizing that hunger could argue strongly in Alexios's favour, Innocent wrote:

> Lest, however, you lack food, we will write to our beloved son in Christ, the emperor of Constantinople, that, just as he promised us in his letter, he will make sure that food is supplied to you. If, perchance, it happen that these supplies are denied to you because you have sworn yourselves to the public service of the Crucified One, to whom the land and its fullness belong (the world and all who live in it), in no way is it absurd to draw an analogy with the earthly emperor, for whom there exists a provision in civil law. If his army lacks food, it may take it anywhere. And you may, without hurting people and with fear of the Lord and the intention of making reparation, take such items when facing a similarly great need.[53]

Here, Innocent cleverly negated the argument of survival. The Crusaders, he implied, had no need of an earthly emperor in order to take provisions from the Greeks, since their allegiance to the heavenly emperor already bestowed on them that right. They could, therefore, head directly to the Holy Land without worrying about provisions. Clever it was, but it was also too late. By June 1203 the Crusade was well on its way to Constantinople. The letter probably did not arrive until after Alexios was installed on the throne.

The Crusaders sailed from Zara in April 1203, making their way down the Adriatic and stopping at Corfu. There they made camp and awaited the arrival of Alexios Angelos. Relations with the Greeks seem to have been peaceful during their three-week stay on the island.[54] With the arrival of the "rightful emperor" it

---

Babyloine ne en Alixandre, quant nous n'avons viande ne avoir par coi nous y puissons aler? Miex nous vient-il, anchois que nous y aillons, que nous conquestonnes viande et avoir par raisnavle acoison, que nous y aillons pour morir de faim."

53 Innocent III, *Crucesignatis*, in Innocent III, *Registra*, VI.102 (vol. 215, col. 109): "Ne autem victualia vobis desint, charissimo in Christo filio nostro Imperatori Constantinopolitano scribemus, ut, juxta quod per litteras suas nobis ipse promisit, victualia vobis faciat exhiberi. «Quod si forsan ea vobis contingeret denegari, cum tamen vos devoveritis ad commune obsequium Crucifixi, cujus est terra et plenitudo ejus, orbis terrarum et universi qui habitant in ea, posset utique non absurdum videri ad similitudinem imperatoris terrae, de quo cautum est in jure civili quod si ejus exercitus indiguerit alimentis, ea possit accipere undecunque, possitis et vos cum timore Domini, sub satisfaciendi proposito, ad necessitatem tantum ea sine personarum accipere laesione.»" [trans. Andrea, *Contemporary sources*, pp. 64-9, here pp. 67-8].

54 The anonymous chronicler of Halberstadt reported that the Greeks attacked the Venetian fleet with petraries, forcing it to withdraw. In retaliation, the Crusaders devastated the island. See *Gesta episcoporum Halberstadensium*, p. 118. No eyewitness reported an

appears that the Crusaders had no problem acquiring food from the locals. Villehardouin described the island as "very rich and bountiful" and no sources referred to any provisioning problem during the sojourn there.[55]

On 24 May 1203 the fleet again set sail, leaving Corfu and travelling around Cape Malea into the Aegean Sea. Landing on Evvoia, a squadron was sent with Alexios to subdue Andros. The remainder headed straight for the Dardanelles. They landed at Abydos in mid-June and waited eight days for the others to catch up. Villehardouin reported that they were again running low on provisions and so helped themselves to the grain harvest, which stood ripe in the fields.[56]

But it was not enough to sustain them for long. When, on 23 June, they cast anchor before the abbey of St Stephen some seven miles southwest of Constantinople, they were again low on food. The leaders discussed their options. More than one knight argued for a general march to the southern land walls of the capital where they could prepare for an attack.[57] Perhaps only the Venetians knew the foolhardiness of such a plan. The mammoth fortifications of Constantinople were legendary, having deflected armies many times the size of the Fourth Crusade. Within the walls the emperor Alexios III Angelos commanded a very large garrison.[58] The aged Dandolo remained silent at the meeting until all had had their say. Then, according to Villehardouin, who was present, he rose and spoke:

> Lords, I know the conditions of this land better than you because I have been here before. You have undertaken the greatest and most perilous enterprise that anyone has ever undertaken. It would therefore be appropriate if you acted prudently. Know this, if we go overland, the country is large and broad, and our people are poor and short of food. Therefore, they will disperse across the land to acquire food, and there are a great many people in this country. We will, therefore, not be able to keep so good a watch that some of ours would not be lost; and we cannot

---

altercation between the Crusaders and the inhabitants and it is impossible to believe that the bombardment of the fleet and the devastation of the island would be so ignored. Nikētas Chōniatēs noted that the Crusaders were on Corfu for 20 days but reported no violence, remarking only that they departed when they realized that the citadel was unassailable. Nikētas Chōniatēs, *Nicetae Choniatae historia*, ed. J.-L. van Dieten (Berlin, 1975), p. 541. It is hard to accept the second-hand testimony of the *Gesta episcoporum Halberstadensium* on this matter. Andrea doubts the whole story. See A. J. Andrea, "Conrad of Krosigk, Bishop of Halberstadt, Crusader and monk of Sittichenbach: his ecclesiastical career, 1184-1225", *Analecta Cisterciensia*, 43 (1987), 11-91, p. 43, n. 133; Andrea, *Contemporary sources*, p. 254, n. 65. Cf. Queller and Madden, *Fourth Crusade*, p. 97, where we accepted it.

55 Geoffrey of Villehardouin, *Conquête de Constantinople*, §113 (vol. 1, p. 116): "... mult ere riche et plenteüroise."

56 Ibid., §126 (vol. 1, p. 128).

57 Ibid., §129 (vol. 1, p. 130).

58 Queller and Madden, *Fourth Crusade*, pp. 106-8.

> afford any losses, for we have very few people for that which we wish to do.
>
> There are islands nearby, which you can see from here, that are inhabited and can furnish grains and victuals and other resources. Let us make port there and take the grain and the provisions of the land. And when we have loaded aboard the provisions, let us go before the city and do that which Our Lord has prepared. For he who has food wages war more securely than he who does not.[59]

The Doge's wise counsel was accepted. The barons ordered the host to board the Venetian ships the following day.[60]

Villehardouin was clearly referring to the Princes' Islands, which are visible from there. He was confused when the next day the fleet did not sail there at all.

> Then God our Lord made them reverse the decision that had been made the day before in the evening to sail to the islands, as if no one had ever spoken of it; and immediately they headed for the mainland, just as directly as they could go.[61]

The fleet made port at Chalkēdōn, a wealthy suburb on the Asian side of the Bosporos, directly across from Constantinople. Why the change in plans? None of the Princes' Islands could have accommodated the fleet or provided much in the way of provisions or water. They could not have been the fleet's destination. Compounding Villehardouin's confusion was the fact that no one seemed to notice or at least mention the change in plans. It is possible, of course, that the entire episode at St Stephen's was a fiction, although that is not at all characteristic of

---

59 Geoffrey of Villehardouin, *Conquête de Constantinople*, §§130-31 (vol. 1, pp. 130-32): "Seignor, je sai plus del convine de cest païs que vos ne faites: car altre foiz i ai esté. Vos avez le plus grant afaire et le plus perillous entrepris que onques genz entrepreïssent: por ce si convendroit que on ovrast sagement. Sachiez, se nos alons a la terre ferme, la terre est granz et large; et nostre gent sont povre et diseteus de la viande: si s'espandront par la terre por querre la vïande: et il i a mult grant plenté de la gent el païs: si ne porriens tot garder que nos n'en perdissiens; et nos n'avons mestier de perdre, que mult avons poi de gent a ce que nos volons faire.

Il a isles ci prés, que vos poez veoir de ci, qui sont habitees de genz et laborees de blez et de viandes et d'autres biens: alons iki prendre port, et recuillons les bleés et les vïandes del païs; et quant nos avrons les vïandes recuillies, alomes devant la ville et fesons ce que Nostre Sires nos avra porveü. Quar plus seürement guerroie cil qui a la vïande que cil qui n'en a point."

60 Geoffrey of Villehardouin, *Conquête de Constantinople*, §131 (vol. 1, p. 132).

61 Ibid., §134 (vol. 1, p. 134): "Ensi lor bestorna Diex nostre Sires le conseil qui fu pris le soir de torner es ysles, ausi con se chascuns n'en aüst onques oï parler; et maintenant traient a la ferme terre plus droit que il onques puent."

this author, particularly when it comes to the business of councils. It is much more likely that Villehardouin simply misunderstood Dandolo. Perhaps the blind doge mistakenly pointed toward the islands while counselling a landfall at Chalkēdōn. The latter is also visible from St Stephen and, indeed, somewhat closer than the Princes' Islands. The names would have meant nothing to the French knight.

Chalkēdōn certainly had everything that Dandolo had described. Villehardouin related that it was beautiful, rich, and well supplied with all good things. The winter wheat was not only plentiful in the fields, but had already been reaped and bound in sheaves. The hungry Crusaders quickly helped themselves. The horses were also disembarked and allowed to graze and exercise after their long journey.[62]

To prepare to cross, the leaders decided to move camp to Chrysopolis, about a mile to the north. The Venetians sailed the fleet there and everyone else went overland. There they waited for the Byzantine people to overthrow their tyrant, Alexios III. That is, after all, what their imperial claimant had assured them would happen. While they waited, food remained a constant concern. With no general stores of provisions, every man was responsible for his own sustenance. That meant that most or all had to forage or purchase provisions. As Villehardouin remarked, at Chrysopolis "... those who had need procured for themselves provisions, and that included everyone in the host."[63] Having to go increasingly further afield as the days passed, foraging parties were protected by a company of some 80 knights. On 1 July these skirmished with a larger force of Greek cavalry. After defeating them decisively, they brought back additional supplies and horses.[64]

On the following day an emissary from Alexios III arrived in the Crusader camp and presented his credentials. He related that the emperor was amazed that such good men, sworn to deliver the True Cross and the Holy Sepulchre, were in his domains making war against fellow Christians. If their problem was insufficient provisions or supplies, he would provide them with everything they needed. Yet if they were there to make war he would be forced, however reluctantly, to crush them. The Crusade leaders, through the eloquent troubadour-knight Conon of Béthune, responded that they had come to place the rightful emperor on the throne of Constantinople. Conon suggested that Alexios III throw himself on the mercy of his nephew and negotiations went nowhere.[65]

From the perspective of the role of food in the dynamics of the Fourth Crusade, the message of the emperor is instructive. We have seen that at Zara the Crusade's lack of provisions was a powerful argument in favour of supporting the

---

62 Ibid., §135 (vol. 1, p. 136).

63 Ibid., §137 (vol. 1, p. 138): "... et qui en avait besoin se procura des vivres, et c'étaient tous ceux de l'armée."

64 Ibid., §§138-40 (vol. 1, pp. 138-42); Nikētas Chōniatēs, *Historia*, p. 542.

65 Ibid., §§141-4 (vol. 1, pp. 142-6); Robert of Clari, *Conquête de Constantinople*, §41 (pp. 40-41).

young Alexios's claims and delaying the expedition to Egypt. Unlike the promised seas of money and military support, the ability to extract food from the rightful emperor's domains was not a luxury or an expedient but rather an immediate necessity. Innocent had already attempted to neutralize this problem by giving the Crusaders permission to forage in Greek lands. Now Alexios III tried something similar, offering provisions and supplies to help them on their way. But it was much too late for that. Once the leaders had sworn to champion the cause of the exiled prince their chivalric honour would not easily allow them to abandon him. In an attempt to keep the faltering enterprise alive, the Crusaders had wrapped themselves up in a web of contracts and oaths that now went far beyond their daily bread.[66]

As long as they could forage, provisions were not a problem. However, they still lacked sufficient stores to feed the host for very long. On 11 July the army of about 10,000 men made camp just outside the northwest walls of Constantinople on a hill near the monastery of Saints Cosmas and Damian. Under constant attack from sallies and missile fire, they were unable to forage more than four bow-shots away from their fortified camp, forcing them to consume their meagre stores. According to Villehardouin, they had flour and salt meat, but little else. There was no fresh meat to be had, save what they could get from horses killed in battle. He estimated that they had only three weeks worth of provisions.[67] In a letter to the Pope, the commanders later reported that "... we faced the immediate necessity of either perishing or conquering because, constrained by an incredible scarcity of all foodstuffs, we could not reasonably extend this siege for 15 days."[68] Hugh of St Pol wrote that he "... had fallen into such poverty that it was necessary for me to exchange my surcoat for bread, although I kept my horses and weapons."[69]

Fortunately for the Crusaders, the city capitulated after a week. Fearing a palace coup, Alexios III had fled. The crown was restored to Isaac II, father of the young Alexios Angelos. Soon Alexios was crowned co-emperor and immediately began to make good on his promises. The Franks' debt to the Venetians was paid off and for once everyone felt happy and reasonably secure. He quartered the

---

66 T. F. Madden, "Vows and contracts in the Fourth Crusade: the Treaty of Zara and the attack on Constantinople in 1204", *International History Review*, 15 (1993), 441-68, pp. 441-3.

67 Geoffrey of Villehardouin, *Conquête de Constantinople*, §165, (vol. 1, p. 166).

68 Innocent III, *Crucesignatis*, in Innocent III, *Registra*, VI.211 (vol. 215, col. 239): " ... in eum statum necessitatis impacti, ut statim necesse haberemus aut perire aut vincere, cum obsidionem ipsam nulla ratione in quindecim dies protelare possemus, ..." [trans. Andrea, *Contemporary sources*, pp. 79-85, here p. 82].

69 Hugh of St Pol, *Epistola* [to R. de Balues], p. 203: "Nocte vero precedenti diem, in qua se civitas reddidit nobis, ad tantam inopiam deveneram, quod oportuit me superticale meum mittere ad panem, retentis tamen equis et armis" [trans. Andrea, *Contemporary sources*, p. 186].

Crusaders in Galata, across the Golden Horn from the city. There he made available to them "a great abundance of good provisions".[70]

For the next five months, the Crusade had no need to worry about its next meal. Alexios IV kept regular supplies coming at all times. However, he was not so regular in keeping his other promises. Having paid about half of the promised 200,000 silver marks, the new emperor requested that the Crusade winter at Constantinople, which would give him time to secure his hold on the throne and gather the remaining funds. He also promised to renew the transportation contract with the Venetians at his own expense. After a great deal of grumbling, the Crusaders at last agreed. However, relations between the Latins and the Greek population were already poor and further familiarity would only breed additional contempt. In August 1203, during a skirmish between Latins and Greeks in the city, one of the most destructive fires in history was set. It raged across Constantinople's most congested and opulent areas, leaving only charred rubble in its wake. Fearing reprisals, the Latin residents in Constantinople fled to Galata. Tensions would remain extremely high.

Early in December, after a spate of particularly horrific anti-Latin massacres, the Crusade leaders gave Alexios IV an ultimatum: either pay what he had promised or they would "pay themselves". Since the emperor could not pay, they spent the winter pillaging Constantinople's suburbs. There was a great deal of wealth to be had, but food was more problematic. Foraging was difficult so close to the city, so they were often forced to purchase provisions in the local markets. Not surprisingly, by January 1204 food prices in Galata had skyrocketed. Alberic of Trois-Fontaines recorded that a three-day-old loaf of bread, normally worth three Parisian *deniers*, was sold for 26. When a ship out of Brindisi sailed into the harbour loaded with food, the leaders purchased everything it had and distributed it among the soldiers.[71] This suggests an abundance of wealth and scarcity of food which would naturally drive up prices. According to Robert of Clari, a *setier* of wine cost as much as 15 sous, a hen 20 sous, and an egg two pence, "... but there was no such scarcity of biscuit; rather they had enough to supply the host for some time."[72]

By February it was almost impossible to find fresh meat or produce in the Crusader camp. Foraging parties had picked clean what was nearby and were now

---

[70] Geoffrey of Villehardouin, *Conquête de Constantinople*, §191 (vol. 1, p. 194): "Ensi sejornerent en pais et en repos en grant plenté de bones viandes."

[71] Alberic of Trois-Fontaines, *Chronica Albrici monachi Trium Fontium, a monacho novi monasterii Hoiensis interpolata*, ed. P. Scheffer-Boichorst, in MGH *SS*, vol. 23, 631-950, Annus 1204 (pp. 882-3).

[72] Robert of Clari, *Conquête de Constantinople*, §60 (p. 178): "Et le kiertés estoit si grans en l'ost que on i vendoit un sestier de vin .xij. saus, .xiiij. saus, .xv.saus tele eure fu, et une geline .xx. saus, et un oef .ij. deniers; mais du bescuit n'i avoit il mie tele kierté, ains en avoient auques a leur ost maintenir une pieche."

forced to travel further distances, making their activities more dangerous and less productive.[73] One such group of foragers was led by Henry of Flanders on 2 February. With 30 knights and a large number of mounted sergeants, he raided the town of Philea on the Black Sea, northwest of Constantinople. They netted a great deal of booty and provisions, much of which was sent by boat down the Bosporos to Galata. The rest, including cattle and other beasts, was brought overland, notwithstanding a failed Byzantine ambush. According to Alberic of Trois-Fontaines, the total take was only enough to feed the army for 15 days,[74] although Alberic's description of the expedition and ambush is so faulty that it is difficult to give this too much credence.[75] Nevertheless, the problem of provisioning was real and the fact that the Crusaders were raiding towns distant a full-day's ride suggests its seriousness. As Gunther of Pairis noted, they were "... suffering from deprivation in a hostile land."[76]

On the night of 8-9 February 1204 Alexios IV was strangled by the newly crowned Alexios V Mourtzouphlos. At once the goal of the Crusaders' war against the Greeks changed from pillaging to conquest. Did the problem of provisions play a role in this decision as well? Perhaps, but there were a great many other factors that certainly weighed more heavily. According to the Anonymous of Soissons, the leaders decided to target Constantinople because after the death of Alexios IV "... they were neither able to enter the sea without danger of immediate death nor delay longer on land because of their impending exhaustion of food and supplies, ...".[77] Yet that is the only source that suggests that food was a consideration. As Robert of Clari noted, the army at least had sufficient biscuit to keep body and soul together.

Much more important in this decision was the desire to punish the Greeks for their faithlessness in dealing with the Crusaders and their rightful lord. This sentiment pervaded all of the contemporary accounts. It was even sanctified by the clergy, who offered the Crusade indulgence to those who took up arms against Constantinople. It is certainly true that the Franks did not have sufficient provisions for a voyage to campaign in Egypt or Palestine yet neither did they have a fleet, given that their lease on the Venetian vessels had long since expired.[78] In other words, their problems were legion. They were disgusted, angry, and yes, sick of biscuit. These were reasons enough to attack the Byzantine capital.

---

73 Baldwin of Flanders, *Literae Balduini imperatoris ad Papam*, in Innocent III, *Registra*, VII.152 (vol. 215, col. 449).

74 Alberic of Trois-Fontaines, *Chronica*, *Annus* 1204 (p. 883).

75 See Queller and Madden, *Fourth Crusade*, p. 278, n. 128.

76 Gunther of Pairis, *Hystoria Constantinopolitana*, §14 (p. 146).

77 Anonymous of Soissons, *De terra Iherosolimitana*, fol. 212v (p. 160): "... cum nec mare sine mortis imminentis periculo intrare possent, nec in terram morari diutius propter victualium et rerum suarum ingruentem egestatem, ..."

78 Queller and Madden, *Fourth Crusade*, pp. 172-5.

Proper provisioning is essential for any army. Successful commanders plan ahead and keep a careful watch on the food stores. The planners of the Fourth Crusade tried to be successful. They were careful to secure a reliable form of transportation and sufficient provisions to free the army from reliance on foraging and local markets for a considerable amount of time. The Venetian execution of this plan was precise, but the plan itself was faulty. Ironically, it was an abundance of supplies and vessels, indeed three times more than was needed, that led to the later shortages. Unable to fulfil their contract, the Franks were forced to spend three months waiting for more arrivals in Venice and an additional seven months waiting for spring to arrive, all the while consuming their provisions. The resulting lack of food played a critical role in their decision to divert the Crusade to Constantinople, where they would become hopelessly trapped in a quagmire of Byzantine intrigue and politics. The army of the Fourth Crusade not only travelled on its stomach, it was in many ways led by it.

Chapter 12

Money and logistics in the forces of the First Crusade: coinage, bullion, service, and supply, 1096-99[1]

Alan V. Murray

How much money did Crusaders take with them? How did they use money during the course of Crusading expeditions? These may seem to be obvious questions to raise in connection with the wider issue of the logistics of Crusading; yet, while considerable research has been carried out into the means by which Crusaders raised money to finance their journeys, much less has been devoted to how such funds were spent, and in particular, how money was used by Crusaders after their departure.[2] As far as the first century of the Crusade movement is concerned, most studies of money have tended to deal with purely numismatic considerations.[3] Any attempt to assess the logistical dimensions of Crusader finances therefore needs to consider all functioning forms of money, as well as how they were used. This paper examines these issues in connection with the forces of the princes during the First Crusade, focusing on three related questions. What sums of money did Crusaders take with them? In what form did they take them? And, how did they use them in the course of their march to the Holy Land?

---

1 Consult Maps 2, 3, 7, 8, 10, 11.

2 For an overview of Crusading finances see F. A. Cazel, jr, "Financing the Crusades", in Setton, *HC*, vol. 6, 116-49. For money in Crusades see K. Leyser, "Money and supplies on the First Crusade", in idem, *Communications and power in medieval Europe: the Gregorian revolution and beyond*, ed. T. Reuter (London, 1994), 77-95; R. Hiestand, "»Precipua tocius christianismi columpna«. Barbarossa und der Kreuzzug", in A. Haverkamp, ed., *Friedrich Barbarossa: Handlungs-spielräume und Wirkungsweisen des staufischen Kaisers* (Sigmaringen, 1992), 51-108. A related, but somewhat different, issue to those treated here is the transfer of sums of money on behalf of Crusaders from the West to *Outremer*, for which see D. M. Metcalf, "The Templars as bankers and monetary transfers between West and East in the twelfth century", in *Coinage in the Latin East: the Fourth Oxford Symposium on coinage and monetary history*, ed. P. W. Edbury and D. M. Metcalf [BAR *IS*, vol. 77] (Oxford, 1980), 1-17.

3 For overviews see J. Porteous, "Crusader coinage with Greek or Latin inscriptions", in Setton, *HC*, vol. 6, 354-420; D. M. Metcalf, *Coinage of the Crusades and the Latin East in the Ashmolean Museum, Oxford*, 2nd ed. (London, 1995).

It has been estimated that less well-off armsbearers needed roughly four times their annual income in order to finance a Crusading journey.[4] This figure is based on investigation of a vast number of charters issued by Crusaders but it does not tell us about the relative amounts of spending before and after departure and thus about how much money may have been carried by Crusaders. The amount of money spent before departure as well as the amount taken on Crusade after departure must have varied according to social status. Many poorer pilgrims, particularly those who joined the so-called "People's Crusade" of 1096-97, may have been prepared to leave their homes with little more than what they stood up in, trusting to charity, foraging, extortion from Jews, and providence. The expenses of those whose normal business was fighting would have been higher, requiring arms, armour, clothing, warhorses, riding horses, pack animals, tents, as well as provisioning for the journey. One might expect that those whose *métier* was fighting would have had most of this gear available anyway but an expedition expected to last a long time may have been seen as an appropriate occasion to replace equipment, at least in part. In the first redaction of his chronicle, Fulcher of Chartres wrote that Crusaders leaving their spouses told them that they would return within three years. Two to three years, therefore, seems a conservative estimate for the amount of time that Crusaders in 1096 thought that they might be away on campaign.[5] Expenses would have increased exponentially according to the number of dependents accompanying a given individual. The retinues of bishops, dukes, counts, and wealthier lords would have included not only knights and grooms but also numbers of household officers, scribes, chaplains, huntsmen, falconers, servants and other personnel, who would have needed to be equipped and supported.

The greatest projected category of expense after departure would have been for food supplies *en route*. Crusaders could, of course, carry food with them but there would be a limit imposed by the perishability of foodstuffs and by the amount that could be transported. Supplies carried by any contingent or individual would not have lasted more than a few weeks at most but, at least until enemy territory was reached, this would not have caused major problems.[6] The overland

---

4 J. Riley-Smith, "Early Crusaders to the East and the costs of Crusading, 1095-1130", in M. Goodich, et al., eds, *Cross cultural convergences in the Crusader period: essays presented to Aryeh Grabois on his sixty-fifth birthday* (New York, 1995), 237-57; idem, *The First Crusaders, 1095-1131* (Cambridge, 1997), p. 112.

5 Fulcher of Chartres, *Historia Hierosolymitana (1095-1127)*, ed. H. Hagenmeyer (Heidelberg, 1913), I.vi (p. 163).

6 R. Hiestand, "Der erste Kreuzzug in der Welt des ausgehenden 11. Jahrhunderts", *Monatshefte für evangelische Kirchengeschichte des Rheinlandes*, 44 (1995), 1-36. Odo of Deuil related that during the Second Crusade the troops of Conrad III of Germany were advised by Byzantine guides to take a week's provisions with them from Nicaea. This amount was supposed to last them as far as Konya and since this march was through enemy

routes taken by the princes' forces would take them through Christian, and thus theoretically friendly, territory as far as Bithynia. From their points of departure as far as the eastern frontiers of the Byzantine Empire, Crusaders might reasonably expect to be provided with a certain amount of supplies by friendly rulers such as the king of Hungary and the Byzantine emperor; however, by and large they would depend on purchase to provide food for themselves and their animals. Unlike many other forms of warfare in medieval Europe, land Crusades to the Holy Land did not really have the option of having additional supplies of money brought to the armies during campaign, at least until they had reached the East. Therefore, although there might well be opportunities to gain plunder *en route*, whatever sums of money were estimated as being needed to purchase supplies would have to be carried on the journey.

We have no precise primary information on what Crusaders in 1095-96 estimated their expenses might be going to be but we can extrapolate from figures dating from the second half of the twelfth century. During preparations for the Third Crusade in 1189-90, Frederick Barbarossa and his advisers ordered that Crusaders should take sufficient funds for a campaign of two years' duration. The actual amount that this implied was given by Otto of St Blasien as three marks.[7] This implied costs of 432 pence, 18 pence per month, or 0.6 pence per day.[8] Yet this amount should probably be regarded as an absolute minimum which was intended to discourage the participation of non-combatants, and Otto stated that Crusaders were urged to take larger sums if they could. A more realistic idea of the costs of military expeditions for armsbearers can be gained from the so-called *Längeres Kölner Dienstrecht*, a set of regulations from around 1165 governing the service of the ministerial knights of the archbishop of Cologne. Paragraph Four described customs in force when knights were obliged to accompany the archbishop to an imperial coronation in Rome. Knights with incomes of less than five marks per annum were not obliged to go. Those with incomes of over five marks were to receive ten marks each from the archbishop, as well as 40 ells of cloth for their retinue. After crossing the Alps, each knight was to receive one mark per month to cover his expenses, which would of course include keep for

territory, it is likely that they took as much as they could carry. In the event this was to prove insufficient. See Odo of Deuil, *De profectione Ludovici VII in orientem*, ed. and trans. V. G. Berry (New York, 1948), bk 5 (p. 90).

7 Otto of St Blasien, *Chronica*, ed. A. Hofmeister, in MGH *ScriptRerGerm.Cont* (Hanover and Leipzig, 1912), §31 (p. 4): "Quibus omnibus imperator sequentis anni Maio tempus profectionis constituit, pauperioribus ad minus trium marcarum expensam, dicioribus pro posse expensis preparari indicens."

8 I have adhered throughout to the important distinction between "pounds", "shillings", and "pence" for monies of account, and "pennies" for actual coins.

grooms and other servants.[9] A journey to an imperial coronation was not a military expedition in the strict sense, but it might involve travel through the territory of hostile, anti-imperialist cities in Lombardy and in reality German emperors often used such expeditions to further their political aims in Italy. Knights would have to be prepared to fight on such journeys, although one might surmise that the amount of fighting involved would be less than on a Crusade. The *Kölner Dienstrecht* gives some useful comparative figures, even though they are likely to have been underestimates when compared with the expenses involved in Crusades. We can say that the journey from Cologne to Regensburg, the starting point of the Third Crusade for the Germans, was roughly equivalent to the journey from Cologne to the Alps, which was not covered by the additional monthly expenses specified in the *Kölner Dienstrecht*. For an expedition lasting two years, a knight with an annual income of at least five marks would require ten marks for his initial outlay and the journey to Regensburg, plus some 24 marks thereafter, a total of 34 marks or just under seven times his annual income.

One could argue that many of the knights who participated in the First Crusade were poorer than the *ministeriales* of Cologne who have been used for the foregoing calculations. However, any lower figures for a proportion of the knights would be offset by the relatively large sums that would have been raised and carried by the households of the princes, bishops, and greater lords, many of whose members would undoubtedly have required a higher level of subsidence than that of knights. These considerations would suggest that Riley-Smith's figure of four times annual income is probably an underestimate for many of the armsbearers on the First Crusade, certainly for the wealthier knights and those higher up the social scale.

Apart from those with ready funds, Crusaders who needed to raise money for a two-year expedition could resort to various possibilities. These included sales of property and movable goods, sales of rights (including renunciations, confirmations, etc.), mortgages and loans against cash, seizure or confiscation of movables, extortion (primarily from Jews), and appeals to charity. In 1096-7, however, there was no dedicated taxation as a form of Crusade finance. The monetary system of western Europe at this time was based on the *libra* or pound of 20 *solidi* or shillings, each *solidus* being worth 12 *denarii* or pence. As well as *marcae* or marks, the pounds and shillings were monies of account only. The basic unit for transactions was the silver penny, which varied in weight and silver content, and consequently in value, throughout western Europe. Some territories issued *oboli* or halfpennies as a smaller unit and it was also common practice to cut pennies into halves or fractions in order to make payments for smaller amounts.

---

9 *Das längere Kölner Dienstrecht (etwa 1165)*, in L. Weinrich, ed., *Quellen zur deutschen Verfassungs-, Wirtschafts- und Sozialgeschichte bis 1250* [AQDGM, vol. 32] (Berlin, 1977), 266-79, §4 (pp. 268-70).

But there were no coins for higher denominations until Venice and other Italian territories began to issue the so-called *grossi* at the end of the twelfth century.[10] This meant that the two principal forms in which the sums of money raised by Crusaders would have been paid over to them would have been coin, that is silver pennies, or bullion, that is unminted silver, valued at a specified sum in monies of account. In practice payments may well have been made in combinations of coin and unminted silver but it is difficult to establish what the typical proportions may have been.

When used in monetary transactions, unminted silver was usually in the form of bars or ingots containing metal of a set weight, representing a multiple of a basic equivalent value in coin.[11] Ingots could be used to make high-value purchases before departure and were a practical means of transporting money on the march since they could be exchanged whole for local currency, or cut up into fractions to provide small or intermediate amounts for payment or exchange.[12] Silver ingots have been found in the so-called "Barbarossa Hoard", a series of finds of coins and treasure which came to light between 1982 and 1985, and which were quickly recognized by specialists as originating from a single hoard in Turkey. Of the total of around 7,700 coins and fragments, none can be dated to later than 1190, suggesting that the hoard represents the funds of a Crusader, or group of Crusaders, from the army of Frederick Barbarossa during the Third Crusade.[13] From the 1070s European silver supplies had begun to go into a decline. The most famous source, the Harz Mountains in Saxony, had been largely worked out, and this situation did not begin to change until the 1160s, with the discovery of new sources of silver at Freiberg in Upper Saxony, Friesach in Carinthia, Montieri in Tuscany, and in Sardinia.[14] The occurrence of silver ingots in the Barbarossa Hoard provides hard evidence of their use in a Crusade army at the end of the twelfth century, a time when there was no shortage of silver available for coin. This would also indicate that ingots were used to an equal or even greater extent during the First and Second Crusades, when silver coins were in far shorter supply than after the 1160s.

---

[10] P. Grierson, "The origins of the grosso and of gold coinage in Italy", *Numismaticky Sborník*, 12 (1971-2), 33-48, pp. 33-5.

[11] A. Luschin von Ebengreuth, *Allgemeine Münzkunde und Geldgeschichte des Mittelalters und der neueren Zeit*, 2nd ed. (Munich and Berlin, 1926), pp. 180-83; H. Van Werveke, "Monnaie, lingots, ou marchandises? Les instruments d'échange aux XIe et XIIe siècles", *AHES*, 4 (1932), 452-68; M. Castaing-Sicard, *Monnaies féodales et circulation monétaire en Languedoc, Xe-XIIIe siècles* (Toulouse, 1961), pp. 63-6.

[12] W. Jesse, *Quellenbuch zur Münz- und Geldgeschichte des Mittelalters* (1924; rpt, Aalen, 1968), pp. 249-52; Luschin von Ebengreuth, *Münzkunde*, p. 180.

[13] G. Stumpf, *Der Kreuzzug Kaiser Barbarossas. Münzschätze seiner Zeit* (Munich, 1993).

[14] P. Spufford, *Money and its use in medieval Europe* (Cambridge, 1988), pp. 98-110.

Silver ingots would have been less practical the further one went down the social scale. The main expense on the march was likely to be the purchase of food supplies, which would happen frequently, but on a small scale in each case. For the majority of Crusaders, this would have been most easily carried out by using coins. This would have meant that large numbers would be aiming to carry silver pennies, which would necessarily have constituted fairly large quantities. On figures established so far, the absolute minimum required for a non-combatant participant in the Third Crusade would have been three marks for a two-year journey and in terms of the actual coinage available this would have amounted to 432 pennies. A knight with a minimum annual income of five marks would require at least 24 marks, which in coin would amount to no less than 3,456 pennies. It is questionable how far Crusaders were actually able to obtain coin even as a proportion of such sums before their departure in 1096. In areas where there had been a strong response, attempts by Crusaders to convert assets such as land and rights into money would have put considerable pressure on the amounts of coinage available in the local economy, remembering of course that at this time huge sectors of the economy functioned without money at all. Were there sufficient supplies of coinage to go around so that all Crusaders were able to realize assets by converting property, rights or movables into coin? It is likely that those who were purchasing land or rights from Crusaders in 1095-6 were not always able to find sufficient coin to pay for their purchases. In 1095 or 1096 the Lotharingian nobleman Warner of Grez sold his allod of Vaux in the county of Huy to the church of Fosses, receiving in return a gold chalice valued at $20^{1}/_{4}$ marks. This single sale would undoubtedly have covered a large proportion of his foreseeable expenses but it would hardly have been possible for him to do this in practical terms unless he was able to convert a significant proportion of its value into coins with which he could make a variety of payments from the lowest denomination upwards.[15]

Difficulties in securing sufficient quantities of coinage suggest that Crusaders were also obliged to carry wealth in less standard forms than coins and unminted silver. Gold was less readily available than silver in the West at this time but could also have been carried in the form of ingots, which would of course have represented a considerably higher value per weight than silver. Gold coins were not issued in the West at the time, but there is some evidence for Byzantine gold coins being used in transactions.[16] We also need to envisage the possibility of hoarded or inherited stocks of Muslim or Byzantine gold coinage which Crusaders may well

---

15 "Documents extraits du Cartulaire du chapitre de Fosses", *Analectes pour servir à l'histoire ecclésiastique de la Belgique*, 4 (1867), 396-8.

16 L. Musset, "Réflexions sur les moyens de paiement en Normandie aux XIe et XIIe siècles", in idem et al., eds, *Aspects de la société et de l'économie dans la Normandie médiévale (Xe-XIIIe siècles)* (Caen, 1988), 65-89, pp. 72-6.

have decided to utilize on the occasion, not only as a means of realizing wealth but precisely because gold currency would have had a greater acceptance than silver in the East, as we shall see. Finally, it is likely that not all Crusaders were able to have silver and gold melted down into ingots or to convert it into coin, but simply took objects which had a high proportion of precious metal, such as plate or jewellery.[17]

The narrative sources indicate that the forces of the First Crusade had plentiful amounts of coin with them. However, discussion of coinage in this context has tended to focus on a single remark made by the chronicler Raymond of Aguilers, who related that while they were besieging the town of 'Arqā in 1099, the Muslim ruler of Tripoli, to whom 'Arqā belonged, tried to persuade them to abandon the siege:

> The king of Tripoli offered us fifteen thousand gold pieces of Saracen money; in addition, he offered horses, mules, many articles of clothing, and many more of these things as tribute in succeeding years. At that time, one gold piece was worth eight or nine solidi of the money of our army. This was our money: [deniers] of Poitou, Chartres, Le Mans, Lucca, Valence, Melgueil, and Le Puy, two of these last being worth one of the others.[18]

The denier coins mentioned were designated according to seven different places of minting. Chartres and Le Mans were in France north of the river Loire. Le Puy in the Auvergne, Melgueil in the county of St Gilles, and Poitou, which could have meant either the mint of Poitiers itself or that at Melle, which was the principal

---

[17] During the Second Crusade, Odo of Deuil described Greek ships with moneychangers aboard, which sold food to the Crusaders at Constantinople, whose "... tables gleamed with gold and were groaning with the silver vessels which they had bought from us.". See Odo of Deuil, *De profectione Ludovici VII*, bk 4 (p. 74). Similarly, when the participants in the Fourth Crusade could not raise the agreed sum of money to pay the Venetians for their sea passage, a further appeal for funds went out which yielded much plate, jewellery, and other valuables that had evidently been intended to be kept as reserves. See Geoffrey of Villehardouin, *La conquête de Constantinople*, ed. and Fr. trans. E. Faral, 2 vols (Paris, 1938-9), §§60-61 (vol. 1, pp. 62-4).

[18] Raymond of Aguilers, *Historia Francorum qui ceperunt Iherusalem*, in RHC *HOcc*, vol. 3, 231-309, c. XVI (p. 278): "Volebat nobis dare rex Tripolis quindecim millia aureos Sarracenae monetae; praeterea equos et mulas, et vestes multas, et tributa deinceps singulis annis, multo his ampliora. Valebat quippe unus aureus eo tempore octo vel novem solidos monetae nostri exercitus. Erat haec nostra moneta: Pictavini, Cartenses, Manses, Luccenses, Valanzani, Melgorienses, et duo Pogesi pro uno istorum.". Cf. Raymond of Aguilers, *Liber Raimundi de Aguilers*, ed. J. H. and L. L. Hill, *Le « Liber» de Raymond d'Aguilers* (Paris, 1969), §17 (pp. 111-12).

mint in the county of Poitou, were in France south of the river Loire.[19] The episcopal mint of Valence, just over the frontier from France, and the nominally royal mint of Lucca in Tuscany were in imperial territory. There was undoubtedly a certain degree of correspondence between the mints named and particular contingents in the Crusade. It could be supposed that the followers of Stephen of Blois used coins of Chartres, those of Bishop Adhemar coins of Le Puy, those of Raymond of St Gilles coins of Melgueil, and so on.[20] However, the correspondences were clearly not representative of the Crusade as a whole. There was no mention of coinage of the dukes of Lower Lotharingia, who are known to have issued coins in the eleventh century.[21] Similarly, Robert of Normandy issued deniers as duke of Normandy and in fact Guibert of Nogent mentioned *Rothomagenses*, that is deniers of Rouen, as being used by Robert's troops at Latakia in 1100.[22]

Both the context and formulations of Raymond of Aguilers' remarks need to be examined closely. They relate to a period between the capture of Antioch and the march on Jerusalem when the Crusade forces had become fragmented. Hugh of Vermandois had been sent in charge of a diplomatic mission to Alexios Komnēnos while Stephen of Blois had left for home. We may assume that most of the followers of Hugh and Stephen attached themselves to other contingents. Baldwin of Boulogne was in the process of establishing a power base around Edessa and had been able to build up a war band of his own which included knights from Lotharingia, Flanders, and Northern France. Bohemond and his Normans had mostly remained in Antioch after securing possession of the city but Tancred and his followers had taken service with Raymond of St Gilles. In January 1099 the remaining leaders marched south from Ma'arrat-al-Nu'mān. Raymond, with Tancred in his service, along with Robert of Normandy, invaded the territories of the *amīr* of Tripoli, and laid siege to the town of 'Arqā. They were joined by Godfrey of Bouillon and Robert of Flanders, who had marched via Jabala, and it was after the

---

19 N. Mayhew, *Coinage in France from the Dark Ages to Napoleon* (London, 1988), p. 45.

20 This is essentially the argument proposed by Porteous, "Crusader coinage", p. 359: "Briefly therefore it can be stated that of the coins mentioned by Raymond of Aguilers, the *pictavini* were contributed originally by the Poitevins, the *cartenses* by the followers of Stephen, the *mansei* by Robert's Normans, the *lucenses* came with the fleet, the *valentinenses* and *pogesi* with the Provençals and those who travelled down the Rhône valley, and the *mergoresi* [sic!] were originally brought by the followers of Raymond of Saint Gilles. All are found in Palestine and Syria."

21 G. Albrecht, *Das Münzwesen im niederlothringischen und friesischen Raum vom 10. bis zum beginnenden 12. Jahrhundert* (Hamburg, 1959), pp. 38-45, 66-7, 79-84.

22 Guibert of Nogent, *Guibert de Nogent Dei gesta per Francos, et cinq autres textes*, ed. R. B. C. Huygens, in CC *ContMed*, vol. 127A, 76-352, VII.xxxvii, (p. 336). See also F. Dumas, "Les monnaies normandes (Xe-XIIe siècles)", *Revue numismatique*, ser. 6, 21 for 1979 (1980), 84-140; Musset, "Moyens de paiement", pp. 76-82.

union of the armies that the *amīr* of Tripoli offered payment in exchange for the lifting of the siege of 'Arqā. The total sum offered was 15,000 "gold pieces of Saracen money" (*aureos Sarracenae monetae*), which must have referred to the gold *dīnārs* which were the standard higher denomination coins in circulation in Muslim Syria, Palestine, and Egypt.[23] This payment, together with 5,000 *dīnārs* from Jabala, was eventually paid over to all of the leaders of the Crusade, rather than to Raymond of St Gilles alone. After detailing the various amounts of the payment, Raymond of Aguilers specified that the exchange rate for the *dīnār* was roughly eight or nine shillings (*solidi*), which would be in the range of 96-108 pence. There was, therefore, only an approximate equivalence between six of the western coinages mentioned while that of Le Puy was outside the accepted range of equivalence, being worth only half of the others. It is actually very surprising that there was not a much greater discrepancy since they all had varying silver contents.

Raymond specified the value of the *dīnār* in terms of "shillings of the money of our army" ("*solidos monetae nostri exercitus*") and then wrote that "this was our money" ("*erat haec nostra moneta*") before going on to identify the individual currencies. Now, the word *exercitus* was ambiguous. Raymond and the other chroniclers sometimes used it in the singular to refer to individual contingents, and sometimes in the singular or plural to the Crusade forces as a whole. It is therefore possible that in this instance it related only to the forces of Raymond of St Gilles.[24] When tribute was received, it had to be split up not only among the different leaders but a good proportion must have been distributed down to their various followers and presumably in turn to the dependents of those in trickle-down fashion. The amount of money involved in this tribute payment was of the order of 20,000 *dīnārs*, which the Crusaders valued as equivalent to at least 160,000 shillings of southern France. It would not have been practical to divide up the tribute received using only high denomination gold *dīnārs*, since we should not assume a democratic division of spoils. It is likely that the lion's share went to the leaders and the more important lords. The distribution of tribute on this scale throughout each of the armies could have functioned smoothly only if it also made use of their existing supplies of western coinage. What Raymond of Aguilers was probably doing, therefore, was specifying which coinages were at that point current in the contingent of Raymond of St Gilles and then giving a rough and ready working tariff as to how gold *dīnārs* were to be calculated in relation to them.

---

23 M. L. Bates and D. M. Metcalf, "Crusader coinage with Arabic inscriptions", in Setton, *HC*, vol. 6, 421-82; Metcalf, *Coinage of the Crusades*, pp. 9-14.

24 For individual armies see Raymond of Aguilers, *Historia* (RHC *HOcc*), cc. I, II, IV, XVI (pp. 236, 238, 241, 277); Raymond of Aguilers, *Liber* (Hill), §§1, 2, 5, 17 (pp. 36, 42, 46, 109). For the Crusade as a whole see Raymond of Aguilers, *Historia* (RHC *HOcc*), cc. III, IV, XVI (pp. 240, 241, 278); Raymond of Aguilers, *Liber* (Hill), §§4, 5, 17 (pp. 44, 46, 110).

Raymond had followers from almost the entire area south of the Loire and one would expect them to have carried coins from throughout this area. Coins from Lucca represent more of a problem but there is evidence that they were in use in other contingents, which suggests that they were circulating freely throughout the forces.

I therefore do not think that the seven different coinages mentioned by Raymond of Aguilers represented the only ones in use by the forces of the Crusade. However, I would agree with Matzke's argument that there was a considerable correspondence between the geographical distribution of the mints involved and the itinerary of Pope Urban II in the years 1095-96.[25] On his journey from Italy through France and back Urban passed through all of the places named, except Chartres, and he did come close to both Chartres and to Blois, towns held by Stephen of Blois. Given the problems that would have been encountered by significant numbers of Crusaders simultaneously trying to convert their assets into coin and precious metal, it is entirely likely that during the preparations for the Crusade, the Papal curia could have conducted negotiations with the lords of mints in those areas where the main recruitment was to be expected in order to ensure that sufficient coinage was produced to meet the needs of Crusaders in the period of fund-raising and preparation. This would have involved managing production so that there was a smooth and continuous supply of silver to the mints and making sure that they were not swamped by the demands of buyers and sellers. Matzke's suggestion is undoubtedly the most plausible to have been put forward to date, even though it is perhaps not entirely satisfactory. There were other currencies in wide circulation in the Languedoc which were not mentioned by Raymond of Aguilers.[26] This may well suggest that a rough parity of value, and thus interchangeability, was the main criterion in the selection of the currencies named by Raymond, which were not the only ones taken to the East by the various contingents but were the ones primarily intended for internal exchange, at least among the southern French and northern Italians.[27] Coins of Lucca and Valence,

---

25 M. Matzke, *Daibert von Pisa. Zwischen Pisa, Papst und erstem Kreuzzug* (Sigmaringen, 1998), pp. 130-31.

26 A. Blanchet and A. Dieudonné, *Manuel de numismatique française*, 4 vols (1912-36; rpt, Paris, n.d.), vol. 4, pp. 234-7. Other significant coinages in the regions from which Raymond's army was recruited included those of Toulouse, Cahors, Narbonne and Rodez. See Castaing-Sicard, *Monnaies féodales* and G. Depeyrot, "Le trésor de Toulouse et le numéraire feudal aux XIIe et XIIIe siècles", *Annales du Midi*, 94 (1982), 125-49.

27 M. Matzke, "Vom Ottolinus zum Grossus — Münzpragung in der Toskana vom 10. bis zum 13. Jahrhundert", *Schweizer Numismatische Rundschau*, 72 (1993), 135-99, pp. 148-50, points out that by 1095 the newer issues of Luccan currency no longer had the same value as the older *denarii veteres* or *rugi* and were worth considerably less than the *denarii* of Pavia, the other main mint in the kingdom of Italy. This would explain why the Luccan, but not the

and to a lesser extent of Le Mans, Chartres and Melle, are known from hoards from the Latin East and from finds *en route*.[28] In fact the need to provide coins in sufficient quantities for Crusaders in the years 1095 and after may well have been a major impetus towards devaluation of currencies such as that of Lucca.

Several of the issues involved in fund-raising and transport of money can be illustrated by the case of Godfrey of Bouillon, Duke of Lower Lotharingia. He raised the greatest funds by the sale or mortgage of his hereditary lands and offices. He sold his rights in the county of Verdun as well as the allods of Stenay and Mouzay to Bishop Richer of Verdun for a sum which is not recorded. He also mortgaged to Bishop Otbert of Liège the castle and territory of Bouillon along with an adjacent fief held from the archbishop of Rheims.[29] The sources agree that the amount realized by this mortgage was at least 1,300 marks of silver.[30] The bishop raised money to pay Godfrey and other Crusaders by ransacking the abbeys and churches of his diocese for jewellery, plate, and precious metals in other forms. These measures went to such extremes that in 1104 Otbert was accused by his own cathedral chapter of having debased the coinage of the bishopric.[31] If paid over as coin, the mortgage of Bouillon would have amounted to a total of between 187,200

Pavian currency had been given the status of rough parity with the southern French currencies mentioned by Raymond.

28 D. M. Metcalf, "Coins of Lucca, Valence and Antioch: some new hoards and stray finds from the time of the Crusades", *Hamburger Beiträge zur Numismatik*, 22-3 for 1968-9 (1972), 443-70; M. Phillips, "A hoard of French feudal coins from the First Crusade", *Numismatic chronicle*, 162 (2002), 402-4.

29 A. V. Murray, *The Crusader Kingdom of Jerusalem: a dynastic history, 1099-1125* (Oxford, 2000), pp. 37-45.

30 For the 1,300 marks of silver see *Triumphus S. Lamberti de castro Bullonio*, ed. W. Arndt, in MGH *SS*, vol. 20, 497-511, §1 (p. 499); *Gesta abbatum Trudonensium continuatoris tertii auctorum*, ed. D. R. Koepke, in MGH *SS*, vol. 10, 361-443, Anno Domini 1096 (p. 387). 1300 marks of silver and three marks of gold are mentioned in other sources. See Renier of St Laurence, *Triumphale Bulonicum*, ed. W. Arndt, in MGH *SS*, vol. 20, 583-92, lib. 1 (p. 584); Giles of Orval, *Aegidii Aureaevallensis Gesta episcoporum Leodiensium*, ed. J. Heller, in MGH *SS*, vol. 25, 1-135, III.14 (p. 91). Laurence of Liège mentions 1,300 marks of silver and one pound of gold. See Laurence of Liège, *Laurentii de Leodio Gesta episcoporum Virdunensium et abbatum S. Vitoni*, ed. G. Waitz, in MGH *SS*, vol. 10, 486-530, §12 (p. 498). Alberic of Troisfontaines mentions 1,500 marks of silver and the *Chronique de Saint-Hubert* 1,500 pounds of silver. See Alberic of Trois-Fontaines, *Chronica Albrici monachi Trium Fontium, a monacho novi monasterii Hoiensis interpolata*, ed. P. Scheffer-Boichorst, in MGH *SS*, vol. 23, 631-950, p. 804; *La chronique de Saint-Hubert dite Cantatorium*, ed. K. Hanquet (Brussels, 1906), p. 206 [ed. L. C. Bethman and W. Wattenbach, as *Chronicon sancti Huberti Andaginensis usque ad a. 1106*, in MGH *SS*, vol. 8, 565-630, §83 (p. 615)].

31 J.-L. Kupper, "Otbert de Liège: les manipulations monétaires d'un évêque d'Empire à l'aube du XIIe siècle", *Le Moyen Age*, 86 (1980), 353-85.

and 216,000 pennies.[32] Duke Godfrey had a mint at Bouillon, the only urban centre in his domains, but its capacity seems to have been fairly limited. The bishopric of Liège was better provided with mints, but a mint simply provided the technological infrastructure where coins could be produced. We should not assume that all mints were permanently staffed or continuously in production. Given the obviously frantic nature of the bishop's fund-raising, it seems doubtful whether the episcopal mints or the ducal one at Bouillon could have processed this large an amount of bullion realized by sales to the bishop in time for the departure of Godfrey and his contingent.[33]

Figure 12.1: Penny of Godfrey of Bouillon with the inscription GODFRIDUS IEROSOLIMITANUS

After V. Tourneur, "Un denier de Godefroid de Bouillon frappé en 1096", *Revue Belge de numismatique et de sigillographie*, 83 (1931), 27-30, p. 27.

It is possible to make a reasonable estimate of the total funds raised by Godfrey. Even if the sales of the rights in Verdun plus Stenay and Mouzay were equivalent to only ten percent of the value of his core allodial territory of Bouillon, this would represent another 18,720 pennies in terms of the coinage available. According to Solomon bar Simson, writing some 45 years later, and whose account of the Crusade is admittedly very garbled, it may have been Godfrey of Bouillon who was responsible for extorting monies from the Rhineland communities of the Jews which amounted to some 1,000-1,500 marks. If so, it is likely that this was composed of a mixture of coin, precious metal, and other valuables.[34] Taking the

---

32 Calculated as 1,300 marks at 144 pence per mark for the lower estimate, and 1,500 at 144 for the higher.

33 Albrecht, *Münzwesen*, pp. 38-45, 66-7, 79-84.

34 Solomon bar Simson, *The chronicle of Solomon bar Simson*, ed. and trans. S. Eidelberg, in *The Jews and the Crusaders: the Hebrew chronicles of the First and Second*

lowest estimates for all of these sums, it would seem that Godfrey had been able to raise funds equivalent to almost 350,000 pence and the total may possibly have been as much as 450,000 pence. Godfrey would have needed to have had at least a proportion of his funds available in the form of coin in order for he and his followers to make low denomination payments and there are two deniers which have been found in hoards in Russia which provide evidence for coins having been minted by him in 1095-6. One of these has an inscription which can be read as GODFRIDUS IEROSOLIMITANUS, which suggests that it was a coin which had been produced for Godfrey by one of the Liège mints before he left.[35] The proportion of Godfrey's funds represented by such coinage was, however, probably limited, as is suggested by the evidence of Albert of Aachen who was well informed about Godfrey's army. Albert was demonstrably familiar with various units of measure of the Meuse region and Flanders but on two occasions in which he specificied currencies, he gave prices in terms of pence of Lucca, a coinage which is unlikely to have been brought from the West by Godfrey's men, thus suggesting that other coinages came to be in greater use amongst his forces during the course of the march than coinages from Lotharingia.[36] Whereas Crusaders from southern France would have been able to secure significant quantities of coin as a result of planning which involved the *curiae* and lords of selected mints, those from Northern France, Flanders, Lotharingia, and other parts of Germany may well have had much greater difficulties in securing sufficient quantities of coin. It is therefore very likely that Crusaders such as Godfrey of Bouillon carried a greater proportion of their wealth in the form of unminted silver than was the case for those of forces recruited in Italy and central and southern France.

The main purpose for which money flowed out of Crusader forces was for the purchase of food *en route*. Crusaders could always try to purchase supplies on an *ad hoc* basis or to take them without payment; however, the numbers involved meant that considerable organization was required to ensure that sufficient quantities could be made available. The authorities in the friendly states through which they passed had to encourage sufficient numbers of producers to come forward and to provide places to buy and sell where their safety would be guaranteed. We repeatedly hear of Crusaders asking for markets to be made available to them and of the authorities granting permission to buy and sell

---

Crusades (Madison, 1977), 21-72, pp. 24-5. See Murray, *Crusader Kingdom of Jerusalem*, pp. 43-5.

35 N. Bauer, "Der Fund von Spanko bei St. Petersburg", *Zeitschrift für Numismatik*, 36 (1926), 75-94; V. Tourneur, "Un denier de Godefroid de Bouillon frappé en 1096", *Revue belge de numismatique et de sigillographie*, 83 (1931), 27-30.

36 Albert of Aachen, *Historia Hierosolymitana*, in RHC *HOcc*, vol. 4, 265-713, III.lii (p. 375), IV.xxxiiii (p. 412).

(*licentia emendi et vendendi*).[37] Large numbers of Crusaders would have an immediate effect on local demand and consequently on prices, and one of the frequent concerns echoed in the chronicles was the demand for "fair prices" and "goods in fair weight and measure".[38]

In Hungary the supply situation was generally straightforward, but it was quite different in Byzantine territory. The Crusaders were confronted with an unfamiliar monetary system in which value was based not on intrinsic silver content, but on gold for high denominations, and on a generally accepted nominal value for lower denominations. According to the currency reform carried out by Alexios Komnēnos in 1092, the standard Byzantine unit of coinage became the gold *hyperpyron nomisma*, which was equivalent to 48 *aspron trachea* of billon (known to the Crusaders as the *staminum*), which had a very low silver content of some 6%. Lower denominations were available in billon and copper.[39] As the Byzantine populace was largely unfamiliar with silver, it is likely that part of the market organization required was the provision of standard weights and measures as well as facilities for money changing. There would have been general problems in agreeing on both fair prices and appropriate rates of exchange.

Food was often difficult to obtain in the quantities required by Crusaders in the western parts of the Byzantine Empire, zones often disputed with Hungarians and Serbs, as well as in the Empire's Anatolian territories, which had been devastated by Turkish invasions. Food supplies seem to have been abundant only in the central parts of the Empire. The further the armies advanced from Constantinople and Edirne and their hinterlands, the more difficult it was to obtain supplies and consequently the higher the prices demanded. Albert of Aachen relates that after crossing over to Asia Minor, Godfrey of Bouillon's followers were obliged to pay higher prices.[40] Of course, Crusaders did not always feel obliged to pay for all the supplies they acquired. Godfrey promised the emperor that his troops would not plunder but reserved their right to take fodder for their horses where it could be

---

37 Albert of Aachen, *Historia Hierosolymitana*, II.x (p. 304).

38 Albert of Aachen, *Historia Hierosolymitana*, II.iii (p. 300): "in pondere et mensura aequitatis"; II.vi (p. 303): "in mensura aequa et justa"; II.xvii (p. 311): "aequo pondere et mensura".

39 This is a simplification of the rather complex monetary situation in the Byzantine empire, for which see M. F. Hendy, *Coinage and money in the Byzantine Empire, 1081-1261* (Washington, 1969), pp. 14-22.

40 Albert of Aachen, *Historia Hierosolymitana*, II.xvii (p. 312). Similarly, Odo of Deuil stated that after Louis VII's army crossed into Byzantine territory from Hungary, the exchange rate it received was five pence to one *stamenon*. At Constantinople, the rate had improved to two pence per *stamenon*. Yet only three days beyond the capital, the exchange rate shot up again to five to six pence. See Odo of Deuil, *De profectione Ludovici VII*, bk 3 (p. 40), bk 4 (p. 66).

found.[41] Crusaders from all contingents indulged in plundering and theft from markets in Byzantine territory. The leaders of the armies also received considerable sums of money from the imperial treasury. Between Christmas 1096 and Pentecost (24 May) 1097, Alexios Komnēnos regularly sent a financial subsidy of gifts and measures of gold bezants (*hyperpyra*) and of copper coinage to the Crusade leaders. Albert of Aachen remarked of this that:

> It was a remarkable thing, though, everything which the Duke distributed to the soldiers out of the Emperor's gift went back straight away to the royal treasury in buying food, and not only this, but even the money which the army had collected there from the whole world. No wonder, for as with wine and oil, so with corn and barley and all the food in the whole kingdom, it was sold at no one's price except the emperor's, and that is why the royal treasury was perpetually overflowing with money and could not be emptied by the presentation of gifts.[42]

Clearly, monies obtained by Crusaders were paid out by them to purchase food and this passage also provides evidence that the Emperor was able to impose price controls on the markets. The essential point, though, is that many Crusaders were able to use gold, copper, and billon coins to buy supplies, and thus could conserve their own stocks of silver for a period of up to five months.

Tendencies evident with regard to supply and expenditure on the march through Byzantine territory became even more pronounced during that through Anatolia to Cilicia and northern Syria. The Crusaders had new opportunities to secure income. After the capture of Nicaea, Alexios made gifts of gold and silver to the princes and distributed copper coins to the rank and file.[43] Considerable booty was taken at the battle of Dorylaion and again after the siege of Antioch.[44] Stephen of Blois, who in a previous letter had described gifts from Alexios and the booty gained at Nicaea, wrote to his wife Adela from Antioch that he now had accumulated twice as much gold, silver, and other riches as she had given him

---

41 Albert of Aachen, *Historia Hierosolymitana*, II.vii (p. 304).

42 Albert of Aachen, *Historia Hierosolymitana*, II.xvi (p. 311): "Mirabile dictu, universa quae ex dono Imperatoris Dux militibus distribuebat, in mutuatione alimentorum ad aerarium Regis protinus redibant, et non solum haec, sed etiam ea quae ab universo orbe illuc congessit exercitus. Nec mirum; nam nullius praeter Imperatoris merces tam in vino et oleo quam in frumento et ordeo omnique esca in toto vendebatur regno. Et ideo Regis aerarium, assidua pecunia habundans, nulla datione vacuari potest."

43 Fulcher of Chartres, *Historia Hierosolymitana*, I.x.10 (pp. 188-9).

44 Albert of Aachen, *Historia Hierosolymitana*, II.xxxxiii (pp. 332-3), IV.lvi (pp. 428-9).

when he had left home.[45] The leaders and knights with horses had the best opportunities to benefit from booty but even ordinary foot-soldiers and non-combatants had the chance to gain money and valuables by robbing enemy prisoners of war and even corpses. Turkish and Egyptian soldiers would swallow coins for security and Crusaders would stop at nothing to recover these from both captives and corpses.[46]

The march from the Byzantine frontier to northern Syria saw widespread privation as a result of lack of food and water and this led to heavy casualties among the Crusaders and their mounts and pack animals, large numbers of which died of thirst and/or were slaughtered for food. Privation grew more acute during the course of the long siege of Antioch as the food supplies of the surrounding areas were gradually exhausted. This situation began to change only when the Armenian rising around Edessa provided new sources of food and money and ships from the West were able to bring in supplies through the port of St Symeon.[47] When the armies resumed their march south towards Jerusalem they moved into relatively fertile areas. In contrast to the previous stage, both food and income were available. Muslim rulers such as the *amīrs* of Jabala and Tripoli were keen to pay tribute to the Crusaders to leave their lands in peace, and were also ready to grant markets and even to supply horses for sale. It was only really after the march through the Judaean highlands towards Jerusalem that food and water again became in short supply.

The siege of Antioch provides numerous examples of the monetary dimensions of the worsening supply situation. Raymond of Aguilers related that bread for one day cost two shillings, the head of an animal cost two to three shillings, and a hen cost eight to nine shillings. The *Gesta Francorum* recorded that at the time of the investment of Antioch an ass's load, which is testimony to supplies having been brought from some distance, cost 120 shillings or eight *hyperpyroi*.[48] Later, when the Crusaders were themselves besieged in the city, those who could afford it paid a penny for a walnut, two shillings for an egg, and

---

45 H. Hagenmeyer, ed., *Epistulae et chartae ad historiam primi belli sacri spectantes quae supersunt aevo aequales ac genuinae: Die Kreuzzugsbriefe aus den Jahren 1088-1100* (1901; rpt, Hildesheim, 1973), nos IV (pp. 138-40), X (pp. 149-52).

46 Anonymous, *Gesta Francorum et aliorum Hierosolimitanorum*, ed. and trans. R. Hill, *The deeds of the Franks and the other pilgrims to Jerusalem* (Oxford, 1962), X.xxxii (p. 80); Fulcher of Chartres, *Historia Hierosolymitana*, I.xxvii.1 (pp. 301-2).

47 For discussions of supply by sea see B. S. Bachrach, "The siege of Antioch: a study in military demography", *War in History*, 6 (1999), 127-46; J. France, "The First Crusade as a naval enterprise", *MM*, 83 (1997), 389-97; J. H. Pryor, " 'Water, water everywhere, Nor any drop to drink.': Water supplies for the fleets of the First Crusade", in M. Balard, B. Z. Kedar and J. Riley-Smith, eds, *Dei gesta per Francos: études sur les croisades dédiées à Jean Richard / Crusade studies in honour of Jean Richard* (Aldershot, 2001), 21-8.

48 *Gesta Francorum*, VI.xiiii (p. 33).

15 shillings for a hen, while a small loaf cost a bezant, by which could be meant either a *hyperpyron* (15 shillings) or a "Saracen bezant" or *dīnār* (eight to nine shillings).[49] The fairly precise details indicate that such items were being bought and sold, even if the prices demanded were greatly inflated. The main problem in evaluating such evidence is the fact that it is not always clear at what time any particular price applied. One of the most useful accounts in this respect is that of Albert of Aachen. Not only did he give some comparative figures for April and May 1098 as compared to the beginning of the investment of the city, but he also specified the currency in which prices were calculated. A small loaf which had cost a penny in the currency of Lucca at the beginning of the siege now cost two shillings, while an ox which had previously cost ten shillings now cost two marks.[50] After the capture of the city by the Crusaders, those who could afford it were paying sixpence for an egg, a bezant for the head of an ass, ox, horse, or camel, and ten shillings for the innards of any of these animals.[51] Several points can be extrapolated from such information. Firstly, all of the sources indicate that any single item such as a loaf or an egg in early 1097 cost considerably more than what was regarded as minimum expenses for an entire day a hundred years later during the Third Crusade. Therefore, it must be concluded, prices were greatly inflated during the siege of Antioch compared to the course of the march up to that point. Secondly, in the period between the commencement of the siege and the time of the greatest privation, there was an even greater spiralling of inflation, fuelled not only by the increasing scarcity of food and the cutting off of the hinterland by the Turks, but also by the greater availability of funds, notably booty gained as a result of the capture of the city. Thirdly, even though food may have been relatively scarce, it was still available and there were some individuals who were prepared to pay relatively high sums to secure it. As Raymond of Aguilers reported, "... to those rich in gold, silver, and clothes it was neither unusual, nor burdensome to pay exorbitant costs."[52] The medieval values of loyalty, honour, and representation meant that lords with households may have felt obliged to pay

---

49 *Gesta Francorum*, IX.xxvi (p. 62).

50 Albert of Aachen, *Historia Hierosolymitana*, III.lii (p. 375): "Nec mirum; nam solus paniculus, qui antea denario Luculensis monetae poterat mutuari, nunc duobus solidis vendebatur indigentibus; bos duabus marcis vendebatur, qui paulo ante decem solidis poterat comparari; agniculus quinque solidis appreciabatur."

51 Albert of Aachen, *Historia Hierosolymitana*, IV.xxxiv (p. 412): "Pro uno namque ovo gallinae, si invenire poterat, sex denarii Luculensis monetae numerabantur; pro decem fabis, denarius; pro capite unius asini, equi, bovis, vel cameli, bisantius unus dabatur; in pede vel aure, sex denarii; in visceribus cujuslibet horum animalium, decem solidi mutuo accipiebantur."

52 Raymond of Aguilers, *Historia* (RHC *HOcc*), c. XI (p. 258): "Nec erat mirum, nec grave esse poterat his qui tam care mercabantur, quum auro et argento et palliis abundarent." Cf. Raymond of Aguilers, *Liber* (Hill), §10 (p. 76).

extortionate rates out of a sense of obligation to provide for their dependents, as for example, when Godfrey of Bouillon paid three marks for a goat and 15 for "a miserable camel" to provide meals for his followers.[53] Yet, such exceptionally high prices were probably the result of the very large amounts of money in the army just as much as of the actual scarcity of food.

In the case of the siege of Antioch, we cannot be certain that the prices in the examples given above actually represented moneys paid by the Crusaders to the local population. It is equally as likely that the foodstuffs traded were obtained by Crusaders through forage or plunder and were then sold on to other Crusaders. If they were indeed purchased, one should then probably assume a profit margin on resale. Some supplies reached the army by ship through Port St Symeon and from there up river or by land, although to what quantities these amounted is a matter for debate. The transactions at Antioch indicate that in terms of logistics and political dynamics the exchange of money within Crusade armies was just as important as its inflow or outflow. From a fairly early stage in the Crusade we can find evidence of considerable amounts of money changing hands. The most obvious example, as we have seen, was for food. However, later, at the siege of Jerusalem, there was "a good supply of eggs and an abundance of water for those who could afford them", but there was a severe lack of water for the poor, so much so that a penny would not buy sufficient water for a man for a day.[54] The essential point to be considered is that whether because of opportunity, equipment, general fitness, or funds, some Crusaders were able to acquire supplies and, in some cases, to sell them on.

The major means of money transfer within Crusade armies, however, was for service. At the siege of Nicaea a siege tower of oak was constructed at their own expense by two lords in Godfrey's contingent: Hartmann of Dillingen and Henry of Esch.[55] Since supplies of oak were available within a day's journey of Nicaea, the expense must have been incurred mainly for labour. It might well be asked why Crusaders were paid for undertaking work which was obviously in their own, and indeed everyone's, interest, but this was the way that things functioned. At the siege of Jerusalem Raymond of St Gilles wished to bring one of his siege towers into position against the city walls. Since it was necessary to fill in the ditch in front of it in order to provide a level surface, Raymond offered a penny each to any pilgrim who brought three stones to fill in this ditch, a payment which seems both very high and also unnecessary. Given the situation they were in, can one imagine Crusaders before the walls of Jerusalem refusing to carry out this kind of labour unless they were paid? Now, there may have been an element of charity in this sort

---

[53] Albert of Aachen, *Historia Hierosolymitana*, IV.xxxiv (p. 412).

[54] Albert of Aachen, *Historia Hierosolymitana*, VI.vii (p. 470); *Gesta Francorum*, X.xxxviii (p. 90).

[55] Albert of Aachen, *Historia Hierosolymitana*, II.xxx (p. 322).

of payment and it would seem to have been in line with inflation since a penny was insufficient for a day's supply of water. The essential point, however, was that this payment would have created or contributed to a relationship of dependence between one of the major leaders and large numbers of ordinary Crusaders. During the First Crusade, service was regularly secured and maintained by means of financial payments. The rates of casualties and desertion between Nicaea and Jerusalem meant that there were ever-increasing numbers who lost or abandoned their lords and who sought out those who could provide protection, patronage, and above all, food. However, during the later stages of the Crusade the workings of the internal market threw such inequalities into even greater relief. Leaders such as Godfrey of Bouillon and Raymond of St Gilles had not only brought considerable funds with them from the West but now had opportunities to acquire huge sums from tribute or subsidy, such as the 20,000 *dīnārs* paid by the rulers of Jabala and Tripoli. When Godfrey's brother Baldwin established control over the lands around Edessa, he not only gained access to their very considerable revenues but he was also able to seize considerable treasuries which had been accumulated by the Armenian princes of the region. He was easily able to subsidize Godfrey, assigning to him the revenues of Turbessel, which were estimated at 50,000 *dīnārs*.[56] However, it is Raymond of St Gilles who provides perhaps the best example of the use of money to purchase services during the Crusade. He is generally regarded by historians as having been one of the wealthiest Crusaders at the time of departure, yet Raymond of Aguilers accused him of being stingy in distributing the huge tribute which he acquired from Tripoli.[57] His funds were so plentiful that even before acquiring this income he offered to pay Godfrey, Tancred, Robert of Flanders, and Robert of Normandy if they would resume the journey south under his leadership, the total sum offered being 31,000 shillings, while unspecified sums were offered to other leaders. Even though Tancred alone took up the offer, the incident suggests that Raymond had funds available to buy service that were equivalent to at least 400,000 pence, and this is unlikely to have represented all of his funds.[58]

The internal market could not cope with all the needs that arose in the course of the expedition. There was clearly a requirement for financial provision in order to pay for services which were meant to benefit the Crusade as a whole, rather than individuals or contingents. Thus, at the siege of Nicaea a Lombard engineer was

---

56 Albert of Aachen, *Historia Hierosolymitana*, IV.ix (pp. 395-6). See G. Dédéyan, "L'argent et le pouvoir chez les chefs arméniens de l'Euphratèse à l'époque de la Première Croisade", in *ΕΥΨΥΧΙΑ: mélanges offerts à Hélène Ahrweiler*, 2 vols (Paris, 1998), 195-208.

57 Raymond of Aguilers, *Historia* (RHC *HOcc*), c. XVIII (p. 289). Cf. Raymond of Aguilers, *Liber* (Hill), §18 (p. 131).

58 Raymond of Aguilers, *Historia* (RHC *HOcc*), c. XIV (p. 271). Cf. Raymond of Aguilers, *Liber* (Hill), §11 (p. 100).

offered 15 pounds in money of Chartres from common funds (*ex communi sumptu*) to construct a siege machine to mine under the walls of Nicaea.[59] From this point we can discern the existence of a common fund for such general expenditure, which was used at Antioch to pay Tancred for garrisoning with his followers a monastery outside the walls of the city. This fund seems to have been maintained by the princes who paid into it as they could afford, although Raymond of St Gilles was evidently the greatest contributor.[60] Yet the predicament of the poor and the clergy, two groups who were presumably less able to benefit from booty and plunder, was severe and a more institutional attempt to provide for them came into being at 'Arqā in 1098 with the establishment of a separate charitable fund for the benefit of the clergy and the poor, financed by voluntary tithes on the spoils of war.[61] Luckily, this was a time when considerable income was coming in to the army and markets were available to spend it on. The financial dynamics of the First Crusade were essentially those of a free market which operated both between and within individual contingents and which was tempered only by a certain amount of charity.

While the First Crusade repeatedly encountered problems in securing supplies of food and water, there never seems to have been a shortage of money. As a result of plunder and tribute, by the time they left Antioch the forces, or at least many individual Crusaders, probably had sums of coinage and bullion available which were far in excess of what they had when they had arrived at Constantinople. A further accumulation of money by individuals within the armies must also have occurred constantly as a result of deaths from battle, disease, or starvation. We must assume that funds held by deceased Crusaders at the time of death would have been either willed to companions or dependents or simply taken by them. However, it was precisely around this time that transport became increasingly problematic. There is little evidence for the use of carts by the Crusaders beyond Nicaea and it is unlikely that carts were employed east of the Anatolian plateau since many of the roads there had sections with steps rather than gradients, and consequently were impassable to wheeled vehicles.[62] From the point that carts

---

59 Albert of Aachen, *Historia Hierosolymitana*, II.xxxv (pp. 325-6).

60 Raymond of Aguilers, *Historia* (RHC *HOcc*), c. VIII (p. 250). Cf. Raymond of Aguilers, *Liber* (Hill), §8 (p. 63).

61 Raymond of Aguilers, *Historia* (RHC *HOcc*), c. XVI (p. 278). Cf. Raymond of Aguilers, *Liber* (Hill), §17 (p. 111).

62 The use of carts by Crusaders deserves a separate logistical study. They were certainly used by more than one contingent as far as Thrace at least. See Albert of Aachen, *Historia Hierosolymitana*, I.viii (pp. 277-8), I.x-xii (pp. 279-82). However, it is doubtful whether they were transported beyond the Bosporos. Many of the roads of eastern Anatolia were stepped, and thus impassable to carts and so the carts mentioned at Nicaea and Antioch by Albert of Aachen were probably obtained locally and not used over long distances. Albert of Aachen, *Historia Hierosolymitana*, II.xxviii (pp. 320-21), III.xxxvi (pp. 364-5).

were abandoned, the transport of money, whether in the form of coin or bullion, would have created ever greater logistical problems. Money previously carried in barrels or chests would have had to be divided into smaller packages and redistributed to be carried on packhorses, mules, camels, or even by Crusaders on foot. The fundamental problem was that there had been great losses among both animals and men as a result of lack of water and food and of exhaustion by the time that the forces reached Northern Syria but, at the same time, large sums of money were flowing into the armies. Many Crusaders must have entered the final stages of their journey with more funds than they had when they left their homes, at a time when there was little to spend their money on and transporting it was probably more difficult than ever before. A much reduced number of Crusaders with a large amount of money, but with greatly reduced numbers of horses, mules, and camels, all of them weakened by exertion and lack of food, must have been increasingly confronted with invidious choices. Could one continue to carry food, weapons, armour, and money, and if not, which should be abandoned? This was one of the logistical paradoxes of the First Crusade.

Chapter 13

The Northern Crusaders: the logistics of English and other Northern Crusader fleets[1]

Richard W. Unger

So great a fleet
together came
for prudent prince,
picked and faithful,
that sixty ships
sailed, gaily planked,
hence by holy
heaven's decree.[2]

That was the way the skaldic poet Thórarin Stuttfeld described the departure in 1107 of King Sigurd of Norway. Sigurd shared the throne with two brothers and four years after his accession at the age of 17 he felt secure enough to sail with his people from Norway for the Holy Land. The great Icelandic historian Snorri Sturluson thought the travels of Sigurd the *Jórsalafari*, as he came to be known, worthy of detailed retelling in his *Heimskringla*, the history of Norway up to 1177. Snorri wrote down the story not long before his death in 1241, at which time the tale was more than a century old. He encapsulated the events as laid out by the earlier Norwegian historian Theodoricus monachus and some skalds.[3]

---

1 Consult Maps 2, 4, 5, 6, 7, 9, 14.

2 Snorri Sturluson, *Heimskringla: History of the Kings of Norway*, trans. L. M. Hollander (1964; rpt, Austin, 1999), p. 689. The description of Sigurd's voyage to *Outremer* and his return is on pp. 688-99. Cf. Snorri Sturluson, *Heimskringla*, ed. B. S. Kristjánsdóttir et al., 3 vols (Reykjavík, 1991), vol. 2, p. 720: "Svo kom fylkis / framt lið saman / margspaks mikið, / mildingi vilt, / að skip við skip / skarfögr við lög / hreins grams héðan / hnigu, sex tigir.". A more colourful and much looser translation is Snorri Sturluson, *Heimskingla. Part two: Sagas of the Norse kings*, trans. S. Laing, rev. P. Foote (1961; rpt, London and New York, 1968), XIII: The sons of Magnus (p. 277). I am indebted to Margaret Clunies Ross for pointing to the similarities in the description of the exploits of Sigurd and of his predecessor, Harald Hardradi, suggesting some structural or literary influence of one recounting on the other.

3 Theodoricus monachus, *Historia de antiquitate regum Norwagiensium / An account of the ancient history of the Norwegian kings*, trans. D. and I. McDougall (London, 1998), c.

Sigurd sailed for England in the autumn and stayed the winter with Henry I. He departed in the spring of 1108 but did not reach Galicia, where he wintered, until the autumn. Because the local ruler failed to make available adequate food supplies for Sigurd and his men to buy, Sigurd attacked him in the best traditions of his Viking ancestors, seizing his castle, the food, and the treasure in it. The king then went on west meeting and defeating a squadron of "heathen" ships and taking eight galleys from them. The next stop was Sintra, where he fought another battle, took the castle and killed every man in it because they would not convert to Christianity. At Lisbon he had his third battle with the "heathen" and took the town and great booty. Alcácer do Sal was the site of the fourth battle and victory for the Norwegians. The pillage proved immense this time and they killed so many Muslims that they left the town empty. Another victory at sea and the defeat of a den of Muslim pirates on an island brought Sigurd and his men more booty. Battles on Iviza and Minorca followed and in these as in the previous six Sigurd proved victorious.

The battles all having taken place in a short period during what must have been winter, Sigurd arrived in Sicily in spring 1109, where Count Roger treated him royally. At the end of a seven-day feast Sigurd responded by recognizing Roger as his equal, saying that he should be king of Sicily and acknowledged as such.[4] The Norwegians finally arrived in the Holy Land in the summer of 1110. There Sigurd met King Baldwin of Jerusalem, who also treated him royally. The two besieged Sidon together and when the town fell, great treasure was taken and Sigurd gifted his share to Baldwin. After the victory he sailed to Cyprus. Snorri gave no reason for Sigurd's short stay in *Outremer* or for what seems to have been a sudden departure. According to the saga, he then sailed on to *Engilsnes*, "Angel's Cape", where he waited for two weeks for the wind to change. The author claimed that he could have easily sailed north with a following wind but that because both sides of the sails were covered with costly stuffs, neither those crew in the bows nor those in the sterns wanted to see the less attractive side of the sails and he therefore waited for a wind from abeam so that the sails could be set fore and aft, and then steered a course for Constantinople. Emperor Alexios Komnēnos opened the Golden Gate, Χρυσεία Πύλη, at the south end of the land walls of the city to Sigurd and his men, who rode to the emperor's magnificent hall in great splendour.[5] Alexios held games in his honour and Sigurd hosted a great banquet in

33 (pp. 52-3).

4 Of course, Count Roger was later crowned King of Sicily in 1130 under the auspices of Antipope Anacletus II and this would have been well known to Snorri Sturluson.

5 Snorri Sturluson, *Heimskringla* (Kristjánsdóttir), Magnússona Saga, §§11-12 (vol. 2, p. 728): "Þeir sigldu norður til eyjar þeirrar er Ky pur heitir og dvaldist Sigurður konungur þar nokkura hríð, fór siðan til Grikklands og lagði öllu liðinu út við Engilsnes og lá þar hálfan mánuð og var hvern dag hraðbyri norður eftir hafinu. Enn hann vildi bíða þess byrjar er þverskytningur væri og seglum mætti aka að endilöngu skipi því að öll segl hans voru sett

his. After a lengthy stay, he gave all his ships to the emperor, who had the figureheads set up in a church.[6] In return, Alexios gave Sigurd horses and guides to make his way north. However, many of his men remained in Byzantine service as mercenaries. That was an old practice but since the ninth century Scandinavians had usually come not by sea but overland down the rivers of Russia and the Ukraine. Just as on his way to the East, so too on his way home, Sigurd did not use the traditional route. He travelled overland through Bulgaria, Hungary, and Bavaria and on his way met Lothar of Supplinburg, later to become Roman Emperor. He was met at Hedeby by the Danish king, who held a feast in his honour, accompanied him through Jutland, and then gave him a fully supplied ship for the final voyage home. He was welcomed joyfully on his return, in fact in 1111 although Snorri said that he returned three years after he had departed, in 1110.[7]

The story of Sigurd illustrates graphically the problems of understanding the voyages of Northern Crusaders, their contributions to the Crusading effort, and logistical problems which they faced. For example, Sigurd was diverted into fighting in Iberia. Crusaders sailing south would be asked again and again to aid their co-religionists in taking towns along the Iberian coast. Often they accepted, delaying their voyages but gaining glory, and more immediately plunder. However,

---

pellum, bæði það er fram vissi og aftur, fyrir því að hvorirtveggju, frambyggjar eða þeir er aftar voru, vildu eigi sjá hið ófegra seglanna. Þá er Sigurður konungur sigldi inn til Miklagarðs sigldi hann nær landi. Þar eru allt á land upp borgir og kastalar og þorp svo að hvergi slítur. Þá sá af landi í bug allra seglanna og bar hvergi í milli svo sem einn garður væri. Allt fólk stóð úti, það er sjá mátti sigling Sigurður konungs."

The whole story reads like a literary conceit and the text appears to be corrupt. The only promontory in all of the Aegean from Cyprus to Constantinople that was ever known as something like "Angel's Cape" was Cape Malea, sometimes known as Cape Saint Angelo. See K. Kretschmer, *Die italienischen portolane des Mittelalters: ein Beitrag zur Geschichte der Kartographie und Nautik* (1909; rpt, Hildesheim, 1962), pp. 633-70. However, it is absurd to imagine Sigurd setting his sails fore-and-aft at Cape Malea and then progressing to Constantinople without changing their setting and, in any case, Cape Malea was far off the route from Cyprus to Constantinople. Surely, the author merely wanted to say that Sigurd set his sails fore-and-aft for onlookers ashore while sailing up to Constantinople and the only place known to him between Cyprus and Constantinople was Cape Saint Angelo/Malea, which was known far and wide for the dangers of rounding it. The story of the costly stuffs on both sides of the sails and of the crew in the bows and sterns makes no sense.

6 The saga said that the figureheads were set up in St Peter's Church but there does not appear to have been a church by that name in Constantinople, although there was a church of Sts Peter and Paul and also some chapels of St Peter. Together with others, the error suggests that the author was not familiar with the eastern Mediterranean.

7 Snorri called Lothar the Emperor of Rome but Henry V was emperor at the time. Presumably the reference was to the man who would succeed him in 1125. Snorri also said that Sigurd had spent three years on his journey but that would have put him back in Norway in 1110, a year earlier than when he must have returned.

the attacks on Muslims were not the only reason for Sigurd's slow progress on his outward voyage. Entertainment offered him by great men along the way may be one explanation for the three years taken to reach *Outremer* but logistical and navigational problems may also help to explain the leisurely pace of the trip.

Sigurd left his ships as gifts for Alexios Komnēnos in Constantinople. They had most probably deteriorated so much that they had become unseaworthy, being unsuited to the warmer waters of the Mediterranean and susceptible to attack by shipworm (*teredo navalis*). It also appears that they were not able to overcome the currents and prevailing westerly winds which made travelling westward through the Straits of Gibraltar all but impossible in the eleventh-twelfth centuries. Sigurd was able to return to Norway overland in a very few months, with the generous hospitality of kings and emperors along the way expediting a swift trip, and his Crusade demonstrates that going by sea was slower than going overland.

In the case of northern Galicia, Snorri Sturluson did write that one reason for difficulties with the local Christian lord was that the land there was barren and unproductive,[8] the one instance where he suggested logistical problems. There may well have been other such problems with decisions not to travel by sea in winter being a product of fear both of rough and unpredictable seas and of shortages of provisions for crews and pilgrims.

If the greatest seafarers of northern Europe in the tenth and eleventh centuries found difficulties in travelling around western Europe and through the Straits of Gibraltar to *Outremer*, then there seems to have been little reason for others to use the sea route; however, through to and beyond the fall of Acre in 1291 some did. That being said, the number of known cases is limited, the number of well-documented cases even fewer. Voyages from England and the Low Countries which were part of major Crusading efforts received more notice than the travels of other pilgrims who apparently went almost continuously to *Outremer* by sea from the eleventh century on. However, a change in the character of the voyages made from northwestern Europe to *Outremer* apparently began in the early thirteenth century, a change generated by technological developments in ship design and increasing knowledge of navigational conditions and political circumstances in Europe and in the East. Voyages up to and through the Fourth Crusade showed common characteristics not fully shared by those after about 1210.

Early Crusade voyages continued previous Norse voyages to Iberia and into the Mediterranean. As early as 843 Vikings had sacked Muslim Lisbon and gone on through the Straits of Gibraltar. They reached Greece in 858. There were further Viking raids in Iberia in 844, 859, 961, 966, 971, 1014 and 1016 and a Viking leader called Ulf the Galician obtained his sobriquet for raids in northwestern

---

[8] Snorri Sturluson, *Heimskringla* (Kristjánsdóttir), Magnússona Saga, §4 (vol. 2, p. 721): "En það entist eigi lengur en til jóla og gerðist þá illt til matar því að landið er skarpt og illt matland."

Spain between 1048 and 1066.[9] For the First Crusade Northerners organized quickly if somewhat informally. In the Viking tradition, a professional pirate named Guynemer of Boulogne had been active in the Mediterranean with a flotilla of vessels from Flanders, Frisia, and Denmark for eight years by the spring of 1097. He put in to the mouth of the Cydnus river downstream from Tarsos soon after Baldwin of Boulogne had taken the town. The Northerners supplied relief and reinforcement at what was a difficult moment for Baldwin's forces and also assisted in the transportation of men and equipment to Antioch.[10] In March of the following year a fleet manned by Englishmen carrying a number of Italian pilgrims put in at the port of Antioch. They brought with them invaluable siege equipment picked up at Constantinople. The flotilla, like that of Guynemer, presumably continued to operate in Levantine waters and some of its ships may have been among the Christian vessels that seized Jaffa in June 1098 after the Muslims had deserted the port.[11] In the last days of May 1102 another fleet said to have numbered some 200 ships arrived at Jaffa with pilgrims and soldiers from England, France, and Germany. Whether they came directly from the North or whether the passengers had gone overland to the shores of the Mediterranean is not clear. In 1107 another large group of pilgrims in many ships arrived in *Outremer* from Flanders and Brabant, the numbers undoubtedly inflated by chroniclers. Obviously in the years around 1100 going by sea from northwestern Europe was feasible and even in some cases practical.[12]

---

9 B. W. Diffie, *Prelude to empire: Portugal overseas before Henry the Navigator* (1960; rpt, Lincoln Nebraska, 1967), pp. 7-8; H. Kennedy, *Muslim Spain and Portugal: a political history of al-Andalus* (London, 1996), p. 47.

10 The expedition of Guynemer, like that of a number of others in the First Crusade, had a Scandinavian flavour, in this case it being the possible presence of Danish ships. The only source for Guynemer is Albert of Aachen, *Historia Hierosolymitana*, in RHC *HOcc*, vol. 4, 265-713, III.xiv (pp. 348-9) and see also III.lix (p. 380), VI.lv (pp. 500-501). See also J. France, *Victory in the East: a military history of the First Crusade* (Cambridge, 1994), p. 217; A. R. Lewis, "Northern European sea power and the Straits of Gibraltar, 1031-1350 A.D.", in W. C. Jordan et al., eds, *Order and innovation in the Middle Ages: essays in honor of Joseph R. Strayer* (Princeton, 1976), 139-64, p. 144.

11 It is said that the fleet which stopped at Constantinople found there the exiled English prince, Edgar Atheling, who then took command of the fleet but the story is spurious since Edgar did not arrive in the East until 1102. France, *Victory in the East*, pp. 215-16; N. Hooper, "Edgar the Ætheling: Anglo-Saxon prince, rebel and Crusader", *Anglo-Saxon England*, 14 (1985), 197-214, pp. 208-210; C. Tyerman, *England and the Crusades 1095-1588* (Chicago: 1988), pp. 19-20.

12 An English pirate, named as Gudericus by Albert of Aachen, was said to have brought king Baldwin I through an Egyptian blockade to Jaffa in early 1102. If the story is true, he may have been among one of the other groups from the North mentioned by chroniclers. Albert of Aachen, *Historia Hierosolymitana*, IX.ix (p. 595). See also H. Hardenberg, *De Nederlanden en de Kruistochten* (Amsterdam, 1941), p. 114; G. Asaert, "Scheepvaart in het

Small contingents continued to arrive but the next well-documented wave of northern European ships to reach the Mediterranean came during the Second Crusade.[13] Possibly the first to make the trip was an earl Rögnvald of the Orkneys who took a small flotilla of 15 ships around Iberia and on to Crete and *Outremer*. After visiting the holy sites, he and his followers also went on to Constantinople.[14] The crowned heads of Latin Europe went by land but many nobles from Flanders, northern Germany, and England chose to travel by sea. Perhaps inspired by the naval expedition of 1097, on 23 May 1147 a mixed force of English, Normans, Flemings, and Germans sailed in 164 ships from Dartmouth on the southwest English coast. However, just like contingents bound for *Outremer* in 1112 and 1140, the fleet was turned aside to help the King of Portugal.[15] The traditional story is that on arrival in Oporto the Crusaders were exhorted by its bishop to join in the siege of Lisbon. The Northerners may have brought siege engines and they certainly brought numbers. If the fleet was not the 164 ships reported by Raol, the priest who is the most likely author of the *De expugnatione Lyxbonensi*, but rather the approximate 200 reported by the priest Winand, author of the so-called Lisbon Letter, it would mean that there was an average of 50 men per ship for a total force of around 10,000, which is a reasonable figure.[16] The ships played an active role in

---

zuiden tot 1400", in *Maritieme Geschiedenis der Nederlanden. Vol. 1: de Boer Maritiem*, eds G. Asaert, J. van Beylen and H. P. H. Jansen (Bussum, 1976), 37-74, p. 44.

13 See S. M. Foster, *Some aspects of maritime activity and the use of sea power in relation to the Crusading states 1096-1169* (D.Phil. thesis, Oxford, 1978).

14 *Orkneyinga Saga*, ed. F. Guðmundsson, in *Íslenzk fornrit*, 35 vols (Reykjavík, 1968-82), vol. 34, 1-300, §§86-8 (pp. 208-34). The same story was repeated by Snorri Sturluson in *Heimskringla* (Kristjánsdóttir), Saga Inga konungs og bræðra hans, §17 (vol. 2, p. 780). So many of Rögnvald's adventures related in the saga were so much like those of Sigurd that there must be doubt about their veracity.

15 Raol, *De expugnatione Lyxbonensi/The conquest of Lisbon*, ed. and trans. C. W. David, rev. J. Phillips (1936; rpt, New York, 2001), pp. 52-7. It took the Rhineland contingent 22 days to travel from Cologne to Dartmouth, an indication of the critical importance of prevailing winds. The Anglo-Norman ships seem to have carried hardly anyone of noble status and it may be that it was a chance for lesser men to avoid the consequences of having been on the wrong side in the recent civil war in England. See F. Barlow, *The feudal kingdom of England 1042-1216*, 4th ed. (London, 1988), p. 223; W. A. Engelbrecht, *Schets der Historische Betrekkingen Portugal-Nederland* (The Hague, 1940), p. 2; Hardenberg, *Nederlanden en de Kruistochten*, p. 136; Tyerman, *England and the Crusades*, pp. 32-4.

16 It is possible that a plan was already in place and that the fleet left early from the North so that its ships and men could join in the attack on Lisbon. See J. Phillips, *The Crusades, 1095-1197* (London, 2002), pp. 68-9. The leaders may have known about the plan but the decidedly pro-English Raol wrote that the Crusaders had to be convinced to join in the effort. The economic incentives offered by the bishop and King Afonso I Henriques appear to have been inadequate. On the number of ships and descriptions of the Lisbon episode compare S. B. Edgington, "The Lisbon letter of the Second Crusade', *HR*, 69 (1996), 328-39, pp. 336-9 [trans. idem, "Albert of Aachen, St Bernard and the Second Crusade", in J.

the siege, patrolling the river and even providing platforms for an attack. The town fell on 24 October after a siege just short of four months.[17] A number of Crusaders, including many Flemish, took advantage of offers of land and freedom from tolls and stayed on in Portugal. The rest sailed for *Outremer* in February 1148.

The fleet had left Dartmouth on 23 May and sighted the Cantabrian mountains on 28 May, despite being becalmed for two days, and reached Oporto on 16 June. Ten days later, after the exhortation from its bishop, it left and took two days to reach the mouth of the Tagus.[18] The speed with which the fleet covered the distance from the English south coast to Iberia was in sharp contrast to the slow progress from landfall at Gozón to reaching Oporto. The elapsed time reflects a storm which dispersed the fleet and the number of stops made along the coast.[19] The decision to hug the coast suggests something about the character of the ships involved as do descriptions of the best documented of all Northern fleets, that of the Third Crusade.

While the English royal house had done little to promote Crusading before 1189, on his accession Richard Cœur de Lion threw himself and the weight of the Angevin government into the support of an expedition. The king hired or purchased ships and had them filled with supplies, including 60,000 horseshoes and 14,000 cured pig carcasses among a broad range of other food and equipment.[20] Chroniclers said of the fleet raised by Richard that it was more to be noted for its strength than its numbers and it is difficult to generate a good estimate, but the total number of ships supplied by the English king, including those from his Continental possessions, was probably between 100 and 150. Of

---

Phillips and M. Hoch, eds, *The Second Crusade: scope and consequences* (Manchester, 2001), 54-70, pp. 61-7] and *De expugnatione Lyxbonensi*, esp. pp. 52-3.

17 Winand described the German contingent as "Lotharingians". They came from the Rhineland and Winand and Raol called them people of Cologne. According to Raol, the Crusaders looted Lisbon, although the English were much more restrained than the Flemings and Germans. A number of Englishmen received offices as rewards for their part in the siege including Gilbert of Hastings, who became the first bishop of Lisbon known since 688. See M. Bennett, "Military aspects of the conquest of Lisbon, 1147", in Phillips and Hoch, *Second Crusade*, 71-89, pp. 71-84; Phillips, *Crusades*, p. 69; Edgington, "Lisbon Letter"; *De expugnatione Lyxbonensi*, pp. 170-81.

Another fleet of Crusaders from the Low Countries commanded by Count Philip of Flanders, who in 1184 married the youngest daughter of King Afonso Enriques, was to stop in Lisbon in 1177, 30 years after its fall. See Engelbrecht, *Historische Betrekkingen Portugal-Nederland*, p. 3.

18 *De expugnatione Lyxbonensi*, pp. 58-69, 86-9; Hardenberg, *Nederlanden en de Kruistochten*, p. 135.

19 *De expugnatione Lyxbonensi*, pp. 60-67.

20 The efficiency of royal administration led to the survival of a number of records of what the king did to supply his force. See Phillips, *Crusades*, p. 141; Tyerman, *England and the Crusades*, pp. 82-3.

those, the Cinque Ports supplied 33 and the king paid two-thirds of the costs for those ships. There is no record of what happened to any of the ships after the Crusade.

When the fleet left England Richard apparently sent the heavier ships on ahead and let the smaller and probably oared vessels leave later. The difference in departure dates probably had more to do with the greater anticipated speed of ships with oars than with any difference in the time or difficulty of fitting out the two different types of ships. The king appointed commanders for the fleet and then travelled overland to Marseilles, arriving at the end of July 1190. He expected to find his fleet waiting for him but it had been delayed in Portugal. Sixty-three ships reached Lisbon and while waiting for another squadron of 30 ships the English sailors ran amok, raping and pillaging in the town. King Sancho I jailed a number of them and it took some time for the English commanders to sort out the problem. The reunited fleet finally left Lisbon on 26 July and reached Marseilles on 22 August, three weeks late for the rendezvous with Richard. Meanwhile, frustrated by the delay, he had put part of the force that had come overland with him on hired ships and sent them directly to Acre while he himself, in 15 galleys according to one report, made his way down the coast of Italy and reached Messina in time to meet the English fleet there on 22 September. The ships wintered in Sicily, undergoing repairs. There were more than 200 vessels in the fleet that left Sicily the following April, some 13 "dromons", 150 *busses*, and 53 galleys.[21] Richard

---

[21] The numbers reported varied. Richard of Devizes gave a grand total of 219 *vasa* or vessels, of which 156 were *naves* or "ships", 24 were *bucee* or *busses*, and 39 *galee* or galleys. There were, he wrote, 180 vessels exclusive of the king's ship and the galleys. Richard of the Holy Trinity, in his revision of the *Itinerarium peregrinorum*, gave a figure of 108 ships. The poet Ambroise wrote that 107 *nes* took to the sea, while Roger of Wendover wrote that the fleet had 13 *buccae*, 100 *naves onerariae*, and 50 *galeiae triremes*. Roger of Hoveden said that Richard left Messina with 50 *magnae naves* and 53 *galeiae*. There were also ships hired by various barons and a contingent of Hollanders and Flemings who sailed in April 1189 from Walcheren to England, where their numbers were strengthened by Frisians. See Ambroise, *L'estoire de la Guerre Sainte: histoire en vers de la Troisième Croisade (1190-1192)*, ed. G. Paris (Paris, 1897), ll. 311, 1179-96; *Itinerarium peregrinorum et gesta regis Ricardi*, in *Chronicles and memorials of the reign of Richard I. Vol. I*, ed. W. Stubbs (Rolls Series, tome 38, vol. 1), II.vii (pp. 147-8); Roger of Hoveden, *Chronica*, ed. W. Stubbs (Rolls series, tome 51), 4 vols (London, 1868-71), vol. 3, pp. 39-41, 45; Richard of Devizes, *Chronicon Richardi Divisensis de tempore regis Richardi Primi / The Chronicle of Richard of Devizes of the time of King Richard the First*, ed. and trans. J. T. Appleby (London, 1963), p. 28; Roger of Wendover, *Rogeri de Wendover liber qui dicitur Flores Historiarum / The Flowers of History by Roger de Wendover*, ed. H. G. Hewlett (Rolls series, tome 84), 3 vols (London, 1886-9), vol. 1, p. 192.

See also F. W. Brooks, *The English naval forces 1199-1272* (1932; rpt, London, 1962), pp. 133-6, 163-5; N. A. M. Rodger, *The safeguard of the sea: a naval history of Britain. Vol. I: 660-1649* (London, 1997), p. 46; J. Gillingham, *Richard I* (New Haven and London, 1999), pp. 129-30, 139, 143-4; S. Painter, "The Third Crusade: Richard the Lionhearted and

sailed to Crete, on to Rhodes, and then to Cyprus where he was reunited with his new queen who had been driven ashore there by contrary winds. He seized the island and then went on to the siege of Acre, his first act being to take a large Muslim sailing ship using his flotilla of galleys. He reached Acre on 8 June with his fleet and much equipment,[22] and on 12 July the city fell.

Richard was to stay in *Outremer* until October 1192 and during that time he used naval superiority gained in part thanks to the Northern ships to carry out an effective campaign against Ṣalāh al-Dīn. In doing so he laid the groundwork for another century of Crusader presence in *Outremer*, a presence which would have been impossible without Christian naval dominance. After concluding a truce he made his way by sea to Corfu and then to the Adriatic coast. Captured by his enemies in Vienna, he was held for ransom. He chose to return overland probably for political reasons but sailing directly back to the North, especially in the winter, apparently was still not an option at the end of the twelfth century.[23]

Richard's fleet was not the only Northern one to make its way to the East to the siege of Acre. From September 1189 onwards a steady stream of Crusaders arrived by ship at Acre and Frisian and Scandinavian ships joined the naval blockade of the city. Most of those had left the North in April and May. Frisians and Flemings appear to have had their progress delayed like their predecessors 40 years earlier by stopping in Portugal to help the king. One fleet of 37 ships that sailed from Dartmouth on 18 May with Crusaders from England, Denmark, and Flanders stopped in Portugal to help King Sancho to capture Silves on 6 September. Other Frisians preceded them, arriving on 1 September and yet another group along with Danes and men of Lübeck reached Acre on 22 September. A Dutch, Flemish, and Frisian fleet arrived even later. Some prominent Londoners left their home port in 1190 but when they reached Silves, local residents scuttled their ship to convince them to stay on to fight the Muslims. The English pilgrims

---

Philip Augustus", in Setton, *HC*, vol. 2, 45-85, pp. 55-6; A. R. Lewis and T. J. Runyan, *European naval and maritime history, 300-1500* (Bloomington, 1985), pp. 115-16.

22 Gillingham, *Richard I*, pp. 144, 157-9. For two descriptions of the taking of the three-masted Egyptian ship from Ibn S̲haddād's *Al-Nawādir al-sulṭāniyya* and from Richard of the Holy Trinity's revision of the *Itinerarium peregrinorum* see Ibn S̲haddād, Bahā' al-Dīn Abu 'l-Maḥāsin, *Al-Nawādir al-sulṭāniyya wa 'l-maḥāsin al-Yūsufiyya* [trans. C. W. Wilson and C. C. Conder, *Saladin; or what befell sultan Yusuf*, in PPTSL, vol. 13], ch. III (pp. 249-50) and *Itinerarium peregrinorum* (Stubbs), II.xlii (pp. 206-9). See also Lewis and Runyan, *European naval and maritime history*, pp. 53, 75-6.

23 H. Mayer *The Crusades*, trans. J. Gillingham, 2nd ed. (Oxford, 1988), pp. 145-9; Gillingham, *Richard I*, pp. 229-32.

reached an agreement for compensation after they promised to assist the Christians.[24]

As had become standard, during the Fourth Crusade a fleet set out from the Low Countries and arrived at Marseilles, presumably a logical landfall along the Christian coast of Europe and also a likely port to gather up any Crusaders who might have travelled overland. There it received instructions from the leaders of the Crusade to join the main force at Zara. However, the commanders decided not to join the principal group but to sail directly to *Outremer*.[25] Their voyage but even more so that of the English fleet fitted out by Richard Cœur de Lion "... indicate a thorough knowledge of the maritime geography of Europe at an early date, as well as testifying to the seaworthiness of the ships used."[26]

From the First to the Fourth Crusade, seaworthy Northern ships appear to have been of two types, with some variations within the two. After the Viking age differentiation in Northern ships increased between cargo vessels, which became more beamy and high sided, and warships which became more slender and low, designed for speed and to carry men and arms.[27] Excavations in Roskilde Fjord in the 1960s unearthed five vessels from about the time of the First Crusade which illustrate the kinds of ships available to Sigurd and other northern Crusaders.[28]

Long, low warships relied on a single square sail for propulsion but also on a number of oars which passed through holes in the gunwale which acted as rowlocks. Their length was some four or five times their beam and they were floored for their entire length. Crews were large. Cargo ships also had a single square sail, partial decks at the bow and stern and a length three and a half to four times their beam. Though the basic design and construction of the two types were the same, and though it was still possible to convert a cargo ship to a warship even at the time of the Third Crusade simply by cutting oarports in the gunwales, there were two different designs of ships used by Northern Crusaders.

Cargo ships in the Baltic and North Sea had various names, the differences possibly reflecting differences in design. They were often called simply *keels* in northwestern Europe, perhaps because until the twelfth century Celtic types did not yet have a true keel while Viking ships had them from the eighth century at least. The *knorr* was the vessel for trade in the North Atlantic to Iceland and Greenland.

---

24 Hardenberg, *Nederlanden en de Kruistochten*, pp. 152, 155, 164; Mayer, *The Crusades*, pp. 143-4; Painter, "The Third Crusade", p. 50; *Itinerarium peregrinorum* (Stubbs), I.xxvii (pp. 63-5). See also Tyerman, *England and the Crusades*, pp. 66, 73-4.

25 E. H. McNeal and R. L. Wolff, "The Fourth Crusade", in Setton, *HC*, vol. 2, 153-85, p. 175; Asaert, "Scheepvaart", p. 45.

26 G. Hutchinson, *Medieval ships and shipping* (London, 1994), p. 164.

27 A. E. Christensen, "Proto-Viking, Viking and Norse craft", in idem, ed., *The earliest ships: the evolution of boats into ships* (London, 1996), 72-88, p. 84.

28 O. Olsen and O. Crumlin-Pedersen, "The Skuldelev Ships (II)", *Acta Archaeologica*, 38 (1967), 73-174; *The Skuldelev Ships*, at http://www.abc.se/~m10354/uwa/skulde.htm.

The largest of the sailing ships from Roskilde Fjord may reflect that type. It was 16.3 metres long and 4.5 metres wide and could be handled by a crew of less than ten men.[29] There were places for a few oars at bow and stern but only for use in manœuvring in and out of harbour and in emergencies. By the twelfth century the term *byrthing* was used for cargo ships and in Latin sources the word became *bucius*. That may have been taken from Old French *bussa* or *buse* or *buce* which appeared in the twelfth century and may have been taken directly from a late Latin word.[30] Albert of Aachen wrote that the English pirate Godric commanded a *buza* off *Outremer* in 1102 and the English chronicler Roger of Hoveden wrote that in October 1189 "... *naves et buciae plusquam quingentæ, exceptis galeis et cursariis* ..." arrived off Acre.[31] He made a distinction between *naves* and *busses*, which possibly reflected a distinction between Mediterranean merchantmen with two lateen rigged masts and Northern cargo ships. Presumably the *busses* looked much like ships on contemporary seals of southern English towns, including the Cinque Ports.[32] Roger, however, also used the word *buss* for ships which were most probably of the Mediterranean type, such as the ten large *buciae* that Richard hired at Marseilles together with 20 well-armed galleys on his way to Messina or the *buscia magna* in which he sailed from *Outremer* to Corfu during his return to England.[33] The *busses* that Richard of Devizes wrote were waiting for Richard when he arrived at Marseilles were undoubtedly not from England but rather Mediterranean ships given a Northern name.[34]

---

29 Christensen, "Proto-Viking, Viking and Norse craft', pp. 85-6.

30 The Old French word *busse* was used for a cask so the original meaning may simply have been a vessel for carrying something. By the fourteenth century in England *buss* meant a vessel of burden. In the fifteenth century the word would come to be used for a Dutch fishing boat of very specific features. *Oxford English Dictionary*, 2nd ed. (Oxford, 1992), "Buss" (vol. 2, p. 698); J. De Vries, *Nederlands etymologisch Woordenboek* (Leiden, 1997), "buis 2" (p. 94); R. W. Unger, "The Netherlands herring fishery in the Late Middle Ages: the false legend of William Beukels of Biervliet", *Viator*, 9 (1978), 335-56, pp. 348-9, 353-4.

31 Albert of Aachen, *Historia Hierosolymitana*, IX.ix (p. 595); Roger of Hoveden, *Chronica*, vol. 3, p. 21.

32 H. Ewe, *Schiffe auf Siegeln* (Berlin, 1972). For example, nos 34 (Dover), 37 (Dunwich), 64 (Hastings), 155-6 (Pevensey), 158 (Portsmouth), 167 (Rye), 170 (Sandwich), 217 (Winchelsea), 221 (Yarmouth).

33 Roger of Hoveden, *Chronica*, vol. 3, pp. 39, 185. He also termed the ship on which the queen travelled to Cyprus a *buscia*, p. 105.

34 Richard of Devizes, *Chronicon*, p. 15. The vessels may have been *taride*, a type which was relatively narrow and combined oars and two sails for propulsion. It was unknown in northern Europe at the time, hence the chronicler's difficulty in finding a word to describe the ships. See J. H. Pryor, "Transportation of horses by sea during the era of the Crusades: eighth century to 1285 A.D.", *MM*, 68 (1982), 9-27, 103-25, pp. 18-21; T. J. Runyan, "Naval logistics in the Late Middle Ages: the example of the Hundred Years' War", in J. A.

English chroniclers put the number of men on board the transports in the fleet that left Dartmouth for the Third Crusade at 80 each. That seems like an upper limit. Smaller ships also made the journey, some with as few as 14 men. Nevertheless 80 is certainly possible and even plausible since a longship excavated at Roskilde and dated to 1025 or later carried perhaps 100 men, 78 of them at any one time able to row the vessel.[35] Richard of Devizes also discriminated between *naves* and *buciae*. The *naves* were to carry 40 warhorses as well as all the arms for 40 knights and footsoldiers and 15 seamen, as well as food for them all. The *buciae*, he wrote, carried twice this payload.[36] The capacity of these *naves* can only be guessed at but it was certainly larger than the 40-50 deadweight tonnes of the *knerrir* that carried families and their goods and animals to Iceland in the late ninth century. The men on board would have consumed around 1.3 kilogrammes of food and 3.7 litres of water each day, a total of 5 kilogrammes. The horses would have needed some 5 kilogrammes of grain, 5 kilogrammes of hay and 32 litres of water, a total of 50 kilogrammes per day or ten times that needed for each man. Such estimates, together with those for the consumption of crews, are notoriously difficult to make accurately, but if the figures of 5 kilogrammes and 50 kilogrammes are close to being accurate then a ship with 40 men and 40 horses would have consumed something like 2.2 tonnes of supplies per day. Carrying supplies for ten days would then have been feasible. In a 100-tonne vessel, supplies would not have amounted to a large percentage of tonnage but doubling or tripling the days at sea would have increased sharply space devoted to food and water and decreased proportionately payload for horses, soldiers, and equipment. For a ship with 80 men, less than 0.5 tonnes of supplies per day would have been needed and therefore remaining at sea for 15-20 days would have been possible.

These calculations are based on weight rather than volume, which might have posed another problem. Fulcher of Chartres wrote that for the Venetian Crusade of 1123, the ships travelling from Corfu to *Outremer* had to put in to port daily for

Lynn, ed., *Feeding Mars: logistics in Western warfare from the Middle Ages to the present* (Boulder, 1993), 79-100, p. 83.

35 J. Bill, "Scandinavian warships and naval power in the thirteenth and fourteenth centuries", in J. Hattendorf and R. W. Unger, eds, *War at sea in the Middle Ages and the Renaissance* (Woodbridge, 2002), 35-51, p. 46.

36 Richard of Devizes, *Chronicon*, p. 15. The text is poorly punctuated and badly translated at this point. Ignoring the editor's misleading punctuation, it reads: "Oneratur navis xl equis de precio exercitatis ad arma et omni armorum genere totidem equitum et quadraginta peditum et quindecim navigantium et victualibus per annum integrum tot hominum et equorum." Clearly Richard of Devizes meant that such *naves* carried 40 horses, 40 knights, 40 footsoldiers, and provisions for a year for the 80 men and 40 horses, as well as the 15 seamen in each crew. See also Tyerman, *England and the Crusades*, pp. 66, 81-2.

water and it seems likely that he was not exaggerating.[37] By the end of the twelfth century it is improbable that Northern ships had to stop that often but once in the warmer South keeping horses watered would have presented a persistent problem.

The vessels would have been open with no protection against the weather for either crews or horses. When Vikings transported animals or when Normans transported horses for the conquest of England in 1066 there appear to have been no special provisions for them. They simply stood in the middle of the ship. So conditions on board, especially for horses and soldiers unfamiliar with the sea, were also such that extended cruises were not advisable. The use of partitions to keep horses from falling and to protect them from pitch and roll certainly were in use by the second half of the thirteenth century in the Mediterranean and it seems likely that similar devices were in place on the ships that carried horses from Messina to Cyprus for Richard of England.[38] Northerners possibly learned about more efficient stowage of horses from their trips to the Mediterranean during the Crusades. Until then having horses on board oared vessels must have made rowing them more of a challenge.

If ships in fleets marshalled at Dartmouth left more or less together, which it seems they did, many of them would have used two or three days' supplies while waiting for others. Additional supplies presumably could be, and were, lightered at high tide in shallow-draft boats out to the ships anchored in the broad estuary of the Dart River. If the ships left three minutes apart then the 164 ships gathered there in May 1147 would have taken the entire day to set out, which would have been impossible because they would have had to sail with the outgoing tide and could not have waited that long. Separation must have been much less than three minutes and departure could only be managed because the vessels were anchored some distance from each other. The fleet did not reach Oporto for 24 days and if the consumption estimates are correct then the 40 horses and 40 men on board would have consumed approximately 52 tonnes of supplies, testing the limits of the capacity of the vessels. However, the ships did make a number of landfalls on the Biscayan coast of Spain, presumably to obtain stores among other things. Those ships with greater tonnage not only offered more space for cargo but also freedom from more frequent resupplying stops along the way. It seems that ships from England and the Low Countries could reach Portugal without a stop but they certainly would have been forced to make more than one stop in the Mediterranean on their way to *Outremer*. Probably, the ships did not actually need to put in

---

37 Fulcher of Chartres, *Historia Hierosolymitana (1095-1127)*, ed. H. Hagenmeyer (Heidelberg, 1913), III.xiv-xv (pp. 656-8).

38 Pryor, "Transportation of horses by sea", pp. 21, 105, 113-14. There, and elsewhere, Pryor implied that horses in slings were actually suspended off their feet by them. He now knows that that was not the case, that the horses' feet were firmly grounded, and that the slings were merely firmed up under their bellies to assist them from being thrown about. See also Runyan, "Naval logistics", p. 90.

anywhere in southern Iberia on their way to France or Italy but, as the case of Sigurd shows, it was preferable and even possibly advantageous to do so. Having safe havens along the coast west of Gibraltar or in Catalonia was apparently not a necessity created by logistics but a valuable asset and insurance should anything go wrong.

King Henry II of England had an *esnecca* which was a fast ship with a crew of 60, strongly suggesting that it was rowed in the tradition of earlier Viking warships. The English word *snacc* was related to the Norse *snekkja* and so presumably were the vessels that carried those names. Such "snakes" were probably more elaborate and larger than other warships. The last voyage for the *esnecca* of the king of England was to the Mediterranean in Richard's Crusading fleet, however, and after the Crusades and especially 1204 and the loss of Normandy the word, and very possibly the ship type, disappeared from English naval forces.[39]

Roger of Hoveden mentioned *galeae* and *cursarii* among the ships before Acre but how many of those were from the North is unclear. It seems that for him and other Northern contemporaries a *galea* was simply a ship with many oars. After he returned to England, Richard built a fleet of *naves cursoriae* in 1196 to defend the Seine. They were suitable for fast travel on rivers and on the sea so presumably they were long, narrow, oared vessels. He also had long ships with crews of 60, which suggests that those were similar to the *esnecca* that went to *Outremer*.[40] How much the *cursarii*, *barges*, and *balingers* of English naval forces in the later Middle Ages owed to the galleys English Crusaders saw in the Mediterranean is at the very best difficult to say.

As kings consolidated their power in Scandinavia after 1000 they began to build larger warships with more oarsmen than the standard fighting ship of 20-25 oarsmen per side. Larger *drekar*, dragon-ships, or *snekkjur* had higher freeboard which made them harder to manœuvre but also more defensible in battle. The oared ships of King Sigurd's expedition, the *esnecca* or *enekes* of Richard, and other oared vessels that went to *Outremer* proved that such ships could make long-distance voyages over the open sea.[41] They were large enough to carry men,

---

39 Brooks, *English naval forces*, pp. 133-5; Rodger, *Safeguard of the sea*, p. 47; R. H. F. Lindemann, "The English *esnecca* in northern European societies", *MM*, 74 (1988), 75-82, pp. 76, 79-80; Christensen, "Proto-Viking, Viking and Norse Craft", p. 85; J. Jesch, *Ships and men in the late Viking age* (Woodbridge, 2001), p. 126.

40 Rodger, *Safeguard of the sea*, p. 47. On changes in English naval administration after 1204 and the development of support for galley forces see F. W. Brooks, "William de Wrotham and the office of Keeper of the King's ports and galleys", *EHR*, 40 (1925), 570-79.

41 A. W. Brøgger and H. Shetelig, *The Viking ships: their ancestry and evolution* (Oslo, 1951), pp. 143-71; Rodger, *Safeguard of the sea*, p. 47; Christensen, "Proto-Viking, Viking and Norse craft", p. 86.

horses, and supplies,[42] but their crews probably never rose above 70, with numbers around 50 probably being more common. In *Outremer* they did patrol work and covered amphibious operations. They were certainly faster than *busses* or *keels* and apparently more manœuvrable than Mediterranean galleys. Like other vessels in the fleets of northern Crusaders, they were open boats with no protection for crew or cargo and, with such large numbers on board, putting in at night to allow the men to sleep on shore must have been appealing. *Barks*, which were also among Sigurd's ships, were probably like other warships carrying both a sail and oars. Versions of them reached the Mediterranean throughout the Middle Ages, suggesting that Scandinavian oared ships had certain features which made them as good as Mediterranean oared vessels, at least for some tasks.[43]

Oared vessels posed serious logistical problems for Northern Crusader fleets. Roger of Hoveden said that galleys, presumably including English *enekes*, could be swamped easily in an open sea and even in the Mediterranean it was best if they stayed close to land. He did say that it was possible, and even advisable, for sailing *naves* to sail directly from Marseilles to Acre, keeping Corsica, Sicily, and Crete to port, but advised against it for galleys. The layovers made by the various detachments of Richard's fleet suggest that weather and design restricted the range and speed of oared vessels. In describing the voyage from Lisbon to Marseilles, Roger of Hoveden listed the names of many different ports. It may be that he was filling in the names from a geographical work he knew but it may be also that the fleet put in to at least some of those ports, again because of the limitations of oared vessels.[44] On oared ships crews might be larger than on cargo ships, perhaps three times the size, but with few if any horses aboard the food and water needed per day would have been the same or even less than on sailing ships. It was not so much, therefore, the total consumption of food and water that forced oared ships to stop but their limited capacity. Cruising for a week, though uncomfortable, was certainly within the capabilities of an *esnecca*. However, stopping at Oporto and Lisbon was sensible given the time taken to reach the Portuguese coast from ports in England or the Low Countries. Involvement in fighting against the Muslims in Iberia was a common by-product of such stops and though the reports are of bribes by Christian rulers to convince Crusaders to help them, those same Crusaders needed the safe harbours and supplies of their hosts. The kings of Portugal had the ability to hinder Crusaders' progress and had some leverage in bargaining with Northerners.

---

42 Lindemann, "English *esnecca*", pp. 77-8.

43 Ibid., pp. 77-8; R. W. Unger, "Portuguese shipbuilding and the early voyages to the Guinea Coast", in *Vice-Almirante A. Teixeira da Mota. In memoriam*, 2 vols (Lisbon, 1987-9), vol. 1, 229-49, p. 231.

44 Roger of Hoveden, *Chronica*, vol 3, pp. 46-50, 160; J. H. Pryor, *Geography, technology, and war: studies in the maritime history of the Mediterranean, 649-1571* (Cambridge, 1988), pp. 36-8, 74.

Whether Muslims in southern Iberia had the same ability and whether they chose to use it is not as obvious as it might seem. Northern Crusaders travelled in convoys, in force. Resistance could yield dramatically negative results for defenders as proven at Lisbon and Silves. Cooperation could yield gain. King Richard's ships, along with men and equipment and horses, carried treasure for hiring ships and buying supplies and anyone along the way, no matter their religion, might receive some of that.[45] By 1094 a single Almoravid ruler had united all of southern Iberia and Morocco but that did not prevent Crusader ships sailing through the Straits of Gibraltar. Allowing Crusaders to pass may well have been a matter of weak Muslim naval power but it may also have been a matter of indifference to, or fear of, the Christians on the part of local governors and rulers. If Christian attacks on Muslims in Iberia could be prevented or deflected, the fate of Muslims in *Outremer* may well have been of little concern to local rulers.[46]

Although the Almoravids became more aggressive at sea in the second decade of the twelfth century, by the 1140s their position had deteriorated badly, thus making it easier for Crusaders to pass through the Straits whenever they wanted to. The ships of the Second Crusade proceeded without incident and the relevance, or rather irrelevance, of Muslim power in Iberia became clear during the Third Crusade. In the mid twelfth century Iberian Muslims had called on the Almohads in Morocco to come to their aid and these pushed the Portuguese northward. From 1190 Abū-Yusūf Ya'qūb, al-Manṣūr, the Caliph of Morocco and southern Iberia, carried out extensive campaigns in Iberia, retaking Alcácer do Sal and Silves in 1191. Al-Manṣūr wanted those river ports to have avenues of resupply for his campaigns by land against the Christian kings. However, they were inland, and his holding them does not seem to have deterred further voyages by Northern fleets to *Outremer* during the Third and Fourth and subsequent Crusades.[47]

The politics of Iberia do not seem to have presented serious constraints to the progress of Crusader ships from the North. Greed was a more important deterrent. So also were problems with the deterioration of the timbers of their ships. When Richard's fleet wintered over in Sicily and there was time available, the ships had to have planking replaced because of shipworm infestation.[48] Shipwrights used Sicilian wood for the repairs, the implication being that either through selection or treatment Mediterranean shipwrights were able to limit the effects of shipworm.

---

45 *Itinerarium peregrinorum* (Stubbs), II.xxx (p. 184); Richard of Devizes, *Chronicon*, p. 36.

46 Kennedy, *Muslim Spain and Portugal*, pp. 161-6, 171-4; Rodger, *Safeguard of the sea*, pp. 44-5.

47 Kennedy, *Muslim Spain and Portugal*, pp. 202-4, 237-47; Lewis, "European sea power", pp. 149-50.

48 Roger of Hoveden, *Chronica*, vol. 3, pp. 71-2: "Rex vero Angliæ interim, dum moram fecit apud Messanam, fecit omnes navigii sui naves duci in terram, et eas refici; nam multæ earum per corrosionem vermium deteriorate erant."

Although their hulls were usually made of strong and durable oak, Northern vessels could not remain serviceable for long in Mediterranean waters. Over a period of a few months, shipworms made questions of range or frequency of resupply irrelevant.

There were many changes in Crusading in the half century from 1175. Transporting horses by sea became common and it also became easier to move large sums of money between Europe and *Outremer*. Vessels of novel design which changed the logistics of voyages dramatically began to appear in Northern fleets. *Cogs* were flat-bottomed coastal traders which were given keels some time during the twelfth century, thus creating vessels that, with modifications and improvements over time, proved capable of long-distance voyages in the open sea. While *cogs* appeared frequently in records in Scandinavia and the Low Countries before 1200, only after that date did they appear in England and France. King John acquired his first known *cog* in 1210.[49] In the thirteenth century *cogs* became common in Northern Crusader fleets and they even appear over time to have come to dominate them.

In response to the call for a new Crusade, Count William I of Holland gathered a fleet in 1217 and, joined by a force from the Cologne region, departed from Vlaardingen on 29 May with a fleet of well-manned *cogs*, said to number 212. It was joined at Dartmouth by a Frisian flotilla reported as more than 80 *keels*, so the older type of cargo ship had not disappeared entirely from Crusader fleets. When the ships reached Lisbon on 25 July the archbishop, following a well-established tradition, tried to convince them to attack Alcácer do Sal; however, the Frisians refused on the grounds that the Pope had forbidden Crusaders to stay in Portugal. They sailed on, though not hesitating to attack Faro and Cadiz on the way, arriving in October at Civitavecchia in Italy, where they wintered before sailing on, reaching Acre on 26 April 1218.[50] Count William and the rest of the expedition stayed in Portugal and with Portuguese forces took Alcácer do Sal on 21 October.

---

49 For discussion of whether the improved *cog* should be differentiated from the earlier Celtic type in use in the Low Countries and Scandinavia see T. Weski, "The Ijsselmeeer type: some thoughts on Hanseatic *cogs*", *IJNA*, 28 (1999), 360-79 and, in reply, O. Crumlin-Pedersen, "To be or not to be a *cog*: the Bremen *cog* in perspective", *IJNA*, 29 (2000), 230-46. See also D. Ellmers, "The cog as cargo carrier", in R. W. Unger, ed., *Cogs, caravels and galleons: the sailing ship 1000-1650* (London, 1994), 29-46, pp. 35, 39. Some scholars insist that even among the fleet of about 200 that marshalled at Dartmouth in 1147 for the Second Crusade there were at least 2 *koggen* from Flanders. How far those ships ventured is uncertain, as is the source for the claim of their presence. See Asaert, "Scheepvaart", p. 45; Crumlin-Pedersen, "Bremen *cog*", p. 238.

50 The *Gesta Crucigerorum Rhenanorum* said that the Count of Holland had 24 *coggones* when he left Lisbon and those were presumably his ships that were part of the fleet that left the Low Countries. See *Gesta Crucigerorum Rhenanorum*, in R. Röhricht, ed., *Quinti belli Sacri scriptores minores* (Geneva, 1879), 27-56, §IV (p. 36). See also Diffie, *Prelude to empire*, p. 32; Engelbrecht, *Historische Betrekkingen Portugal-Nederland*, p. 4.

They then wintered in Lisbon and in the spring made their way first to Seville, which was still in Muslim hands, where they waited for a favourable wind before entering the Mediterranean. From there they moved straight to the mouth of the Ebro, which was said by a chronicler to mark the limit of Muslim territory. More plausible is the chronicler's suggestion that the fleet stopped there to take on water. They had free access there to "*aquas potabiles*" but also took aboard other drinks, including wine and beer. The fleet remained there for two days and then left either because it had acquired the supplies it needed or more probably because a favourable wind sprang up. The decision to make a stop not long after a stay in Seville, where presumably they would have resupplied their ships in any case, would have been dictated more by prudence than by necessity. Given the disastrous consequences of running out of water at sea, especially with valuable horses on board, opportunities to obtain water would have been rarely passed up. After the stop at the Ebro, the fleet then went on to *Outremer*. After Damietta fell to the Crusaders in the autumn of 1219, Count William followed the common practice of returning home overland through Germany.[51] Less than a decade later in May 1227, in response to a new call for a Crusade, a fleet of ships from Frisia set sail from Borkum but storm, sickness, and hunger took their toll and the few surviving vessels joined other Crusaders at Brindisi. They then sailed from there to Jaffa.[52]

The last major Northern effort to save *Outremer* came during the Crusade of 1270. King Louis IX of France, who led and financed the campaign, decided that attacking Tunis would be the most effective way to defend the few remaining Christian strongholds in *Outremer*. A force of fifty *cogs* left Frisia to join the new Crusade, stopping first at the mouth of the Zwin river where they were greeted by the countess of Flanders. They then sailed to Marseilles, only to find that Louis IX had already left for Tunis. The Frisians followed him to Africa and after his death and the failure of the siege of Tunis some sailed on to Tyre but many also made their way back to the North overland since their ships were no longer seaworthy.[53] Prince Edward of England, the future Edward I, also joined the Crusade, fulfilling the 1252 plan of his father, Henry III.[54] He took at least 225 knights with him by land across France to Marseilles where they boarded 13 ships for Tunis. Edward also arrived too late and with many of the others in the Crusading force retired to Sicily where they wintered. In the spring he hired ships and proceeded to Acre, where he joined in the fighting, leaving only in September 1272 after the death of

---

51 The report of watering at the Ebro is from the *De itinere Frisonum*. See *De itinere Frisonum*, in Röhicht, *Scriptores minores*, 57-70, pp. 56-7. See also Hardenberg, *Nederlanden en de Kruistochten*, pp. 175-80; Kennedy, *Muslim Spain and Portugal*, p. 259.

52 Hardenberg, *Nederlanden en de Kruistochten*, p. 182.

53 A number of Crusaders died on the way home. Hardenberg, *Nederlanden en de Kruistochten*, pp. 189-90; Engelbrecht, *Historische Betrekkingen Portugal-Nederland*, p. 5.

54 Tyerman, *England and the Crusades*, p. 116.

his father made his return to England urgent. Travelling overland, he took almost two years to reach London.[55] Edward's Crusade demonstrated the change in the character of Crusading between the twelfth and thirteenth centuries. In England especially, before the Third Crusade, Crusading had been an enterprise of lesser nobles and urban dwellers. After that it became a royal enterprise involving extensive preparations and financing from the highest political level. One reason lying behind the change was the exorbitant cost of sending well-equipped soldiers and all the support they needed in supplies and personnel; although, another was the developing political structures of northern Europe. Changes in the ships used played a role as well.

Cogs were larger than *keels*. They might not be longer but they were deeper and that increased their carrying capacity sharply. They had crews which were as large or larger because their size demanded a large sail and enough human muscle power to raise it. They were much less manœuvrable than their predecessors and could not move as fast, or make way against wind and tide as well as *keels*, which therefore retained a place in Crusader fleets, *cogs* seeing perhaps more specialized use. Large *cogs* could not be beached and therefore needed ports with quays to function. Being larger with greater carrying capacity for each unit of length, however, *cogs* had significantly greater range than their predecessors and could stand out from the coasts making less frequent landfalls. They also had better accommodation since their depth in hold made it possible to build enclosed spaces safe from the elements to protect people and horses.[56]

Cogs' greater range may have eliminated a number of logistical problems; however, it was also precisely at the time when *cogs* came to be used that the problem of ports of call between Lisbon and Marseilles or Sicily was resolved. The Christians took the offensive in Iberia in 1209 and in 1212 won a great victory at Las Navas de Tolosa, after which the Almohads withdrew from Iberia to the Maghrib. Although the Christians were slow to capitalize on their advantage, they had broken Almohad power. In 1248 Seville fell and the Portuguese Algarve followed in the next year,[57] leaving only a strip of the coast of Iberia about 200 kilometres long from Gibraltar to Almeria under the control of the Naṣrid sultans of Granada. Even though it may have had little influence before, from the mid thirteenth century Muslim naval power was no longer a factor affecting the logistics of Northern fleets.

---

[55] Edward had borrowed heavily from Louis IX to finance his expedition. However, despite promises made when the loan was agreed, Edward reached Marseilles too late to join the main force. See M. Prestwich, *Edward I* (London, 1988), pp. 71-5, 79, 85.

[56] The decks typically were not watertight but even so provisions for protection were still much greater than with earlier vessels. See Ellmers, "*Cog* as cargo carrier", pp. 41-4.

[57] Kennedy, *Muslim Spain and Portugal*, pp. 249-56, 265-6; Diffie, *Prelude to empire*, p. 51.

In 1277 Genoese galleys arrived in Flanders, the first Italian ships known to have gone from the Mediterranean to the Low Countries. Venetian galleys followed soon after and established a pattern of regular voyages between the Mediterranean and northern Europe.[58] And, at the same time as Mediterranean galleys became capable of regular voyages back and forth through the Straits of Gibraltar, improvements in *cogs* made it possible for them also to make similar regular trips. The addition of a second mast at the stern carrying a lateen sail created the *cocha* in the Mediterranean. The type came to be called a *carraca* in northern Europe when it made its way there in the fourteenth century.[59] Mediterranean shipwrights emulated the improved *cog* in part because of its superior design but also because of lower labour requirements at a time when seamen's wages were rising. The new type was much more manœuvrable than the earlier *cog* and could carry as much cargo but more reliably and in all likelihood more quickly. By the end of the fourteenth century carracks were sailing back and forth through the Straits of Gibraltar.

The problem of sailing west into the Atlantic from the Mediterranean had plagued Northerners since the time of King Sigurd and before. The passage was over 70 kilometres long, the prevailing winds typically adverse and the strong contrary surface currents running at around 6 knots made going west through the Straits difficult. It may have been possible to creep through but it would have required knowledge of local waters. The presence of hostile naval forces in ports such as Malaga, Gibraltar, Cadiz, and along the Algarve, as well as in Ceuta and Tangier, only increased the difficulty. Even with fluctuations in Muslim naval power, natural conditions still made movement of ships out of the Mediterranean all but impossible. Despite the large numbers of ships from northern Europe making their way into the Mediterranean from the Vikings through all the major expeditions of the Crusades there appears to be no record before the later thirteenth century of a ship sailing out again.[60] Only when builders fitted a rig combining a mizzen lateen sail and a main square sail on a modified *cog* and built galleys large enough with high enough freeboard to weather Atlantic swells, did vessels become capable of making round trips between the North and the South. The ultimate logistical constraint on Northern Crusaders was that their vessels could make one-way trips only. Ironically by the time that Europeans did develop vessels that could go back and forth between the Mediterranean and the Atlantic, there was no need for them because the Crusader States were on their last legs.

---

58 Hardenberg, *Nederlanden en de Kruistochten*, p. 223; Diffie, *Prelude to empire*, pp. 49-53; R. S. Lopez, "Majorcans and Genoese on the North Sea route in the thirteenth century", *Revue Belge de philologie et d'histoire*, 29 (1951), 1163-79.

59 I. Friel, *The Good Ship: ships, shipbuilding and technology in England 1200-1520* (London, 1995), pp. 157-60.

60 Lewis, "European sea power", pp. 140, 151.

The principal difficulty in evaluating the action of Northern Crusaders is the paucity of evidence. The lack of information even by standards of the study of the Middle Ages is remarkable. Knowledge about the kinds and types of ships used, their characteristics and dimensions, is shaky enough. Material on what was in the ships from the North is even harder to find. While it is hard to establish the dimensions and capacity and sailing characteristics of the Crusaders' ships, it is at least possible to say that Northerners had no advantages over Southerners. The Northern *busses* and long ships and *esneccas* that entered the Mediterranean from the time of King Sigurd had no inherent superiority over the ships in use in the South. Indeed Northern ships may have been inferior for most purposes. Once in the Mediterranean the English in the Third Crusade, and for that matter all other Crusaders from the North, fell into patterns of behaviour, action, and deployment like those of the people they met, often using locally produced ships in preference to those brought with them. They did that because their ships had no inherent advantages and also because shipworm often destroyed the seaworthiness of their own ships.

In contrast to all the problems Northern Crusaders faced on the way to *Outremer*, their counterparts along the shores of the Baltic had a very simple life. Everything in the Baltic was different. As part of the Second Crusade, a combined army of Danes and Saxons attacked pagan settlements by land in the summer of 1147 but had limited success. The Papal bull *Divini dispensatione* of 11 April 1147 gave Christians fighting Slavs remission of sins as if they had gone to *Outremer*.[61] By that time ships of Scandinavian monarchs had already started to transport horses and this gave them greater range on land in amphibious attacks. They continued to carry horses on board but nothing they had in their armaments or vessels could overcome the adverse weather conditions which stymied the King of Denmark's naval expedition in 1147. In the second half of the twelfth century the Danes began to take the Crusade to the Baltic Sea,[62] and from 1198 the German bishop Albert of Buxhövden took the lead in the Northern Crusade. In 1202 he set up the small order of the Sword Brothers and in the course of the following 30 years they subdued Livonia, the start of a series of successes in Poland, Lithuania, and Estonia. Letters of Pope Innocent III, especially *Etsi verba* of 10 October 1204 addressed to the archbishop of Bremen, had the effect of creating what was in essence a perpetual Crusade against the Slavs though the idea and practice did not

---

[61] Eugenius III, Pope, ed. Migne, *Eugenii III Pontificis Romani epistolae et privilegia*, in PL, vol. 180, coll. 1013-1614, no. 166 (coll. 1203-4). See also E. Christiansen, *The Northern Crusades: the Baltic and the Catholic frontier 1100-1525* (London, 1980), p. 51.

[62] Otto of Freising, *Ottonis et Rahewini Gesta Friderici I. imperatoris*, 3rd ed., ed. G. Waitz, in MGH *ScriptRerGerm.Cont* (Hanover, 1912), I.xlii-xliii (pp. 60-63). See also Phillips, *Crusades*, pp. 71-2.

become fully developed for another 40 years.[63] The Crusade became perpetual in the sense that it became unnecessary to ask the Pope for remission of sins for Crusaders each time an expedition was mounted. Albert laid the foundation of conquest when he transferred the centre of Christian power in Livonia from Üxküll down the Dvina River to Riga because *cogs* from Lübeck could reach the port.[64] As with *cogs* that made the trip to *Outremer* those in the Baltic had much greater carrying capacity than the old *keels* and by the mid thirteenth century were able to take five or even ten times the normal load of the typical *keel*. The *cog* was the perfect transport for carrying military supplies and reinforcements and its high freeboard made it almost impregnable at sea. After losing a battle to the Slavic tribe of Curonians in Riga Bay in 1210 the Sword Brothers gave up using small ships and instead turned to *cogs* and drove the pagans away from Riga in 1215. Not only were *cogs* superior but they were also beyond the technical capabilities of their enemies to build so the Christian knights feared no competition. In the conquest of Livonia the Crusaders set up a series of stone blockhouses along the Dvina and supplied them with food, arms, and reinforcements from Riga using river boats like the *bolskip* and other forms of lighters. So long as they worked along rivers Christian success followed but once away from their lifeline to supplies brought by *cogs*, for example in the more forested Lithuania to the south of Livonia, the Crusaders had greater difficulty.

From 1309 the Teutonic Knights had their headquarters at Marienburg in Prussia but they had already been operating against pagans in the region for some time. When the Order first arrived in Poland in the 1230s it followed the example of the Sword Brothers, setting up river forts and using them as bases from which it could recruit or intimidate people along the banks. In the beginning it worked down the river Vistula but by the end of the thirteenth century and especially after it acquired Danzig in 1308-9 it also relied on supplies brought by sea and carried upriver.[65] The lifelines that ran up the rivers of northeastern Europe were the avenues of Crusader conquest and the *cog* gave them an unassailable advantage over their Slavic opponents. The Christians used a new form of warfare which allowed them to dominate river valleys and to give soldiers virtually endless supplies all the time, except in winter when ice closed ports. After the introduction of the *cog*, logistics proved to be the least of the worries of Crusaders in northeastern Europe. Ease of supply, more comfortable travel by *cog*, and chances

---

63 Innocent III, Pope, *Innocentii III Romani pontificis Regestrorum sive epistolarum libri*, eds E. Baluze et al., in PL, vols 214-16, and 217, coll. 9-308, 239 (vol. 215, coll. 428-30). See also J. Riley-Smith, *The Crusades: a short history* (New Haven and London, 1987), p. 131.

64 Christiansen, *Northern Crusades*, p. 94-6; E. N. Johnson, "The German Crusade on the Baltic", in Setton, *HC*, vol. III, 545-85, pp. 556-65.

65 Christensen, *Northern Crusades*, pp. 86, 96-7, 101; Johnson, "German Crusade on the Baltic", pp. 570-75; Riley-Smith, *Crusades*, p. 213.

of success may help to explain why knights and other fighters in northern Europe preferred to fight for their beliefs in the valleys of rivers leading into the Baltic, especially after the fall of Acre.

Chapter 14

Digest[1]

John H. Pryor

The inspiration for the workshop that led to this volume came from a realization that in the context of the Crusades in particular, but also of medieval warfare in general, most historians have written about military campaigns as though they took place in a geographical, meteorological, and oceanographic vacuum. In most books military forces move from one place to another without the slightest difficulty. Actual routes followed, if discussed at all, are almost never done so from the perspective of trying to understand why they were chosen from a logistical point of view, sometimes from a tactical point of view but almost never from a logistical one. Naval forces are given even less consideration. They leave the West and arrive in the Holy Land as though their commanders had merely engaged the engines on their cruise ships and set course for the Holy Land by the shortest possible route. Almost any book about the Crusades showing the supposed routes of those who came by sea will demonstrate this lack of appreciation of the technological and geographical realities of the Middle Ages. In fact far more attention has been paid to logistical infrastructures, primarily military rather than naval, by ancient historians. Some ancient sources are more informative than most medieval ones and it will have become apparent to anyone reading this book that most contributors have utilized the research of ancient historians considerably. A cursory glance at the Consolidated Bibliography will reveal just how few publications have been devoted to the logistics of medieval military campaigns.

Moreover, as John Haldon in particular has pointed out, logistical infrastructures on which the movement of military forces depended have been almost entirely neglected: the condition of roads, where they existed at all; systems of communications; the capability of the land, population, and technological mix to produce surpluses capable of sustaining extraneous military forces; climatic and weather variations. Occasionally it has been appreciated that excessive rain or

1 This "Digest" was compiled at the request of the publishers after the volume had been completed from a reading of all of the papers. Ideas, and even quotations, from the various contributors have been assumed tacitly and not cross-referenced. It has not been indexed since matters discussed should be able to be found easily through the index. All references to the contributors are to their papers in this volume only. I have adduced a few pieces of evidence, and some other matters for consideration, in addition to those adduced elsewhere in the volume, which have seemed to me to be relevant; however, these have not been documented or indexed either.

excessive heat did impact on the movement of military forces, but rarely has the analysis been much more sophisticated than that. Yet Tom Madden has been able to demonstrate how the issue of food supplies drove the entire course of the Fourth Crusade. The whole argument over the outcome of the fifty-years' war in Syria between the Īlkhāns of Persia and the Mamlūks of Egypt hinges on the issue of the seasons, the carrying capacity of the grasslands of Syria, and to a lesser degree the easing off of water flows in the few rivers, through the summer. In the case of the First Crusade, it is important to consider the fact that Peter the Hermit and the other contingents of the People's Crusade left the West at the end of the annual food cycle, before the winter corn would have been harvested, but that the baronial forces did not leave until August-October, after the harvest. It was probably no logistical accident that Urban II had nominated the Feast of the Assumption of the Blessed Mary, 15 August, as Adhemar of Le Puy's departure date. Leaving in early autumn, and hoping to reach Constantinople before deep winter set in, a not unrealistic hope, forces might have expected to have had access to recent harvests all the way. Furthermore, the horses should have been able to graze on summer grass throughout the march. Grass was of paramount importance. Military campaigns by large cavalry forces were normally impossible before the summer grass had grown or after it had died off in late autumn. Basil II's refusal to be bound by seasonal constraints on campaigning was so unusual as to invite comment by Michael Psellos. As it eventuated, only Hugh of Vermandois and Godfrey of Bouillon reached Constantinople before the winter; however, the expectation may have been there. Leaving too late for that, Bohemond of Taranto almost certainly took a circuitous route through the valley of the Drino because of its fertility and because by then the autumn harvest would have been in. As Yaacov Lev has pointed out, the Crusaders abandoned the siege of 'Arqā on 13 May 1099 at least partly because the harvest was about to begin and they could hope to live off the land to Jerusalem.

In the analysis of such matters some of us have had recourse to modern geographical, meteorological, animal-physiological, population, and other data and in most cases, as long as it is used with due caution, this can be defended. There is nothing in the historical record to suggest, for example, that climatic variations in historical times have produced any variation in prevailing wind patterns in Europe. However, Yaacov Lev is surely correct to emphasize that smaller medieval populations placed much less pressure on the environment than do modern ones and that in some cases at least it may have been much easier for medieval armies to obtain the necessary water, for example, than modern data may suggest. Both Lev and Haldon are optimistic that progress can be made in attempts to reconstruct the medieval human and geographical environment within which historical events were played out.

With few exceptions medieval historians have not considered applying to the past what can be learned about logistics from the modern era. Even if medieval animals were not the same as modern ones, even nineteenth-century ones, and data from nineteenth-century military manuals must be used with caution, they nevertheless complement the fragmentary historical record. There is no doubt that even where logistical problems did exist and were encountered, medieval authors rarely commented on them because they were part of the ordinary and of little interest except where they impacted upon the extraordinary, which was what medieval authors were interested in. Before the twelfth century documentary sources such as those used here by me, or the letters used by John France, are very rare. That Caffaro recorded no ports of call for the Genoese flotilla which sailed for the First Crusade, even though it took four months to reach the Levant, or that Western chroniclers passed over Byzantine efforts to supply the First Crusade, is hardly surprising. As Yaacov Lev points out, Muslim sources contain few references to logistical problems and medieval people obviously had the skills and means to provide. But that does not mean that the problems did not exist. For obvious reasons, authors such as Odo of Deuil, attempting to explain failure rather than record success, often provide more information about such matters than do those glorifying success. There is the further consideration that almost without exception the sources on which we depend were literary creations rather than objective accounts of what actually occurred. Their purposes were rarely to record. Without recourse to alternative sources of information, argument over the interpretation of the occasional references in medieval texts will ultimately prove futile. Resources such as the sun and moon data of the United States Naval Observatory's Astronomical Department can be of enormous use in modelling military campaigns. For example, assuming, with good authority, that the climate in 1800 was not too dissimilar to that of 1187, we can know that Guy of Lusignan and the army of the Kingdom of Jerusalem would have had 13.86 hours of daylight between sunrise and sunset on 3 July when they left the Springs of Saffuriya to march to the relief of Tiberias, and there would have been around another 57 minutes of twilight at either end. The distance was only around 30 kilometres, requiring a rate of march of only 2.15 kilometres per hour, perhaps not to have been considered an impossibility, even in the face of Ṣalāh al-Dīn's opposition.

Benjamin Kedar reminds us forcefully that even if we may properly have recourse to modern data, in his case accurate geographical maps, in order to better comprehend historical events such as the marches of military forces, we should not forget that that gives us an ability to make logistical and other calculations that medieval people would not have been able to make themselves. It is perhaps significant that *The atlas of the Crusades* (London, 1991) contains a great many maps but none are medieval *mappemundi* such as might have been viewed by someone with an education leaving for the First Crusade. That is not to criticize, merely to suggest another way of thinking. We should bear in mind that medieval people perceived their physical world differently to us. How did they in fact view the way ahead? To take the ferry from Istanbul out to the Princes' islands and to observe how the Gulf of Nicomedia recedes into the haze of infinite distance is to give one a perspective on the First Crusade that modern maps in books falsify. What must have been the emotions of those crossing the Bosporos and then setting out from Chalkēdon to march into the unknown? What did people know of the geography of the land stretching away beyond the horizon? In some cases they had personal experience, in other cases local guides, and in the case of the Crusaders facing a march across Asia Minor more than a millennium of Romano-Byzantine experience. The way ahead must have been well known to general Tatikios. But it is curious that no Byzantine maps survive from between the sixth and thirteenth centuries, although some literary texts allow the surmise that they did once exist. In the case of Western and Muslim *mappemundi*, which literary sources make clear were available to military commanders as embroideries either on cloth or on the walls of tents, Kedar's exploration of the fact that one such was known to Stephen of Blois, and his hypothesis that that explains some curious statements in his letters to his wife Adela, leads to the tantalizing idea that he was appointed *provisor atque gubernator* of the expedition because of a singular ability to see the way ahead.

Benjamin Kedar also refers to the first-known reference, in William of Nangis's account of the Tunis Crusade of St Louis, to what is usually considered to have been a portolan chart; that is, one of the new generation of "accurate" maritime charts of which the oldest surviving example is the *Carta Pisana* of ca 1290. Compass bearings were marked on such charts and in the accompanying *portolans*, the books of sailing directions, distances were measured in miles. This raises the intriguing issue of how medieval people envisaged maritime space. By land there were surveying techniques for measuring distances, or distances could be paced out. By sea that was not possible. Around the world in pre-modern times journeys through maritime space were normally reckoned in time: a half-day's sailing, three days' sailing, etc. How could anyone measure distance across open sea out of sight of land at both ends? Some medieval Muslim juridical texts declared that freight and passenger rates could not be calculated for voyages out of sight of land because the distances could not be measured, as they could be for coastal voyages. It should be borne in mind that from the masthead of a medieval galley the horizon was no more than around 12 kilometres distant. The realities of traversing maritime space raise questions concerning the relations between that and the texts which survive to us.

Such considerations also raise issues of planning. To what degree was it even possible for political authorities and military commanders to plan logically for projected campaigns. In the case of the Crusades, as Charles Glasheen and John France have emphasized, these expeditions were almost completely unplanned. In the case of the First Crusade, its totally inchoate nature surely meant that there was virtually no logistical planning except at the most elemental level; that is, individual lords' retinues. The issue of whether the Pope or individual leaders announced their coming to Alexios Komnēnos so that the Byzantines could provide for their arrival is an open one. Probably Bohemond did. Probably both Peter the Hermit and the other baronial leaders did not, at least there is little evidence to suggest it. In the case of the Second Crusade, commanders controlled the direction only with the acquiescence of participants and, as John France has clarified, Louis VII was responsible only for his vassals from the royal domain. Others had to fend for themselves. In the case of the Fourth Crusade, the entire outcome was precipitated by the erroneous guesstimates of numbers made by the French lords, compounded by Venetian ambition to conquer Egypt which led to the preparation of totally unprecedented battle and transport fleets. No one had ever done anything like attempt to convey 4,500 horses to the Levant on *huissier* transport galleys of the *tarida* type before. Around 150 would have been needed. The closest attempt to it had been the Sicilian attack on Alexandria in 1174 when 1,500 horses were transported, probably on 50 *taride*. The Venetians certainly did plan well, but on false assumptions. There was no Papal direction until the Fifth Crusade and the inchoate nature of Crusading impacted radically on any logistical planning. No doubt Byzantine and Muslim military expeditions were rather better planned than Crusades, as the evidence of Byzantine military treatises suggests, but even in their case we should not be deceived into regarding their military forces as "armies" in any modern sense of the word. Both Muslim and Byzantine forces were either provided with money to purchase provisions *en route* or were expected to provide at least some for themselves, as both John Haldon and Yaacov Lev have pointed out. Reuven Amitai uses the expression "nation-in-arms" for the military forces of the Īlkhāns of Persia, and that is revealing.

Debate and discussion between participants here has focused on a number of central issues, in particular: the width and condition of Roman roads in the Balkans and Asia Minor by the tenth to twelfth centuries; standard marching rates for infantry and distances covered in a normal day's march; the volume and weight of provisions needed per man per day; the volume and weight of fodder and/or the amount of grazing needed by animals and time necessary to consume it; the productivity of pasture; the amount of water needed by animals and, to a lesser extent, men; loads that could be carried by animals and men or in carts and waggons; the range that the equation of loads which could be carried to daily consumption gives before resupply became necessary; the establishment of supply depots along frequently used routes at distances no more than maximum ranges; the provision and acquisition of money or convertible wealth in other forms; the provision of markets when in friendly territory as opposed to the need or desirability for foraging or pillaging; and the logistical capabilities of naval forces of various kinds and in particular of horse transports. It will have become apparent to readers who have worked their way through the various papers that we have not agreed on all such matters, although in most cases some agreement on broad parameters has been reached. In some cases it has not been, however, and that must be acknowledged. In what follows I have not attempted to present a consensus point of view but rather my own. In most cases it is in agreement with that of most of the participants.

Medieval forces did not necessarily remain confined to the width of Roman roads and where the terrain made it possible no doubt spread out to either side of the roads; however, for the main part that would not have been possible. During the wet months from late autumn to early spring sodden ground either side of roads would have made it difficult for those not on them to keep pace with those on the pavement. In the Balkans and the Tauros and anti-Tauros, hilly terrain to either side would have made it impossible. So the width and condition of the roads was particularly important. Except near cities, the Roman roads were in fact quite narrow, and even where they had been wider in antiquity, as John Haldon has pointed out, Byzantine repairs to them narrowed them considerably. Moreover, by the tenth-twelfth centuries they were in a state of considerable disrepair. The *Via militaris* from the Danube seems to have been particularly badly deteriorated, but the *Via Egnatia* was probably not much better and there is evidence that the road through the Cilician gates was in complete disrepair. Bernard Bachrach believes that the Byzantines had maintained the roads and that between Nicaea and Dorylaion they were six to seven metres wide. That is probably an optimistic assessment given the evidence for the Byzantines increasing use of pack animals as opposed to carts and waggons for military baggage trains. In hilly country some roads became stepped, which would have meant that they would have become impassable to carts and waggons.

Of course, in the context of the cavalry warfare in Syria-Palestine discussed by Reuven Amitai, remnants of Roman roads would have been irrelevant. Just exactly how 325,000 Mongol ponies accompanied by 50,000 camels managed to move from one point to another is a question to excite the imagination. One can only envisage a vast cloud of men and animals moving across the grasslands spread out from horizon to horizon. For the Fāṭimids, Ayyūbids, and Mamlūks moving up from Egypt to Palestine there were some remnants of Roman roads in places but it would not have been necessary to remain confined to them and no doubt columns spread out considerably.

Overall rates of march were very variable, being related to the type of terrain, the composition of forces, their size, the season, and urgency. Where columns were forced into fewer files by bottlenecks, bridges, narrow passes, etc., that would also have slowed them down considerably. Standard marching rates for infantry, however, have always been around three miles (4.83 kilometres) per hour, dropping to two miles (3.2 kilometres) on broken, uneven, or hilly ground. The question then becomes for how long each day forces could march given the assumption that they always encamped together, an assumption which may not always have been justified. Did forces sometimes split into divisions to extend daily route marches? Modelling shows how daily marching times would have been related to the size of divisions camping together. But, apart from the splitting of the forces of the First Crusade into two divisions between Nicaea and Dorylaion, there is no evidence for splitting forces. In hostile territory it would have been suicidal, although in friendly territory it would seem to have been only sensible. Nevertheless, no sources refer to the practice and the Byzantine military treatises assume that forces encamped together. That being said, it may well have been the case that, when passing through the Empire, Crusader columns may well have encamped in extended camp zones stretching along the line of march for many kilometres, as long as access to water and pasture made that possible. In modelling daily route marches one of the critical variables is the distance to be allowed for between the ranks in columns. But what modelling can do effectively is to relate movement to space realistically and to provide insights into what meagre information the medieval sources provide. The emperor Nikēphoros Phōkas's assessment that 16 Byzantine *milia* or approximately 24 kilometres was a long and tiring day's march for both men and animals in the terrain of the Tauros mountains is a very significant piece of evidence which can be comprehended by modelling. In fact forces of any size at all containing infantry must rarely have averaged more than 20 kilometres per day. The slowest we have encountered was Bohemond's winter march from Avlona to Kastoria at a rate of only 6-7 kilometres per day.

Ancient and late antique sources provide some useful information on provisions for military forces but the only hard medieval data known to us are those which I collected for the oarsmen of thirteenth-century galleys and for Crusaders being transported by sea. Most of the contributors here have assumed a weight of around 1.33 kilogrammes of provisions per man per day but that may be rather high. What medieval data we have suggests that men could be maintained in good physical condition for fairly heavy labour on a diet weighing around one kilogramme, which would have included around 750 grammes of twice-baked bread, biscuit. But that would have needed only around 560 grammes of winnowed grain. If carried as grain, the total weight needed per man per day may have been as low as around 800 grammes. This is not dissimilar to some conclusions reached by John Haldon for Byzantine forces.

There is even more debate and less certainty about the amount of fodder needed by animals, in particular by horses. There is considerable variation between the ancient, medieval and nineteenth-century data, a result, no doubt, of differing circumstances and perhaps also of differing types and sizes of animals. By and large the figures for nineteenth- and early twentieth-century cavalry horses can be up to twice as much as those which can be deduced from ancient and medieval sources. What is certain, however, is that, with the exception of the Mongols' ponies, the horses of all military forces considered here required some portion of grain in their diet to remain in good condition. Obviously circumstances did arise in which they simply had to survive for some periods on a diet of hay or grazing alone. But that would not have been desirable for very long. For transporting horses by sea, the medieval barley or other grain ration appears to have been around 2.7 litres or 1.6 kilogrammes, to be increased by a third to 2.4 kilogrammes for animals in moderate work such as on a route march. The hay allowance of around 2.7-3.7 litres for horses being transported by sea cannot be converted to a weight because the medieval techniques for baling, or indeed "barrelling", hay are unknown. However, data from elsewhere suggests that hay allowances should be roughly equivalent to, or a little higher than, the hard fodder allowance and therefore, for horses in moderate work a figure of around 2.5-3 kilogrammes may be appropriate if hay had to be transported. A total weight of provisions per horse per day of around 5-6 kilogrammes or 11-13 pounds would probably be about right. On an extended route march some hay, or simply cut grass, may have been carried when necessary but in normal circumstances grazing would have substituted for it. Assuming that marches were made in seasons of green grass when the water content of grass was around two-thirds, horses would need to graze around 8-9 kilogrammes of green grass. That is somewhat less than what nineteenth- and twentieth-century cavalry manuals would suggest was desirable and would probably be a minimum figure. Note that J. M. Smith's calculation for Mongol ponies, which were only about a half the weight of medieval Western warhorses, is 14.2 kilogrammes, but that is for a diet of grass alone.

The total amount of provisions for men and fodder for horses required for large military forces compels consideration of the productive capacity of populations and land to provide it. John Haldon calculates that a Byzantine army of 15,000 would need 288.4 tonnes of provisions for 2-3 weeks. My own calculations suggest that a modelled force of 3,860 men and 7,730 animals would require around 2,425 tonnes of provisions for a 110-day march from Avlona to the Vardar river, and that does not include any hay or cut grass. Bernard Bachrach estimates that 600 metric tonnes of grain would have been needed by the First Crusade for the six days between 26 June and 1 July 1097. Tom Madden calculates that the Venetians undertook to provide 17,000 metric tonnes of food for the men and horses of the Fourth Crusade for a year, as well as provisions for their own forces. This imposed such a strain on the region that the contract with the Frankish Crusaders forbade them from purchasing provisions in the Po valley from Cremona eastwards in order to prevent prices being driven up. Smith's calculations for Mongol ponies, if they were consuming hay and grain rather than solely grazing grass, would be that the 325,000 ponies of the 1299 Īlkhānid expedition to Syria would have needed around 735 tonnes each of grain and hay per day. The ability of pre-modern economies to produce such enormous surpluses and of pre-modern governments to collect and supply them to transient military forces consideration of entire economic and organizational systems. It is not a wonder that there were skirmishes over food between Crusader contingents, local inhabitants, and Byzantine authorities and shepherding forces but, rather, a wonder that there were so few of them.

A collateral problem associated with grazing is the time taken by horses to do it. By nature horses are grazing animals. They have small stomachs and cannot eat a great deal at a time but need to eat frequently. Figures of around an hour per kilogramme and a total necessary grazing time of around 10 hours per day are about right. Smith's calculation for Mongol ponies is 10.5 hours, which gives a figure of 1.35 kilogrammes per hour, which is rather on the high side. When such figures are considered against the productivity of pasture the size of encampments needed for cavalry forces of any size, merely for a single night, becomes enormous. The 325,000 ponies of the Mongol expedition of 1299 would have consumed an area of 2,500 hectares or 2.5 square kilometres per day. A Crusader contingent with a modelled 3,860 men and 7,730 horses or mules would have needed 62.5 hectares per day. Problems associated with simply finding such large encampments every night while on a route march and then cutting grass and bringing it to the picketed animals must have been enormous.

Water requirements for men and animals were even more important than food and fodder. Both men and horses can struggle on for a considerable time without adequate food, but running out of water will bring them to a stop in a few days. That being said, of course, horses that became emaciated or unfit for battle during route marches could create major problems, as for the Mongols before the battle of Wādī 'l Khaznadār. Western horses in moderate work in temperate weather would have needed around 8 gallons or 35 litres of water per day and men around 8 litres. Mongol ponies would have needed around 19 litres per day. In harder work or in hot weather those needs would have risen by at least 50%. Watering would normally have been done by men bringing water to the animals in leather buckets rather than taking them to water, simply because of the enormous length of bank which would have been required. Horses also drink slowly. When dependent on wells, enormous numbers of well-lifts would have been needed. Crossing the desert between Egypt and Palestine, or Egypt and Eilat, where wells and springs were the only water sources, must have made watering cavalry forces of any size extremely time consuming. No wonder that Ṣalāḥ al-Dīn built a dam at Qalat Ṣadr to hold the water from the spring. Watering the 325,000 ponies of the 1299 Mongol expedition defies belief. One can imagine that it could have been done only where there was a very long stretch of river bank available and even then it would have had to have been done in continuous rotating shifts. The time needed to water horses certainly would have influenced, and in some cases probably determined, entire orders of march.

One text on imperial military expeditions produced for emperor Constantine VII noted that cavalry horses carrying men could carry an additional 4 *modioi* of barley each for their fodder, unridden saddle horses could carry 8 *modioi* as well as their saddles, and pack animals 10 *modioi* each. These figures translate to 40, 80, and 100 kilogrammes respectively and accord quite well with the general principle that horses can carry around 20% of their own weight. Bernard Bachrach's conclusions that four-wheeled waggons could carry around 650 kilogrammes and two-wheeled carts around 500 kilogrammes have been generally accepted. The range that the equation produced by loads which could be carried to daily consumption can then be modelled with various ranges of variables. It is significant that both my own and John Haldon's calculations, proceeding independently by different methods, come down to a maximum range of around three weeks' march before total resupply would become necessary and that this accords closely with what the Anonymous tenth-century treatise on campaign organization, *'Ανωνύμου Βιβλίον τακτικόν*, claimed was the maximum range of Byzantine cavalry forces as determined by the barley requirement of horses: 24 days.

The Byzantines developed a system of marching camps, *aplēkta*, in Asia Minor from the mid seventh century, distances apart being related to average daily rates of march and load capabilities. John Haldon calculates that *aplēkta* would need to be no more than 400-475 kilometres apart but in reality they would have been much closer together than that whenever possible to avoid having to use so many animals. Their locations were also determined by natural factors such as water supplies, fertility, and proximity to major military routes, of course. The major Byzantine *aplēkton* in Asia Minor was at Malagina, which was only around 150 kilometres from Constantinople. In the ninth century the *aplēkton* for expeditions to reconquer Crete was at Kepoi at the mouth of the Maeander river. For the desert crossings from Egypt to Palestine or Eilat depots were determined by oases where water could be found: especially Al-Arish in the case of Palestine and 'Uyūn Mūsā, Qalat Sadr, and Qalat Nakhl in the case of the Eilat crossing. The Fāṭimids made Ascalon into a great storehouse for military campaigns and after its loss in 1153 attacking the Kingdom of Jerusalem from the south became much more difficult because Egyptian forces entering Palestine would enter hostile territory immediately after crossing the desert. The Mamlūks also used supply depots along the routes to Syria from Egypt for provisions, fuel, and animals and there were also areas such as Shaqḥab south of Damascus where animals could be grazed.

Alan Murray above all has drawn attention to the fact that all Crusaders had to carry money or negotiable wealth in other forms. John France has also drawn attention to the circulation of money in the forces of the Second Crusade, Tom Madden to it in those of the Fourth Crusade, and Yaacov Lev to it in Muslim armies. Even those of us who have not actually pointed to it would agree that in all the forces we have discussed individuals carried money for the purchase of provisions *en route* in particular as well as for other purposes. That being said, Crusades did differ from other military expeditions in that participants could not expect to be resupplied with money during the expeditions and had to take what was considered necessary from the outset. There is no evidence for the transfer of money to *Outremer* from the West for individual Crusaders until the Sixth

Crusade, after 1249, although it is true that Henry II of England did build up a war chest in Jerusalem and Frederick Barbarossa arranged for the transfer of money to *Outremer* via a Venetian financier. Jonathan Riley-Smith's estimate is that Crusaders needed roughly four times their annual income. Frederick Barbarossa ordered that men take funds for two years, calculated by Otto of St Blasien at three marks, or 432 pence. This was no doubt a minimum and Murray suggests that a knight with a minimum annual income of five marks would have required at least 24 marks, 3,456 pence. His analysis of the *Längeres Kölner Dienstrecht* suggests that for "armsbearers" on the First Crusade four times annual income is probably conservative. The statement in the first redaction of Fulcher of Chartres's *Historia Hierosolymitana* that some Crusaders expected to be away for three years is very significant.

Money was raised for Crusades by sales of property, movables, and rights, mortgages, confiscations, extortion from Jews, charity, and the evidence of Stephen of Blois's letters to his wife Adela suggests that women's jewellery and perhaps dowries were important. "Money" was carried either in plate, jewellery, silver ingots, or in silver pennies, vast numbers of which would have been needed. In the case of the First Crusade, for which the lead time was minimal, problems must have been experienced at mints in the West as Crusaders tried to convert assets to pennies. The mortgage of Bouillon alone by Godfrey of Bouillon would have required between 187,200 and 216,000 silver pennies if paid in coin and there is evidence that the Church paid for some sales of land to it with gold and silver plate. Transporting such large numbers of coins may have been a real problem, especially after Nicaea, when there is no evidence for carts and coin would have had to have been carried by men or animals.

When Stephen of Blois wrote to Adela from Antioch that "... I now have double the gold, silver, and many other riches which, when I left you, your beloved self had assigned to me", he was almost certainly telling the truth. The forces of the First Crusade acquired money and other wealth in many ways. Alexios Komnēnos had given the leaders money and again gave gifts of money to all after the fall of Nicaea. It was also acquired as booty and through plundering enemy corpses. Loss of lives in Asia Minor also meant that those who survived acquired the wealth of those who died. Later, the Muslim rulers of Jabala and Tripoli proved willing to buy off the Crusaders with tribute money.

Alan Murray has exposed the complex issues raised by the existence of the internal market in the forces of the First Crusade. The inflation that the chroniclers commented on would have occurred even without scarcity, simply through oversupply of money. Prices were not a product of supply and demand only. Scarcity of provisions combined with an abundance of coins both contributed to the inflation outside Antioch, but even though food was scarce, those who had enough money could still buy it. Provisions obtained by forage or pillage could be sold on the market. Moreover, the internal market was so pervasive that money came to form the basis for many transactions. Some services became paid for with money. Some lords used money to attract followers, the most famous example being Raymond of St Gilles, who offered 31,000 shillings to the other leaders if they would accept his leadership. Rank and file Crusaders might be attracted to the service of leaders by offers of money to avoid starvation. There is evidence that even services such as building siege engines and filling in moats were paid for at times and the establishment of a "common fund", at least by the time of Nicaea, appears to have been used to pay for such expenditures. Exchange of money within the forces was just as important as its inflow and outflow from them to others.

The First Crusade was unusual in that to even characterize its forces as "armies" in any understood sense of the word would be to misrepresent it. Nevertheless, the operation of similar factors can be discerned elsewhere. John France has shown that even when provisions were abundant there could be pillaging and rioting when those without money could not buy them. Tom Madden suggests that it was a combination of scarcity plus much wealth in the army that drove up prices outside Constantinople in 1204 and Reuven Amitai points to Muslim merchants in Syria driving up the price of supplies to the Mamlūk forces before the battle of Wādī 'l-Khaznadār.

Charles Glasheen has analyzed the various stages of Peter the Hermit's march to Constantinople during which his forces were supplied either by provisions taken with them at the beginning, or by those bought in markets, or by those foraged or pillaged, or by those provided by Byzantine authorities. The issue of foraging or pillaging is a complex one. Obviously it would have been inefficient because it would slow down the progress of any forces. It was also undesirable when in one's own territory, even if it did have to be resorted to at times. In friendly territory, but territory not one's own, it might be difficult to locate what was needed, certainly time-consuming to do so, and then dangerous to forage. Resistance would have to be expected. Even the Mongols might avoid pillage when in territory whose populations they did not wish to alienate. However, as John Haldon points out, foraging might be done if there was good local knowledge and scouting and it could extend ranges of march considerably, an important consideration, even if it would always have remained an inefficient option. After running out of the provisions supplied by the Venetians, the French Crusaders of the Fourth Crusade purchased food at Corfu, then pillaged on Andros and again at Abydos, but were running out of food again by the time that they reached Constantinople.

There can be no denying that all Crusader forces occasionally foraged or pillaged. Northern Crusaders had little difficulty combining a taste for plunder and pillage with the religious motives that induced them to assist the Portuguese in attacks on Muslim towns. In Byzantine territory, in some cases there would have been no alternative. For example, by the time that Bohemond's forces reached Kastoria, they would certainly have exhausted the supplies gathered in the valley of the Drino and when the inhabitants refused them markets, it would have been a question of either taking what they needed by force or starving. Given the circumstances, however, the reports of pillage are less to be wondered at than the paucity of them. Provisions to be sold by local populaces to transient forces in markets would in normal circumstance have to have been gathered from narrow corridors either side of a line of march. Bernard Bachrach suggests one to two days' journey either side but it may well rarely have been more than a half a day's journey if the evidence of local markets in the medieval West is anything to judge by. Forces of the size of the First Crusade demanded such huge amounts of provisions that they must have placed severe demands on the economies of such corridors to supply them and there is clear evidence that problems were created by the passage of such forces for those coming after them. For example, Raymond of St Gilles's forces attacked and sacked the town of Roussa when the inhabitants refused to provide provisions, no doubt because there was little surplus left after the passage of Bohemond of Taranto's forces only a couple of weeks earlier. The passage of the German forces along the *Via militaris* during the Second Crusade created problems of supply for the French coming behind them.

In one's own or friendly territory, purchase of provisions would always have been preferable to foraging or pillaging. In some cases it would have been the only alternative. For example in the desert where, as Yaacov Lev has shown, Bedouin provided markets for Fāṭimid and Zangīd armies. The evidence of the Crusades is instructive. Careful diplomatic and political preparations were necessary to create good supply and there are suggestions that these were made for the First Crusade and clear evidence of them for all subsequent Crusades. The money changers and merchants hovering around the Second Crusade armies reveal clearly the provision of markets. However, the authorities *en route* had to encourage sufficient producers to come forward and to provide markets where safety could be assured. Fear of lack of discipline, for example, might determine the willingness of rulers to allow markets. Witness the demand by King Coloman of Hungary for hostages from Godfrey of Bouillon in the person of Godfrey's brother Baldwin and his family before he would allow passage and provide markets or Alexios Komnēnos's provision of Patzinak escorts for forces coming via the *Via Egnatia*.

There is also evidence that private provision of supplies through markets organized by Byzantine authorities was insufficient to provide for the needs of large Crusader forces. According to Anna Komnēnē, Alexios ordered large stockpiles to be gathered along the route of the *Via Egnatia*. There is good evidence to suggest, in fact, that Byzantine authorities acted very well throughout the twelfth century in supplying Crusader forces. Manuel certainly collected and supplied provisions for Louis VII.

There is a collateral issue to be addressed. In 1092 Alexios Komnēnos had reformed the Byzantine currency, creating as its base the gold *hyperpyron* 20.5 carats fine. Lower denominations were an electrum *trachy* containing mostly silver and worth one-eighth of a *hyperpyron*, a billon trachy containing less than 10%

silver, and a billon *tetartēron* with only a trace of silver and a copper *tetartēron*, together with their halves. The standard was the gold *hyperpyron* and by and large the state was only interested in gold. Taxation was paid in gold *hyperpyra* or electrum *trachea*. There is a very significant issue, therefore, of what use large amounts of Western silver coins would have been to Byzantine producers or merchants providing provisions to transient Crusader forces and whether this would in fact have been a disincentive for them to sell provisions. At the very least it may have led to differences of opinion over the relative values of gold to silver.

One of the most important processes associated with the entire Crusading movement which has yet to be resolved satisfactorily is that of the extension of the range of Western naval forces into Levantine waters. For the First Crusade the only Crusader fleet to attempt to make the voyage to the East in one passage was the small Genoese flotilla of only 12 galleys and a transport galley, a *sandanum*, a Latinization of the Greek χελάνδιον for a transport galley. The fleets of Pisa and Venice did not attempt to make the voyage in one passage but rather wintered over in the Ionian islands and on Rhodes respectively. And even the Genoese took four months to make the voyage.

In fact voyages such as these, across the length of the Mediterranean far from one's own territory, by war fleets had not been seen before. Not since Belisarios's voyage from Constantinople to Vandal Africa had there been anything like them and in his case he had had Byzantine territory all the way to the Ionian islands and then the hospitality of the Ostrogothic regent Amalasuntha who provided in Sicily a large number of the horses which he landed in Africa. In all the age of naval warfare between the Muslim and Christian worlds ca 650-1100, fleets always coasted their own territory before making short passages to targets of attack. The Byzantines never attempted to reconquer Crete directly from Constantinople. They always brought their forces overland to one of the *aplēkta* on the south-west coast of Asia Minor before attempting a short crossing to Crete. In 960 Nikēphoros Phōkas marshalled his forces at Phygela, Kuş Adası, for the final, successful attack. In spite of Alexios Komnēnos's efforts to rebuild something of a Byzantine fleet, Byzantine naval forces remained weak well into the twelfth century. In 1125 they were no match for the Venetian fleet returning from the Crusade of 1122-4 and even as late as 1147 did not have the ability to oppose the Norman fleet which raided into the Aegean. Not until the 1160s did Byzantine naval forces become really formidable again.

What was undertaken by the Genoese, Pisans, and Venetians between 1096 and 1098 was unprecedented in the Mediterranean and it is no wonder that they took between one and a half and three years to prepare for their voyages. It is also important to remember, as John Dotson reminds us, that the naval forces of the maritime republics remained *ad hoc* assemblies, even into the thirteenth century. Fleets were composed of privately-owned ships and it would take time to gather them if they were scattered across the Mediterranean.

Even in 1122-3, the Venetians still wintered over in Corfu and then took over two months to make the Holy Land from Crete in the spring of 1123. In their case, however, it was probably because they were the first to attempt to transport horses from the West to *Outremer*. But already by then matters were evolving. As early as 1100 a larger Genoese fleet of 26 galleys and either four or six *naves* reached Latakia from Genoa between 1 August and the onset of winter. All subsequent Genoese Crusading expeditions appear to have made the voyage in a single passage in one season, as did the Venetian fleet of 1109. By the time of the Third Crusade and that of Frederick II, voyages to the Levant were being made regularly by large fleets, even carrying horses, in a matter of three to four weeks. Something had changed radically during the preceding century. What that was remains unexplored.

Between the Venetian Crusade of 1122-3 and the Third Crusade, there were no actual Crusader fleets which reached *Outremer* except for the Northern fleet for the Second Crusade. If it really was meant to provide transportation to *Outremer* for the entire French army for the Second Crusade, Roger II's offer to Louis VII of a fleet and logistical support would have been a major innovation and it must be questionable whether the Kingdom of Sicily in fact had the necessary naval resources as early as this, although it certainly did by 1174 when the fleet which attacked Alexandria was said to have included 150 galleys as well as 50 horse transport galleys. Throughout the twelfth century, however, it was the sailing *naves* discussed by John Dotson, rather than galleys, by which Western naval power was projected into the Levant and which sustained the economic and human lifelines which maintained the Crusader states. Fāṭimid, and later Ayyūbid, galley forces operating out of Egypt proved ineffectual against them. Dotson argues that only *naves* could engage *naves* and that was true, as long as the wind held up. Becalmed *naves* were a different matter.

If seasonal factors influenced campaigns by land, by sea they were infinitely more important. The Mediterranean remained closed to winter navigation through into the thirteenth century, the Genoese, according to James of Vitry, being the first to build ships large enough and seaworthy enough to cross the sea in winter, thus making possible more than one round trip per year to *Outremer*. Even Northerners knew about the seasonal nature of navigation. During the Second Crusade, William Veil knew that participating in the Portuguese siege of Lisbon could mean delaying the arrival of the fleet in *Outremer* by a whole year.

The impact of sea voyages on land-lubber Crusaders and horses explored by Ruthy Gertwagen must have been serious. It is a commonplace, of course, that the sea in which seamen delight, even if they also respect it for its dangers, can be terrifying to landsmen. Some complete nonsense has occasionally been written as a result; for example, that all Byzantines feared and avoided the sea on the basis of the stories of aristocrats and ecclesiastics. This for a civilization which was among the most sea-faring of all Mediterranean civilizations. Perhaps if we actually had any testimony from Byzantine seamen, such nonsense would not be written. That being said, Ambroise, John of Joinville, and pseudo Brocardus or William Adam, all have graphic descriptions of the debilitating effects of Crusader voyages on men and horses. Approaching particularly dangerous and stormy stretches of sea such as the Ionian, the south coast of Crete, and the Gulf of Antalya must have aroused fearful emotions in many a Crusader. The effect on horses of weeks on end confined in stalls and prevented from lying down by underbelly slings must have been severe. As for those carried in the open boats of Northern fleets in the twelfth century before the introduction of *cogs*, the logistics of making such long voyages with such technology defy the imagination. However, transporting horses in such a way would have had at least one advantage over transporting them on Mediterranean *naves* and, especially, *huissier* transport galleys. Horses need plenty of fresh air and well-ventilated stables. They suffer badly in cramped, stuffy conditions and can die of suffocation in them. Even at rest they use up 2.4 cubic metres of oxygen and produce 2.0 cubic metres of carbon dioxide every 24 hours, all of which would have had to have been replaced and disposed of respectively. This is an issue which has not specifically been addressed by any of the papers here but it is very probably reflected in the extremely slow voyages of the Fourth Crusade. Ruthy Gertwagen has emphasized that mooring, or better beaching, ships would have allowed removing the slings, cleaning stalls, pumping bilge water, and feeding and watering the animals. One should add that it would also have allowed the hatches to be opened in the case of sailing ships and, if beached, the stern quarter ports to be opened in the case of transport galleys so that the animals could be aired. Just as the Venetians made port whenever possible to water the crews, passengers, and horses in 1122-3, in 1202-3 they no doubt did so for the additional purpose of airing the now much larger number of horses. Their *taride* transport galleys must have had some form of forced ventilation into the holds by means of wind-sails sucking it in and cowls or something else allowing it out, but frequent complete airing would have been highly desirable in addition. A wind-sail was a cloth funnel which could be rotated to face the wind and used to deflect it through a hatch to below deck.

One of the variables which may possibly have been improved during the twelfth century and which may have contributed to speeding up voyages was watering, at least in the case of galleys. Sailing *naves* could carry sufficient water to last them the entire voyage, perhaps not early in the century but at least by its end when voyage times had speeded up greatly. But galleys would have had to take on water every few days, at least every five days or so, just to meet the needs of their crews. If they were horse transports, that figure should be halved. The question to be asked is whether the development of strong maritime commerce between the West and *Outremer* during the century may have generated improvements in port facilities *en route* which facilitated the watering process. How many ports had developed quays in the twelfth and thirteenth centuries and how many were merely protected roadsteads? Chandax was in ruins. What was the *porporella* in Garitsas Bay at Corfu? Methone, Korone, and Monemvasia were just roadsteads. How maintained was Rhodes? There is some evidence for the ancient harbour being dilapidated. Paphos was in ruins. Limassol was just an open roadstead but did have a wooden quay by the thirteenth century at least. Famagusta had an anchorage protected by natural reefs. Tyre did have a functioning developed port. Quays would have speeded up watering ships a great deal by comparison to having to use boats to take water out to anchored ships. Very few ports of call *en route* to *Outremer* had streams from which water casks could be filled quickly. Chandax and Phoinikous were two, although Chandax was not *en route* in the twelfth century, but almost everywhere else wells would have to be relied upon. If local populations increased the numbers of these for supply to commercial shipping, perhaps that was another factor which impacted on voyage times. Or did Crusading fleets learn to sail in divisions so that watering could be done successively rather than all together, thus also speeding the progress of fleets? Was it that which explains Richard of Devizes's apparently inexplicable comment on the divisions in which Richard's fleet sailed from Messina?

One would like to know the answers to other questions also. Richard Unger has shown that Northern ships leaving Dartmouth for the Third Crusade and carrying 40 horses and 40 men each would have needed total resupply in Portugal. The introduction of the *cog* in Northern Crusader fleets in the thirteenth century must have made a considerable difference since they had much deeper and capacious hulls and horses in particular must have been able to be transported much more easily. Their much greater cargo capacity must also have increased their range greatly. Nevertheless, the Kings of Portugal remained essential to the progress of Northern Crusaders into the Mediterranean because the Almoravids, and later the Almohads controlled both sides of the Gibraltar Channel. They could not prevent the ingress of large fleets but they could deny them landfalls. In the case of Mediterranean fleets, total resupply would also have been necessary for those that wintered *en route* and probably also for the Genoese flotilla for the First Crusade. It is doubtful whether it could have carried provisions for four months. Galley fleets perhaps needed resupply even later. But what sort of food surpluses would island populations have had to sell? Is it this which explains why the Pisans pillaged the Ionian islands in the winter of 1098-9 and why there are some reports of trouble between the Venetians and islanders on Corfu in 1122-3. Were there simply not sufficient surpluses on such islands to satisfy their needs?

None of us would argue for logistical determinism, or reductionism, as Reuven Amitai perhaps most strongly points out. But he also accepts that logistical study supports the source references pointing to the fact that the Mongols did have great difficulty maintaining their forces in Syria. Acknowledging the importance of other factors such as leadership, morale, ideology and attitudes to enemies, internal socio-political makeup, and international relations, as Yaacov Lev points out good leadership also meant paying attention to logistics. Leadership was not only a matter of brilliant tactics or the ability to inspire loyalty and raise morale. It could have much more mundane aspects than those. The study of logistics in the context of the Crusades and related military activities is essential to understanding their courses and outcomes.[2]

---

2 An international project based at the University of Birmingham has now been set up to study such issues, with several participating institutions: The Centre for Byzantine, Ottoman, and Modern Greek Studies and the Institute of Archaeology and Antiquity of the University of Birmingham, the Departments of History, Art & Archaeology, and nearEastern Studies of Princeton University, the Institut für Byzantinistik und Neogräzistik of the University of Vienna, the *Tabula imperii Byzantini* of the Austrian Academy of Sciences, the British School of Archaeology at Rome, the British Institute of Archaeology at Ankara, and the University of Athens. The project's website is: www.medievallogistics.bhm.ac.uk. An initial volume has appeared: J. F. Haldon, ed., *General issues in the study of medieval logistics: sources, problems and methodologies* (Leiden, Brill, 2005).

Glossary of Technical Terminology[1]

abnā' (*al-dawla*): term applied during the early 'Abbāsid Caliphate to Khurāsāni and other *mawālī* (q.v.) who entered its military service.

amīr: Originally a commander or governor of a province primarily concerned with maintaining order and collecting taxes. The title was later given to military officers of all ranks and to lesser princes under the Saljūqids, Ayyūbids, and Mamlūks.

amīr al-juyūsh: "commander of the armies", title assumed by Badr al-Jamālī, effective ruler of Egypt 1073-94.

aplēkton: literally a fortified camp but in Byzantine terms a staging area, a rendezvous for forces, a magazine where resources were stockpiled.

archōn abydikos/ἄρχων ἀβυδικός: title derived from the port of Abydos, originally designating the Byzantine official controlling maritime traffic through the Dardanelles. Later it became a generic for officials in control of maritime traffic in any port.

biscuit: *biscoctum* (med. Lat.), *ka'k* (Ar.): twice-baked bread used by military forces, and also travellers, in antiquity, the Middle Ages, and through to the twentieth century.

bromegrass, smooth (*bromus inermis*): a perennial grass native to Eurasia found in temperate to cool climates.

bucius, pl. *bucii* (Lat.) / *buce*, pl. *buci* (It.): term used especially at Genoa in the late twelfth and thirteenth centuries for certain types of transport galleys. It was also used both at Genoa and elsewhere for sailing ships. The relationship of the Mediterranean term to the Northern terms *byrthing* and *buss* is uncertain.

buss (Eng.), *bussa/buse/buce* (Old Fr.): twelfth-thireteenth-century Northern merchant ship descended from the Norse *knorr* (q.v.).

byrthing (corr. *byrðing*): lit. "burden carrier", Norse merchant ship.

chelandra, see sandanum

cog: merchant ship of a new type employing straight keel, stempost, and sternpost, a sternpost rudder, and a single square sail which appeared in the North, apparently in Germany, in the twelfth to thirteenth centuries. The relationship of the new type to older types known by the same name is much debated.

Compasso de navegare, Lo: portolan compiled in Sardinia ca. 1260.

esnecca/esneke/esneche: term used in England and France in the twelfth to thirteenth centuries for long warships of a type descended from Norse *snekkjur* (q.v.).

---

[1] Compiled by John Pryor.

gorabius, pl. gorabii (Lat.) and variants: term for a galley of some kind which appeared in Latin sources towards the end of the eleventh century and was common for a while in the early twelfth, probably a Latinization of the Greek κάραβος via the Arabic *ghurāb*.

ḥalqa: term used in Ayyūbid and Mamlūk times for a socio-military unit attached to a commander, in Ayyūbid times often including mamlūks but in Mamlūk times composed of non-mamlūks.

iqṭā': lit. "allotment". Revenue from land granted to military commanders by Muslim rulers to support them and their retinues.

knorr, pl. *knerrir*: Norse merchant ship, cargo carrier, primarily a sailing ship.

mamlūks: military slaves, frequently of Turkish origin, raised as elite troops of Muslim commanders.

mawālī, sing. *mawlā:* lit. "client". In early centuries non-Arab converts to Islam were adopted into Arab tribes, becoming clients or *mawālī* of Arabs who became their protectors. The system was eventually abolished by the Umayyad Caliph 'Umar ibn 'Abd al-'Azīz (717-20 C.E.).

navis, pl. *naves* (Lat.) / *nave*, pl. *navi* (It.): generic term for a ship in general but used in the twelfth to fourteenth century with specific reference to large, lateen-rigged Mediterranean sailing ships.

Outremer, Ultra mare (med. Lat.): "over the sea". Term used in the medieval West for the Crusader states across the Mediterranean, often synonymous for the Holy Land.

Optimatoi,' Οπτιμάτοι: Byzantine theme of N.W. Asia Minor opposite Constantinople, including both shores of the Gulf of Nikomedeia and running E. to the Sangarios.

panzone, pl. *panzoni* (It.), *panzonus* (Lat.): term used especially at Genoa for a type of large cargo ship, exact import unknown.

portolans: books of sailing directions giving directions and distances from one port to another.

qāḍi, pl. *quḍāh*: judge appointed by a Muslim ruler in recognition of his knowledge of Muslim law.

Romania: term used in the West from the twelfth century for territories belonging to, or once belonging to, the Byzantine Empire.

Sachkritik: methodological technique used by historians, especially ancient historians, to critique sources by showing in detail that what they say happened could not have been true, at least in the way they say it happened. Objective criteria set limits on the activity of armies and these must be respected as the basic building blocks for historical reconstruction on the one hand and source criticism on the other.

sagita, pl. *sagitae* (Lat.)/*saetta*, pl. *saette* (It.): type of galley, presumably unusually fast, the term, derived from Latin *sagitta*, "arrow", appeared in a variety of forms in medieval documents, referring to a hull form that could be found in a range of sizes, including *sagittina* of only 24 oars.

salandrium, see sandanum

sandanum/chelandra/salandrium (Lat.): Latinizations of the Byzantine Greek χελάνδιον, *chelandion*, used for both war and transport galleys, but became a term for transport galleys in the West by the twelfth century, sometimes horse transports.

snekkjur: Norse longship, "dragon ship", warship primarily to be rowed.

sultan: "power, authority", hence "holder of power, authority". Title assumed by the Great Saljūqids when they seized real power in Baghdad from the 'Abbāsid Caliphs.

tarida, pl. *taride* (Lat.), from Arabic *ṭarrīda*: term used especially in the thirteenth century for transport galleys, particularly horse transports.

theme, *thema* (θέμα), pl. *themata* (Byz. Gr.): military division and territorial unit in the Byzantine empire administered by a *stratēgos*, military commander, who combined both military and civil authority.

transhumance: movement of flocks and herds between summer and winter pastures

tümen: standard formation of Mongol armies, theoretically a division of 10,000 men

Turcopoles: light cavalry among the Franks of *Outremer*, perhaps the offspring of mixed marriages.

wazīr: in earlier times the highest-ranking civil functionary in Muslim states, the director of civil administration, transformed in the later Fāṭimid Caliphate into a military institution exercising supreme political and military authority.

Zanj: name given by Arabs to black tribes inhabiting the coasts of East Africa who were sources of slaves.

Selective Gazetteer[1]

al-Abbāsiyya	al-Abbāsiyya (Ar.): town built by Ibrāhīm ibn al-Aghlab three miles S.E. of al-Qayrawān (q.v.), Tunisia.
Abydos	*Abydus* (cl. Lat.), Ἄβυδος (Gr.), *Abydus* and variants (med. Lat.): town, now abandoned, on the Dardanelles, near Çanakkale, Turkey.
Antioch of Pisidia	Ἀντιόχεια τῆς Πισιδίας (Gr.), *Antiochia/Antiochia Minor/Antiochia Pisidiae* and variants (med. Lat.): town, now abandoned in favour of Yalvaç, Turkey.
Aquileia	*Aquileia* (cl. Lat.), Ἀκυληία (Byz. Gr.), *Aquilegia* and variants (med. Lat.): town and port, **Aquileia**, Italy.
Aradus	*Aradus* (cl. Lat.), ἡ Ἄραδος (Gr.), *Aradus/Aradiensis* (med. Lat.), Arwād (Ar.): island and port off Tortosa, **Arwād**, Syria.
'Arqā	*Arca/Irkata* (cl. Lat.), Ἄρκη/Ἄρκαι (Gr.), *Archas/Archados* and variants (med. Lat.): town and fortress, **'Arqā** or **'Irqā**, Lebanon.
Arsuf	*Apollonia/Sozusa* (cl. Lat.), *Arsur* (med. Lat.), Arsūf (Ar.): town and port, **Tel Arshaf (Apollonia)**, Israel.
Avarinos	Ἀβάρινος (Byz. Gr.), Πύλος (cl. Gr.), *Pylus* (cl. Lat.), *Abarinum/Ionchium/Pylus* and variants (med. Lat.): bay, **Pilos/Navarino**, Greece. [Classical Πύλος lay at what is now Palaiokastro on the northern promontory of the bay. Byzantine Ἀβάρινος and its successors lay at the south of the bay.]
Avlona	Αὐλών (Byz. Gr.), *Aulona/Valona* and variants (med. Lat.), Valona (med. It.): town and port, superseding classical

---

1 Compiled by John Pryor with the assistance of the contributors. Special thanks to John Haldon, Benjamin Kedar, and Yaacov Lev.

The gazetteer is intended as a guide only to those places which may not be known commonly and which cannot be found in the widely-available English language atlas: J. C. Bartholomew, *The world atlas*, 10th ed. (Edinburgh, 1975). By and large the style of the Wisconsin *History of the Crusades* [Setton, *HC*] has been followed with the exception that for Turkish and Arabic names a modified version of that of *The Encyclopedia of Islam* [EI.2] has been followed and for Byzantine names that of *The Oxford Dictionary of Byzantium* [ODB]. Where places had different names in different languages, the practice has been to use the name as it would have been used by the occupying power at the time.

Where places have a modern equivalent, that equivalent is given in **bold.**

	Apollonia, **Vlorë**, Albania.
'Ayn Jālūt	'Ayn Jālūt (Ar.): "Spring of Goliath", spring, **'Eyn Ḥarōd**, Jezreel valley, Israel.
al-A'waj	al-A'waj (Ar.): river flowing W.-E., south of Damascus, Syria.
Babylon	*Babylon* (cl. Lat.), Βαβυλών (Gr.), *Babylon/Babilonia* and variants (med. Lat.); fortress in the Nile delta, south of Fusṭāṭ (q.v.), Egypt.
Baradā	*Abana* (cl. Lat.), *Farfar* (med. Lat.), Baradā (Ar.): river of Damascus, **Barada**, Syria.
al-Bijāya	*Saldae* (cl. Lat.), Σάλδαι (Gr.), al-Bijaya (Ar.), Bougie (Fr.): port and town, **Béjaïa**, Algeria.
Bilbays	Phelbès (Copt.), Bilbays (Ar.): town, **Bilbays**, Egypt.
al-Bīra	*Birtha* (cl. Lat.), Βίρθα (Gr.), Bir (Arm.), Bireçik (Turk.), al-Bīra (Ar.): town, **Bireçik**, Turkey.
al-Buqay'a	*La Bochea/La Boquea* (med. Lat.), al-Buqay'a (Ar.): river valley, Lebanon.
Chalkēdōn	*Chalcedon* (cl. Lat.), Χαλκηδών (Gr.), *Calcedonia* and variants (med. Lat.): town, **Kadıköy**, Turkey.
Chandax	*Al-Khandaq* (Ar.), Χάνδαξ (Byz. Gr.), *Candia/Chandax/Megalocastrum/Heracleum* and variants (med. Lat.): port, **Iraklion**, Crete.
Chastellet	*Castelletum/Vadum Jacob* (med. Lat.), Bayt al-Aḥzān (Ar.): castle now destroyed, **Metsad 'Ateret**, Israel.
Chrysopolis	*Chrysopolis* (cl. Lat.), Χρυσόπολις (Byz. Gr.), *Scutarium* (med. Lat.); town, **Üsküdar**, Turkey.
Cilician Gates	*Pylae Ciliciae* (cl. Lat.), Πύλαι Κιλικίας (Byz. Gr.), Gülek Boghaz (Turk.): pass through the Tauros mountains north of Namrun, **Gülek Boğaz**, Turkey, through which the *Via Tauris* (q.v.) ran.
Cydnus	*Cydnus/Cidnus* (cl. Lat.), Κύδνος (Gr.); river of Tarsos, **Tarsus**, Turkey.
Cyrenaica	*Cyrenaica* (cl. Lat.), Κυρήνη (Gr.): prov., **Barqa**, Libya.
Danzig	*Danesse* (med. Lat.): town and port, **Gdansk**, Poland.
Dorostolon	Δορόστολον (Byz. Gr.), *Durostorum/Silistra/Distra* and variants (med. Lat.): city and fortress on the Danube, **Silistra**, Bulgaria.
Dorylaion	*Dorylaeum* (cl. Lat.), Δορύλαιον (Byz. Gr.): fortress town near Eskişehir, Turkey.
Döşeme pass	Pass from Pamphylia to Lycia, through which the *Via*

	Sebaste (q.v.) ran.
Drakōn	*Draco* (cl. Lat.), Δράκων (Byz. Gr.): river, **Yalak**, Turkey.
Drizipera	Δριζίπερα (Byz. Gr.), *Drizipera/Drizipara* and variants (med. Lat.); town near modern Büyük Karıştıran between Lüleburgaz and Corlu, Turkey.
Drynopolis	*Andrinopolis* (med. Lat.), Byzantine name unknown: town on the river Drino near Gjirokaster, Albania.
Dyrrachion	*Epidamnus/Dyrrachium* (cl. Lat.), Δυρράχιον (Byz. Gr.), Durazzo (It.): port and town, **Durrës**, Albania.
Edessa	Ἔδεσσα (Byz. Gr.), *Rohas* and variants (med. Lat.), Ouṛha (Arm.), al-Ruhā' (Ar.): town, **Urfa**, Turkey.
Engilsnes	Probably *Mallea promontorium* (cl. Lat.), Maleo Sancto Angelo and variants (med. It.), Μαλέα/Μαλέαι (Gr.): **Cape Malea**, Greece.
Falkonera	Falconera/Farconare and variants (med. It.): islet, **Falkonera**, between Karavi and Parapola, east of the Peloponnesos, Greece.
Fāqūs	Classical and Byzantine name unknown: town, **Fāqūs**, Egypt.
al-Faramā'	*Pelusium* and variants (Lat.), Πηλούσιον (Gr.), al-Faramā' (Ar.): town, **al-Faramā'**, Egypt.
Fermia	*Fermia/Ferminia/Thermia/Cythnus* and variants (cl. & med. Lat.), Ferminia and variants (med. It.), Κύθνος (Gr.): island, **Kythnos**, Greece.
Fusṭāṭ	Al-Fusṭāṭ (Ar.), *Babylon/Babilonia* and variants (med. Lat.): Arab town founded north of Babylon (q.v.), later Ḳaṣr al-Sham'a, now Miṣr, Old Cairo, Egypt.
Galata	*Sycae* (cl. Lat.), Γαλατᾶς/τὰ Γαλάτου (Byz. Gr.), *Calata/Galatas* (med. Lat.): suburb of Constantinople N. of the Golden Horn, **Galata**, Istanbul, Turkey.
Gallos	*Gallus* (cl. Lat.), Γάλλος (Gr.): river, tributary of the Sangarios (q.v.), **Lefke**, Turkey.
Gozón	*Gozzem/Gozzim, Portus Sancti Salvatoris* (med. Lat.): town, **Luanco**, Spain.
Ḥarrān	*Carrhae* (cl. Lat.), *Carrhae* and variants (med. Lat.), Κάρραι (Gr.), Ḥarrān (Ar.): town now abandoned, near Altinbasak, Turkey.
Ḥaṭṭīn, Horns of	*Madon* (Lat.), Ḥaṭṭīn (Ar.): hilltop and battle site, **Qarney Ḥiṭṭīn**, Israel.
Ḥaydarān	Site of a battle in 1052 C.E., probably between 30-60

kilometres S.-S.W. of al-Qayrawān (q.v.), Tunisia.

Jacob's Ford — *Vadum Iacob* (med. Lat.), Jisr Banāt Ya'qūb (Ar.): ford across the Jordan N. of the Sea of Galilee, **Gesher Bnot Ya'qov**, Israel.

Jurjīr — Town of medieval Egypt on a peninsula jutting into Lake Manzala, apparently now abandoned.

al-Karak — Characmoba, Cracum and variants (med. Lat.), Χάρακα and variants (Gr.), al-Karak (Ar.): fortress and prov., **Karak**, SE of Dead Sea, Jordan.

Karavi — Mauro Caravi/Caravio and variants (med. It.); three small, barren islets off the east coast of the Peloponnesos, Greece.

Kassiopi Bay — Caxopo/Sancta Maria di Casopoli and variants (med. It.): small bay south of Cape Kassiopi, Corfu.

Koukousos — *Cucusus* (cl. Lat.), ἡ Κουκουσσός (Byz. Gr.): fortress, **Göksun**, W. of Elbistan, Turkey.

Krak de Montréal — *Mons Regalis* (med. Lat.), al-Shaubak (Ar.): castle, **Shaubek**, SSE of the Dead Sea, Jordan.

Kuch — *Mansio Apsos XIV* (cl. Lat.): town at the crossing of the southern branch of the *Via Egnatia* (q.v.) over the Seman river, **Kuch**, Albania.

Lepanto — *Naupactus* (cl. Lat.), Ναύπακτος (Gr.), *Naupactus/Lepan-tum* (med. Lat.), Lepanto (It.): harbour and town, **Navpaktos**, Greece.

Levkas — Λευκάς (Byz. Gr.); place at which there was a bridge over the river Gallos (q.v.), Turkey.

Lido — *Malamoccum/Methamaucum, Litus* (med. Lat.): island separating the Venetian lagoon from the Adriatic.

Litani — *Leontes* (cl. Lat.), Λέοντος (Gr.), al-Līṭānī (Ar.): river, **Litani**, Lebanon.

Lopadion — *Lopadium* (cl. Lat.), Λοπάδιον (Byz. Gr.): town, **Ulubad**, Turkey.

Lotharingia — *Lotharingia* (med. Lat.): med. duchy of the Empire, now in France, Belgium, and Germany.

Maeander — *Maeander* (Lat.), Μαίανδρος (Gr.): river, **Menderes**, Turkey.

al-Mahdiyya — al-Mahdiyya (Ar.): town and port founded by the Fāṭimid 'Ubayd Allāh al-Mahdī in 912-13 C.E., **Mahdia**, Tunisia.

Mainē — Μαΐνη (Byz. Gr.), *Maina* (med. Lat.), Le Grand Magne (med. Fr.): castle and town, unidentified, Mani region,

	Peloponnesos, Greece.
Malagina	Μαλάγινα (Byz. Gr.): district and imperial *aplēkton* on the left bank of the river Sangarios (q.v.) a few kilometres upstream from the junction with the river Gallos (q.v.), Turkey.
Margat	*Margat/Margatum* (med. Lat.), al-Marqab (Ar.): castle, County of Tripoli, **Qal'at al-Marqab**, Syria.
Marienburg	*Marienburg/Marienborg* (med. Lat.): town and fortress, Marienburg, island near Welun in the river Memel, **Malbork**, Poland.
Marj al-Ṣuffār	Marj al-Ṣuffār (Ar.): plain south of the al-'Awaj river (q.v.), **Marj al-Şuffar**, Syria.
Melgueil	*Melgorium* and variants (med. Lat.): Mauguio, med. county, déps Hérault and Gard, France.
Mont Gisard	Gezer (Heb.), *Gezer/Gazara/Gadara* (cl. Lat.), *Mons Gisardi* (med. Lat.), Γαζέρ (Gr.), Tall al-Jazar (Ar.): hill, **Tel Gezer**, eastern Shephelah, Israel.
Mouzay	*Muzacum/Mosagum* (med. Lat.): town, **Mouzay**, France.
Myriokephalon	Μυριοκέφαλον (Byz. Gr.): pass in Phrygia, **Çardak Boğazî**, Turkey.
Mysia	*Mysia* (cl. Lat.), Μυσία (Byz. Gr.): region of NW Asia Minor between Bithynia and Phrygia, prov. Balıkesir, Turkey.
Nakhl, Qala'at	Fortress in southern Sinai on the route taken by Ṣalāḥ al-Dīn in 1169.
Nicaea	*Nicaea* (cl. Lat.), Νίκαια (Byz. Gr.): city, **Iznık̦**, Turkey.
Nikomedeia	Νικομήδεια (Byz. Gr.), *Nicomedia* (cl. Lat.): city, **Izmit**, Turkey.
Palaiokastron	*Portus Fraschee* (med. Lat.), Policastro/Paleocastro and variants (med. It.): port, **Eremopolis**, Crete.
Patara	Πάταρα (Gr.), *Patara* (cl. Lat.), *Patara/Patera* and variants (med. Lat): town and port, Lycia, now abandoned, near Kalkan, Turkey.
Pelagonia	Πελαγονία (Byz. Gr.), *Herakleia Lynci* (cl. Lat.): town and region on the *Via Egnatia* (q.v.), south of Bitola, Greece.
Pelekanos	Πελεκάνος (Byz. Gr.), *Pelecanum/Pantichium* (med. Lat.): town, **Pendik**, Turkey.
Pella	Πέλλα (Gr.), *Pella* (cl. Lat.): capital of ancient Macedonia. After the decline of the city, the name became given to its region, between Yiannitsa and Yefira, Greece.
Philea	Φιλέα (Gr.), *Affileta/Philea/Phileas* (med. Lat.), La Philée

	(med. Fr.): town, **Karacaköy**, Turkey.
Philopation	Φιλοπάτιον (Byz. Gr.): *Philopation*, a hunting preserve outside the land walls of Constantinople in the N.W. quarter from the gate of St Romanus to the gate of Charisius/Adrianople.
Phoinikous	Φοινικοῦς (Gr.), *Phoenicus* (cl. Lat.), *Finica/Phoenix* (med. Lat.): port, now abandoned, near Finike, Turkey.
Portovenere	*Portus Veneris* (cl. Lat.), *Portusveneris* (med. Lat.): port and town, **Portovenere**, Italy.
Portus Guiscardus	*Portus Guiscardus/Portus Wiscardi* and variants (med. Lat.), Porto Biscardo/Viscardo/Fiscardo (med. It.): bay and town of **Fiskhardo**, Keffalinia, Greece.
Princes' islands	Πριγκίπιοι νῆσοι (Byz. Gr.), *Principeae insulae* (med. Lat.): group of nine islands in the Sea of Marmara, **Kızıl adalar**, Turkey.
al-Qādisiyya	Site of a Muslim victory over the Sāsānids west of al-Ḥīrah (now abandoned) in Iraq in 637 C.E.
al-Qayrawān	al-Qayrawān (Ar.): Ag͟hlabid capital of Tunisia founded by 'Uḳba ibn Nāfi' in 670: town, **Kairouan**, Tunisia.
Quallie	Qualle/Qualie/Laquaie/Gualles and variants (med. It.): anchorage, perhaps **Porto Asomato**, Peloponnesos, Greece.
Quwayq	Al-Quwayq (Ar.): river of Aleppo, **Quwayq**, Syria.
Rhēgion	Ῥήγιον (Byz. Gr.), *Regio/Rhegium* (med. Lat.): town, **Küçük Çekmece**, Turkey.
Rialto	*Rivus Altus* (med. Lat.): "High Bank", central island of Venice, synonym for Venice.
Riviera di Levante	Coastal reach east of Genoa to the mouth of the Arno River.
Romania	*Romania* (med. Lat.): for all territories in and around the Aegean Sea which were or had been part of the Byzantine Empire.
Roussa	*Rossa* and variants (med. Lat.): town, **Keşan**, Turkey.
Sabā'	*Saba/Sabaea* (cl. Lat.), Σάβη/Σαβαί (Gr.), Sabā' (Ar.): Sheba, pre-Muslim polity in the Yemen.
Ṣadr, Qala'at	Fortress in southern Sinai on the route taken by Ṣalāḥ al-Dīn in 1169.
Sağlıklı	Town, **Sağlıklı**, Turkey.
Sangarios	*Sangarius* (cl. Lat.), Σαγγάριος (Gr.): river, **Sakarya**, Turkey.
S͟haqḥab	S͟haqḥab (Ar.): town on the Marj al-Ṣuffār (q.v.), **S͟haqḥab**

Syria.

Shayzar	*Larissa* (cl. Lat.), *Cesara* (med. Lat.), Σέζερ (Byz. Gr.), Shayzar (Ar.): town and fortress, **Saijar**, Syria.
Siberis	*Siberis* (cl. Lat.), Σίβερις (Byz. Gr.): river, **Kirmir**, Turkey.
St Symeon	*Sancti Symeonis Portus* and variants (med. Lat.), al-Suwaydīya (Ar.): town and port, **Samandağ**, Turkey.
Stenay	*Sathanagium* and variants (med. Lat.): canton, **Stenay**, France.
Tinnīs	*Tenessus* (cl. Lat.), Θιννησός (Gr.), Tinnīs (Ar.): town on an island in Lake Manzala, **al-Tina**, Egypt.
Trajan's gate	Pass on the *Via Militaris* (q.v.) between Ikhtiman and Pazardčik, Bulgaria.
Turbessel	*Turbessel* and variants (med. Lat.), Tall Bā<u>sh</u>ir (Ar.), Ṭlbashar (Arm.): town, **Tilbeşar**, Turkey.
Üxküll	*Ykescola* (med. Lat.), Ikškile (Lettish): village on the Dvina river, Latvia.
'Uyūn Mūsa	Springs of Moses: oasis on the Gulf of Suez, Egypt.
Via Egnatia	Roman road from Dyrrachion (q.v.) and Avlona (q.v.) to Constantinople via Ohrid, Florina, and Thessalonikē.
Via Militaris	Roman road from Aquileia via Belgrade on the Danube to *Viminacium*/Οὐιμινάκιον (near Smederevo) and thence to Constantinople via Niš Sofia, Plovdiv, Edirne, and Tekirdağ.
Via Sebaste	Roman road from the *Via Tauris* (q.v.) south of the Cilician Gates (q.v.) to Sebaste [near Silifke, Turkey], continuing through Pamphylia into Lycia.
Via Tauris	Roman road from Tyana (Kemerhisar, Turkey) through the Cilician Gates (q.v.) in the Tauros Mts to Tarsos.
Wādī 'l-<u>Kh</u>aznadār	Site of a battle on the plain N. of Ḥomṣ, Syria, unidentified.
Zompos/Zompē	Ζόμπος/Ζόμπη (Byz. Gr.): site of a bridge over the Sangarios river (q.v.).

Consolidated Bibliography

(a) Manuscripts

'Abd al-Rahīm, continuation of Ibn Wāṣil, Abū 'Abd Allāh Jamāl al-Dīn Muḥammad, *Mufarrij al-kurūb fī akhbār Banī Ayyūb*, Paris, Bibliothèque Nationale, MS. Ar. 1702.

Cristoforo Buondelmonti, *Isolario*, Venice, Biblioteca Nazionale Marciana, MS. Ital. Class. VI, Cod. XIX (= 6087).

Ibn al-Furāt, Nāsir al-Dīn Muḥammad ibn 'Abd al-Raḥīm, *Ta'rīkh al-duwal wa 'l-mulūk*, Vienna, Staatsbibliothek, MS. 814.

al-Jazarī, Shams al-Dīn Muḥammad, *Ḥawādith al-ẓamān*, Paris, Bibliothèque Nationale, MS. Ar. 6739.

Venice, Archivio di Stato, MS. *Maggior Consiglio, Clericus civicus.*

Venice, Archivio di Stato, MSS *Senato mar*, Regs 2, 4, 12, 14.

Venice, Archivio di Stato, MSS *Senato misti*, Regs 40, 43.

Venice, Archivio di Stato, MSS *Senato misti secreta*, Reg. 40.

Venice, Biblioteca Nazionale Marciana, MS. Ital. Class. XI, No. 87 (colloc. 7353), fols 1r-7r.

(b) Published Primary Sources

Abrahams, P., ed., *Les oeuvres poétiques de Baudri de Bourgueil (1046-1130)* (Paris, 1926).

Abū Shāma, Shihāb al-Dīn Abu 'l-Qāsim 'Abd al-Raḥmān, *Kitāb al-rawḍatayn fī Akhbār al-Dawlatayn*, ed. I. al-Zaybaq, 2 vols (Beirut, 1997). Extracts trans. A. C. Barbier de Meynard as *Le livre des deux jardins*, in RHC *HOr*, vols 4-5.

Alberic of Trois-Fontaines, *Chronica Albrici monachi Trium Fontium, a monacho novi monasterii Hoiensis interpolata*, ed. P. Scheffer-Boichorst, in MGH *SS*, vol. 23, 631-950. Pp. 870 (*Annus* 1193) – 886 (*Annus* 1205) part. trans. Andrea, *Contemporary sources*, 291-309.

Albert of Aachen, *Historia Hierosolymitana*, in RHC *HOcc*, vol. 4, 265-713.

Amadi, Francesco, *Chronique*, ed. R. de Mas Latrie, in *Chroniques d'Amadi et de Strambaldi*, 2 vols (Paris, 1891-3), vol. 1.

Ambroise, *L'estoire de la Guerre Sainte: histoire en vers de la Troisième Croisade (1190-1192)*, ed. G. Paris (Paris, 1897). Trans. M. J. Hubert, ed. J. L. La Monte, *The Crusade of Richard Lion-Heart by Ambroise* (New York, 1941).

Ammianus Marcellinus, *Historia*, in J. C. Rolfe, trans., *Ammianus Marcellinus*, 3 vols (London, 1935-40).

Andrea, A. J., *Contemporary sources for the Fourth Crusade* (Leiden, 2000).

Andrea Dandolo, *Chronica per extensum descripta, aa. 46-1280 d. C.*, ed. E. Pastorello, in RISS *NS*, tome 12, parte 1.

Anna Komnēnē, *Alexiade*, ed. and Fr. trans. B. Leib, 3 vols (Paris, 1937-45). Trans. E. R. A. Sewter, *The Alexiad of Anna Comnena* (Harmondsworth, 1969).

Annales Cavenses, ed. G. H. Pertz, in MGH *SS*, vol. 3, 185-97.

Annales Fuldenses, ed. F. Kurze, in MGH *ScriptRerGerm.Cont* (Hanover, 1891). Trans. T. Reuter, *The Annals of Fulda: ninth-century histories, volume II* (Manchester and New York, 1992).

Annales Ianuenses, ed. G. H. Pertz, in MGH *SS*, vol. 18, 1-356.

Annales Ianuenses, eds L. T. Belgrano and C. Imperiale, *Annali genovesi di Caffaro e de' suoi continuatori dal MXCIX al MCCXCIII*, 5 vols [FStI] (Rome and Genoa, 1890-1929).

Anonymous, *Ἀνωνύμου Βιβλίον τακτικόν* in G. T. Dennis, ed. and trans., *Three Byzantine military treatises*, 241-335.

Anonymous, *L'Estoire de Eracles empereur et la conqueste de la terre d'Outremer*, in RHC *HOcc*, vols 1.1, 1.2, 2. Different manuscript tradition ed. P. Paris, *Guillaume de Tyr et ses continuateurs: texte français du XIIIe siècle, revu et annoté*, 2 vols (Paris, 1879-80).

Anonymous, *Li Estoire de Jérusalem et d'Antioche*, in RHC *HOcc*, vol. 5, 621-48.

Anonymous, *Tudebodus imitatus et continuatus Historia peregrinorum*, in RHC *HOcc*, vol. 3, 165-229.

Anonymous monk of the Lido, *Monachi anonymi Littorensis historia de translatione sanctorum Magni Nicolai, terra marique miraculis gloriosi, ejusdem avunculi, alterius Nicolai, Theodorique, martyris pretiosi, de civitate Mirea in monasterium S. Nicolai de Littore Venetiarum. 6 Dec. 1100*, in RHC *HOcc*, vol. 5, 253-81.

Anonymous of Soissons, *De terra Iherosolimitana*, eds and trans A. J. Andrea and P. I. Rachlin, "Holy war, holy relics, holy theft: the Anonymous of Soissons's *De terra Iherosolimitana*: an analysis, edition, and translation", *Historical reflections*, 18 (1992), 147-75. Also trans. Andrea, *Contemporary sources*, 223-38.

al-Anṭākī, Yaḥyā ibn Sa'īd ibn Yaḥyā, *Kitāb al-ta'rīkh*, ed. L. Cheikho et al., in CSCO *ScriptAr*, 3rd ser., vol. 7 (Paris, 1909). Part. Fr. trans. I. Kratchkovsky and A. Vasiliev, *Histoire de Yaḥyā ibn Sa'īd d'Antioche continuateur de Sa'īd-ibn-Biṭrīq*, 2 vols, in PO, 18.5, 23.3.

al-Balādhurī, Abu 'l-'Abbās Aḥmad ibn Yaḥyā ibn Jābir, *Futūḥ al-Buldān*, ed. M. J. de Goeje, *Liber expugnationis regionum, auctore Imámo Ahmed ibn Jahja ibn Djabir al-Beládsorí* (1866; rpt, Leiden, 1968). Trans. P. K. Hitti and F. C. Murgotten, *The origins of the Islamic state*, 2 vols (New York, 1916, 1924).

al-Balawī, Abū Muḥammad 'Abd Allāh, *Sīrat Aḥmad ibn Ṭūlūn*, ed. M. K. 'Alī (Damascus, 1939).

Basilicorum libri LX, ser. A, ed. H. J. Scheltema et al., 8 vols (Groningen, 1953-88).

Baudri of Bourgueil, *Baldricus Burgulianus, Carmina*, ed. K. Hilbert [Editiones Heidelbergenses, 19] (Heidelberg, 1979).

Baudri of Bourgueil, *Baldrici, episcopi Dolensis, Historia Jerosolimitana*, in RHC *HOcc*, vol. 4, 1-111.

Baybars al-Manṣūrī al-Dawādār, Rukn al-Dīn, *Kitāb al-tuḥfa al-mulūkiyya fī 'l-dawla al-mamlūkiyya*, ed. 'Abd al-Ḥamīd S. Ḥamdān (Cairo, 1987).

Baybars al-Manṣūrī al-Dawādār, Rukn al-Dīn, *Zubdat al-fikra fī ta'rīkh al-hijra: History of the early Mamluk period*, ed. D. S. Richards (Beirut, 1998).

Belgrano, L. T., ed., *Documenti inediti riguardanti le due Crociate di San Ludovico IX* (Geneva, 1859).

Bernard of Breydenbach, *Peregrinatio in Terram Sanctam* (Speyer, 1502).

Bernard of Clairvaux, St, *Epistolae*, in PL, vol. 182, coll. 67-716. Trans. B. S. James, *Letters of Bernard of Clairvaux*, 2nd ed. (Stroud, 1998).

Bernardo Maragone, *Annales Pisani*, ed. M. L. Gentile, *Gli Annales Pisani di Bernardo Maragone*, in RISS *NS*, tome 6, parte 2, 1-74.

Beševliev, V., *Spätgriechische und spätlateinische Inschriften aus Bulgarien* [BBA, 30] (Berlin, 1964).

Blockley, R. C., *The fragmentary classicising historians of the Later Roman Empire: Eunapius, Olympiodorus, Priscus and Malchus. Vol. 2: text, translation and historiographical notes* (Liverpool, 1983).

Breve chronicon de rebus Siculis a Roberti Guiscardi temporibus inde ad annum 1250, ..., in J.-L.-A. Huillard Bréholles, *Historia diplomatica Friderici Secundi*, 6 vols (Paris, 1852-61), vol. 1, part 2, 887-908.

Caffaro de Caschifelone, *De liberatione civitatum Orientis liber*, ed. L. T. Belgrano, in *Annales Ianuenses* (FStI), vol. 1, 95-124.

Chronica monasterii Casinensis, ed. H. Hoffmann, in MGH *SS*, vol. 34.

Chronicle of the Morea, The/Τὸ Χρονικὸν τοῦ Μορέως, ed. J. Schmitt (1904; rpt, New York, 1979).

Chronique de Saint-Hubert dite Cantatorium, La, ed. K. Hanquet (Brussels, 1906). Also ed. L. C. Bethman and W. Wattenbach, as *Chronicon sancti Huberti Andaginensis usque ad a. 1106*, in MGH *SS*, vol. 8, 565-630.

Cobham, C. D., trans., *Excerpta Cypria: materials for a history of Cyprus* (Cambridge, 1908).

Codice diplomatico della repubblica di Genova, ed. C. Imperiale di Sant' Angelo, 3 vols [FStI, nos. 77, 79, 89] (Rome, 1936-42).

Columella, Lucius Junius Moderatus, *Res Rustica*, in H. B. Ash, ed. and trans., *Lucius Junius Moderatus Columella on agriculture*, 3 vols (London, 1941-55).

Constantine VII Porphyrogennētos, *De administrando imperio*, vol. 1, ed. G. Moravcsik, trans. R. J. H. Jenkins (Budapest, 1949); R. J. H. Jenkins, *Volume II: commentary* (London, 1962).

Constantine VII Porphyrogennētos, *Constantine Porphyrogenitus: three treatises on imperial military expeditions*, ed. and trans. J. F. Haldon [CFHB, vol. 28] (Vienna, 1990).

Continuation de Guillaume de Tyr de 1229 à 1261 dite du manuscrit de Rothelin, in RHC *HOcc*, vol. 2, 483-639. Trans. J. Shirley, *Crusader Syria in the thirteenth century: the* Rothelin *continuation of the* History *of William of Tyre with part of the* Eracles *or* Acre *text* (Aldershot, 1999), 13-120.

Cristoforo Buondelmonti, *Descriptio insule Crete et liber insularum, cap. XI: Creta*, ed. M.-A. van Spitael (Iraklion, 1981).

Daniele di Chinazzo, *Cronica de la guerra da veneciani a zenovesi*, ed. V. Lazzarini, in *Monumenti storici pubblicati dalla Deputazione di Storia Patria per le Venezie*, n.s., XI (Padua, 1958).

De Boislisle, A., "Projet de Croisade du premièr Duc de Bourbon (1316-33)", *Annuaire-bulletin de la Société de l'histoire de France*, 9 (1872), 230-36, 246-55.

Dennis, G. T., ed. and trans., *Three Byzantine military treatises: text, trans. and notes* [CFHB, vol. 25] (Washington, 1985).

Devastatio Constantinopolitana, ed. and trans. A. J. Andrea, "The *Devastatio Constantinopolitana*, a special perspective on the Fourth Crusade: an analysis, new edition, and translation", *Historical reflections*, 19 (1993), 107-49. Also trans. Andrea, *Contemporary sources*, 205-21.

De velitatione, ed. and trans. G. T. Dennis, *Περὶ παραδρομῆς*, in *Three Byzantine military treatises*, 137-239. Also eds and Fr. trans. G. Dagron and H. Mihăescu, *Le traité sur la guérilla (De velitatione) de l'empereur Nicéphore Phocas (963-969)* (Paris 1986).

"Documents extraits du Cartulaire du chapitre de Fosses", *Analectes pour servir à l'histoire ecclésiastique de la Belgique*, 4 (1867), 396-8.

Dosio, G. B., ed., *Ragioni antique spettanti all'arte del mare et fabriche de vasselli: manoscritto nautico del sec. XV* (Venice, 1987).

Dotson, J. E., trans., *Merchant culture in fourteenth century Venice: the Zibaldone da Canal* (Binghamton, 1994).

Du Chesne, C., *Historiae Francorum scriptores*, 5 vols (Paris, 1636-49).

Ernoul, *Chronique d'Ernoul et de Bernard le Trésorier*, ed. L. de Mas Latrie (1871; rpt, Famagusta, 1974).

Eugenius III, Pope, *Epistolae Eugenii III Papae*, in RHGF, vol. 15, 426-83; also ed. Migne, *Eugenii III Pontificis Romani epistolae et privilegia*, in PL, vol.

180, coll. 1013-1614. *Variorum ad Eugenium Papam epistolae*, in PL, vol. 180, coll. 1613-42.

Felix Fabri, *Evagatorium*, ed. C. D. Hassler, *Fratris Felicis Fabri evagatorium in Terrae Sanctae, Arabiae, et Egypti peregrinationem*, 3 vols [BLVS, Bde 2-4] (Stuttgart, 1843-9).

Felix Fabri, *Wanderings*, trans. A. Stewart, *The book of the wanderings of brother Felix Fabri*, 2 vols, 4 parts, in PPTSL, vols 7-10 (London, 1893-6).

Filangieri, R., ed., *I registri della cancelleria angioina*, 33 vols (Naples, 1950-81).

Flodoard, *Annales*, ed. G. Pertz, in MGH *SS*, vol. 3, 363-408.

Fulcher of Chartres, *Historia Hierosolymitana (1095-1127)*, ed. H. Hagenmeyer (Heidelberg, 1913). Trans. F. R. Ryan and H. S. Fink, *A history of the expedition to Jerusalem, 1095-1127* (New York, 1973).

Geoffrey Malaterra, *De rebus gestis Rogerii Calabriae et Siciliae comitis et Roberti Guiscardi ducis fratris eius*, in RISS *NS*, tome V, parte I (Bologna, 1928).

Geoffrey of Villehardouin, *La conquête de Constantinople*, ed. and Fr. trans. E. Faral, 2 vols (Paris, 1938-9). Trans. M. R. B. Shaw, *Joinville and Villehardouin: chronicles of the Crusades* (Harmondsworth, 1963), 27-160.

Gesta abbatum Trudonensium continuatoris tertii auctorum, ed. D. R. Koepke, in MGH *SS*, vol. 10, 361-443.

Gesta episcoporum Halberstadensium, ed. L. Weiland, in MGH *SS*, vol. 23, 73-123. Pp. 116-121 trans. Andrea, *Contemporary sources*, pp. 239-64.

Gesta Francorum et aliorum Hierosolimitanorum, ed. and trans. R. Hill, *The deeds of the Franks and the other pilgrims to Jerusalem* (Oxford, 1962).

Giles of Orval, *Aegidii Aureaevallensis Gesta episcoporum Leodiensium*, ed. J. Heller, in MGH *SS*, vol. 25, 1-135.

Guibert of Nogent, *Guibert de Nogent Dei gesta per Francos, et cinq autres textes*, ed. R. B. C. Huygens, in CC *ContMed*, vol. 127A, 76-352; also ed. *Historia quae dicitur Gesta Dei per Francos*, in RHC *HOcc*, vol. 4, 113-263. Trans., or rather loose paraphrase, in R. Levine, *The Deeds of God through the Franks: a translation of Guibert de Nogent's* Gesta Dei per Francos (Woodbridge, 1997). Fr. trans., M.-C. Garand, *Geste de Dieu par les Francs. Histoire de la Première Croisade* (Turnhout, 1998).

Gunther of Pairis, *Hystoria Constantinopolitana: Untersuchung und kritische Ausgabe*, ed. P. Orth (Hildesheim and Zurich, 1994). Trans. [and independently ed.] A. J. Andrea, *The capture of Constantinople: the* Hystoria Constantinopolitana *of Gunther of Pairis* (Philadelphia, 1997).

Hagenmeyer, H., ed., *Epistulae et chartae ad historiam primi belli sacri spectantes quae supersunt aevo aequales ac genuinae: Die Kreuzzugsbriefe aus den Jahren 1088-1100* (1901; rpt, Hildesheim, 1973).

Haldon, J. F., "Theory and practice in tenth-century military administration: chapters II, 44 and 45 of the *Book of Ceremonies*", *Travaux et Mémoires*, 13 (2000), 201-352 [bk II, chs 44, 45, Ed.]

Hausmann, F., ed., *Die Urkunden Konrads III. und seines Sohnes Heinrich*, in MGH *DipReg*, vol. 9.

Helmold of Bosau, *Helmoldi presbyteri Chronica Slavorum*, 3rd ed., ed. B. Schmeidler, in MGH *ScriptRerGerm.Cont.* Trans. (from the 2nd ed.) F. J. Tschan, *The Chronicle of the Slavs by Helmold, priest of Bosau* (1935; rpt, New York, 1966).

Heṭoum, *La Flor des estoires de la Terre d'Orient*, in RHC *DArm*, vol. 2, 111-363.

Historia ducum Veneticorum, ed. H. Simonsfeld, in MGH *SS*, vol. 14, 72-89.

Hugh of St Pol, *Epistola*, in *Annales Colonienses maximi*, ed. G. Waitz, in MGH *SS*, vol. 17, 723-847, pp. 812-24.

Hugh of St Pol, *Epistola*, ed. R. Pokorny, in "Zwei unedierte Briefe aus der Frühzeit des lateinischen Kaiserrecihs von Konstantinopel", *Byzantion*, 55 (1985), 180-209; pp. 203-9. Trans. Andrea, *Contemporary sources*, 177-201.

Ibn 'Abd al-Ẓāhir, Muḥyī 'l-Dīn 'Abd Allāh, *Al-Rawḍ al-zāhir fī sīrat al-malik al-ẓāhir*, ed. 'A.-'A. al-Khuwayṭir (Riyadh, 1976).

Ibn al-'Adīm, Kamāl al-Dīn Abū 'l-Qāsim 'Umar ibn Aḥmad, *Bughyat al-ṭalab fī ta'rīkh Ḥalab*, ed. S. Zakkār, 11 vols (Damascus, 1988-9).

Ibn al-'Adīm, Kamāl al-Dīn Abū 'l-Qāsim 'Umar ibn Aḥmad, *Zubdat al-ḥalab fī ta'rīkh Ḥalab*, ed. S. Dahan, 3 vols (Damascus, 1951-68).

Ibn al-Athīr, 'Izz al-Dīn Abū 'l-Ḥasan 'Alī, *Al-Kāmil fī 'l-ta'rīkh*, ed. C. J. Tornberg, 13 vols (1851-76; rpt, Beirut, 1965-6). Part. Fr. trans., *Extrait de la chronique intitulée Kamel-Altevarykh par ibn-Alatyr*, in RHC *HOr*, vol. 1, 189-744; vol. 2, 1-180.

Ibn al-Dawādārī, Abū Bakr ibn 'Abd Allāh, *Kanz al-durar wa-jāmi' al-ghurar. Vol. 9: Al-Durr al-fākhir fī sīrat al-malik al-Nāṣir (Der Bericht über den Sultan al-Malik an-Nāṣir Muḥammad ibn Qalā'ūn*, ed. H. R. Roemer (Freiburg and Cairo, 1960).

Ibn al-Furāt, Nāṣir al-Dīn Muḥammad ibn 'Abd al-Raḥīm ibn 'Alī al-Miṣrī al-Ḥanafī, *Ta'rīkh al-duwal wa 'l mulūk*, trans. U. and M. C. Lyons, *Ayyubids, Mamlukes and Crusaders: selections from the Tārīkh al-Duwal wa'l Mulūk*, 2 vols (Cambridge, 1971).

Ibn al-Ma'mūn al-Baṭā'iḥī, Abū 'Alī Mūsā, *Akhbār Miṣr*, ed. A. F. Sayyid (Cairo, 1983).

Ibn al-Qalānisī, Abū Ya'lā Ḥamza ibn Asad, *Dhayl ta'rīkh Dimashq*, ed. H. F. Amedroz (Beirut, 1908). Part. trans. H. A. R. Gibb, *The Damascus chronicle of the Crusades* (1932; rpt, London, 1967).

Ibn al-Ṭuwayr, Abū Muḥammad 'Abd al-Salām, *Nuzhat al-muqlatayn fī akhbār al-dawlatayn*, ed. A. F. Sayyid (Beirut, 1992).

Ibn 'Arabshāh, Aḥmad ibn Muḥammad ibn 'Abd Allāh ibn Ibrāhīm Shihāb, *Kitāb 'Ajā'ib al-maqdūr fī nawā'ib Tīmūr* (Calcutta, 1841-2). Trans. J. H. Sanders, *Tamerlane or Timur, the Great Amir* (London, 1936).

Ibn Baṭṭūṭa, Shams al-Dīn Abū 'Abd Allāh Muḥammad, *Riḥla*, eds and Fr. trans C. Defrémery and B. R. Sanguinetti, 4 vols, *Voyages d'Ibn Batoutah* (Paris, 1853-9). Trans. H. A. R. Gibb, *The travels of Ibn Baṭṭūṭa A.D. 1325-1354*, 3 vols, in HSW, 2nd ser., vols 110, 117, 141 (London, 1956-71).

Ibn Ḥawqal, Abū 'l-Qāsim ibn 'Alī al-Naṣībī, *Kitāb ṣūrat al-arḍ*, 3rd ed., ed. J. H. Kramers in De Goeje, *BGA*, vol. 2. Fr. trans. J. H. Kramers and G. Wiet, *Ibn Hauqal, Configuration de la terre (Kitab surat al-ard)*, 2 vols paginated continuously (Paris, 1964).

Ibn Jubayr al-Kinānī, Abū 'l-Ḥusayn Muḥammad ibn Aḥmad, *Riḥla*, Fr. trans. M. Gaudefroy-Demombynes, *Ibn Jobair - Voyages*, 4 parts (Paris, 1949-65).

Ibn Jubayr al-Kinānī, Abū 'l-Ḥusayn Muḥammad ibn Aḥmad, *Riḥla*, trans. R. J. C. Broadhurst, *The travels of Ibn Jubayr* (London, 1952).

Ibn Khurradādhbīh, Abū 'l-Qāsim 'Ubayd Allāh ibn 'Abdallāh, *Kitāb al-Masālik wa 'l-mamālik*, ed. and Fr. trans. in De Goeje, *BGA*, vol. 6, 3-183 (trans. 1-144).

Ibn Muyassar, Taj al-Dīn Muḥammad ibn Yūsuf, *Akhbār Miṣr*, ed. A. F. Sayyid as *Choix de passages de la chronique d'Egypte d'Ibn Muyassar ... selection faite par Taqiyy ad-Dīn Aḥmad ibn 'Alī al-Maqrīzī en l'anneé 814 H.* (Cairo, 1981).

Ibn Shaddād, Bahā' al-Dīn Abū 'l-Maḥāsin, *Al-Nawādir al-sulṭāniyya wa 'l-maḥāsin al-Yūsufiyya* (published under the title *Sīrat Ṣalāḥ al-Dīn*) (Cairo, n. d.). Ed. and Fr. trans. as *Anecdotes et beaux traits de la vie de sultan Youssof (Salâh ed-Dîn)*, in RHC *HOr*, vol. 3, 1-370. Trans. C. W. Wilson and C. C. Conder, *Saladin; or what befell sultan Yusuf*, in PPTSL, vol. 13.

Ibn Shaddād, 'Izz al-Dīn Abū 'Abd Allāh Muḥammad ibn 'Alī al-Ḥalabī, *Ta'rīkh al-Malik al-Ẓāhir*, ed. A. Khuṭayṭ (Wiesbaden, 1983).

Ibn Taghrī Birdī, Abū 'l-Maḥāsin, *Al-Nujūm al-zāhira fī mulūk miṣr wa'l-qāhira*, 16 vols (Cairo, 1930-56). Trans. from his own ed. W. Popper, *History of Egypt 1382-1469 A.D.*, 8 vols (Berkeley and Los Angeles, 1954-63).

Ibn Wāṣil, Abū 'Abd Allāh Jamāl al-Dīn Muḥammad, *Mufarrij al-kurūb fī akhbār Banī Ayyūb*, eds M. Shayyal, H. Rabie, and S. Ashour, 5 vols (Cairo, 1954-77).

Ibn Ẓāfir, Jamāl al-Dīn Abū 'l-Ḥasan 'Alī, *Akhbār al-Duwal al-munqati'a*, ed. A. Ferré (Cairo, 1972).

Idrīs 'Imād al-Dīn, *'Uyūn al-akhbār*, ed. M. Ghālib, 6 vols (Beirut, n.d.).

al-Idrīsī, Abū 'Abd Allāh Muḥammad ibn Allāh, al-Sharīf, *Kitāb Nuzhat al-mushtāk fi 'khtirāq al-āfāq*, Fr. trans. P.-A. Jaubert, *La géographie d'Edrisi*, 2 tomes (1836-40; rpt in 1 vol., Amsterdam, 1984).

al-Idrīsī, Abū 'Abd Allāh Muḥammad ibn Allāh, al-Sharīf, *Kitāb Nuzhat al-mushtāk fi 'khtirāq al-āfāq*, part. Fr. trans. H. Bresc and A. Nef, *La première géographie de l'Occident* (Paris, 1999), from G. Tucci, et al., eds, Al- Idrīsī, *Opus geographicum sive «Liber ad eorum delectationem qui terras peragrare studeant»*, 9 fasc. (Leiden, 1970-84).

Innocent III, Pope, *Innocentii III Romani pontificis Regestrorum sive epistolarum libri*, eds E. Baluze et al., in PL, vols 214-16, and 217, coll. 9-308.

Itinerarum peregrinorum, Anonymous, *Das Itinerarium peregrinorum: eine zeitgenössische englische Chronik zum dritten Kreuzzug in ursprünglicher Gestalt*, ed. H. E. Mayer (Stuttgart, 1962).

Itinerarium peregrinorum et gesta regis Ricardi, in *Chronicles and memorials of the reign of Richard I. Vol. I*, ed. W. Stubbs (Rolls Series, tome 38, vol. 1). Trans. H. J. Nicholson, *Chronicle of the Third Crusade: a translation of the Itinerarium peregrinorum et Gesta regis Ricardi* (Aldershot, 1997).

James of Vitry, *Historia Orientalis*, in *Iacobi de Vitriaco ... libri duo. Quorum prior Orientalis sive Hierosolymitanae, alter, Occidentalis Historiae nomine inscribitur*, ed. F. Moschius (Douai, 1597), 1-258.

James of Vitry, *Lettres*, ed. R. B. C. Huygens, in CC *ContMed*, vol. 171, 491-652 [rpt of *Lettres de Jacques de Vitry (1160/1170-1240) évêque de Saint Jean-d'Acre* (Leiden, 1960)].

Jean le Bel, *Chronique*, eds J. Viard and E. Déprez, *Chronique de Jean le Bel*, 2 vols (Paris, 1904-5).

John Kinnamos, *Ἱστοριῶν βιβλία ϛ´*, ed. A. Meineke [CSHB, vol. 15] (Bonn, 1836). Trans. C. M. Brand, *Deeds of John and Manuel Comnenus* (New York, 1976).

John of Joinville, *Vie de saint Louis*, ed. J. Monfrin (Paris, 1995). Trans. M. R. B. Shaw from the ed. of Natalis de Wailly (Paris, 1874), in Joinville & Villehardouin, *Chronicles of the Crusades* (Harmondsworth, 1963), 161-353.

Khan, G., ed. and trans., *Arabic legal and administrative documents in the Cambridge Genizah collections* (Cambridge, 1993).

Längere Kölner Dienstrecht (etwa 1165), Das, in L. Weinrich, ed., *Quellen zur deutschen Verfassungs-, Wirtschafts- und Sozialgeschichte bis 1250* [AQDGM, vol. 32] (Berlin, 1977), 266-79.

Laurence of Liège, *Laurentii de Leodio Gesta episcoporum Virdunensium et abbatum S. Vitoni*, ed. G. Waitz, in MGH *SS*, vol. 10, 486-530.

Leo VI, *Leonis imperatoris tactica*, ed. R. Vari, 2 vols (Budapest, 1917-22) [incomplete].

Leo VI, *Leonis imperatoris tactica*, ed. J. Meursius, in PG, vol. 107, coll. 669-1094.

Lombardo, A., and R. Morozzo della Rocca, eds, *Nuovi documenti del commercio veneto nei secoli XI-XIII* (Venice, 1953).

Ludolph of Suchem, *Description of the Holy Land and of the way thither. Written in the year A.D. 1350*, trans. A. Stewart in PPTSL, vol. 12.C (London, 1895), from the ed. of F. Deycks, *De itinere Terrae Sanctae liber* ..., in BLVS, 25 (1851), 1-104.

Lupus Protospatharios, *Annales*, ed. G. Pertz, in MGH *SS*, vol. 5 (Hanover, 1844), 51-63.

al-Maghribī, Abū 'l-Ḥasan 'Alī ibn Mūsā ibn Sa'īd, *Kitāb al-Mughrib fī ḥulā 'l-Maghrib*, ed. and trans. K. Vollers as *Fragmente aus dem Muġrib des Ibn Sa'īd*, in *Semitistische Studien*, Heft 1 (Berlin, 1894).

al-Maghribī, Abū 'l-Ḥasan 'Alī ibn Mūsā ibn Sa'īd, *Al-Nujūm al-zāhira fī ḥulā ḥaḍrat al-Qāhira*, ed. Ḥ. Naṣṣār (Cairo, 1969).

Malchos of Philadelphia, *Byzantiaka*, ed. and trans. Blockley, *Fragmentary classicising historians*, vol. 2, 402-70.

Al-Maqrīzī, Taqī al-Dīn Abū 'l-'Abbās Aḥmad ibn 'Alī, *Itti'āẓ al-ḥunafā'*, ed. J. al-Dīn al-Shayyal and M. H. M. Aḥmad, 3 vols (Cairo, 1967-73).

al-Maqrīzī, Taqī al-Dīn Abū 'l-'Abbās Aḥmad ibn 'Alī, *Al-Mawā'iẓ wa 'l-i'tibār fī dhikr al-khiṭat wa 'l-āthār*, 4 vols (Būlāq, 1906). Fr. trans. U. Bouriant, *Description topographique et historique de'Egypte*, 2 pts (Paris, 1895, 1900).

al-Maqrīzī, Taqī al-Dīn Abū 'l-'Abbās Aḥmad ibn 'Alī, *Al-Sulūk li-ma'rifat al-mulūk*, ed. M. M. Ziyādah and S. 'A. 'Ashūr, 10 vols (Cairo, 1934-73). Part. Fr. trans. E. M. Quatremère, *Histoire des sultans mamlouks de l'Egypte*, 2 vols (London, 1837, 1845); E. Blochet, "Histoire d'Égypte de Makrizi", *ROL*, 6 (1898), 435-89; 8 (1900-1901), 165-212, 501-53; 9 (1902), 6-163, 466-530; 10 (1903-4), 248-371; 11 (1905-8), 192-239; part. Eng. trans. R. J. C. Broadhurst, *A History of the Ayyūbid sultans of Egypt* (Boston, 1980).

Martino da Canale, *Les estoires de Venise: cronaca veneziana in lingua francese dalle origini al 1275*, ed. A. Limentani [Civiltà veneziana, fonti e testi, XII, serie terza, 3] (Florence, 1972).

Matthew of Edessa, *Chronicle*, trans. A. E. Dostourian, *Armenia and the Crusades tenth to twelfth centuries: the chronicle of Matthew of Edessa* (Lanham, 1993).

Matthew Paris, *Mathaei Parisiensis monachi Sancti Albani, Chronica Majora*, ed. H. R. Luard (Rolls series, tome 57), 7 vols (London, 1872-83). Trans. J. A. Giles, *Matthew Paris's English history from the year 1235 to 1273*, 3 vols (1852-4; rpt, New York, 1968).

Maurice, *Das Strategikon des Maurikios*, eds and Germ. trans. G. T. Dennis and E. Gamillscheg [CFHB, vol. 17] (Vienna, 1981). Trans. G. T. Dennis, *Maurice's Strategikon: handbook of Byzantine military strategy* (Philadelphia, 1984).

Mayer, P., "La chanson de Doon de Nanteuil. Fragments inédits", *Romania*, 13 (1884), 1-26.

Mayer, P., "Un récit en vers français de la première croisade fondé sur Baudri de Bourgeuil", *Romania*, 5 (1876), 1-63.

Michael Attaleiates, *Michaelis Attaliotae Historia*, eds W. Brunet de Presle and I. Bekker [CSHB, vol. 34] (Bonn, 1853).

Michael Psellos, *Chronographie ou histoire d'un siècle de Byzance (976-1077)*, ed. and Fr. trans. E. Renauld, 2 vols (Paris, 1926-8). Trans. E. R. A. Sewter, *Fourteen Byzantine rulers: the* Chronographia *of Michael Psellus* (Harmondsworth, 1966).

Morozzo della Rocca, R., and A. Lombardo, eds, *Documenti del commercio veneziano nei secoli XI-XIII*, 2 vols (Turin, 1940).

Motzo, B. R., ed., *Il compasso da navigare: opera italiana della metà di secolo XIII* (Cagliari, 1947).

al-Musabbiḥī, al-Amīr al-Mukhtār 'Izz al-Mulk, *Al-Juz' al-arba'ūn min Akhbār Miṣr*, eds A. F. Sayyid and T. Bianquis (Cairo, 1978).

Nāṣir-i Khusraw, Abū Mu'īn, *Safar-Nāmah*, trans. W. M. Thackston, *Nāser-e-Khosraw's Book of Travels (Safarnama)* (New York, 1985).

Nicolas de Martoni, *Liber peregrinationis ad Loca Sancta*, ed. L. Legrand, "Relation du pèlerinage a Jérusalem de Nicolas de Martoni notaire italien (1394-1395)", *ROL*, 3 (1895), 566-669.

Nikēphoros I, Patriarch, *Ἱστορία σύντομος*, ed. and trans. C. Mango, *Short history* [CFHB, vol. 13] (Washington, 1990).

Nikēphoros Bryennios, *Nicephori Bryenni Historiarum libri quattuor*, ed. and Fr. trans. P. Gautier [CFHB, vol. 9] (Brussels, 1975).

Nikēphoros Phōkas, *Praecepta militaria* [*Στρατηγικὴ Ἔκθεσις καὶ Σύνταξις Νικηφόρου δεσπότου*], ed. and trans. E. McGeer, *Dragon's teeth*, 12-59. Also ed. J. Kulakovskij, "Nikephori praecepta militaria ex codice Mosquensi", *Mémoires de l'Académie impériale des sciences de St.-Pétersbourg, VIII sér. Classe historico-philologique, No. 9* (St Petersburg, 1908), 1-21.

Nikētas Chōniatēs, *Nicetae Choniatae historia*, ed. J.-L. van Dieten (Berlin, 1975). Trans. H. Magoulias, *O city of Byzantium, annals of Niketas Choniatēs* (Detroit, 1984).

Odo of Deuil, *De profectione Ludovici VII in orientem*, ed. and trans. V. G. Berry (New York, 1948).

Oliver of Paderborn, *Historia Damiatina*, ed. H. Hoogeweg, *Die Schriften des Kölner Domscholasters, späteren Bischofs von Paderborn und Kardinal-Bischofs von S. Sabina, Oliverus* (Tübingen, 1894), 159-282. Trans. J. J. Gavigan, *The capture of Damietta by Oliver of Paderborn* (1948; rpt, New York, 1980).

Orderic Vitalis, *The Ecclesiastical History of Orderic Vitalis*, ed. and trans. M. Chibnall, 6 vols (Oxford, 1968-80).

Orkneyinga Saga, ed. F. Guðmundsson, in *Íslenzk fornrit*, 35 vols (Reykjavík, 1968-82), vol. 34, 1-300.

Otto of Freising, *Ottonis et Rahewini Gesta Friderici I. imperatoris*, 3rd ed., ed. G. Waitz, in MGH *ScriptRerGerm.Cont* (Hanover, 1912). Trans. C. C. Mierow,

The deeds of Frederick Barbarossa by Otto of Freising and his continuator, Rahewin (New York, 1953).

Otto of St Blasien, *Chronica*, ed. A. Hofmeister, in MGH *ScriptRerGerm.Cont* (Hanover and Leipzig, 1912).

Peter Tudebode, *Historia de Hierosolymitano itinere*, in *RHC HOcc*, vol. 3, 1-117.

Peter Tudebode, *Historia de Hierosolymitana itinere*, ed. J. H. and L. L. Hill (Paris, 1977). Trans. J. H. and L. L. Hill, *Peter Tudebode: Historia de Hierosolymitana itinere* (Philadelphia, 1974).

Philip of Novara, *Estoire de la guerre qui fu entre l'emperor Frederic & Johan d'Ibelin*, in G. Raynaud, ed., *Les gestes des Chiprois. Recueil de chroniques françaises écrites en Orient aux XIIIe et XIVe siècles (Philippe de Navarre & Gérard de Monréal)* (Geneva, 1887), 25-138. Trans. J. L. LaMonte and M. H. Hubert, *The wars of Frederick II against the Ibelins in Syria and Cyprus* (New York, 1936).

Pietro Casola, *Viaggio a Gerusalemme di Pietro Casola*, ed. A. Paoletti (Alessandria, 2001). Trans. M. M. Newett [from the ed. of 1855 by G. Porro], *Pietro Casola's pilgrimage to Jerusalem in the year 1494* (Manchester, 1907).

Pliny, *Natural history,* Gaius Plinius Secundus, *Historia naturalis*, in H. Rackham, trans., Pliny, *Natural History*, 10 vols (London, 1938-62).

Polybios, *The Histories*, in W. R. Paton, trans., *Polybius: the histories*, 6 vols (London, 1922-7).

Prokopios of Caesarea, *Buildings*, in H. B. Dewing, ed. and trans., *Procopius*, 7 vols (London, 1914-40), vol. 7.

Prokopios of Caesarea, *History of the wars*, in H. B. Dewing, ed. and trans., *Procopius*, 7 vols (London, 1914-40), vols 1-6.

Pseudo-Brocardus, *Directorium ad Passagium faciendum*, in RHC *DArm*, vol. 2, 365-517.

al-Qalqashandī, Shihāb al-Dīn Abu 'l-'Abbās Aḥmad ibn 'Alī, *Ṣubḥ al-a'shā fī ṣina'at al-inshā'*, 14 vols (Cairo, n.d.).

Qudāma ibn Ja'far al-Katīb al-Baghdādī, Abū 'l-Faraj, *Kitāb al-Kharaj*, ed. and Fr. trans. in de Goeje, *BGA*, vol. 6, 184-266 (trans. 144-208).

Ralph of Caen, *Gesta Tancredi in expeditione Hierosolymitana, auctore Radulfo Cadomensi, ejus familiari*, in RHC *HOcc*, vol. 3, 587-716.

Ralph of Coggeshall, *Chronicon Anglicanum*, ed. J. Stevenson [Rolls Series, tome 66] (London, 1875), 1-208.

Raol, *De expugnatione Lyxbonensi/The conquest of Lisbon*, ed. and trans. C. W. David, rev. J. Phillips (1936; rpt, New York, 2001).

Rashīd al-Dīn Faḍl Allāh ibn 'Imād al-Dawla, Abū 'l-Khayr, *Jāmi' al-tawārīkh*, in K. Jahn, ed., *Geschichte Gāzān-Hān's aus dem Ta'rīh-i-Mubārak-i-Gāzāni des Rasīd al-Dīn Faḍlallāh b. 'Imād al-Daula Abūl-Hair* (London, 1940).

Rashīd al-Dīn Faḍl Allāh ibn 'Imād al-Dawla, Abū 'l-Khayr, *Jāmi' al-tawārīkh*, trans. W. M. Thackston, *Rashiduddin Fazlullah's Jami'u 't-tawarikh:*

Compendium of chronicles: a history of the Mongols, 3 vols (Cambridge Mass., 1998-9).

Raymond of Aguilers, *Historia Francorum qui ceperunt Iherusalem*, in RHC *HOcc*, vol. 3, 231-309.

Raymond of Aguilers, *Liber Raimundi de Aguilers*, ed. J. H. and L. L. Hill, *Le* «Liber» *de Raymond d'Aguilers* (Paris, 1969). Trans. J. H. and L. L. Hill, *Raymond d'Aguilers: Historia Francorum qui ceperunt Iherusalem* (Philadelphia, 1968).

Renier of St Laurence, *Triumphale Bulonicum*, ed. W. Arndt, in MGH *SS*, vol. 20, 583-92.

Riant, P. E. D., "Inventaire critique des lettres historiques des croisades", *Archives de l'Orient Latin*, 1 (1881), 1-224.

Richard, J., *Chypre sous les Lusignans. Documents chypriotes des archives du Vatican (XIVe et XVe siècles)* (Paris, 1962).

Richard of Devizes, *Chronicon Richardi Divisensis de tempore regis Richardi Primi / The Chronicle of Richard of Devizes of the time of King Richard the First*, ed. and trans. J. T. Appleby (London, 1963).

Riley-Smith, L. and J., *The Crusades: idea and reality, 1095-1274* (London, 1981).

Robert of Auxerre, *Chronicon*, ed. O. Holder-Egger, in MGH *SS*, vol. 26, 219-76.

Robert of Clari, *La conquête de Constantinople par Robert de Clari*, Fr. trans. A. Micha, texte de P. Lauer [1924] (Paris, 1991). Trans. E. H. McNeal, *The conquest of Constantinople: translated from the Old French of Robert of Clari* (New York, 1936).

Roger of Hoveden, *Chronica*, ed. W. Stubbs [Rolls series, tome 51], 4 vols (London, 1868-71).

Roger of Wendover, *Rogeri de Wendover liber qui dicitur Flores Historiarum / The Flowers of History by Roger de Wendover*, ed. H. G. Hewlett [Rolls series, tome 84], 3 vols (London, 1886-9).

Röhricht, R., ed., *Quinti belli Sacri scriptores minores* (Geneva, 1879).

Roman de Thèbes, Le, ed. G. Raynaud de Lage, 2 vols (Paris, 1968-69).

Romans d'Alixandre par Lambert li Tors et Alexandre de Bernay, Li, ed. H. Michelant [BLVS, vol. 13] (Stuttgart, 1846).

Saewulf, in R. B. C. Huygens, ed., *Peregrinationes tres: Saewulf, John of Würzburg, Theodericus* [CC *ContMed*, vol. 139], 58-77.

Sanudo Torsello, Marino, *Liber secretorum fidelium crucis super Terrae Sanctae recuperatione et conservatione* (1611; rpt, Jerusalem, 1972).

Sathas, C. N., *Documents inédits relatifs à l'histoire de la Grèce au moyen-âge*, 9 vols (Paris, 1880-90).

Shāmī, Niẓām al-Dīn, *Ẓafar-nāmah*, ed. F. Tauer in *Histoire des conquêtes de Tamerlan intitulée Ẓafarnāmah*, 2 vols (Prague, 1937, 1956).

Snorri Sturluson, *Heimskringla*, ed. B. S. Kristjánsdóttir et al., 3 vols (Reykjavík, 1991).

Snorri Sturluson, *Heimskringla: History of the Kings of Norway*, trans. L. M. Hollander (1964; rpt, Austin, 1999).

Snorri Sturluson, *Heimskingla. Part two: Sagas of the Norse kings*, trans. S. Laing, rev. P. Foote (1961; rpt, London and New York, 1968).

Solomon bar Simson, *The chronicle of Solomon bar Simson*, ed. and trans. S. Eidelberg, in *The Jews and the Crusaders: the Hebrew chronicles of the First and Second Crusades* (Madison, 1977), 21-72.

Stussi, A., ed., *Zibaldone da Canal: manoscritto mercantile del sec. XIV* (Venice, 1967).

Suger of St Denis, *Epistolae Sugerii abbatis S. Dionysii*, in RHGF, vol. 15, 483-532.

Tafel, G. L. Fr., and G. M. Thomas, eds, *Urkunden zur älteren Handels- und Staatsgeschichte der Republik Venedig*, 3 vols (1856-7; rpt, Amsterdam, 1964).

Taj al-Dīn Shāhanshāh ibn Ayyūb, *Kitāb al-ta'rīkh*, in Ibn Shaddād, *Al-Nawādir al-sulṭāniyya*, 253-310.

Theodoricus monachus, *Historia de antiquitate regum Norwagiensium / An account of the ancient history of the Norwegian kings*, trans. D. and I. McDougall (London, 1998).

Theophanes continuatus, in I. Bekker, ed., *Theophanes continuatus, Ioannes Cameniata, Symeon magister, Georgius monachus*, [CSHB], 3 vols (Bonn, 1838), vol. 1.

Theophilus, *De diversis artibus / The various arts*, ed. and trans. C. R. Dodwell (London, 1961).

Theophylakt Simokattēs, *Theophylacti Simocattae historiae*, ed. P. Wirth (Stuttgart, 1972). Trans. M. and M. Whitby, *The History of Theophylact Simocatta* (Oxford, 1986).

Triumphus S. Lamberti de castro Bullonio, ed. W. Arndt, in MGH *SS*, vol. 20, 497-511.

al-'Umarī, Aḥmad ibn Yaḥyā ibn Faḍl Allāh, *Masālik al-abṣār fī 'l-mamālik al-amṣār: mamālik Miṣr wa 'l-Shām wa 'l-Ḥijāz wa 'l-Yaman*, ed. A. F. Sayyid (Cairo, 1985).

al-'Umarī, Aḥmad ibn Yaḥyā ibn Faḍl Allāh, *Al-Ta'rīf fī 'l-muṣṭalaḥ al-sharīf*, (Cairo, 1312 A.H. [1894-5 C.E.]).

Usāma ibn Munqidh, ibn Murshid ibn 'Alī ibn Muqallad ibn Naṣr, *Kitāb al-I'tibār*, ed. P. Hitti (Princeton, 1930). Trans. P. Hitti, *An Arab-Syrian gentleman and warrior in the period of the Crusades* (1929; rpt, London, 1987).

Vegetius Renatus, Flavius, *Epitoma rei militaris*, ed. C. Lang (Stuttgart, 1872). Trans. N. P. Milner, *Vegetius: Epitome of military science* (Liverpool, 1993).

Vita de S. Petro episcopo Anagniae in Italia, in AASS, Augusti, tome I (Venice, 1750), 230-41.

Voyage de la sainte cyté de Hierusalem avec la description des lieux, portz, villes, citez et aultures passaiges fait l'an mil quatre cens quatre vingtz, Le, ed. C. Schefer (Paris, 1882).

Walter of Châtillon, *Galterus de Castellione, Alexandreis*, ed. M. L. Colker (Padua, 1978). Trans. D. Townsend, *The Alexandreis of Walter of Châtillon. A twelfth-century epic* (Philadelphia, 1996).

Waṣṣāf al-Ḥaḍrat, *Ta'rīkh-i Waṣṣāf* (1852-3; rpt, Teheran, 1959-60).

William of Nangis, *Vita Sancti Ludovici regis Franciæ / Vie de Saint Louis*, in RHGF, vol. 20, 309-465.

William of Tyre, *Willelmi Tyrensis archiepiscopi chronicon*, ed. R. B. C. Huygens, 2 vols, in CC *ContMed*, vols 63 & 63A paginated continuously. Trans. A. Babcock and A. C. Krey [from the ed. in RHC *HOcc*, vol. 1], *A history of deeds done beyond the sea* (New York, 1943).

Yaqūt al-Rūmī, Shihāb al-Dīn Abū 'Abd Allāh, *Mu'jam al-buldān*, ed. F. 'Abd al-'Aziz, 5 vols (Beirut, 1990).

Yazdī, Sharaf al-Dīn 'Alī, *Ẓafar-nāma*, in F. Petis de La Croix, the younger, trans., *Histoire de Timur-Bec connu sous le nom du Grand Tamerlan, empereur des Mongols et Tartares* (Paris, 1723).

al-Yūnūnī, Quṭb al-Dīn Mūsā ibn Muḥammad, *Dhayl mir'āt al-zamān fī ta'rīkh al-a'yān*, in Li Guo, *Early Mamluk Syrian historiography: al-Yūnūnī's Dhayl Mir'āt al-zamān*, 2 vols (Leiden, 1998).

Zetterstéen, K. V., *Beiträge zur Geschichte der Mamlükensultane in den Jahren 690-741 der Hig) ra, nach arabischen Handschriften* (Leiden, 1919).

(c) Reference Works and Websites

Anonymous, *Baedeker's Turkey*, trans. W. Bell et al. (Basingstoke, 1994).

Calder, W. M., and G. E. Bean, eds, *A classical map of Asia Minor* (London, 1958).

Cornu, G., *Atlas du monde Arabo-Islamique à l'époque classique, IXe-Xe siècles* (Leiden, 1985).

daylight, darkness, moonlight: http:/aa.usno.navy.mil.data/docs/RS_OneDay. html/

De Vries, J., *Nederlands etymologisch Woordenboek* (Leiden, 1997).

Doursther, H., *Dictionnaire universel des poids et mesures anciens et modernes contenant des tables des monnaies de tous les pays* (1840; rpt, Amsterdam, 1976).

Great Britain, Admiralty, Hydrographic Office, *Mediterranean Pilot. Volume I: the coasts of Spain ...*, 10th ed. (London, 1978).

Great Britain, Admiralty, Hydrographic Office, *Mediterranean Pilot. Volume III: the Ionian Sea bordering the western coast of Greece ...*, 10th ed. (Taunton, 1988).

Great Britain, Admiralty, Hydrographic Office, *Mediterranean Pilot. Volume IV: the islands of the Grecian archipelago with the adjacent coasts ...*, 9th ed. (London, 1968).

Great Britain, Admiralty, Hydrographic Office, *Mediterranean Pilot. Volume V: the coasts of Libya, Egypt, Syria, Lebanon and Israel; ...*, 6th ed. (London, 1976).

Hagenmeyer, H., *Chronologie de la première Croisade* (1898-1901; rpt, Hildesheim, 1973).

Heikell, R., *Italian waters Pilot: a yachtsman's guide to the west and south coasts of Italy with the islands of Sardinia, Sicily and Malta*, 6th ed. (London, 2002).

Heikell, R., *Greek waters Pilot: a yachtsman's guide to the coasts and islands of Greece*, 8th ed. (London, 2001).

Heikell, R., *Turkish waters and Cyprus Pilot: a yachtsman's guide to the Mediterranean and Black Sea coasts of Turkey with the island of Cyprus*, 6th ed. (London, 2001).

Hooper, N. and M. Bennett, *Cambridge illustrated atlas of warfare: the Middle Ages, 732-1487* (Cambridge, 1996).

Jal, A., *Glossaire nautique: répertoire polyglotte de termes de marine anciens et modernes*, 1 vol. in 2 tomes (Paris, 1848).

Kemp, P., ed., *The Oxford companion to ships and the sea* (Oxford, 1976).

Kretschmer, K., *Die italienischen portolane des Mittelalters: ein Beitrag zur Geschichte der Kartographie und Nautik* (1909; rpt, Hildesheim, 1962).

Martini, L., *Manuale di metrologia ossia misure, pesi e monete* (Turin, 1883).

Oxford dictionary of Byzantium, The, ed. A. Kazhdan, 3 vols (New York and Oxford, 1991).

Salvati, C., *Misure e pesi nella documentazione storica dell'Italia del mezzogiorno* (Naples, 1970).

Smith, W., ed., *Dictionary of Greek and Roman geography*, 2 vols (London, 1854, 1857).

The Skuldelev Ships: http://www.abc.se/~m10354/uwa/skulde.htm.

United States War Department, *Manual for stable sergeants* (Washington, 1917).

Zupko, R. E., *Italian weights and measures from the Middle Ages to the nineteenth century* (Philadelphia, 1981).

(d) Secondary Works, Published and Unpublished

Abels, R. P. and B. S. Bachrach, eds, *The Normans and their adversaries at war: essays in memory of C. Warren Hollister* (Woodbridge, 2001).

Abulafia, D., "Henry Count of Malta and his Mediterranean activities, 1203-1230", in A. T. Luttrell, ed., *Medieval Malta: studies on Malta before the Knights*

(London, 1975), 104-25. Rpt in his *Italy, Sicily and the Mediterranean 1100-1400* (London, 1987), No. III.

Adams, C. E. P., "Supplying the Roman army: O. Petr. 245", *Zeitschrift für Papyrologie und Epigraphik*, 109 (1995), 119-24.

Ahrweiler, H., *Byzance et la mer: la marine de guerre, la politique, et les institutions maritimes de Byzance aux VIIe-Xve siècles* (Paris, 1966).

Airaldi, G., "Roger of Lauria's expedition to the Peloponnese", *MHR*, 10 (1995), 14-23.

Aksan, V., "Feeding the Ottoman troops on the Danube, 1768-1774", *War and Society*, 13 (1995), 1-14.

Albrecht, G., *Das Münzwesen im niederlothringischen und friesischen Raum vom 10. bis zum beginnenden 12. Jahrhundert* (Hamburg, 1959).

Amitai, R., "Whither the Ilkhanid army? Ghazan's first campaign into Syria (1299-1300)", in N. Di Cosmo, ed., *Warfare in inner Asian history*, 221-64.

Amitai-Preiss, R., " 'Ayn Jālūt revisited", *Tārīḫ: papers in Near Eastern studies*, 2 (1991), 119-50.

Amitai-Preiss, R., *Mongols and Mamluks: the Mamluk-Īlkhānid war, 1260-1281* (Cambridge, 1995).

Amitai-Preiss, R., and D. O. Morgan, eds, *The Mongol Empire and its legacy* (Leiden, 1999).

Anderson, J. G. C., "The campaign of Basil I against the Paulicians in 872 A.D.", *Classical review*, 10 (1896), 136-40.

Andrea, A. J., "Conrad of Krosigk, Bishop of Halberstadt, Crusader and monk of Sittichenbach: his ecclesiastical career, 1184-1225", *Analecta Cisterciensia*, 43 (1987), 11-91.

Andrews, M. C., "The study and classification of medieval mappae mundi", *Archaeologia or miscellaneous tracts relating to antiquity*, 75 (1926), 61-76.

Asaert, G., "Scheepvaart in het zuiden tot 1400", in *Maritieme Geschiedenis der Nederlanden. Vol. 1: de Boer Maritiem*, eds G. Asaert, J. van Beylen, and H. P. H. Jansen (Bussum, 1976), 37-74.

Ashkenazi, T. B., *The Turkmans in Palestine (Ha-Turkmanim be-ereṣ-yisra'el)* (Jerusalem, 1931).

Ayalon, D., "Discharges from service, banishments and imprisonments in Mamluk society", *Israel Oriental studies*, 3 (1972), 25-50. Rpt in his *Mamlūk military society*, No. V.

Ayalon, D., *Eunuchs, Caliphs and sultans: a study in power relationships* (Jerusalem, 1999).

Ayalon, D., *Gunpowder and firearms in the Mamluk kingdom: a challenge to a medieval society*, 2nd ed. (London, 1988).

Ayalon, D., *Islam and the Abode of War: military slaves and Islamic adversaries* (Aldershot, 1994).

Ayalon, D., "Preliminary remarks on the Mamlūk military institution in Islam", in V. J. Parry and M. E. Yapp, eds, *War, technology and society*, 44-58.

Ayalon, D., "The mamlūks of the Seljuks: Islam's military might at the crossroads", *JRAS*, 6 (1996), 305-33.

Ayalon, D., *The Mamlūk military society: collected studies* (London, 1979).

Ayalon, D., "Notes on the *Furūsiyya* exercises and games in the Mamluk sultanate", *Scripta Hierosolymitana*, 9 (1961), 31-62. Rpt in his *Mamlūk military society*, No. II.

Ayalon, D., "The Nubian dam", *JSAI*, 12 (1989), 372-90. Rpt in his *Islam and the Abode of War*, No. XII.

Ayalon, D., *Outsiders in the lands of Islam: mamluks, Mongols and eunuchs* (London, 1988).

Ayalon, D., "Studies on the structure of the Mamluk army", *BSOAS*, 15 (1953), 203-28, 448-76; 16 (1954), 57-90. Rpt in his *Studies on the Mamlūks*, No. I.

Ayalon, D., "The system of payment in Mamluk military society", *JESHO*, 1 (1958), 37-65, 257-96. Rpt in his *Studies on the Mamlūks*, No. VIII.

Ayalon, D., *Studies on the Mamlūks of Egypt (1250-1517)* (London, 1977).

Ayalon, D., "The Great *Yāsa* of Chingiz Khān. A re-examination", *Studia Islamica*, 33 (1971), 97-140; 34 (1971), 151-80; 36 (1972), 113-58; 38 (1973), 107-56. Rpt in his *Outsiders in Islam*, Nos IV (a)-(d).

Bacharach, J. L., "African military slaves in the medieval Middle East: the cases of Iraq (869-955) and Egypt (868-1171)", *IJMES*, 13 (1981), 471-95.

Bachrach, B. S., *The anatomy of a little war: a diplomatic and military history of the Gundovald affair (568-586)* (Boulder, 1994).

Bachrach, B. S., "Animals and warfare in early medieval Europe", in *L'uomo di fronte al mondo animale nell'alto medioevo* [SSCISAM, XXXI] (Spoleto, 1985), 707-64.

Bachrach, B. S., "Some observations on the role of the Byzantine navy in the success of the First Crusade", *Journal of medieval military history*, 1 (2002), 83-100.

Bachrach, B. S., "*Caballus et caballarius* in medieval warfare", in H. Chickering and T. H. Seiler, eds, *The study of chivalry: resources and approaches* (Kalamazoo, 1988), 173-211. Rpt in his *Warfare and military organization*, No. II.

Bachrach, B. S., *Early Carolingian warfare: prelude to empire* (Philadelphia, 2001).

Bachrach, B. S., "Early medieval military demography: some observations on the methods of Hans Delbrück", in D. Kagay and L. J. Andrew Villalon, eds, *The circle of war in the Middle Ages* (Woodbridge, 1999), 3-20.

Bachrach, B. S., "The Hun army at the battle of Chalons (451): an essay in military demography", in K. Brunner and B. Merta, eds, *Ethnogenese und*

Überlieferung: Angewandte Methoden der Frühmittelalterforschung (Vienna and Munich, 1994), 59-67.

Bachrach, B. S., "Logistics in pre-Crusade Europe", in J. A. Lynn, ed. *Feeding Mars*, 57-78. Rpt in his *Warfare and military organization*, No. V.

Bachrach, B. S., "Magyar-Ottonian warfare: à-propos a new minimalist interpretation", *Francia*, 27 (2001), 211-30.

Bachrach, B. S., "Medieval siege warfare: a reconnaisance", *The journal of military history*, 58 (1994), 119-33. Rpt in his *Warfare and military organization*, No. IV.

Bachrach, B. S., "The siege of Antioch: a study in military demography", *War in History*, 6 (1999), 127-46.

Bachrach, B. S., "Some observations on administration and logistics of the siege of Nicaea", forthcoming in *War in history*.

Bachrach, B. S., *Warfare and military organization in pre-Crusade Europe* (Aldershot, 2002).

Bachrach, B. S., "William Rufus's plan for the invasion of the mainland in 1101", in Abels and Bachrach, *Normans and their adversaries*, 31-63.

Bachrach, B. S., "On the origins of William the Conqueror's horse transports", *Technology and culture*, 26 (1985), 505-31. Rpt in his *Warfare and military organization*, No. XIII.

Bachrach, D. S., *Religion and the conduct of war c. 300-c. 1215* (Woodbridge, 2003).

Balard, M., B. Z. Kedar, and J. Riley-Smith, eds, *Dei gesta per Francos: études sur les croisades dédiées à Jean Richard / Crusade studies in honour of Jean Richard* (Aldershot, 2001).

Balard, M., ed., *Autour de la Première Croisade: actes du Colloque de la Society for the Study of the Crusades and the Latin East (Clermont-Ferrand, 22-25 juin, 1995)* (Paris, 1996).

Barlow, F., *The feudal kingdom of England 1042-1216*, 4th ed. (London, 1988).

Barral i Altet, X., "Architecture, sculpture et mosaïque", in *Le monde roman 1060-1220: Les royaumes d'Occident*, ed. F. Avril, X. Barral i Altet, and D. Gaborit-Chopin (Paris, 1983), 8-158.

Barral i Altet, X., "Poésie et iconographie: un pavement du XIIe siècle décrit par Baudri de Bourgueil", *DOP*, 41 (1987), 41-54.

Bartlett, R., "Provisioning war in the twelfth century", in A. McKay with D. Ditchburn, eds, *Atlas of Medieval Europe* (London, 1997), 125-7.

Bartolomucci, N., "Il registro epico di Baudri de Bourgueil. *Adelae comitissae*, vv. 207-582", *Annali della Facoltà di Lettere e Filosofia della Università di Bari*, 39 (1996), 73-87.

Bartolomucci, N., "Tecnica e poesia in Balderico de Bourgueil: c. CXCVI", *Annali della Facoltà di Lettere e Filosofia della Università di Bari*, 17 (1974), 137-57.

Bates, M. L., and D. M. Metcalf, "Crusader coinage with Arabic inscriptions", in Setton, *HC*, vol. 6, 421-82.

Bauer, N., "Der Fund von Spanko bei St. Petersburg", *Zeitschrift für Numismatik*, 36 (1926), 75-94.

Beeston, A. F. L., *Qahtan: studies in old South Arabian epigraphy. Fasc. 3: Warfare in ancient South Arabia (2nd – 3rd centuries A.D.)* (London, 1976).

Belke, K., "Von der Pflasterstrasse zum Maultierpfad? Zum kleinasiatischen Wegenetz in mittelbyzantinischer Zeit", in N. Oikonomides, ed., *Byzantine Asia Minor*, 267-84.

Belke, K., "Roads and travel in Macedonia and Thrace in the middle and late Byzantine period", in R. Macrides, ed., *Travel in the Byzantine world*, 73-90.

Belke, H., and N. Mersich, eds, *Phrygien und Pisidien* (TIB, vol. 7) (Vienna, 1990).

Bennett, M., "La *Règle du Temple* as a military manual, *or* how to deliver a cavalry charge", in C. Harper-Bill et al., eds, *Studies in medieval history presented to R. Allen Brown* (Woodbridge, 1989), 7-19.

Bennett, M., "Military aspects of the conquest of Lisbon, 1147", in J. Phillips and M. Hoch, eds, *Second Crusade*, 71-89.

Bennett, M., "The Crusaders' 'fighting march' revisited", *War in history*, 8 (2001), 1-18.

Bennett, M., "The myth of the military supremacy of knightly cavalry", in M. Strickland, ed., *Armies, chivalry and warfare*, 304-16.

Berggren, J. L., and A. Jones, *Ptolemy's Geography: an annotated translation of the theoretical chapters* (Princeton, 2000).

Berry, V. G., "The Second Crusade", in Setton, *HC*, vol. 1, 463-512.

Bill, J., "Scandinavian warships and naval power in the thirteenth and fourteenth centuries", in J. Hattendorf and R. W. Unger, eds, *War at Sea*, 35-51.

Blake, E. O., and C. Morris, "A hermit goes to war: Peter and the origins of the First Crusade", in W. J. Shiels, ed., *Monks, hermits and the ascetic tradition* [SCH, 22] (Oxford, 1985), 79–107.

Blanchet, A., and A. Dieudonné, *Manuel de numismatique française*, 4 vols (1912-36; rpt, Paris, n.d.).

Bombaci, A., "The army of the Saljuqs of Rum", *Annali del Instituto orientale di Napoli*, 38 (1978), 343-69.

Bon, A., *Le Péloponnèse byzantin jusqu'en 1204* (Paris, 1951).

Boswell, J., *Christianity, social tolerance, and homosexuality: gay people in Western Europe from the beginning of the Christian era to the fourteenth century* (Chicago and London, 1980).

Bosworth, C. E., "Abu 'Amr 'Uthmān al-Ṭarsūsī's *Siyar al Thughūr* and the last years of Arab rule in Tarsus (fourth/tenth century)", *Graeco-Arabica*, 5 (1993), 183-95.

Bosworth, C. E., "Ghaznavid military organization", *Der Islam*, 36 (1960), 37-77.

Bosworth, C. E., "Military organization under the Būyids of Persia and Iraq", *Oriens*, 18-19 (1965-6), 143-67.

Bowlus, C. R., *Franks, Moravians, and Magyars: the struggle for the middle Danube, 788-907* (Philadelphia, 1995).

Bowlus, C. R., "Tactical and strategic weaknesses of horse archers on the eve of the First Crusade", in Balard, *Autour de la Première Croisade*, 159-66.

Boyle, J. A., "The dynastic and political history of the Īl-Khāns", in *The Cambridge history of Iran*, vol. 5, ed. J. A. Boyle (Cambridge, 1968), 303-421.

Braudel, F., *The Mediterranean and the Mediterranean world in the age of Philip II*, trans. S. Reynolds, 2 vols (London, 1972).

Breeze, D. J., "The logistics of Agricola's final campaign", *Talanta*, 18-19 (1986-7), 7-28.

Brett, M., "The military interest of the battle of Ḥaydarān", in V. J. Parry and M. E. Yapp, eds, *War, technology and society*, 78-88.

Brett, M., "The Battles of Ramla (1099-1105)", in U. Vermulen and D. De Smet, eds, *Egypt and Syria*, 17-39.

Brett, M., *The rise of the Fatimids: the world of the Mediterranean and the Middle East in the fourth century of the Hijra, tenth century CE* (Leiden, 2001).

Brice, W. C., "Compasses, compassi and *Kanābīs*", *Journal of Semitic studies*, 29 (1984), 169-78.

Brøgger, A. W., and H. Shetelig, *The Viking ships: their ancestry and evolution* (Oslo, 1951).

Brooks, F. W., *The English naval forces 1199-1272* (1932; rpt, London, 1962).

Brooks, F. W., "William de Wrotham and the office of Keeper of the King's ports and galleys", *EHR*, 40 (1925), 570-79.

Brundage, J. A., "An errant Crusader: Stephen of Blois", *Traditio*, 16 (1960), 380-95.

Brunt, P. A., *Italian manpower, 225 B.C. – A.D. 14* (Oxford, 1971).

Bulliet, R. W., *The camel and the wheel* (Cambridge Mass., 1975).

Byrne, E. H., *Genoese shipping in the twelfth and thirteenth centuries* (Cambridge Mass., 1930).

Cameron, A., et al., eds, *The Byzantine and early Islamic Near East*, 3 vols (Princeton, 1992-5). Vol. 3: *States, resources and armies*.

Campbell, J., "The significance of the Anglo-Norman state in the administrative history of Western Europe", in W. Paravicini and K. F. Werner, eds, *Histoire comparée*, 117-34.

Caro, G., *Genua und die Mächte am Mittelmeer, 1257-1311*, 2 vols (Halle, 1895, 1899); trans. O. Soardi, *Genova e la supremazia sul Mediterraneo (1257-1311)*, 2 vols [ASLSP, n.s., 14-15] (Genoa, 1974-5).

Carrié, J.-M., and S. Janniard, "L'armée romaine tardive dans quelques travaux récents, 1: l'institution militaire et les modes de combat", *L'antiquité tardive*, 8 (2000), 321-41.

Castaing-Sicard, M., *Monnaies féodales et circulation monétaire en Languedoc, Xe-XIIIe siècles* (Toulouse, 1961).

Castiñeiras González, M. A., "Topographie sacrée, liturgie pascale et reliques dans les grands centres de pèlerinage: Saint-Jacques de Compostelle, Saint-Isidore-de-Léon et Saint-Étienne-de-Ribas-de-Sil", *Cahiers de Saint-Michel de Cuxa*, 34 (2003), 26-49.

Cazel, F. A., jr, "Financing the Crusades", in Setton, *HC*, vol. 6, 116-49.

Chenevix-Trench, C., *A history of horsemanship* (Harlow, 1970).

Christensen, A. E., "Proto-Viking, Viking and Norse craft", in idem, ed., *The earliest ships: the evolution of boats into ships* (London, 1996), 72-88.

Christiansen, E., *The Northern Crusades: the Baltic and the Catholic frontier 1100-1525* (Minneapolis, 1980).

Ch'i-ch'ing, H., *The military establishment of the Yuan dynasty* (Cambridge Mass., 1978).

Constable, G. "The Second Crusade as seen by contemporaries", *Traditio*, 9 (1953), 213-79.

Contamine, P., *War in the Middle Ages*, trans. M. Jones (Oxford, 1984).

Craig, G. A., "Delbrück: the military historian", in P. Paret, ed., *Makers of modern strategy: from Machiavelli to the nuclear age* (Princeton, 1986), 326-53.

Crone, P., "The 'Abbāsid abnā' and Sāsānid cavalrymen", *JRAS*, 8 (1998), 1-19.

Crone, P., *Slaves on horses: the evolution of the Islamic polity* (Cambridge, 1980).

Crumlin-Pedersen, O., "To be or not to be a *cog*: the Bremen *cog* in perspective", *IJNA*, 29 (2000), 230-46.

David, C. W., *Robert Curthose, Duke of Normandy* (Cambridge Mass., 1920).

De Bruyne, E., *Etudes d'esthétique médiévale. Vol. II: l'époque romane* (Bruges, 1946).

Dédéyan, G., "L'argent et le pouvoir chez les chefs arméniens de l'Euphratèse à l'époque de la Première Croisade", *ΕΥΨΥΧΙΑ: mélanges offerts à Hélène Ahrweiler*, 2 vols (Paris, 1998), 195-208.

Delbrück, H., *History of the art of war within the framework of political history*, trans. W. J. Renfroe, jr, 4 vols (Westport, 1975-85).

Delbrück, H., *Numbers in history: how the Greeks defeated the Persians, the Romans conquered the world, the Teutons overthrew the Roman Empire, and William the Norman took possession of England. Two lectures delivered before the University of London on October 6 and 7, 1913* (London, 1913).

Depeyrot, G., "Le trésor de Toulouse et le numéraire feudal aux XIIe et XIIIe siècles", *Annales du Midi*, 94 (1982), 125-49.

Destombes, M., *Mappemondes A.D. 1200-1500* [*Imago Mundi*, Supplement no. IV] (Amsterdam, 1964).

Di Cosmo, N., ed., *Warfare in inner Asian history (500-1800)* [Handbuch der Orientalistik: Abt. 8, Zentralasien, vol. 6] (Leiden, 2002).

Diffie, B. W., *Prelude to empire: Portugal overseas before Henry the Navigator* (1960; rpt, Lincoln Nebraska, 1967).

Dixon, K. R., and P. Southern, *The Roman cavalry from the first to the third century AD* (London, 1992).

Donner, F. M., "Centralized authority and military autonomy in the early Islamic conquests", in Cameron, *States, resources and armies*, 337-60.

Donner, F. M., *The early Islamic conquests* (Princeton, 1981).

Donovan, L. G., *Recherches sur le Roman de Thèbes* (Paris, 1975).

Dotson, J. E., "Economics and logistics of galley warfare", in J. Morisson, ed., *The age of the galley*, 217-23.

Dotson, J. E., "Fleet operations in the first Genoese-Venetian War, 1264-1266", *Viator*, 30 (1999), 165-80.

Dotson, J. E., "Merchant and naval influences on galley design at Venice and Genoa in the fourteenth century", in C. L. Symonds, ed., *New aspects of naval history: selected papers presented at the Fourth naval history Symposium, United States Naval Academy* (Annapolis, 1981), 20-32.

Duffy, C., *Siege warfare: the fortress in the Early Modern world, 1494-1660* (London, 1979).

Dumas, F., "Les monnaies normandes (Xe-XIIe siècles)", *Revue numismatique*, ser. 6, 21 for 1979 (1980), 84-140.

Duncalf, F., "The First Crusade: Clermont to Constantinople", in Setton, *HC*, vol. 1, 253-79.

Duncalf, F., "The Peasants' Crusade", *AHR*, 26 (1920-21), 440-53.

Duncan-Jones, R., "The choenix, the artaba and the modius", *Zeitschrift für Papyrologie und Epigraphik*, 21 (1976), 43-52.

Eadie, J., "The transformation of the eastern frontier, 260-305", in R. W. Mathisen and H. S. Sivan, eds, *Shifting frontiers in late antiquity* (Aldershot, 1996), 72-82.

Edbury, P. W., ed., *Crusade and settlement: papers read at the First Conference of the Society for the Study of the Crusades and the Latin East and presented to R. C. Smail* (Cardiff, 1985).

Eddé, A.-M., "Saint Louis et la Septième Croisade vus par les auteurs arabes", *Cahiers de recherches médiévales (XIIIe-XVe s.)*, 1 (1996), 65-92.

Edgington, S. B., "Albert of Aachen, St Bernard and the Second Crusade", in J. Phillips and M. Hoch, eds, *Second Crusade*, 54-70.

Edgington, S. B., "The Lisbon letter of the Second Crusade", *HR*, 69 (1996), 328-39.

Edgington, S. B., "The First Crusade: reviewing the evidence", in J. Phillips, ed., *First Crusade*, 57-77.

Edson, E., *Mapping time and space: how medieval mapmakers viewed their world* [The British Library studies in map making, 1] (London, 1999 [1997]).
Ehrenkreutz, A. S., "Byzantine tetartera and Islamic dīnārs", *JESHO*, 7 (1964), 183-90.
Ehrenkreutz, A. S., "Studies in the monetary history of the Near East in the Middle Ages II. The standard of fineness of western and eastern dīnārs before the Crusades", *JESHO*, 6 (1963), 243-77.
El-Azhari, T. K., *The Saljuqs of Syria during the Crusades 463-549 AH / 1070-1154 AD* (Berlin, 1997).
Ellmers, D., "The cog as cargo carrier", in Unger, *Cogs, caravels and galleons*, 29-46.
Elton, H., *Warfare in Roman Europe, AD 350-425* (Oxford 1996).
Engelbrecht, W. A., *Schets der Historische Betrekkingen Portugal-Nederland* (The Hague, 1940).
Engels, D. W., *Alexander the Great and the logistics of the Macedonian army* (Berkeley, 1978).
Epstein, H., *Domestic animals of China* (Farnham Royal, 1969).
Epstein, S. A., *Genoa and the Genoese, 958-1528* (Chapel Hill, 1996).
Evans, J. W., et al., *The horse* (San Francisco, 1977).
Ewe, H., *Schiffe auf Siegeln* (Berlin, 1972).
Finley, M. I., *Ancient history: evidence and models* (London, 1986).
Fischel, W. J., *Ibn Khaldūn and Tamerlane. Their historic meeting in Damascus, 1401 A.D. - 803 A.H.* (Berkeley, 1952).
Fischel, W. J., *Ibn Khaldūn in Egypt. His public functions and his historical research, 1382-1406* (Berkeley, 1967).
Forey, A. J., "The Second Crusade: scope and objectives", *Durham University journal*, 86 (1994), 165-75.
Forsyth, J. H., *The Byzantine-Arab chronicle (938-1034) of Yaḥyā b. Sa'īd al-Antākī*, 2 vols (Ph.D. thesis, University of Michigan, 1977).
Foster, S. M., *Some aspects of maritime activity and the use of sea power in relation to the Crusading states 1096-1169* (D.Phil. thesis, Oxford University, 1978).
Fotheringham, J., "Genoa and the Fourth Crusade", *EHR*, 25 (1910), 26-57.
Foxhall, L., and H. A. Forbes, "Σιτομετρεία: the role of grain as a staple food in classical antiquity", *Chiron*, 12 (1982), 41-90.
France, J., "Crusading warfare and its adaptation to eastern conditions in the twelfth century", *MHR*, 15 (2000), 49-66.
France, J., "The First Crusade as a naval enterprise", *MM*, 83 (1997), 389-97.
France, J., "La stratégie arménienne de la première croisade", in C. Mutafian, ed., *Les Lusignans et l'Outremer: actes du colloque Poitiers-Lusignan, octobre 1993* (Poitiers, 1995), 141-9.

France, J., *Victory in the East: a military history of the First Crusade* (Cambridge, 1994).

France. J., "The Western Mediterranean powers and the First Crusade", *Journal of Mediterranean studies*, 10 (2000), 265-74.

France. J., *Western warfare in the age of the Crusades, 1000-1300* (Ithaca, 1999).

French, D. H., "A road problem: Roman or Byzantine?", *Istanbuler Mitteilungen*, 43 (1993), 445–54.

French, D. H., *Roman roads and milestones of Asia Minor. Fasc. 1: the Pilgrim's Road* [BAR *IS*, 105] (London, 1981).

French, D. H., "The Roman road-system of Asia Minor", in ANRW, 7.2 (1980), 698-729.

French, D. H., "1989 Roma Yolları Mıltaşları ve Yazıtları Araştırması", *Araştırma Sonuçları Toplantısı*, 8 (1990), 229–40.

Friel, I., *The Good Ship: ships, shipbuilding and technology in England 1200-1520* (London, 1995).

Gautier Dalché, P., *La "Descriptio mappe mundi" de Hugues de Saint-Victor: texte inédit avec introduction et commentaire* (Paris, 1988).

Gautier Dalché, P., and J.-Y. Tilliette, "Un nouveau document sur la tradition du poème de Baudri de Bourgueil à la comtesse Adèle", *BEC*, 144 (1986), 241-57.

Gazzola, P., *Ponti Romani. Vol. 2: contributo ad un indice sistematico con studio critico bibliografico* (Florence, 1963).

Gertwagen, R., "The concept of ports in the medieval eastern Mediterranean: construction and maintenance on Crete to the end of the fifteenth Century", *IJMH*, 12 (2000), 177-241.

Gertwagen, R., "The Crusader port of Acre: layout and problems of maintenance", in Balard, *Première Croisade*, 553-82.

Gertwagen, R., "Maritime activity concerning the ports and harbours of Cyprus from the late 12th to the 16th centuries (1191-1571)", in N. Coureas and J. Riley-Smith, eds, *'Η Κύπρος καὶ οἱ Σταυροφορίες/Cyprus and the Crusades* (Nicosia, 1995), 511-38.

Gertwagen, R., "Does naval activity – military and commercial – need artificial ports? The case of Venetian harbours and ports in the Ionian and Aegean till 1500", forthcoming in *Graeco-Arabica.*

Gertwagen, R., "The Venetian colonies in the Ionian and Aegean Seas in Venetian defensive policy in the fifteenth century", *JMS*, 12 (2002), 351-84.

Gertwagen, R. "Venetian Modon and its port, 1358-1500", in A. Cowan, ed., *Mediterranean urban culture, 1400-1700* (Exeter, 2000), 125-48, 248-54.

Gertwagen, R., "The Venetian port of Candia, Crete, 1299-1363: construction and maintenance", *MHR*, 3 (1988), 141-58.

Gibb, H. A. R., "The armies of Saladin", *Cahiers d'histoire égyptéenne*, ser. 3, 4 (1951), 304-20. Rpt in his *Studies on the civilization of Islam* (London, 1962), 74-90.

Gillingham, J., *Richard I* (New Haven and London, 1999).

Gladitz, C., *Horse breeding in the medieval world* (Dublin 1997).

Goffart, W., *The narrators of barbarian history (A.D. 550-800): Jordanes, Gregory of Tours, Bede, and Paul the Deacon* (Princeton, 1988).

Golden, P. B., "War and warfare in the Pre-Cinggisid western steppes of Eurasia", in di Cosmo, *Warfare in inner Asian history*, 105-72.

Goldsworthy, A. K., *The Roman army at war, 100 BC – AD 200* (Oxford, 1996).

Goodich, M., et al., eds, *Cross cultural convergences in the Crusader period: essays presented to Aryeh Grabois on his sixty-fifth birthday* (New York, 1995).

Grabar, O., "Imperial and urban art in Islam: the subject matter of Fatimid art", in *Colloque international sur l'histoire du Caire, 27 mars-5 avril 1969* (Cairo, 1972), 1-57. Rpt in his *Studies in medieval Islamic art* (London, 1976), No. VII.

Greatrex, G., "The dates of Procopius' works", *BMGS*, 18 (1994), 101-14.

Grierson, P., *Byzantine coinage* (Washington, 1982).

Grierson, P., "The origins of the grosso and of gold coinage in Italy", *Numismaticky Sborník*, 12 (1971-2), 33-48.

Grousset, R., *The Empire of the steppes: a history of central Asia*, trans. N. Walford (New Brunswick, 1970).

Haldon, J., *Byzantium in the seventh century: the transformation of a culture*, 2nd ed. rev. (Cambridge, 1997).

Haldon, J., "The organisation and support of an expeditionary force: manpower and logistics in the Middle Byzantine period", in N. Oikonomides, ed., *Τό εμπόλεμο Βυζάντιο (9ος - 12ος αι.)/Byzantium at war (9th-12th c.)* (Athens, 1997), 111-51.

Haldon, J., *Warfare, state and society in the Byzantine world 565-1204* (London, 1999).

Halm, H., *Ägypten nach den mamlukischen Lehensregistern*, 2 vols (Wiesbaden, 1979-82).

Hamblin, W. J., *The Fatimid army and the early Crusades: 1099-1124* (Ph.D. thesis, University of Michigan, 1985).

Harari, Y., "The military role of the Frankish Turcopoles: a reassessment", *MHR*, 12 (1997), 75-116.

Hardenberg, H., *De Nederlanden en de Kruistochten* (Amsterdam, 1941).

Harley, J. B., and D. Woodward, eds, *The history of cartography. Vol. One: cartography in prehistoric, ancient and medieval Europe and the Mediterranean* (Chicago and London, 1987). *Vol. Two, Book One:*

cartography in the traditional Islamic and South Asian societies (Chicago and London, 1992).
Harvey, A., *Economic expansion in the Byzantine empire, 900-1200* (Cambridge 1989).
Harvey, P. D. A., "Medieval maps: an introduction", in Harley and Woodward, *History of cartography. 1*, 283-5.
Hasson, I., "Les mawālī dans l'armée musulmane sous les premièrs Umayyades", *JSAI*, 14 (1991), 176-213.
Hattendorf, J., and R. W. Unger, eds, *War at sea in the Middle Ages and the Renaissance* (Woodbridge, 2002).
Hayes, M. H., *Horses on board ship: a guide to their management* (London, 1902).
Hendy, M. F., *Coinage and money in the Byzantine Empire, 1081-1261* (Washington, 1969).
Hiestand, R., "The Papacy and the Second Crusade", in J. Phillips and M. Hoch, eds, *Second Crusade*, 32-53.
Hiestand, R., "»Precipua tocius christianismi columpna«. Barbarossa und der Kreuzzug", in A. Haverkamp, ed., *Friedrich Barbarossa: Handlungsspielräume und Wirkungsweisen des staufischen Kaisers* (Sigmaringen, 1992), 51-108.
Hiestand, R., "Der erste Kreuzzug in der Welt des ausgehenden 11 Jahrhunderts", *Monatshefte für evangelische Kirchengeschichte des Rheinlandes*, 44 (1995), 1-36.
Hild, F., *Das byzantinische Strassensystem in Kappadokien* [Veröffentlichungen der Kommission für die *TIB* 2. Denkschriften der Österreichischen Akad. der Wiss., phil.-hist. Kl., 131] (Vienna, 1977).
Hill, R., "The role of the camel and the horse in the early Arab conquests", in V. J. Parry and M. E. Yapp, eds, *War, technology and society*, 32-43.
Hocquet, J.-C., "Tonnages ancien et moderne: botte de Venise et tonneau anglais", *Revue historique*, 281 (1989), 349-60.
Hodge, A. T., *Roman aqueducts and water supply*, 2nd ed. (London, 2002).
Hooper, N., "Edgar the Ætheling: Anglo-Saxon prince, rebel and Crusader", *Anglo-Saxon England*, 14 (1985), 197-214.
Howard-Johnston, J., "Anna Komnene and the *Alexiad*", in M. Mullett and D. Smythe, eds, *Alexios I Komnenos. I: Papers* (Belfast, 1996), 260-302.
Howorth, H. H., *History of the Mongols from the 9th to the 19th century*, 4 vols (London, 1876-1927).
Hurter, F., *Storia di Papa Innocenzio III*, trans. T. G. Gliemone, 4 vols (Milan, 1857-8).
Hutchinson, G., *Medieval ships and shipping* (London, 1994).
Hyland, A., *Equus: the horse in the Roman world* (New Haven, 1990).
Inalcık, H., ed., with D. Quataert, *An economic and social history of the Ottoman empire, 1300-1914* (Cambridge, 1994).

Irwin, R., *The Middle East in the Middle Ages: the early Mamluk sultanate 1250-1382* (London, 1986).

Irwin, R., "Gunpowder and firearms in the Mamluk sultanate reconsidered", in A. Levanoni and M. Winter, eds, *The Mamluks in Egyptian and Syrian politics and society* (Leiden, 2004), 117-39.

Jacoby, D., "A Venetian manual of commercial practice from Crusader Acre", in G. Airaldi and B. Z. Kedar, eds, *I comuni Italiani nel Regno Crociato di Gerusalemme: atti del Colloquio "The Italian communes in the Crusading Kingdom of Jerusalem" (Jerusalem, May 24 – May 28, 1984)* [Collana storica di fonti e studi, 48] (Genoa, 1986), 401-28.

Jandora, J. W., "The battle of the Yarmūk: a reconstruction", *JAH*, 19 (1985), 8-21.

Jandora, J. W., "Developments in Islamic warfare: the early conquests", *SI*, 64 (1986), 101-13.

Jesch, J., *Ships and men in the late Viking age* (Woodbridge, 2001).

Jesse, W., *Quellenbuch zur Münz- und Geldgeschichte des Mittelalters* (1924; rpt, Aalen, 1968).

Johns, J., and E. Savage-Smith, "The book of curiosities: a newly discovered series of Islamic maps", *Imago Mundi* (forthcoming).

Johnson, E. N., "The German Crusade on the Baltic", in Setton *HC*, vol. III, 545-85.

Kaegi, W. E., "The contribution of archery to the Turkish conquest of Anatolia", *Speculum*, 39 (1964), 96-108.

Kaegi, W. E., *Byzantium and the early Islamic conquests* (Cambridge, 1992).

Kalligas, H., *Byzantine Monemvasia: the sources* (Monemvasia, 1990).

Kalligas, H., "Monemvasia, seventh-fifteenth centuries", in A. E. Laiou, ed. *Economic history of Byzantium*, vol. 2, 879-97.

Kedar, B. Z., "The subjected Muslims of the Frankish Levant", in J. M. Powell, ed., *Muslims under Latin rule, 1100-1300* (Princeton, 1990), 135-74. Rpt in his *The Franks in the Levant, 11th to 14th centuries* (Aldershot, 1993), No. XVIII.

Kellner, M. G., *Die Ungarneinfälle im Bild der Quellen bis 1150: von der "gens detestanda" zur "gens ad fidem Christi conversa"* (Munich, 1997).

Kennedy, H., *The armies of the Caliphs: military and society in the early Islamic state* (London, 2001).

Kennedy, H., "The financing of the military in the early Islamic state", in Cameron, *States, resources and armies*, 361-78.

Kennedy, H., *Muslim Spain and Portugal: a political history of al-Andalus* (London, 1996).

King, D. A., *World-Maps for finding the direction and distance to Mecca: innovation and tradition in Islamic Science* (London and Leiden, 1999).

Köhler, M. A., *Allianzen und Verträge zwischen frankischen und islamischen Herrschern im Vorderen Orient* (Berlin, 1991).

Krueger, H. C., *Navi e proprietà navale a Genova, seconda metà del sec. XII* (Genoa, 1985).

Kupfer, M., "Medieval world maps: embedded images, interpretive frames", *Word & image*, 10 (1994), 262-88.

Kupper, J.-L., "Otbert de Liège: les manipulations monétaires d'un évêque d'Empire à l'aube du XIIe siècle", *Le Moyen Age*, 86 (1980), 353-85.

Laiou, A. E., ed., *The economic history of Byzantium from the seventh through the fifteenth century*, 3 vols paginated continuously (Washington, 2002).

Laiou, A. E., "Byzantine trade with Christians and Muslims and the Crusades", in idem and R. P. Mottahedeh, eds, *The Crusades*, 157-96.

Laiou, A. E., "Exchange and trade, seventh–twelfth centuries", in idem, ed., *Economic history of Byzantium*, vol. 2, 681-770.

Laiou, A. E., and R. P. Mottahedeh, eds, *The Crusades from the perspectives of Byzantium and the Muslim world* (Washington, 2001).

Landau-Tasseron, E., "Features of the pre-conquest Muslim armies in the time of Muḥammad", in Cameron, *States, resources and armies*, 299-336.

Lane, F. C., "Diet and wages of seamen in the early fourteenth century", in his *Venice and history: the collected papers of Frederic C. Lane* (Baltimore, 1966), 263-8 [trans from his "Salaires et régime alimentaire des marins au début du XIVe siècle", *Annales: économies, sociétés, civilizations*, 18 (1963), 133-8].

Lane, F. C., *Venetian ships and shipbuilders of the Renaissance* (Baltimore, 1934).

Lane, F. C., *Venice: a maritime republic* (Baltimore, 1973).

Lane Poole, A., "Germany, 1125-1152", in CMH, vol. 5, 334-59.

Langdon, J., *Horses, oxen and technological innovation: the use of draught animals in English Farming from 1066-1500* (Cambridge, 1986).

Larin, I. V., *Pasture economy and meadow cultivation*, trans from the Russian ed. of 1956 (Jerusalem, 1962).

Lefebvre des Noëttes, R., *L'attelage: le cheval de selle à travers les âges*, 2 vols (Paris, 1931).

Leighton, A. C., *Transport and communications in early medieval Europe AD 500-1100* (Newton Abbot, 1972).

Lev, Y., "Army, regime, and society in Fatimid Egypt, 358-487/968-1094", *IJMES*, 19 (1987), 337-65.

Lev, Y., "Regime, army and society in medieval Islam, 9th-12th centuries", in idem, *War and society in the Eastern Mediterranean*, 115-52.

Lev, Y., *Saladin in Egypt* (Leiden, 1999).

Lev, Y., ed., *War and society in the Eastern Mediterranean, 7th-15th centuries* (Leiden, 1997).

Lewis, A. R., "Northern European sea power and the Straits of Gibraltar, 1031-1350 A.D.", in W. C. Jordan et al., eds, *Order and innovation in the Middle Ages: essays in honor of Joseph R. Strayer* (Princeton, 1976), 139-64.

Lewis, A. R., *Naval power and trade in the Mediterranean, 500-1100* (Princeton, 1951).

Lewis, A. R., and T. J. Runyan, *European naval and maritime history, 300-1500* (Bloomington, 1985).

Lewis, B., *Race and color in Islam* (N. Y., 1970).

Lewis, B., *Race and slavery in the Middle East: an historical enquiry* (New York, 1990).

Leyser, K., "Money and supplies on the First Crusade", in idem, *Communications and power in medieval Europe: the Gregorian revolution and beyond*, ed. T. Reuter (London, 1994), 77-95.

Lilie, R.-J., *Byzantium and the Crusader states 1096-1204*, trans. J. C. Morris and J. E. Ridings (Oxford, 1993).

Lilie, R.-J., *Handel und Politik zwishen dem byzantinischen Reich und den italienischen Kommunen: Venedig, Pisa, und Genua in der Epoche der Komnenen und der Angeloi (1081-1204)* (Amsterdam, 1984).

Lindemann, R. H. F., "The English *esnecca* in northern European societies", *MM*, 74 (1988), 75-82.

Livermore, H., "The Conquest of Lisbon and its author", *Portuguese Studies*, 6 (1990), 1-16.

Livermore, H., *A new history of Portugal* (Cambridge, 1966).

Lomax, D. W., *The reconquest of Spain* (London, 1978).

Lopez, R. S., *Storia delle colonie genovesi nel Mediterraneo* (Bologna, 1938).

Lopez, R. S., "Majorcans and Genoese on the North Sea route in the thirteenth century", *Revue Belge de philologie et d'histoire*, 29 (1951), 1163-79.

LoPrete, K. A., "Adela of Blois as mother and countess", in J. C. Parsons and B. Wheeler, eds, *Medieval mothering* (New York and London, 1996), 313-33.

LoPrete, K. A., "The Anglo-Norman card of Adela of Blois", *Albion*, 22 (1990), 569-89.

LoPrete, K. A., *Adela of Blois, countess and lord* (Dublin, 2003).

Luschin von Ebengreuth, A., *Allgemeine Münzkunde und Geldgeschichte des Mittelalters und der neueren Zeit*, 2nd ed. (Munich and Berlin, 1926).

Lynn, J. A., ed., *Feeding Mars: logistics in Western warfare from the Middle Ages to the present* (Boulder, 1993).

Lynn, J. A., "Food, funds, and fortresses: resource mobilization and positional warfare in the campaigns of Louis XIV", in idem, *Feeding Mars*, 137-59.

Macrides, R., ed., *Travel in the Byzantine world: papers from the Thirty-fourth Spring Symposium of Byzantine studies, Birmingham, April 2000* (Aldershot, 2002).

Madden, T. F., *Enrico Dandolo and the rise of Venice* (Baltimore, 2003).

Madden, T. F., "Vows and contracts in the Fourth Crusade: the Treaty of Zara and the attack on Constantinople in 1204", *International History Review*, 15 (1993), 441-68.

Magdalino, P., "The grain supply of Constantinople, ninth–twelfth centuries", in C. Mango and G. Dagron, eds, *Constantinople and its hinterland*, 35-47.

Malamut, E., *Les îles de l'Empire Byzantin VIIIe – XIIe siècles*, 2 vols paginated continuously (Paris, 1988).

Mango, C. "The water supply of Constantinople", in idem and G. Dagron, eds, *Constantinople and its hinterland*, 9-18.

Mango, C., and G. Dagron, eds, *Constantinople and its hinterland: papers from the Twenty-seventh Spring Symposium of Byzantine studies, Oxford, April, 1993* (Aldershot, 1995).

Manitius, M., *Geschichte der lateinischen Literatur des Mittelalters*, 3 vols (1911-31; rpt, Munich, 1965-74).

Marshall, C. J., "The use of the charge in battles in the Latin East, 1192-1291", *HR*, 63 (1990), 221-6.

Martin, E. E., *The transport of horses by sea* (Calcutta, 1901).

Martinez, A. P., "Some notes on the Īl-Xānid army", *Archivum Eurasiae medii aevi*, 6 (1986), 129-242.

Matz, S. A., *The chemistry and technology of cereals as food and feed* (Westport, 1959).

Matzke, M., *Daibert von Pisa. Zwischen Pisa, Papst und erstem Kreuzzug* (Sigmaringen, 1998).

Matzke, M., "Vom Ottolinus zum Grossus — Münzpragung in der Toskana vom 10. bis zum 13. Jahrhundert", *Schweizer Numismatische Rundschau*, 72 (1993), 135-99.

Maurice, F., "The size of the army of Xerxes in the invasion of Greece, 480 B.C.", *Journal of Hellenic studies*, 50 (1930), 210-35.

Mayer, H., *The Crusades*, trans. J. Gillingham, 2nd ed. (Oxford, 1988).

Mayhew, N., *Coinage in France from the Dark Ages to Napoleon* (London, 1988).

McCormick, M., *Origins of the European economy: communications and commerce, A.D. 300–900* (Cambridge, 2001).

McGeer, E., *Sowing the dragon's teeth: Byzantine warfare in the tenth century* (Washington, 1995).

McNeal, E. H., and R. L. Wolff, "The Fourth Crusade", in Setton, *HC*, vol. 2, 153-85.

Metcalf, D. M., *Coinage of the Crusades and the Latin East in the Ashmolean Museum, Oxford*, 2nd ed. (London, 1995).

Metcalf, D. M., "Coins of Lucca, Valence and Antioch: some new hoards and stray finds from the time of the Crusades", *Hamburger Beiträge zur Numismatik*, 22-3 (for 1968-9; 1972), 443-70.

Metcalf, D. M., "The Templars as bankers and monetary transfers between West and East in the twelfth century", in *Coinage in the Latin East: the Fourth Oxford Symposium on coinage and monetary history*, eds P. W. Edbury and D. M. Metcalf [BAR *IS*, vol. 77] (Oxford, 1980), 1-17.

Meyvaert, P., "An unknown letter of Hulagu, Il-Khan of Persia, to king Louis IX of France", *Viator*, 11 (1980), 245-60.

Miller, K., *Mappae arabicae: Arabische Welt- und Länderkarten des 9.-13. Jahrhunderts*, 6 vols (Munich, 1926-31) [also rpt, Frankfurt am Main, 1994].

Miller, K., *Mappaemundi: die ältesten Weltkarten*, 6 fasc. (Stuttgart, 1895-8).

Morgan, D. O., "The Mongols in Syria, 1260-1300", in Edbury, *Crusade and settlement*, 231-5.

Morris, C., "Peter the Hermit and the chroniclers", in J. Phillips, ed., *First Crusade*, 21-34.

Morrison, J., ed., *The age of the galley: Mediterranean oared vessels since pre-classical times* (London, 1995).

Mouton, J.-M., et al., "La route de Saladin (tariq Sadr wa Ayla) au Sinaï", *Annales Islamologiques*, 30 (1996), 41-71.

Mouton, J.-M., *Le Sinaï médiéval: un espace stratégique de l'Islam* (Paris, 2000).

Mulinder, A., "Albert of Aachen and the Crusade of 1101", in A. V. Murray, ed., *Clermont to Jerusalem*, 69-77.

Murphey, R., *Ottoman warfare, 1500-1700* (London 1999).

Murray, A. V., *The Crusader Kingdom of Jerusalem: a dynastic history, 1099-1125* (Oxford, 2000).

Murray, A. V., ed., *From Clermont to Jerusalem: the Crusades and Crusader societies 1095-1500. Selected proceedings of the International Medieval Congress, University of Leeds 10-13 July, 1995* (Turnhout, 1998).

Musset, L., "Réflexions sur les moyens de paiement en Normandie aux XIe et XIIe siècles", in L. Musset et al., eds, *Aspects de la société et de l'économie dans la Normandie médiévale (Xe-XIIIe siècles)* (Caen, 1988), 65-89.

Nesbitt, J. W., "The rate of march of Crusading armies in Europe", *Traditio*, 19 (1963), 167-81.

Nicolle, D., "The impact of the European couched lance on Muslim military tradition", *The journal of the arms and armour society*, 10 (1980), 6-40.

Northrup, L., *From slave to sultan: the career of al-Manṣūr Qalāwūn and the consolidation of Mamluk rule in Egypt and Syria (678-689 A.H./1279-1290 A.D.)* (Stuttgart, 1998).

O'Connor, C., *Roman bridges* (Cambridge, 1993).

Oikonomides, N., ed., *Byzantine Asia Minor (6th-12th cents)/Η Βυζαντινή Μικρά Ασία· 6.-12. αι*, ed. S. Lampakis (Athens, 1998).

Olsen, O., and O. Crumlin-Pedersen, "The Skuldelev Ships (II)", *Acta Archaeologica*, 38 (1967), 73-174.

Ortalli, G., "Venezia e Creta. Fortune e contraccolpi di una conquista", in idem, ed., *Venezia e Creta: Atti del Convegno internazionale di studi Iraklion-Chanià, 30 settembre – 5 ottobre 1997* (Venice, 1998), 9-31.

O'Sullivan, F., *The Egnatian Way* (Newton Abbot, 1972).

Painter, S., "The Third Crusade: Richard the Lionhearted and Philip Augustus", in Setton, *HC*, vol. 2, 45-85.

Pallucchini, R., *La pittura veneziana del Trecento* (Venice and Rome, 1964).

Paravicini, W., and K. F. Werner, eds, *Histoire comparée de l'administration (IVe-XVIIIe siècles): actes du XIVe colloque historique franco-allemand, Tours, 27 mars – 1er avril 1977 organisé en collaboration avec le Centre d'études supérieures de la Renaissance par l'Institut historique allemand de Paris* (Munich, 1980).

Parry, V. J., and M. E. Yapp, eds, *War, technology and society in the Middle East* (London, 1977).

Paviot, J., "Comment reconquérir la Terre sainte et vaincre les Sarrasins?", in Balard, *Dei gesta per Francos*, 79-85.

Peddie, J., *Invasion: the Roman invasion of Britain in the year AD 43 and the events leading to their occupation of the West Country* (New York, 1987).

Perjés, C., "Army provisioning, logistics and strategy in the second half of the 17th century", *Acta historica Academiae Scientiarum Hungaricae*, 16 (1970), 1-51.

Phillips, J., *The Crusades, 1095-1197* (London, 2002).

Phillips, J., "Ideas of Crusade and Holy War in *De expugnatione Lyxbonensi (The Conquest of Lisbon)*", in R. N. Swanson, ed., *The Holy Land, holy lands and Christian history* [SCH, vol 36] (Woodbridge, 2000), 123-41.

Phillips, J., "Papacy, Empire and the Second Crusade", in J. Phillips and M. Hoch, eds, *Second Crusade*, 15-31.

Phillips, J., ed., *The First Crusade: origins and impact* (Manchester, 1997).

Phillips, J., and M. Hoch, eds, *The Second Crusade: scope and consequences* (Manchester, 2001).

Phillips, M., "A hoard of French feudal coins from the First Crusade", *Numismatic chronicle*, 162 (2002), 402-4.

Porteous, J., "Crusader coinage with Greek or Latin inscriptions", in Setton, *HC*, vol. 6, 354-420.

Prestwich, M., *Edward I* (London, 1988).

Pryor, J. H., "The Crusade of Emperor Frederick II, 1220-29: the implications of the maritime evidence", *The American Neptune*, 52 (1992), 113-32.

Pryor, J. H., "From dromōn to galea: Mediterranean bireme galleys AD 500-1300", in J. Morrison, ed., *Age of the galley*, 101-16.

Pryor, J. H., "The galleys of Charles I of Anjou, King of Sicily: ca. 1269-84", *Studies in medieval and Renaissance history*, n.s., 14 (1993), 33-103.

Pryor, J. H., *Geography, technology, and war: studies in the maritime history of the Mediterranean, 649-1571* (Cambridge, 1988).

Pryor, J. H., "The maritime republics", in NCMH, vol. 5, 419-46.

Pryor, J. H., "The Mediterranean round ship", in R. W. Unger, ed., *Cogs, caravels and galleons*, 59-76.

Pryor, J. H., "The naval architecture of Crusader transport ships and horse transports revisited", *MM*, 76 (1990), 255-73.

Pryor, J. H., "Stephen of Blois: sensitive new-age Crusader or victim of history?", *Arts: the journal of the Sydney University Arts Association*, 20 (1998), 26-74.

Pryor, J. H., "Transportation of horses by sea during the era of the Crusades: eighth century to 1285 A.D.", *MM*, 68 (1982), 9-27, 103-125.

Pryor, J. H., "The Venetian fleet for the Fourth Crusade and the diversion of the Crusade to Constantinople", in M. Bull and N. Housley, eds, *The experience of* Crusading, 2 vols (Cambridge, 2003). *Vol. 1: Western approaches*, 103-123.

Pryor, J. H., "The voyages of Saewulf", in R. B. C. Huygens, ed., *Peregrinationes tres: Saewulf, John of Würzburg, Theodericus*, [CC *ContMed*, vol. 139] (Turnhout, 1994), 34-57.

Pryor, J. H., " 'Water, water everywhere, Nor any drop to drink.': water supplies for the fleets of the First Crusade", in Balard, *Dei gesta per Francos*, 21-8.

Pryor, J. H., "Winds, waves, and rocks: the routes and the perils along them", in K. Friedland, ed., *Maritime aspects of migration* (Cologne, 1989), 71-85.

Queller, D. E., and T. F. Madden, *The Fourth Crusade: the conquest of Constantinople*, 2nd ed. (Philadelphia, 1997).

Raepsaet, G., "Attelage antiques dans le nord de la Gaule: les systèmes de traction par équides", *Trierer Zeitschrift*, 42 (1982), 215-73.

Raepsaet, G., "La faiblesse de l'attelage antique: la fin d'un mythe?", *L'antiquité classique*, 48 (1979), 171-6.

Ramsay, W. M., *The historical geography of Asia Minor* (London, 1890).

Ratkowitsch, C., *Descriptio Picturae. Die literarische Funktion der Beschreibung von Kunstwerken in der lateinischen Großdichtung des 12. Jahrhunderts* [Wiener Studien, 15] (Vienna, 1991).

Reuter, T., "Carolingian and Ottonian warfare", in M. Keen, ed., *Medieval warfare: a history* (Oxford, 1999), 13-35.

Reuter, T., *Germany in the early Middle Ages, 800-1056* (London and New York, 1991).

Reynolds, P. J., *Iron-Age farm: the Butser experiment* (London, 1979).

Riley-Smith, J., *The Crusades: a short history* (New Haven and London, 1987).

Riley-Smith, J., "Early Crusaders to the East and the costs of Crusading, 1095-1130", in Goodich, *Cross cultural convergences*, 237-57.

Riley-Smith, J., *The First Crusade and the idea of Crusading* (London, 1986).

Riley-Smith, J., *The First Crusaders, 1095-1131* (Cambridge, 1997).

Riley-Smith, J., *What were the Crusades*, 3rd ed. (Houndmills, 2002).

Robbert, L. B., "Venetian participation in the Crusade of Damietta", *Studi Veneziani*, n.s., 30 (1995), 15-33.

Roberts, I. P., *The horse* (New York, 1905).

Rodger, N. A. M., *The safeguard of the sea: a naval history of Britain. Vol. I: 660-1649* (London, 1997).

Röring, C. W., *Untersuchungen zu römischen Reisewagen* (Koblenz, 1983).

Roth, J. P., *The logistics of the Roman army at war (264 B.C.-A.D. 235)* (Leiden, 1999).

Rousset, P., "Étienne de Blois, croisé fuyard et martyr", *Genava*, n.s., 11 (1963), 183-95.

Runciman, S., *A history of the Crusades*, 3 vols (1951-54; rpt, Harmondsworth, 1965).

Runyan, T. J., "Naval logistics in the Late Middle Ages: the example of the Hundred Years' War", in Lynn, *Feeding Mars*, 79-100.

Russell, J. C., *Late ancient and medieval population* [TAPS, 43, no. 3] (Philadelphia, 1958).

Sato, T., *State and rural society in medieval Islam: sultans, muqtās and fallahun* (Leiden, 1997).

Savage, E., *A gateway to hell, a gateway to paradise: the North African response to the Arab conquest* (Princeton, 1997).

Sayyid, A. F., *Al-Dawla al-fāṭimiyya fī Miṣr. Tafsīr jadīd*, 2nd ed. (Cairo, 2000).

Searle, J., *The construction of social reality* (New York, 1995).

Schulze, I. and W., "A coin hoard from the time of the First Crusade, found in the Near East", *Revue numismatique*, 159 (2003), 323-53.

Sezgin, F., *The contribution of the Arabic-Islamic geographers to the formation of the world map* (Frankfurt, 1987).

Shepard, J., "Cross-purposes: Alexius Comnenus and the First Crusade", in J. Phillips, ed., *First Crusade*, 107-29.

Shirley, A., *Remarks on the transport of cavalry and artillery* (London, 1854).

Sinor, D., "Horse and pasture in Inner Asian history", *Oriens extremus*, 19 (1972), 171-83. Rpt in his *Inner Asia and its contacts with medieval Europe* (London, 1977), No. II.

Sitwell, N. H. H., *Roman roads of Europe* (London, 1981).

Slicher van Bath, B. H., *The agrarian history of western Europe, A.D. 500-1850*, trans. O. Ordish (London, 1963).

Smith, C. D., "Geography or Christianity? Maps of the Holy Land before AD 1000", *Journal of theological studies*, 42 (1991), 143-52.

Smith, F., *A manual of veterinary hygiene* (London, 1905).

Smith, J. M., jr, " 'Ayn Jālūt: Mamlūk success or Mongol failure", *HJAS*, 44 (1984), 307-45.

Smith, J. M., jr, "Mongol nomadism and Middle Eastern geography: Qīshlāqs and Tümens", in Amitai-Preiss and Morgan, *The Mongol Empire and its legacy*, 39-56.

Smith, J. M., jr, "Nomads on ponies vs. slaves on horses", *JAOS*, 118 (1998), 154-62.

Spruytte, J., *Early harness systems. Experimental studies: a contribution to the history of the horse*, trans. M. Littauer (London, 1983).

Spufford, P., *Money and its use in medieval Europe* (Cambridge, 1988).
Strickland, M., ed., *Armies, chivalry and warfare in medieval Britain and France: proceedings of the 1995 Harlaxton Symposium* (Stamford, 1998).
Stumpf, G., *Der Kreuzzug Kaiser Barbarossas. Münzschätze seiner Zeit* (Munich, 1993).
Tahir, G. M. A. al-, "The Nubian archers in pre-Islamic and Islamic periods", *Graeco-Arabica*, 5 (1993), 139-52.
Talbi, M., "Law and economy in Ifriqiya (Tunisia) in the third Islamic century: agriculture and the role of slaves in the country's economy", in A. L. Udovitch, ed., *The Islamic Middle East, 700-1900: studies in economic and social history* (Princeton, 1981), 209-49.
Talbi, M., *L'émirat Aghlabide* (Paris, 1966).
Teall, J. L., "The grain supply of the Byzantine Empire, 330-1025", *DOP*, 13 (1959), 89-139.
Tibbetts, G. R., "The beginnings of a cartographic tradition", "The Balkhī school of geographers", "Later cartographic developments", in Harley and Woodward, eds, *History of cartography. 2.1*, 90-107, 108-36, 137-55 respectively.
Thomas, R. D., "Anna Comnena's Account of the First Crusade", *BMGS*, 15 (1991), 269-312.
Tilliette, J.-Y., "La chambre de la comtesse Adèle: savoir scientifique et technique littéraire dans le c. CXCVI de Baudri de Bourgueil", *Romania*, 102 (1981), 145-71.
Tourneur, V., "Un denier de Godefroid de Bouillon frappé en 1096", *Revue Belge de numismatique et de sigillographie*, 83 (1931), 27-30.
Tucci, U., "L'alimentazione a bordo delle navi veneziane", *Studi veneziani*, n. s., 13 (1987), 103-45.
Tucci, U., "L'impresa marittima: uomini e mezzi", in G. Cracco and G. Ortalli, eds, *Storia di Venezia*, (Rome, 1995), vol. 2, 627-59.
Tyerman, C., *England and the Crusades 1095-1588* (Chicago, 1988).
Tyerman, C., *The invention of the Crusades* (London, 1998).
Unger, R. W., "The Netherlands herring fishery in the Late Middle Ages: the false legend of William Beukels of Biervliet", *Viator*, 9 (1978), 335-56.
Unger, R. W., "Portuguese shipbuilding and the early voyages to the Guinea Coast", in *Vice-Almirante A. Teixeira da Mota. In memoriam*, 2 vols (Lisbon, 1987-9), vol. 1, 229-49.
Unger, R. W., *The ship in the medieval economy 600-1600* (London and Montreal, 1980).
Unger, R. W., ed., *Cogs, caravels and galleons: the sailing ship 1000-1650* (London, 1994).
Van Creveld, M., *Supplying war: logistics from Wallerstein to Patton* (Cambridge, 1977).

Van Creveld, M., *Technology and war: from 2000 B.C. to the present* (New York and London, 1989).

Van den Brincken, A.-D., "Die Ausbildung konventioneller Zeichen und Farbgebungen in der Universalkartographie des Mittelalters", *Archiv für Diplomatik, Schriftgeschichte, Siegel- und Wappenkunde*, 16 (1970), 325-49.

Van Werveke, H., "Monnaie, lingots, ou marchandises? Les instruments d'échange aux XIe et XIIe siècles", *AHES*, 4 (1932), 452-68.

Verbruggen, J. F., *The art of warfare in Western Europe during the Middle Ages: from the eighth century to 1340*, trans. S. Willard and Mrs R. W. Southern, 2nd ed. rev. (Woodbridge, 1997).

Vermulen, U., and D. de Smet, eds, *Egypt and Syria in the Fatimid, Ayyubid, and Mamluk eras. I: Proceedings of the 1st, 2nd, and 3rd international colloquium organized at the Katholieke Universiteit Leuven in May 1992, 1993, and 1994* (Louvain, 1995).

Vidier, A., "La mappemonde de Théodulphe et la mappemonde de Ripoll (IXe-XIe siècles)", *Bulletin de géographie historique et descriptive*, 26 (1911), 285-313.

Von Hagen, V. W., *The roads that led to Rome* (London, 1967).

Von Schlosser, J., "Ein veronesisches Bilderbuch und die höfische Kunst des XIV. Jahrhunderts", *Jahrbuch der kunsthistorischen Sammlungen des Allerhöchsten Kaiserhauses*, 16 (1895), 144-230. Italian trans., *L'arte di corte nel secolo decimoquarto* (Cremona, 1965).

Vryonis, S., jr, *The decline of medieval Hellenism in Asia Minor and the process of Islamization from the eleventh through the fifteenth century* (Berkeley, 1971).

Waas, A., *Geschichte der Kreuzzüge*, 2 vols (Freiburg, 1956).

Watson, A. M., *Agricultural innovation in the early Islamic world: the diffusion of crops and farming techniques, 700-1100* (Cambridge, 1983).

Watson, G., *The Roman soldier* (London, 1969).

Werckmeister, O. K., "The political ideology of the Bayeux Tapestry", *Studi Medievali*, ser. 3, 17 (1976), 535-96.

Werner, K. F., "Heeresorganization und Kriegsführung im deutschen Königreich des 10. und 11. Jahrhunderts", [SSCISAM, 15] (Spoleto, 1968), 791-856.

Werner, K. F., "*Missus-Marchio-Comes*: entre l'administration centrale et l'administration locale de l'Empire Carolingien", in Paravacini and Warner, *Histoire comparée*, 191-239.

Weski, T., "The Ijsselmeeer type: some thoughts on Hanseatic *cogs*", *IJNA*, 28 (1999), 360-79.

Whitby, M., "Justinian's bridge over the Sangarius and the date of Procopius' *De Aedificiis*", *JHS*, 105 (1985), 129-48.

White, K. D., *Roman farming* (Ithaca, 1970).

Williams, J., *The illustrated Beatus: a corpus of the illustrations of the commentary on the Apocalypse*, 5 vols (London, 1994-2002).

Williams, J., "Isidore, Orosius, and the Beatus map", *Imago mundi*, 49 (1997), 7-32.

Wolseley, G. J. [General Viscount], *The soldier's pocket-book for field service*, 5th ed. (London, 1886).

Woodward, D., "Medieval *mappaemundi*", in Harley and Woodward, *History of cartography. 1*, 286-370.

Wright, J. K., *The geographical lore of the time of the Crusades: a study in the history of medieval science and tradition in Western Europe* (New York, 1925).

Yewdale, R. B., *Bohemond I, Prince of Antioch* (Princeton, 1924).

Index

'Abbāsids, Caliphal dynasty, 133, 186-8
  see also military forces
al-Abbāsiyya, 186, 297
'abd, pl. *'abīd*, black slave soldiers, 186-8 & n. 9, 192
  see also military forces
abnā' (al-dawla), 186, 293
Abrahams, P., 180 n. 53, 181 n. 59
Abū Muḥammad 'Abd al-Wāḥid II al-Rashīd, Almohad Caliph, 69-70
Abū Yazīd, 188
Abū-Yusūf Ya'qūb, al-Manṣūr, Almohad Caliph, 266
Abydos, 79, 222, 293, 297
Acre, 39, 63, 65, 70-71, 73, 97-8, 100-102, 111, 114-16, 198, 200, 254, 258, 260, 265, 267-8, 273
  bay, 114
  port, 114
  siege of (1189-91), 195, 205, 259
Adam and Eve, 172 & n. 24
Adela, countess of Blois, 177-81, 243
Adhemar, bishop of Le Puy, 236
al-'Adil, Ayyūbid *amīr* of Damascus, 200
Adour, river, *see* maps
Adriatic Sea, 72, 79-80, 213, 218, 221
  see also maps
Adrastus, king of Argos, 160
Aegean Sea, 102-4, 109-11, 222
al-Afḍal Shāhānshāh, Fāṭimid *amīr al-juyūsh*, 191
Afonso I Henriques, king of Portugal, 88-92, 256-7 nn. 16 & 17
Africa, *see* maps
Africa, sub-Saharan, 187
Aghlabids, *amīrs* of Tunisia, 186-7
agricultural output, *see* environment
aḥdāth, *see* military forces
Aḥmad ibn Ṭūlūn, *see* Ibn Ṭūlūn
Aigues-Mortes, 161
Aksungur al-Ḥājib, Saljūqid governor of Aleppo, 195
Albania ("Bulgaria"), 1, 19, 106
Alberic of Trois-Fontaines, chronicler, 226-7
Albert, bishop of Buxhövden, 271-2
Albert of Aachen, historian, 2, 58, 60 n. 87, 119-21, 123-5, 127, 129, 190, 241-3, 245, 248 n. 62, 255 n. 10, 256 n. 12, 261
Alcácer do Sal, 252, 266-7
Aleppo, 29-30, 42, 144, 194, 200
  see also maps
Alexander the Great, 49, 57, 160
Alexandria, 102, 189, 202, 220
  see also maps
Alexios I Komnēnos, see Komnēnos
Alexios III Angelos, Byzantine emperor, 210, 221-2, 224-5
Alexios V Angelos, Mourtzouphlos, 227
Alexios Angelos, Byzantine pretender [Alexios IV], 219-22, 224-7
Aliakmon, river, 3-4, 9
'Allāqa, 188
Almeria, 269
Almohads, Caliphs of Morocco and Iberia, 266, 269
Almoravids, *amīrs* of Morocco and Iberia, 266
Alps, mountains, 231
  see also maps
Alvisus, bishop of Arras, 84
Amalfi, 1-2
Amalric I, king of Jerusalem, 200
Ambroise, poet, 99, 112, 258 n. 21
Amorgos, 103, 111
Āmū Daryā, river, *see* maps
Anacletus II, Antipope, 252 n. 4
Anatolia, 242-3, 248 & n. 62
  Byzantine control of, 85
  frontier, 158, 244
Andikithera, 99
Andrea, A. J., 222 n. 54
Andrew, king of Hungary, 95

Andrew, leader of men of London, 87 n. 35
Andros, 222
Angevin registers, 10-14 & Tables 1.4 & 1.5, 10 n. 24, 11 n. 28, 15 & n. 39, 16-17 & Table 1.6 & n. 47
Anonymous of Halberstadt, historian, 218, 221-2 n. 54
Anonymous of Soissons, historian, 218, 227
anonymous treatise on campaign organization (*Ἀνωνύμου Βιβλίον τακτικόν*), 149
Antalya, 86
 Gulf of, 113
Anti-Tauros, mountains, 143
Antioch, 43-4, 50, 54, 113, 137, 248 n. 62, 255
 siege of (1097-8), 49, 111-12, 182-3, 196, 236, 243-4, 246, 248 & n. 62
Antioch-Aleppo/Chalkis road, 140
Antioch of Pisidia, 85, 297
Antipodes, 172
Apennines, mountains, 215
aplēkta, *see* magazines
apostles, 172
Apulia, 104-5
'Aqaba, Gulf of, 202
aqueducts, 137
Aquileia, 297, 303
Arabia, 201, 203
 pre-Islamic military forces, 185
Aradus, 114, 297
Archipelago, *see* Duchy of the
Argolis, Gulf of, 71
Argos Orestikon, 4
Argostoli, Gulf of, 107
al-'Arīsh, 201
Armenians, 244
Arnissa, 4 n. 13
Arno, river, 66
Arnold, count of Aerschot, 87
Arnulf, bishop of Lisieux, 83
'Arqā, and siege of, 206, 235, 297
Arsuf, battle (1191), 205, 297
Arta, 2
Ascalon, 53, 201-2
 battle of (1099), 190-91, 202
 Muslim infantry of, 196
 see also maps
al-Ashraf, Ayyūbid sultan of Damascus, 199
Asia, *see* maps
Asia Minor, 132, 161, 193, 200, 243
 see also maps
asini, (asses or donkeys), 18 n. 51
'askar, Ayyūbid standing armies, 198
Astipalaia, 111
Atlantic Ocean, 261, 270-71
Auvergne, 235
Avarinos, 108, 297
Avlona, 1-3, 4 & n. 13, 23, 297-8
al-A'waj, river, 31, 298
Ayla, *see* Eilat
Ayalon, D., 25-6, 38-9, 200 n. 60
'Ayn Jālūt, battle (1260), 34, 298
Ayyūbids, dynasty of sultans, 114
 see also military forces, Ṣalāḥ al-Dīn
al-'Azīz, Fāṭimid Caliph, 192 & n. 27

Bachrach, B. S., 9 n. 22, 146
Babylon (Fusṭāṭ), 218, 220, 298
Badr al-Jamālī, Fāṭimid *amīr al-juyūsh*, 193, 293
Baghdad, 34, 186, 192
 see also maps
al-Balkhī, geographer, 162-3
Baldwin, abbot, 177
Baldwin (of Boulogne) I, king of Jerusalem, 236, 247, 252, 255 n. 12
Bālis, *see* maps
Balkhī school map set, *see* maps
Baltic Sea, 260, 271-3
Banū Hilāl, 192
al-Baqqāra, 201
Baradā, river, 31, 298
"Barbarosssa Hoard", 233
Barral i Altet, X., 179 n. 44
Bartholomew Boel, vice-lord of Chartres, 2 & n. 6
Basgöz, river, 112
Basil I, Byzantine emperor, 137-8, 143
Basil II, Byzantine emperor, 139, 144 & n. 38
Baudri, abbot of St-Pierre de Bourgueil,

archbishop of Dol, historian and poet, 159, 177, 178 & n. 35, 179-83
Bavaria, 253
Baybars, Mamlūk sultan, 31-4, 39
Baybars al-Manṣūrī, Mamlūk commander, historian, 40-41
Bayonne, *see* maps
Bazas, *see* maps
Beatus of Liébana, 171-7 *passim* & n. 23
Bedouin, 33, 190, 192, 194-5, 198-9, 201-2
  Kinaniyya tribe, 202
Beirut, 113
Beit Shean, *see* maps
Bela Pelanka, 128
Belgrade, 121, 127-8, 136
Belisarios, Byzantine general, 98, 108
Benevento, *see* maps
Bernard of Breydenbach, 116
Bernard, St, abbot of Clairvaux, 78, 86-9, 93
Berry, 119
al-Bijāya, 68, 298
Bilbays, 200-201, 203, 298
Bilisht, 3-4
al-Bīra, 39, 298
biscuit, *see* provisions
Bithynia, 137, 231
Bitola, 2-3, 4 & n. 14
blockhouses, 272
Blois, 179, 238
Boccanegra, Guglielmo, 70
Bohemond of Taranto, lord of Apulia, 1-24 *passim*, 58, 59 n. 84, 60, 125 n. 41, 126 n. 42, 127, 236
Bohemond II, prince of Antioch, 105, 108, 113
Bologna, 215
Boniface, marquis of Montferrat, 220
Bonifacio, 73
booty, 243-5
Borborino, Lanfanco, Genoese commander, 73
Boris, Hungarian pretender, 81
Borkum island, 268
Bosporos, 81, 84, 138 n. 17, 182 & n. 61, 223, 227, 248 n. 62
  see also maps
Boucicault, Marshal, Genoese admiral, 108
Bourgeuil, monastery, 179
Bouillon, 239-40
Brabant, 255
bran, 14 & n. 36
Branas, Michael, Byzantine governor of Niš, 83
Bremen, archbishop of, 271
bridges, 80, 137-8
  maintenance of, 138
  pre-fabricated, 81-2
bridging rivers, 80, 138 & n. 17
Brindisi, 97, 105, 111, 226, 268
Brocardus, Pseudo, chronicler, 99-100, 103
Brundage, J. A., 182 n. 61
Bryennios, Nikēphoros, historian, 120 n. 16, 126
Bulgaria, 253
bullion, 233-4, 241, 248
al-Buqay'a, valley, 196, 298
buss, *see* ships
Byzantium/Byzantines/Byzantine Empire
  in general, 53, 80-81, 83, 104-5, 110, 119-45, 196, 198, 199 n. 56, 209, 242 & n. 40
  and Second Crusade, 79-81
  assistance to First Crusade, 51
  fodder for cavalry horses, 18
  gifts of money, 243
  guides for Crusader forces, 161 & n. 9, 230 n. 6
  money changers, 84-5
  of Nicaea, 103
  supplying Crusaders, 50-51, 53-4, 57-9, 80-81, 83-5, 121, 124, 125 & n. 41, 126-9, 242
  see also military forces, money, naval forces
Byzantine-'Abbāsid frontiers, 186

Cadiz, 267, 269
Caesarea, *see* maps
Cagliari, 68, 161-2
Cairo, 41, 188-9, 193, 203
  see also maps
Calabria, 104, 105
  see also maps
Calcurini, Christian mercenaries, 69
calms at sea, 103
camels, 28 n. 15, 29, 39-40, 61, 185, 198,

203
Cantabrian mountains, 257
Cape Andreas, 113
Cape Arnauti, 115
Cape Asprokavos, 108
Cape Akritas, 108
Cape Atheras, 107
Cape Chelidonia, 112-13, 115
Cape Gata (Cyprus), 116
Cape Greco, 115
Cape Ince, 112
Cape Malea, 110, 222, 253 n. 5
Cape Santa Maria de Leuca, 105
Cape Sidhero, 106-7
Cape Spartivento (Calabria), 104
Cape Tainaron, 110
Carmel, mount, *see* maps
Caro, G., 64
Carrara, *see* maps
Carrocia, Genoese *nave*, 65-7, 75
Cartagena, *see* maps
Carthage, 49, 98
  see also maps
carts and wagons
  in general, 134, 140-41, 148-50, 248 & n. 62
  People's Crusade, 123, 127-8, 159
  Second Crusade, 84
  length of supply train, 56
  load capacity, 54-7, 146
  Mamlūk, 39
  numbers required, 54-7
Caspian Sea, 186
Catalonia, 264
Ceuta, 69-70, 73, 270
  amīr of, 69
Chaghatayids, *see* Mongols
Chalkēdōn, 223-4, 298
Chandax, 99, 104, 298
  gulf of, 97, 104
Changeon (*Cambio Burgullii*), river, 179
charitable fund, *see* money
Charlemagne, *see* maps
Charles I of Anjou, king of Sicily, 66, 72-3
Charlie Chaplin, *The Great Dictator*, 160
Chartres, 238
Chastellet, castle, 198, 298
Chios, 117
Christian, castellan of Ghistelles, 87
Christians under Muslim rule, 51, 53
Chronicle of the Morea, 109
Chrysopolis, 224, 298
Cilicia, 113, 137, 243
  see also coasts
Cilician Armenia, 113
Cilician Gates, pass, 2, 298
Cinque Ports, 258, 261
cisterns, 137
Civitavecchia, 267
Clermont, Council of, 1, 51, 119, 124
climate, *see* environment
Clunies Ross, M., 251 n. 2
coasts
  Adriatic, 259
  Bay of Biscay, 263
  Cilicia, 103
  Iberia, 254, 257, 264-5, 269
  Lycia, 103, 112
  North Africa, 102, 117-18
Codex Theodosianus, 136, 147
Coimbra, 90
coins, *see* money
Cologne, 86, 119, 121, 122 & n. 24, 232, 256 n. 15, 257 n. 17, 267
  archbishop of, 90, 231
Coloman I, king of Hungary, 78, 80-81, 124
Columella, Lucius Junius Moderatus, 23 & n. 70
columns, length of
  cavalry, 142-3
  cavalry and infantry together, 7-9, 142-3
  infantry, 143-4
  supply train, 56
common fund, *see* money
compasso de navegare, Lo, 101-2, 109-10, 112-17 *passim*, 293
Conon of Béthune, troubadour, 224
Conrad III, king of Germany, 78-83, 88, 92 n. 51, 230 n. 6
Constable, G., 77, 90
Constantine VII, Byzantine emperor, 18
Constantinople, 1, 53, 58, 64-5, 71, 80-81, 83-5, 88, 109, 119-21, 125, 127, 129,

137, 144, 209, 220-28, 235 n. 17, 242 n. 40, 248, 252, 253 n. 5, 254-6
Golden Horn, 226
St Peter's Church, 253 n. 6
see also maps
Contractus navigii domini regis cum Venetis factus anno domini MCCLXVIII, 10-12 & Table 1.4, 10 n. 25, 16-17 & Table 1.6
contubernium, tent group, 147 n. 50
Copts, 201
Corfu
in general, 2, 98, 105-9, 221-2 & n. 54, 259, 261-2
Corfu Channel, 107
Corfu town, 106-7
corsairs, 109-12
Corsica, 64, 73, 100, 117, 265
see also maps
Cosmas and Damian, Sts, monastery, 225
Cremona, 215
Crete, 97, 99, 102-4, 111, 133 n. 2, 256, 259, 265
see also Chandax, maps
Cristoforo Buondelmonti, geographer, 103-4
Crusader states, *see Outremer*
Crusaders, as pilgrims, 82
Crusades/Crusaders
in general, 185
financing, 230-32, 234
First
in general, 64, 77, 79, 82, 92, 111, 170, 173, 176-7, 182, 193, 200, 229-49 *passim*, 255
oaths of leaders to Alexios I, 54, 80
route of northern French, 79
size of forces, 49 & n. 34, 54, 55 n. 61, 82 n. 16
People's, 51, 54, 80-81, 119-29 *passim*, 159, 230
of 1101, 54
of 1123, 262
of 1227, 268
Second, 77-93 *passim*, 196, 230 n. 6, 233
against Wends, 87, 271
baggage fleets on Danube, 81, 84
baggage trains, 86
composition of French army, 79, 82, 84
English, 83, 87
Flemish, 87, 258 n. 21, 267 n. 49
Frisians, 258 n. 21
French, 138-9
Germans, 87
Hollanders, 258 n. 21
naval forces, 92, 256-7
Normans, 83
oaths to Manuel I, 84-5
size of forces, 82 n. 16, 92 n. 51
Third, 95-6, 102, 195, 205, 231-3, 245, 257-62, 266, 269, 271
Fourth, 64, 106-7, 154, 209-28 *passim*, 235 n. 17, 254, 260, 266
Treaty of Partition, 109, 111
Venetian fleet, 213-14 & n. 24, 217-18 & n. 40
horse transport galleys, 210, 213, 217 & n. 40
Fifth, 91 n. 50
Sixth, 101
Seventh, 268
Curonians, 272
currents, 113
Straits of Gibraltar, 254
Cyclades islands, 103
Cydnus, river, 255, 298
Cyprus, 97, 99-100, 103-4, 111, 113-15, 252, 253 n. 5, 259, 263
Byzantine naval base, 52 & n. 50, 111
see also maps
Cyrenaica, 39, 298

Da Costa, Alamanno, Genoese corsair, count of Syracuse, 66-8 & n. 15, 73, 75
Damascus, 31, 37-8, 40-42, 91-2, 196, 199, 203
see also maps
Damietta, 100 n. 28, 101, 116, 199 n. 56, 202, 268
Dandolo, Enrico, Doge of Venice, 210, 216-17, 219-20, 222-4

Dandolo, Jacopo, Venetian admiral, 73
Dandolo, Ranieri, Venetian admiral, 109
Danes, *see* Denmark
Danube, river, 79-81, 121, 137 n. 12
  baggage fleets for Second Crusade, 81, 84
  route of First and Second Crusades, 79-80, 88
Danzig, 272, 298
Dardanelles, 79, 222
  see also maps
Darius, king of Persia, 160
Dartmouth, 87-8, 92, 256 & n. 15, 257, 259, 262-3, 267 & n. 49
Dax, *see* maps
Daylamites, 186 & n. 5, 192, 195
daylight, hours of, 6 & Table 1.1, 7
De administrando imperio, 136
Dead Sea, *see* maps
Dean, Forest of, 78
De expugnatione Lyxbonensi, 89-93, 256
Delbrück, H., 46-8
Della Turca, Rosso, Genoese commander, 71
demography, *see* environment
Denmark/Danes, 88, 253, 255 & n. 10, 259, 271
Destombes, M., 171 n. 20
Devastatio Constantinopolitana, 216
De velitatione, 126, 134 n. 46, 149
Devol, river, 3-4
divisions, sizes of, 7-9
Divini dispensatione, 271
Diyār Bakr, 194
Dodecanese islands, 96, 103, 111
Doliana, 4
Donau, river, 121
donkeys
  feed requirements, 145, 146 Table 8.1
  load capacity, 146 Table 8.2
Dorostolon, 137 n. 12, 298
Dorylaion, 43 n. 3, 44 & n. 9, 53, 59-61, 298
  battle (1097), 44-5, 50, 243
  battle (1147), 83, 85
Döşeme pass, 140, 298-9
Drakōn, river, 137, 299
Drava, river, 84
Drino, river, 3-4, 19
Drizipera, 137 n. 12, 299
Drynopolis, 1, 3, 4 & n. 13, 299
Duncalf, F., 49 n. 34
Duchy of the Archipelago, 111
Dutch, *see* Holland
Dvina, river, 272
Dyrrachion, 1-2, 4 & n. 13, 72, 80, 127, 299

Ebro, river, 268
ecology, *see* environment
Edessa, 236, 244, 247, 299
Edgar Atheling, English pretender, 255 n. 11
Edgington, S. B., 256 n. 16
Edhessa, 3, 4 n. 13
Edict of Prices of Diocletian, 146
Edirne, 121, 242
Edremit, 138
Edward, prince of England, 268-9 & n. 55
Egypt/Egyptians, 91 n. 50, 117-18, 185-8, 195, 197-8, 200-203, 207, 209, 214, 218, 220, 224, 227, 237, 243
  ḥalqa, 39
  see also Fāṭimids, Mamlūks, maps, military forces, naval forces
Eilat, 202-3
  see also maps, *Ayla*
Elafonisi Channel, 110
Elbasan, 3
Elbistan, battle (1276), 32
Elburz mountains, 186
Empire, Western Roman, 80, 209
Engels, D. W., 57, 148 n. 54, 150
Engilsnes, "Angel's Cape", 252 & n. 5, 299
England/English, 113, 252, 254, 257, 258 n. 21, 259-61, 263-5
  and Second Crusade, 83, 256 & n. 15
  expedition to Ireland, 78
  see also naval forces
environment, 28, 38, 43, 51, 92, 104, 121, 122 & n. 21, 131-4, 139, 158, 200, 205-6, 213
Eracles, chronicle, 97, 126
Erdkamp, P., 15 n. 40
Eşen/Xanthus, river, 112

Estoire de Jérusalem et d'Antioche, Li, chronicle, 3, 126
Estonia, 271
Ethiopia, *see* maps
Etsi verba, 271
Eugenius III, Pope, 78-80, 87-8, 91, 93
Eumathios Philokalēs, governor of Cyprus, 52 n. 50
Euphrates, river, 27, 33, 38-9, 194
see also maps
Europe, *see* maps
Evvoia, 222
exchange rates, 84 & n. 26, 85, 124 n. 32, 242 n. 40
see also money
extortion, *see* money

Faenza, 215
Falcone, Venetian *nave*, 67
Falkonera, 111, 299
Famagusta, 114-17
port of, 116-17
Fāqūs, 201, 299
al-Faramā', 201, 299
Faro, 92, 267
fasting, 46
Fāṭimids, 188, 192, 197-9, 206
farḥiyya, 193
juyūshiyya, 193
qādi, chief, 189
rayḥāniyya, 193
wazīrs, 189, 192, 295
see also Egypt, maps, military forces, naval forces
Felix Fabri, pilgrim, 99 n. 20, 103
Fermia, 111, 299
Fidenzio of Padua, 39
Flanders/Flemish
in general, 86, 236, 241, 255-7, 258 n. 21, 259, 270
count of, 85
Florence, 65
Florina, 3, 4 & n. 14
fodder
for donkeys, 145
for horses
in general, 28 n. 15, 29, 40, 136, 201, 242-3, 262
barley, 15-16 & nn. 40 & 41, 17, 19, 28-9, 38-9, 122 n. 25, 125, 129, 144-5 & n. 40, 149, 211 & n. 5, 243
bran, 15-16 n. 41
bromegrass, 29-30, 293
carrots, 15-16 n. 41
chaff, 15 n. 41
grain, 15-16 & nn. 40 & 41, 27, 29, 33, 55-6, 123, 145-6, 211, 213, 262
grass/grazing/pasturage, 15, 16, 19-20 & n. 60, 21-3, 27-30, 36 n. 45, 38, 55, 123, 139 n. 21, 145, 213
hay, 15-16 & nn. 40 & 41, 16-17, 28-9, 55, 145, 213, 262
mowing, 22-3
oats, 15-16 n. 41, 16
water content of, 16, 20 n. 60, 21 & n. 63, 30, 145
see also pasture, markets
for mules, 145
for oxen, 145
foraging, 9, 19, 37, 61, 80, 201, 222, 224-7, 230, 241-2, 246
see also pillaging
Foxhall, L. and H. A. Forbes, 14
France, 176, 235-8, 241, 264, 268
see also maps
France J., 44 nn. 8 & 11, 49, 51 n. 46, 52 n. 49, 54, 55 nn. 61 & 63, 61 n. 91, 121 n. 19, 126 n. 44, 128-9, 206, 264
Franks, *see* maps
Frankish Crusaders/Franks, of Fourth Crusade, 209-28 *passim*
Franks, of *Outremer*, 26, 35, 161, 185, 190-92, 196, 198, 199 & n. 56, 201-6
horses, 39
see also military forces, Turcopoles
Fraschea, bay of (Crete), 104
Frederick I, Barbarossa, Western Roman emperor, 231, 233
Frederick II, king of Prussia, 144 n. 35
Frederick II, Western Roman emperor, 95, 97, 99, 103-4, 107-9, 112, 117
Free passage, 3, 80-81, 126 & n. 43
Freiberg silver mines, Upper Saxony, 233

Friesach silver mines, Carinthia, 233
Frisia/Frisians, 255, 258 n. 21, 259
Fulcher of Chartres, historian, 53, 58, 98, 113, 119-20 & n. 7, 190-91, 230, 262
Fusṭāṭ, 188, 201, 299

Galata, 226-7, 299
Galatia, 137
Galicia, 252, 254
galleys, *see* ships
galley oarsmen
  numbers of, 65-6
  provisioning of, 9, 10-13 & Tables 1.4 & 1.5, 15 n. 39
  roles of, 74
Gallos, river, 60 n. 85, 299
Garitsas Bay (Corfu), 107
Gascony, *see* maps
Gaston IV, viscount of Béarn, 82
Gaza, 33, 42, 200
  see also maps
Gazopolis islands, 105
Geza II, king of Hungary, 80
Genoa/Genoese
  in general, 63-75 *passim*, 103, 106, 108, 117, 211, 270
  and First Crusade, 64, 96
  and Byzantine Empire, 65
  civic chronicle, 64
    anonymous author, 70, 72
    Lanfranco Pignolo, 72
    Ogerio Pane, 66-8
  financing expeditions, 69
  see also maps, naval forces
Geoffrey III, count of Roussillon, 2 & n. 6, 4, 7
Geoffrey Malaterra, chronicler, 2
Geoffrey of Villehardouin, historian, 216, 219, 222-5
German Empire, *see* Empire
Germany, 121, 124 & n. 31, 241, 256, 268
Gesta Francorum, chronicle, 1 & n. 2, 9, 51, 56 n. 65, 127, 190, 200, 206, 244
g̱hilmān, *see* military forces
g̱huzāh, *see* military forces
Gibraltar, 264, 269
  Straits of, 69, 254, 266, 270
Gifts of money, *see* money, gifts of
Gilbert of Hastings, bishop of Lisbon, 257 n. 17
Godfrey Burel, 120
Godfrey of Bouillon, duke of Lower Lotharingia, 80-82, 87, 125, 129, 159-61, 177, 236, 239-43, 246-7
Godric, English pirate, 261
Goffart, W., 47 n. 25
gold, African, 187
Golden Gate, Χρυσεία Πύλη, of Constantinople, 252
Golden Horde, *see* Mongols
Gozón, 257, 299
grain, *see* fodder, provisions
Greece, 254, *see also* maps
Greenland, 260
Grillo, Simone, Genoese admiral, 67, 72, 74-5
Gudericus, English pirate, 255 n. 12
Guerre de course, *see* naval warfare
Guibert, abbot of Nogent, 120, 124 n. 32, 125 & n. 37, 159, 161, 181, 236
guides, *see* Byzantium
"Gulf of Constantinople", *see* maps
Gunther of Pairis, historian, 227
Guy, knight of the household of the viscount of Melun, 100 n. 28
Guynemer of Boulogne, 255 & n. 10

Hagenmeyer, H., 43 n. 2, 182 n. 61
ḥajjārūn, *see* military forces
al Ḥākim, Fāṭimid Caliph, 188 & n. 14
Haldon, J, 9 n. 22, 15 n. 40, 18 n. 55, 20 n. 59, 21-3
ḥalqa, 39, 195, 294
Ḥamā, 31, 34-5, 199
ḥarāfish, urban riff-raff, 199
Harald Hardradi, king of Norway, 251 n. 2
Harari, Y., 204
harbours and ports, 95-118 *passim*
Harim, fortress, 191
harness for draft animals, 55 n. 60
Ḥarrān, battle (1104), 194, 299
Hartmann, count of Dillingen, 246
Harz mountains, Saxony, 233
Hastings, battle (1066), 47 n. 28, 177
Ḥaṭṭīn, Horns of, battle (1187), 206, 299
Ḥaydarān, battle (1052), 192, 299-300

Hayes, captain, 20 n. 59
Hebron, *see* maps
Hedeby, 253
Hellenistic kingdoms, 132
Helmold of Bosau, chronicler, 87
Henry I, king of England, 252
Henry II, king of England, 264
Henry III, king of England, 268
Henry V, Western Roman emperor, 253 n. 7
Henry, castellan of Esch, 246
Henry, count of Flanders, 227
Henry, bishop of Olmütz, 88
Henry of Louvain, 220
Henry Pescatore, count of Malta, Genoese corsair, 67, 68 & n. 15, 69, 104
Herakleios, Byzantine emperor, 126 n. 17
Hervey of Glanvill, 87 & n. 35, 90-91
Heṭoum, chronicler, 36, 38
Hohenstaufen, house of, Western Roman emperors, 65
Holland/Hollanders/Dutch, 91 n. 50, 258 n. 21, 259
Holy Cross, Augustinian priory of, 90
Holy Land, 78, 87, 176, 221, 229, 251-2
    see also maps, *Outremer*
Holy Sepulchre, 224
Holy War, 93
homage, oaths of, 80
Ḥomṣ, 34-5, 40-41, 199
    battle of (1281), 31, 35, 40-41
Honorius III, Pope, 91 n. 50
horses
    aboard ship, 15-16, 100 & n. 23
    Arabians, 39
    feed requirements, 134 Table 8.1
    feeding and watering, 19-22
    harness, 39
    load capacity, 18, 19 n. 56, 144, 146 & Table 8.2, 147-58 *passim*
    numbers per mounted warrior, 7 & n. 19, 55 & n. 62
    picketing, 22
    remounts, 147, 151-2
    resting of, 144
    stomach, 19-20
    transportation by sea, 95-6, 98, 100, 107, 133 n. 2, 262-3, 267-8, 271
    slings, 263 n. 38
    warhorses, 7, 18
    see also fodder, pack animals, ponies, water
Howard-Johnston, J., 120-21 n. 16
Hugh, count of St Pol, 219-20, 225
Hugh, count of Vermandois, 79, 236
Hugh of St Victor, 176 n. 31, 177
Hūlāgū, Īlkhān of Persia, 36
Hungary, kingdom of/Hungarians 78, 80-81, 121, 124 n. 32, 125, 127, 129, 231, 242 & n. 40, 253
Huns, 15
Hyland, A., 15 n. 40, 20 n. 59, 21-2, 145 n. 40

Ibelin, battle (1123), 191, 202
Iberia, 253-7, 264-6, 269
    see also coasts, maps, Spain
Ibn al-'Adīm, 186 n. 4
Ibn al-Athīr, historian, 160-61
Ibn al-Qalānisī, historian, 190
Ibn Baṭṭūṭa, pilgrim and traveller, 203
Ibn Ḥawqal, geographer, 163, 167, 168-9 & Fig. 9.4, 170, 176 & n. 31, 179
Ibn Jubayr, pilgrim, 98-9, 100 n. 24, 102, 114
Ibn Muyassar, 191
Ibn Shaddād, historian, 205
Ibn Taghrī Birdī, historian, 42
Ibn Ṭūlūn, Aḥmad, *amīr* of Egypt, 114, 187 & n. 8
Ibrāhīm ibn al-Aghlab, 186
Iceland, 260, 262
al-Idrīsī, geographer, 4 n. 13, 101, 106, 110-11, 170 & n. 18
Imola, 215
India, 197
    see also maps
Indus, river, *see* maps
inflation, *see* money
Informationes pro passagio transmarino, of Marseilles, 10-12 & Table 1.4, 11 n. 27, 16-17 & Table 1.6
ingots, *see* bullion
Innocent III, Pope, 210, 220-21, 225, 271
Ioannina, 2, 4 n. 13
Ionian Sea, 98, 102-4, 109
    islands, 96

iqṭā'(s), 198-9, 207, 294
Isaac II Angelos, Byzantine emperor, 225
Isidore of Seville, St, 172, 178, 181
map in an eighth-century MS. of the *Etymologiae* of, 173 n. 28, 176, n. 29
al-Iṣṭakhrī, geographer, 163, 164-5 & Figs 9.1 & 9.2, 166, 167 & Fig. 9.3, 170
Italy, 164, 232, 238, 264
see also maps
Ithaka, 107-8
Itinerarium peregrinorum, chronicle, 99, 112
Iviza, 252

Jabala, 236
amīr of, 244, 247
Jacob's Ford, 198, 300
Jacoby, D., 11 n. 29
Jaffa, 52, 107, 255 & n. 12, 268
James of Vitry, historian, 102, 162
al-Jazīra, 194, 199
Jericho, *see* maps
Jerusalem, 50, 52-4, 82, 160-61, 182-3, 236, 244, 246
garrison of in 1099, 53 n. 52
Kingdom of, 114, 203, 206 & n. 74
objective of First Crusade, 51, 54, 159
objective of Second Crusade, 93
siege of (1099), 246
see also maps
jewellery, *see* money
Jews, 240
John, king of England, 267
John, archbishop of Braga, 90
John of Ibelin, lord of Beirut, 116-17
John of Joinville, hagiographer, 99, 101
John Sarrasin, 101
Jordan, river, 33
see also maps
Judaea, 244
Julianus, *kandidatos*, 137
al-Jundī, Qa'lat, 203
Jurjīr, 201, 300
Justinian I, Byzantine emperor, 5-6, 137
Jutland, 253

ka'k, see provisions
al-Kāmil, Ayyūbid sultan of Egypt, 199
al-Karak, 33, 300
Karavi, 111, 300
Karpathos, 103-4
Kassiopi Bay (Corfu), 105-6, 300
Kastoria, 1-3, 4 & n. 13, 9, 19, 127
Kastron/Hagios Georgios, 107-8
Keffalinia, 97, 107-8
see also Portus Guiscardus/ Fiskhardo, Kastron/Hagios Georgios, Argostoli, gulf of
Kerbogha, *Saljūqid amīr* of Mosul, 165
Khurāsān/Khurāsānians, 186, 195-6
Kh^wārazmians, 194 & n. 33, 199
Kilāb, bedouin tribe, 195
Kitbuga Noyon, Mongol commander, 34
Kölner Dienstrecht, Längeres, 231-2
Komentiolos, Byzantine general, 136-7
Komnēnē, Anna, historian, 58, 120, 126 & n. 44
Komnēnos, Alexios I, Byzantine emperor, 2-3, 52 n. 47, 53-4, 58, 60, 111, 120-21, 124, 125 & n. 41, 126 & n. 42, 127, 129, 236, 242-3, 252-4
Komnēnos, Isaac Ducas, Byzantine emperor of Cyprus, 97
Komnēnos, John, Byzantine governor of Dyrrachion, 3
Komnēnos, Manuel I, Byzantine emperor, 78-81, 83, 88, 124 n. 32
Konya, 230 n. 6
Korone, 108-10
Kos, Byzantine naval base, 111
Koukousos, 143, 300
Krak de Montréal, 202, 300
Kuch, 3, 300
Kupfer, M., 177
Kurds, 33, 193-4, 195 n. 39, 197, 200, 207
Kutāma Berbers, 188

La Coruña, 91
Lakonia, Gulf of, 110
Lambert of St-Omer, mapmaker, *see* maps
Lampedusa, 118
landfalls, *see* ports of call
Lane, F. C., 63
Langres, bishop of, 85
Las Navas de Tolosa, battle (1212), 269

Latakia, 114, 194, 236
Latin Empire of Constantinople, 63
  Empress of, 115
"Laws of the camp", 83
Lebanon, 33
Leo VI, Byzantine emperor, *Taktika* of, 147 & n. 50
Leo, abbot of St Bertin, 84
Leo Vetrone, Genoese corsair, 109
Leopardo, Pisan *nave*, 65-8, 73
Lepanto, battle (1571), 63, 300
Lerinska gorge, 3-4
Lesbos, 103
Lescar, *see* maps
Levant, 104
  see also Holy Land, *Outremer*
Levkas, 60 n. 85, 300
Levkas island, 107
Lewis, A. R., 51 n. 47
Libya, *see* maps
Lido, *see* Venice
Liège, bishopric of, 240
Liguria, *see* maps
Limassol, 97, 101, 115-16
  port of, 116
Linz, 121
Lisbon, 86, 88-90, 91 & n. 50, 92-3, 252, 254, 258, 265-7, 267-8 & n. 50, 269
  siege of, (1147), 256-7 & nn. 16, 17
Lisbon Letter, 256
Litani, river, 33, 300
Lithuania, 271-2
Livermore, H., 88
Livonia, 271-2
logistical infrastructures, *see* environment
Loire, river, 179, 235-6, 238
  see also maps
Lombard engineer, 247-8
London/Londoners, 259, 269
Lopadion, 138, 300
Lot, F., 48
Lothar of Supplinburg, duke of Saxony, 253 & n. 7
Lotharingia, 89, 236, 241, 257 n. 17, 300
  Lower, dukes of, 236
  see also Rhinelanders
Louis, count of Blois, 215
Louis VII, king of France, 78-88, 91, 124 nn. 31 & 32, 242 n. 40
Louis IX, king of France, 17, 36, 99, 101-2, 116-18, 161, 177 n. 32, 268, 269 n. 55
  see also Contractus navigii domini regis cum Venetis
Low Countries, 254, 257 n. 17, 260, 263-5, 267 n, 50, 270
  see also Flanders, Holland, Brabant
Lübeck, 259, 272
Lucca, 65
Ludolph of Suchem, pilgrim, 113, 117 & n. 96
Lupus Protospatharios, chronicler, 2
Lycia, 140
  see also coasts

Madden, T., 9 n. 22
Ma'arrat-al-Nu'mān, 236
Maeander, river, 85, 300
al-Mahdiyya, 300
  see also maps
Maḥmūd G͟hāzān, Īlk͟hān of Persia, 28, 36-8, 42
magazines
  Byzantine, 55, 60, 126, 142 & n. 32, 149, 293
  Mamlūk, 38-9
Magdalino, P., 125 n. 41
Mag͟hrib, 70, 72, 269
  see also maps
Mainē, 110 & n. 70, 300-301
Majorca, *see* maps
Malaga, 69, 270
Malagina, 60, 143-4, 155, 301
Malchos of Philadelphia, historian, 136
Malea, 110 & n. 70
Malta, 72, *see also* maps
mamlūks, slave soldiers, 25, 36, 39, 294
Mamlūks/Mamlūk sultanate of Egypt, 25-43 *passim*
  see also military forces, naval forces
al-Ma'mūn, 'Abbāsid Caliph, 162 & n. 13
Manente di Segalari, Ranieri, Pisan count, 68 n. 15
Manfred of Hohenstaufen, king of Germany, 72
al-Manṣūra, battle (1249), 199

mappemundi, *see* maps
maps, 159-83 *passim*
Adour, river, 172, 179
Adriatic Sea (*Sinus Adriaticus*), 170, 176
Africa, 160, 167, 17-2, 173 n. 28, 176 n. 31, 178-9
Āmū Daryā, river, 163
Aleppo, 166
Alexandria, 166
al-Mahdīyya, 163
Alps, mountains, 176
Arabian gulf, 172
Ascalon (*'Asḳalān*), 166
Asia, 160, 171, 172, 176 n. 31, 178
Asia Minor, 183
Ayla, 166
Baghdad (*Bogdada*), 176 n. 31
Bālis, 166
Balkhī school map set, 162-3, 170, 171 n. 21
Bayonne (*Laburdum*), 172
Bazas (*Basata*), 172
Beit Shean (*Baisān*), 166
Benevento, 176
Bosporos, 171 n. 21
Caesarea, 173
Cairo, 176 n. 31
Calabria, 170, 176
Carmel, mount, 173
Carrara, 176
Cartagena, 176
Carthage, 173, 176
Caspian Sea, 163
Charlemagne, 170
Constantinople, 172, 173
Corsica, 170, 172
Crete, 166, 172
Cyprus, 163, 166, 172
Damascus (*Dimashq*), 166, 173
Dardanelles, 171 n. 21
Dax (*Aquis*), 172
Dead Sea (*Buḥayrat Zu'ar*), 166, 173, 178
Egypt, 175, 176
Ethiopia/Ethiopians, 160, 179
Euphrates, river, 163, 166
Europe, 160, 166, 170-72, 175-6, 178-9
Fāṭimid Egypt, 163, 170
France, 170, 176
Franks, 176 n. 31
Gascony (*Wasconia*), 173
Gaza (*Ghazza*), 166
Genoa, 170, 176 n. 31
Greece, 170
"Gulf of Constantinople", 166
Hebron (*Masjid Ibrāhīm*), 166
Holy Land (*Iudea*, *Palestina*, *Galilea Superior*, *Galilea Inferior*), 173 & n. 29, 175-6, 178, 183
Iberia, 173
India, 160
Indian Ocean, 163
Indus, river, 163
Italy, 170, 176
Jericho (*Rīḥa*), 166, 173
Jerusalem (*Bayt al-Quds*), 166, 173
Jordan, river, 166, 173
Lambert of St-Omer, mapmaker, 179-80
Lescar, 172
Libya (*Libia*), 172, 176, 178-9
Liguria, 160
Loire, river, 172
Maghrib, 170
Majorca, 170
Malta, 170
mappemundi, 159 & n. 2, 160 & n. 3, 161, 162, 170-71 & n. 20, 177 & n. 32, 178, 181
Marseilles, 172
Mediterranean Sea (*Baḥr al-Rūm*), 163, 166, 171-3, 176, 179
Minorca, 170
Mont-de-Marsan (*Sancta Maria Mirsanensis*), 172
Morocco (*Maroch*), 176 n. 31
Mount Etna, 160
Nablus (*Nābulus*), 166
Nile, river, 160, 163, 166, 171 n. 21, 172-3
North Africa, 175
of Adela of Blois, 177-81, 183
Oloron, 172
Orontes, river, 173

O-T/T-O maps, 171, 175-6, 180
Pantelleria, 170
Paradise, 160, 172, 178
Paris, 172
Persian Gulf, 172
Portugal (*Lusitania*), 176
Pyrenees, mountains, 176
Ramlah (*al-Ramla*), 166, 176 n. 31
Red Sea (*baḥr al-qulzum*), 166, 172
Reggio, 176
Rhône, river, 176
Rome, 166, 170, 172-3, 176
"Samaria", 173
Sanudo Torsello, Marino, 177 n. 32
Sardinia, 170, 172
Sea of Tiberias/Galilee (*Buḥayrat Ṭabariyya*), 166, 173
Seine, river, 172
Sicily, 163, 166, 170, 172, 176
Sidon, 173
Sinai, mount, 173
Spain, 166, 170, 176
Sri Lanka, 172
Stephanus Garsia, map maker, 171, 173, 176 & n. 31, 179
St Sever, monastery, & Beatus map, 171-6 *passim* & nn. 24 & 28 & 29 & Fig. 9.5, 178-9, 181, 183
Syria, 161, 166
Tanais (Don), river, 171-2
Tarbes (*Bigorra*), 172
Teutons, 160
Tiberias (*Ṭabariyya*), 166
Tigris, river, 163
Tinnīs, 163
Toflit, 176 n. 31
Tuscany, 176
Tyre, 173
Zachary, Pope, 170
zones, 160
al-Maqrīzī, historian, 163 n. 16, 187
Maraş, 143
Marcus Antonius, 57
Margat, castle, 114, 301
Marienburg, 272, 301
Marj al-Ṣuffār/Shaqḥab, battle (1303), 34 n. 39, 38, 301
markets for provisions and fodder, 40-41, 55, 59 & n. 84, 61, 80-81, 83, 85, 121, 124 & n. 32, 125 & n. 41, 127-9, 201, 203, 214-15, 231, 241-6
internal market, 244-7
Marmara, Sea of, 182 n. 61
Marne, river, 182
Marseilles, 102, 258, 260-61, 265, 268-9
Crusades launched from, 95, 100, 102, 269 & n. 55
see also maps
Martin, abbot of Pairis, 215
Martin, Lieutenant, 20 n. 59
Maṣāmida/Maṣmūda Berbers, 189
al-Mas'ūdī, geographer, 162
Matzke, M, 238
Maurice, *Stratēgikon* attributed to, 146 n. 46, 147 & n. 50
mawālī, 187 & n. 8, 294
Mayenne, river, 179
Mediterranean Sea, 197, 254, 270
see also maps
Melgueil, 301
Melos, 111
merchants, supplying forces, 41, 78, 83-4
Messina, 68, 97-9, 104, 258 & n. 21, 261, 263
Straits of, 98, 102, 117 n. 96
Messinia, Gulf of, 110
mētata, imperial Byzantine ranches, 152
Methone, 99, 108-10
metrology, 12, 24
Metsovon pass, 3
Metz, 81-3
Meuse, river, 241
Michael Psellos, historian, 139
Milan, 215
military forces
'Abbāsid, 186
Ayyūbid
cavalry, 198, 199, 200
infantry, 199, 200
mujāhidūn, 199
rajjāla, 199, 200
behaviour of, 132
Byzantine, in general, 132-3, 205
exercitus, 237
Fāṭimid, 188-93, 194 & n. 31, 195, 199, 200-201, 203

'abd, pl. *'abīd*, 188-91, 193, 197, 206-7
archers, 189
Armenians, 193
camels and horses, 198
cavalry, 191, 193
infantry, 188-91, 193-4, 202, 206-7
Franks of *Outremer*, in general, 205
archers, 206
cavalry and infantry cooperation, 204-5
infantry, 206 & n. 74
knights, charge of, 204
Turcopoles, 204, 206, 295
Macedonian, size of, 49, 57
Mamlūk
archery, 31
armies, 27, 38-9
as troops, 31-2
cavalry, 31, 39
firearms, 26
horses, 27, 39-40
Mongol
archery, 31
as troops, 27, 32
cavalry, 31-2
numbers, 27-8, 33
ponies, 27-8, 37
Muslim in general
archers, 204-5 & n. 70
cavalry (heavy), 206 (light), 204
infantry, 206
resources for, 134
Roman
rates of march, 144
size of, 49, 57, 134-5
Ṣalāḥ al-Dīn in Egypt, 197
bedouin, 198, 199
qarāghulams, 197-9
rajjāla, 198
ṭawāshī, 197-9
sizes of, 48, 120, 121 & n. 19, 131-5, 204-5
Syria
aḥdāth, urban militia, 195-6
'āmma, 196
archers, 196
cavalry, 196-7
ghilmān, 196
ghuzāh, 196
ḥajjārūn, 196
mujāhidūn, 195-6, 198
rajjāla (infantry), 195 & n. 41, 196, 198
sūqa, 196, 198
Turkish,
archers, 205
Miller, K., 162, 166
Military Orders, 206
ministeriales, 232
Minorca, 252
see also maps
mints, *see* money
Monemvasia, 71, 110-11
money, 124-5, 129 & n. 63, 229-49
passim, 267
Byzantine, 129 & n. 63, 234, 242-5
hyperpyron nomisma, 242-5
stamenon/staminum (*aspron trachea*), 242 & n. 40
charitable fund, 248
coins of Godfrey of Bouillon, 240-41
common fund, 248
denominations, 232-3
extortion, 240
Fourth Crusade, 216-17, 219-20, 225-6
gifts of, 243
gold, 234
grossi, 233
inflation, 245
in Muslim armies, 39-41, 194, 198, 201
see also nafaqat al-safar
jewellery, 235 & n. 17
mark, money of account, 231-2, 234, 239-40, 245-6
mints
Bouillon, 240
Chartres, 248
Chartres, Le Mans, Le Puy, Melgueil (& 301), Poitiers, Melle, Valence, Lucca, 235-6 & n. 20, 237
Le Mans, Chartres, Melle, 239

Lucca, 238 & n. 27, 239, 241, 245
Pavia, 238 n. 27
Toulouse, Cahors, Narbonne & Rodez, 238 n. 26
Valence, 238
Muslim, 234, 237
oboli, halfpennies, 232
pennies, silver, 232-4, 240, 244-7
plate, 85, 235
"Saracen" *dīnārs*, 237, 247
Second Crusade, 84
transportation of, 248-9
tribute, 237, 247, 248
see also bullion; Crusades, financing; gifts, treasure
money changers, 84, 235 n. 17
see also exchange rates
Mongols, 15, 25-42 *passim*
transhumance, 34
see also military forces
Mongols/Mongol, Chaghatayid Khānate, 36
Mongols/Mongol, Īlkhānate of Persia, 26-43 *passim*
Mongols/Mongol, Khānate of the Golden Horde, 36
Mont-de-Marsan, *see* maps
Mont Gisard, battle (1177), 202, 301
Montieri silver mines, Tuscany, 233
Morgan, D. O., 27-8, 32
Morocco, *amīr* of, 69
see also maps
Morris, C., 124 n. 32
Mosul, 194
Moulia Rocks, 115
Mount Cadmus, battle of (1148), 85
Mount Etna, 104
see also maps
Mountains of Asia, ten, 178
Mouzay, 239-40, 301
al-Mu'azzam, Ayyūbid *amīr* of Damascus, 199
al-Muhallabī, geographer, 201
Mu'īn al-Dīn Sökmen I, *amīr* of Mardin, 194
al-Mu'izz, Fāṭimid Caliph, 163
al-Mujāhid Asad al-Dīn, *amīr* of Ḥomṣ, 199
mujāhidūn, *see* military forces
mules, 18, 39, 123, 134 n. 50
feed requirements, 133, 134 Table 8.1
load capacity, 134 Table 8.2
muleteers, 156
numbers required, 151-7
munawwirūn, "burners", 38
al-Muqaddasī, geographer, 163
al-Mustanṣir, 'Abbāsid claimant to the Caliphate, 34
Myra, 113
Myriokephalon, 139, 301
Mysia, 139, 301

Nablus, *see* maps
nafaqat al-safar (campaign allowance), 39-41
naffāṭūn (naphtha-throwers), 188, 196
Nakhl, Qalat, 203, 301
naqqābūn (sappers), 196
Nāṣir-i Khusraw, Persian traveller, 189
Naṣrids, sultans of Granada, 269
naval forces
in general, 95-118 *passim*
Almoravid, 266
Byzantine, 51-3, 64, 110-11, 121
Dutch/Hollander, 258 n. 21, 259, 267 & n. 50
English, 52 & n. 49, 89, 95, 255, 257 n. 20, 258-60, 262-3, 264 & n. 40, 267
Fāṭimid, 51 n. 37, 113-14, 190, 202
First Crusade, 51-2 & nn. 46 & 47, 52
Flemish, 256-7, 258 n. 21, 259
fleet organization, 98
France, 267-8
Frisian, 255, 259, 267-8
Frederick II, 95, 97, 99
Genoa, 52, 63-75 *passim*, 96, 106-7, 114
German, 256, 257 n. 17
in Baltic, 272
Kingdom of Sicily, 64, 72-3, 106, 109
Mamlūk, 116
Marseilles, 95
Muslim, in Morocco and Iberia, 266, 269

Norman, 256, 262-3
Northern in general, 52 & n. 49, 95, 253, 255, 267, 269-71
Pisa, 52, 63-4, 66-7, 73, 75, 95-6
Second Crusade, 92, 266-7
Third Crusade, 262
Venice, 52, 63-75 *passim*, 95 & n. 3, 98, 105, 107-9, 111, 114, 263
Fourth Crusade, 210, 213-14, 217-18 & n. 40, 227
water supplies, 51 n. 47
watering, 97, 105, 108, 212, 263, 268
wintering *en route*, 96, 98, 115, 218, 258, 266-8
see also ports of call, ships, voyages
naval warfare
guerre de course, 67, 68, 73
marines, 66
missiles, 63
siege artillery, 70
naves, *navi*, *see* ships
Naxos, 103, 111
Neckar, river, 121
Nesbitt, J. W., 119
Nestorion, 4
Nicaea, 43, 44 & n. 9, 45, 49, 54, 58-60, 120 n. 16, 182-3, 230 n. 6, 243, 247, 248 n. 62, 301
siege of (1097), 49-50, 53, 246-8 & n. 62
see also Byzantines
Nicholas, Saint, 113
Nicholaus de Marthono, 110, 111
Nicolle, D., 197
Nikēphoros II Phōkas, Byzantine emperor, 141
Nikētas, Byzantine governor of Belgrade, 128
Nikētas Chōniatēs, Byzantine historian, 106, 221-2 n. 54
Nikomedeia, and Gulf of, 294, 301
Nile, river, 218
see also maps
Niš, 121, 127-9
non-combatants, 120
Normans,
and Second Crusade, 83, 92, 256
and Sigurd of Norway, 96-7
see also Sicily, Kingdom of
Normandy, 264
North Africa, *see* coasts, maps, Maghrib
North Atlantic ocean, 260, 270
Norway/Norwegians, 251-4
Nubia, 189
Nūr al-Dīn, Zangid *amīr*, 191, 202-3, 205

oarsmen, *see* galley oarsmen
Odo of Deuil, historian, 78-9, 81-5, 88, 230 n. 6, 235 n. 17, 242 n. 40
Ogerio Pane, chronicler, *see* Genoa, civic chronicle
Oghuz Turks, *see* Turkmen
Oloron, *see* maps
Oporto, 89, 91, 256-7, 263, 265
bishop of, 256
Optimatoi, 155-6, 294
Orontes, river, 30-31, 34
see also maps
Osbert of Bawdsey, 90
Otbert, bishop of Liège, 239
Othonoi, 105
Otranto, 105
Straits of, 104
O-T/T-O maps, *see* maps
Otto of St Blasien, chronicler, 231
Ottomans/Ottoman sultans, 26, 132-3, 140
Outremer, 68, 71, 91-2 & n. 51, 96, 98, 105, 109, 112, 115, 161, 229 n. 2, 251-2, 254-7, 259, 260-68, 271-2, 294
see also Holy Land

pack-animals, 7, 123, 134-5, 140, 147-50
numbers required, 151-7
see also donkeys, horses, mules
Palaiokastron (Crete), 104, 301
Palermo, 68
Palestine, 33, 35, 113-14, 185, 194-5, 201-3, 220, 227, 237
Pamphylia, 140
Pantelleria, 118,
see also maps
Papacy, 88, 91, 109, 209, 238
Paphos, 115
port of, 115
Saranda Kolonnes, castle, 115

Paradise, *see* maps
Paris, *see* maps
Parma-Magliabecchi portolan, 117
Paros, 111
pasture/pasturage
  Arabia, 185
  Egypt, 39
  productive capacity of, 21-3, 30, 145 & n. 43
  setting fire to, 38
    see also munawwirūn
  Syria, 40
Patara, 112-13, 301
Pavia, 215
Paxos/Andipaxos, 107
Pelagonia, 4 n. 14, 302
Pelekanos, 50, 58, 302
Pella, 3, 302
Peloponnesos, 71-2, 103
Perjés, C., 22 n. 68, 23 n. 70
Permarino, Rogerio, Venetian admiral, 109
Persian Gulf, *see* maps
Peter, bishop of Oporto, 90
Peter, brother of Afonso I, king of Portugal, 88-9
Peter the Hermit, 80, 119-29 *passim*
Pevkofiton, 4
Philea, 227, 302
Philip, count of Flanders, 257 n. 17
Philip II Augustus, king of France, 95-7, 99, 102-3, 106, 110-11, 113-14, 159, 211-12
Philip, prince of France, 161
Philopation, pleasure park at Constantinople, 83, 302
Phoinikous, 97, 99, 112, 302
Piacenza, 215
Pilgrims' Route, *see* roads
pillaging, 80, 82-3, 85, 92, 121, 123, 125, 126 & n. 42, 126-7 & n. 46, 128, 222, 224-7, 241-2, 246
  see also foraging, plunder
Pindos, mountains, 9, 22
pirates, 105-6, 111, 255 n. 12
  see also corsairs
Pisa/Pisans
  in general, 63-5, 70-71
  and First Crusade, 96
  see also corsairs, naval forces
plate, *see* money
Pliny the Elder, 14-15
Plovdiv, 121, 124, 129
plunder, 231, 242-3, 246-8, 253
Po, river and valley, 214-15
Poitou, 235-6
Poland, 271-2
ponies, Mongol, 27-30
  numbers of 28-9
population, *see* environment
Porteous, J., 236 n. 20
Portolan of Gratiosus Benincasa, 117
portolans, 101, 294
  portolan charts, 161, 177 n. 32
Portovenere, 66, 302
ports of call, 96, 101, 103-4, 109, 265-6, 269
Portugal/Portuguese/Kings of, 92, 256-9, 263-8
  Algarve, 269-70
  see also maps
Portus Guiscardus/Fiskhardo, 107-8, 302
Portus Pisanorum, *see* Phoinikous
prevailing winds, *see* Winds
Princes' Islands, 223-4, 302
Prokopios of Caesarea, historian, 98, 136-8, 144, 147
Provence, 68, 73
provisioning naval forces, 97
provisions/provisioning:
  for horses, *see* fodder
  for men, 9-15, 38-9
    in general, 121, 127-8 147-8, 150-57, 200, 230, 262
    as gifts, 129
    biscuit/*ka'k*, 10-13 & Tables 1.4, 1.5, 14-15, 59, 96, 148, 201, 203, 226-7, 293
    bread, 14, 125, 135, 148, 211-12, 225-6
    by sea, 244 & n. 47, 246
    charity, 85
    cheese, 12 & Table 1.4, 148-9
    droving live animals, 123 & n. 28
    duration of, 19, 121-3, 146-8, 230

fish, 148-9
flour, 211, 225
for seamen and Crusaders, 9-12, 96, 214
Fourth Crusade, 209-28 *passim*
grain, 13 & Table 1.5 & n. 33, 19, 50, 54, 59, 122 & n. 25, 125, 128-9 135, 148 & n. 54, 149, 202, 211, 222, 224, 243
legumes, 12 & Table 1.4, 211-12
meat (pork, mutton), 9-13 & Table 1.4, 122 n. 25, 123 n. 27, 125, 128, 148-9, 201-2, 225
oil, 129, 148-9, 243
prices of, 41, 215-16, 226, 242-6
spoilage, 123
supply in saddle bags, 147
weight of, per day, 9-13, 54, 122, 148
wine, 10-12 & Table 1.4, 125, 129, 148-9, 211-12, 226, 243
individual/private, 84, 85
see also bran, wheat, markets, water
Pryor, J. H., 183 n. 67, 214, 217 n. 40, 263 n. 38
Ptolemy, 162 n. 13, 176 n. 30
Puységur, Marquis de, 21 & n. 63, 22 & n. 68
Pyrenees, mountains, *see* maps

al-Qādisiyya, battle (637), 185, 302
al-Qalqas͟handī, Mamlūk administrator, 187
al-Qanṭara, 201
al-Qayrawān, 186, 302
Qïlïj Arslan, Saljūqid *amīr* of Rum, 50, 61
Quallie, 110, 302
Quantum praedecessores, 79, 88
Qutlug͟h-S͟hāh Noyan, 34, 37
Quwayq, river, 30-31, 302

Radulf, Cistercian monk, 86
Rainald of Broyes, knight, 120
rajjāla, *see* military forces
Ramlah
Second battle (1102), 182
Third battle (1105), 191
see also maps
Ramon Berenguer IV, count of Barcelona, king of Aragon, 87-8
Ramsay, W. M., 60 n. 90
Raol, historian, 87 n. 35, 90, 256-7 & nn. 16 & 17
see also De expugnatione Lyxbonensi
Ras͟hīd al-Dīn, historian, 37
rates of march
breaks in, 20
distance of, per day, 8 & Table 1.3, 121, 142-4, 182-3
distances between resupply, 19, 149
rates of, 141-2
cavalry, 142-3, 203
infantry, 8, 142, 144
cavalry and infantry together, 1-2, 43, 44 & nn. 8 & 9, 45, 81, 125 n. 12, 141-4, 149, 201
individual horseman, 136 & n. 12
times of, per day, 7-9
Ratkowitsch, C., 179 n. 44
Raymond of Aguilers, historian, 58, 235-8, 244-5, 247
Raymond of St Gilles, count of Toulouse, 50, 58, 80, 82, 127, 236-8, 246-8
Red Sea, *see* maps
Regensburg, 81, 84, 232
Reggio (Calabria), 104
see also maps
Reuter, T., 48 n. 31
Rhēgion, 125-6, 302
Rheims, archbishop of, 239
Rhine, river, 121
Rhinelanders/Rhenish, 91 n. 50, 256 n. 15
Rhodes, 97, 99, 103-4, 110-12, 116-17, 259, 288, 291
archōn abydikos, 111, 293
Emborikos harbour, 112
Mandraki harbour, 111-12
Rhône, river, *see* maps
Rialto, 302, *see also* Venice
Richard I, Cœur de Lion, king of England, 96-100, 103-5, 112, 114-17, 160, 257, 258-9 & n. 21, 260-61, 263-6
Richard of Devizes, chronicler, 98 & n. 15,

258 n. 21, 261-2 & n. 36
Richard of the Holy Trinity, chronicler, 258 n. 21
Richer, bishop of Verdun, 239
Richigona, Genoese *cog*, 67
Riḍwān, Saljūqid *amīr* of Aleppo, 191, 195
Riga/Riga Bay, 272
Riley-Smith, J., 45 n. 16, 232
Riviera di Levante, 66, 73, 302
roads, Romano-Byzantine, 43, 44 n. 9, 56, 135-41
 Antioch-Aleppo/Chalkis road, 140
 degrading of, 5-6, 136, 138-41
 Egypt to Palestine, 201
  see also Via maris
 maintenance of, 136-8, 140-41
 Pilgrims' Route, 182-3
 stepped, 140-41, 248 n. 62
 strata Diocletiana, 136
 Tarsos-Cilician Gates road, 140
 width of, 5, 56, 138, 140
 see also Via (various)
Robert, count of Flanders, 236, 247
Robert Guiscard, duke of Apulia, 2, 105, 107, 121 n. 19
Robert of Auxerre, chronicler, 214 n. 27
Robert of Clari, historian, 216, 218-20, 226-7
Robert, duke of Normandy, 52 n. 50, 236, 247
Roccafortis, Venetian *nave*, 67-8, 72, 74-5
Roger II, count of Sicily, king of Sicily, 79-80, 252 & n. 4
Roger of Hoveden, chronicler, 96, 102, 106, 110-11, 114, 117, 258 n. 21, 261 & n. 33, 264-5
Roger of Lauria, Aragonese admiral, 108
Roger of Wendover, chronicler, 258 n. 21
Rögnvald, earl of the Orkneys, 256 & n. 14
Roland, "son" of Bernard of Clairvaux, 88
Roman de Thèbes, 160
Romania, 64-5, 67, 71-2, 294, 302-3
Roman Empire, 132
Romans d'Alixandre, 160
Rome, 72
 see also Papacy, maps
Rosa, Venetian *nave*, 67
Roskilde Fjord, 260-62
Roth, J. P., 15 n. 40, 135 n. 6, 146, 151-7
Roussa, 127, 303
Rrogozhinë, 3
Rum, white Byzantine or European military slaves in Egypt, 187
Runciman, S., 49 n. 34
Russia/Ukraine, rivers of, 253

Sabā', 185, 303
Sachkritik, 46, 54, 294
Ṣadr, Qalat, 203, 303
Saewulf, pilgrim, 113
Ṣafavids, dynasty of Shāhs of Persia, 132
sagita, *see* ships
Sağlıklı, 140, 303
Saharan trade routes, 187 n. 7
Saher, lord of Archelle, 87
Ṣalāḥ al-Dīn, sultan, 165, 193, 195, 197-8, 199 & n. 56, 202-3, 205-6, 259
 qāḍīs, 199
 see also military forces
al-Ṣāliḥ Najm al-Dīn, Ayyūbid sultan of Damascus, 199
Saljūqids, *amīrs* of Syria, 200
Saljūqids (great), sultans of Bagdhad, 36, 193
Saljūqids, sultans of *Rum*, 36, 85, 193, 200
"Samaria", *see* maps
Sancho I, king of Portugal, 258-9
Sangarios, river, 60, 137, 294, 303
San Remo, 73
Santa Carenta, 106
Santarém, 89
Sanudo, Marco, 111
Sanudo Torsello, Marino, *Liber secretorum fidelium crucis*, 10-14 & Tables 1.4, 1.5, 10 n. 26, 13 n. 32, 15, 114
 see also maps
Sapienza, 109
 battle (1403), 108
Saranda Kolonnes, castle, *see* Paphos
Sarandaporos, river, 4
Sardinia, 64, 70-71, 100
 silver mines, 233
 see also maps
Sarrī ibn al-Ḥakam, governor of Egypt, 188

Sāsānids, kings of Persia, 163, 186 n. 5
Sava, river, 121, 127-8
Saxons, 88, 271
Scandinavia/Scandinavians, 253, 259
scouts, 139 & n. 22
sea lanes, 101-3, 104-5, 107, 113-15
Sea of Galilee, *see* maps
seasons, *see* environment
Seine, river, 182, 264
    see also maps
Seman, river, 3, 4
Serbs, 242
service, payment for, 246-8
Seville, 268-9
    amīr of, 69
Sezgin, F., 162 n. 13
Shaqḥab, 40, 303
Sharaf al-Dawla al-Mu'izz, Zīrid sultan of Tunisia, 192
Shayzar, 196-7, 303
Shepard, J., 52 n. 48, 54 n. 55
ships
    balingers, 264
    barca, pl. *barce* (Lat.), *barcha*, pl. *barche* (It.), 65, 69, 71
    barges, 264
    barks, 265
    bolskip, 272
    bucius, pl. *bucii*, also *buza* (Lat.), *bucio*, pl. *buci* (It.), *buss* (Engl.), *bussa/buse/buce* (O. Fr.), 65-6, 69, 73, 97, 258 & n. 21, 261 & n. 30, 261-2, 265, 271, 293
    byrthing, 261, 293
    carraca/carrack, 270
    Celtic, 260
    cocha, 270
    cogs, 68, 267 & n. 49, 267 & nn. 49 & 50, 269-70, 272-3, 293
    cursarii, 261, 264
    δρόμωνες, 95 n. 3
    "dromons", 258
    esneccas, *enekes*, 264-5, 271, 293
    galea, pl. *galeae*, galleys, 63, 65-75, 95 & n. 3, 101, 104, 213, 258 & n. 21, 261, 264-5
        cost of, 75
        crews of, 75
        Genoese, 270
        great galleys, 270-71
        horse transports, 95-6, 112, 210, 213, 217 & n. 40
        manœuvrability of, 74, 265
        sailing qualities, 98-9
        sizes of, 65, 68-70,
        speed of, 74
        Venetian, 270
    gorabius, 106, 294
    keels, 260, 265, 267, 269, 272
    knorr, pl. *knerrir*, 260-62, 294
    legno, pl. *legni* (It.), 65, 73
    longships (Norse), 95, 260, 270
        drekar, dragon ships, 264
        snekkja, pl. *snekkjur*, 264, 295
    Low Countries, 267
    management of, 96
    navilia, 116
    navis, pl. *naves* (Lat.), *nave*, pl. *navi* (It.), *nef*, pl. *nes* (Fr.), 63, 65, 67-74, 95, 97, 102, 112, 114, 116-17, 258 n. 21, 262, 265, 294
        as horse transports, 96, 262 & n. 36
        defensive capabilities, 67, 74
        offensive capabilities, 68, 74
        Paradiso, Venetian *navis*, 213 n. 19
        sailing qualities, 98-9
        speed of, 74
        tonnage of, 214
    Northern, in general, 259-60, 263-4, 266-7
        size of, 262
    panzone, pl. *panzoni* (It.), 72, 294
    sagita, pl. *sagitae* (Lat.), *saetta/saettia*, pl. *saette/saettie* (It.), 63, 65, 66 & n. 8, 71-3, 294-5
    sandanum/chelandra/salandrium, 95-6, 116, 295
        horse transports, 116-17
    Scandinavian, 259, 267
    tarida, pl. taride (Lat.), 63, 65, 68, 71-3, 95-6, 100-101, 107, 261 n. 34, 295
    three-masted Egyptian, 259 n. 22
    triremes, 63

Viking, 261
see also naval forces, naval warfare, voyages
shipworm (*teredo navalis*),97,254,266-7
Shirley, lieutenant colonel, 20 n. 59
Shkumbi, river, 3
shoals, 108, 116
Siberis, river, 137, 303
Sicilian Channel, 73, 102, 117
Sicily, 64, 73, 96, 98, 100, 252, 258, 265-6, 268
Kingdom of, 66, 97, 106, 109
see also maps
Sidon, 113-14
siege of (1110), 252
see also maps
Sifnos, 111
Sigurd the *Jórsalafari*, king of Norway, 95-6, 251 & n. 2, 252-4, 256 n. 14, 260, 264-5, 270-71
silver mines, 233
Silves, 259, 266
Simon of Dover, 87 n. 35
Sinai, mount, *see* maps
Peninsula, 202
Sinor, D., 28 n. 13
Sintra, 252
Siros, 111
Sirocco, 112
Slavs, 87, 271-2
see also Curonians, Wends
Smith, C. D., 173 n. 28
Smith, J. M., jr, 26-43 *passim*
Snorri Sturluson, historian, 251, 252 & n. 4, 253-4 & n. 7
Heimskringla, 251, 256 n. 14
Sofia, 121, 128 & n. 56, 129, 137
Sögüt, 60 n. 90
Solinus, 90
Solomon bar Simson, historian, 240
Sopron, 121
Spain, in general, 70, 87, 255
see also Iberia, maps
Spetsai, battle (1263), 71
Sri Lanka, *see* maps
Stenay, 239-40, 303
Stephanus Garsia, mapmaker, *see* maps
Stephen, count of Chartres, Blois, and Meaux, 181-3, 236, 238, 243-4
St Gilles, 235
St George of Sestos, 79
Stravopotamos, river, 4
Struga, 3, 5
St Sabas, monastery, 63, 70
War of, 63, 70-74
St Sever, monastery, *see* maps
St Stephen, abbey, 222-4
St Symeon, 96, 113, 244, 246, 303
storms at sea, 99 n. 22, 99-101
Libecchi, 104
Tormenta, 104
Straits of Gibraltar, *see* Gibraltar
Sudan, 189
Suez, 202-3
Suger, abbot St Denis, 78, 84
supply
problems of, 93
systems, 45
trains, 55-7, 60, 144, 149
Sword Brothers, 271-2
Symeon, patriarch of Jerusalem, 52 n. 50
Syracuse, 68
Syria, 25-43 *passim*, 113, 182, 185, 193-5, 197, 203, 207, 218, 237, 243-4, 249
resources of, 27-35, 36 & n. 45, 37
water resources, 30-35, 37
see also maps, military forces
Syro-Palestinian coast, ports of, 51 n. 47, 53, 64, 114-15, 166, 202

Taghrī Birdī, Mamlūk *amīr*, 42
Tagus, river, 257
Taktika of Leo VI, 147 & n. 50
Tanais (Don), river, *see* maps
Tancred, 2 & n. 6, 4, 82, 236, 247-8
Prince of Antioch, 191
Tangier, 270
Tara, river, 112
Taranto, Gulf of, 105
Tarbes, *see* maps
Tarida, *see* ships
Tarsos, 2, 255
Tarsos-Cilician Gates road, 140
Tatikios, Byzantine general, 58, 61
Tauros, mountains, 27, 113, 142
Templar, Knights, 86

tent group, *see contubernium*
temperature, *see* environment
Teutons, *see* maps
Teutonic Knights, 272
themes/*themata*, 120, 149, 295
Theodoricus monachus, historian, 251
Theophilus, 180
Theophylaktos Simokattēs, historian, 137 n. 12
Thessalonikē, 1-3, 23, 125 n. 41, 136
Thierry, count of Flanders, 87
Thórarin Stuttfeld, skaldic poet, 251
Thrace, 137 n. 12, 248 n. 62
Tibbetts, G. R., 163
Tiber, river, 72
Tiberias, *see* maps
Tigris, river, 194
 see also maps
Tīmūr Lang, 41-2
Tinnīs, 303
 see also maps
Tortosa, 115
Trajan's Gate, pass, 136, 303
transhumance, 295
 see also Mongols
Trapani, battle (1266), 73
treasure, 234-5 & n. 17, 247, 266
treatise on imperial military expeditions, 18, 149
Treaty of Venice (1201), 16-17 & Table 1.6, 210-15
tribute, *see* money
Trikkala, 3
Tripoli, 51 n. 47, 53, 111, 113-15, 183, 202, 235
 amīr of, 236-7, 244, 247
"troop", knight's, 7, 8 & Table 1.2, 19 & Table 1.7
 resupply, 19
 size of, 55 n. 62
True Cross, 224
Tudebodus imitatus et continuatus Historia peregrinorum, chronicle, 1
Ṭūlūn, Aḥmad ibn, *see* Ibn Ṭūlūn
Tunis, 69, 73, 268-9
Tunisia, 186-8
Turbessel, 247, 303
Turcopoles, *see* military forces
Turkmen, 33, 35-6, 193, 194 & n. 33, 195, 197, 200, 207
Turks, 25, 85, 188-9, 193, 195, 200, 205, 207, 242, 244-5
 see also military forces
Tuscany, 236
 see also maps
Tyerman, C., 52 n. 49
Tyre, 67, 71, 73, 113-14, 188, 202, 268
 port of, 114
 see also maps
Tyrrhenian Sea, 63, 64
Tzachas, Turkish corsair, 111

Ulf the Galician, Viking, 254
al-'Umarī, Mamlūk administrator, 38
'Umar ibn 'Abd al-'Azīz, Umayyad Caliph, 294
Umayyads, Caliphal dynasty, 114
Urban II, Pope, 3, 51, 79, 119, 120 n. 7, 126 n. 44, 127, 238
 letter to the Flemish, 119
Usāma ibn Munqidh, 196-7
Üxküll, 272, 303
'Uyūn Mūsā, 203, 303

Van Creveld, M., 22 & n. 68
Vandals, 98, 132
Vardar, river, 1-2, 4, 19, 125 n. 41
Vegetius Renatus, Flavius, 144, 148
Veil, William, 92
Venice/Venetians
 in general, 63-75 *passim*, 103, 105-11, 209-28 *passim*, 233, 235 n. 17, 262, 270
 chroniclers of, 64
 contract of 1268 with Louis IX, 17
 Lido, 215-16 & n. 33, 217, 300
 maritime traffic, 104
 population of, 214 n. 24
 see also Crusades (Fourth), naval forces, Treaty of Venice
Verdun, 82, 239-40
Verona, 215
Via Egnatia, Roman road, 2-4 & nn. 13 & 14, 5-6, 8-9, 126, 136-7, 140-41, 303
Via maris, 201-2

Via militaris, Roman road, 136, 141, 303
Via Sebaste, Roman road, 140, 303
Via Tauris, Roman road, 298, 303
Vienna, 49, 121, 259
Vienne, river, 179
Vikings, 254, 263
Viola, Genoese *nave*, 68
Vistula, river, 272
Vjosa, river, 3, 19
Vlaardingen, 267
voyages, duration of, 92, 97-9
Voidhomatis, river, 4

Waas, A., 49 n. 34
Wādī 'l-Khaznadār, battle (1299), 28, 32, 35, 37-8, 40-41, 303-4
Walcheren, 258 n. 21
Walter of Châtillon, 160
Walter SansAvoir, knight of Poissy, 80, 119-20, 124, 127
Warner, count of Grez, 234
al-Warrāda, 201
Waṣṣāf (al-Ḥaḍrat), historian, 28-9, 37
water
    in general, 43, 56, 200-204, 216, 244, 246, 248, 268
    for horses, 19-20 & n. 59, 28, 30, 37, 98, 145, 185, 212 & n. 14, 262-3
    for seamen, 96, 98, 121 n. 2, 262
    water bags/skins (*rāwiya/qirba*), 203 & n. 67
watering, *see* military forces, naval forces
weather, *see* environment
weights and measures, 242
Wellington, Duke of, 145
wells, 20
Wends, 87, 88
Werner, K. F., 47-8
wheat, 14, 123
Wibald, abbot of Corvey, 83
Wilbrand of Oldenburg, pilgrim, 115
William I, count of Holland, 267-8 & n. 50
William Adam, propagandist, 100 n. 21
William of Nangis, hagiographer, 161, 177 n. 32
William Porco, Genoese corsair, 109
William the Conqueror, king of England, 179 n. 44
William, archbishop of Tyre, historian, 200
Williams, J., 172 n. 23
Winand, German priest, 90, 256, 257 n. 17, *see also* Lisbon letter
winds, prevailing, 99, 101-4, 109-10, 113, 254, 270
Wolseley, G. J. [General Viscount], 7-8 n. 20, 16, 18 n. 56, 19, 20 & n. 59, 21 n. 63, 22 n. 67
world maps, *see* maps
Worms, 78, 83

Yāqūt al-Rūmī, geographer, 201
Yarmūk, river, battle (636), 185

Zachary, Pope, *see* maps
Zangī, 'Imād-al-Dīn, Saljūqid *amīr*, 165
    Zangīds, 200
Zanj, 189, 295
Zara, 106, 217-19, 221, 224, 260
Zemun, 120-21, 123, 126 n. 42, 127, 128 & n. 56, 129
Zeno, Carlo, Venetian admiral, 67
Zibaldone da Canal, merchant's manual, 11-12 n. 29, 14 n. 36
Zompos/Zompē bridge, 137, 304
zones of the earth, *see* maps
Zwin, river, 268